Regenerative Strategies for Maxillary and Mandibular Reconstruction

James C. Melville · Jonathan W. Shum
Simon Young · Mark E. Wong
Editors

Regenerative Strategies for Maxillary and Mandibular Reconstruction

A Practical Guide

Springer

Editors
James C. Melville
Department of Oral and Maxillofacial Surgery
University of Texas Health Science Center Department of Oral and Maxillofacial Surgery, Oral, Head & Neck Oncology and Micro-vascular Reconstructive Surgery
Houston
TX
USA

Jonathan W. Shum
Department of Oral and Maxillofacial Surgery
University of Texas Health Science Center Department of Oral and Maxillofacial Surgery, Oral, Head & Neck Oncology and Micro-vascular Reconstructive Surgery
Houston
TX
USA

Simon Young
Department of Oral and Maxillofacial Surgery
University of Texas Health Science Center Department of Oral and Maxillofacial Surgery
Houston
TX
USA

Mark E. Wong
Department of Oral and Maxillofacial Surgery
University of Texas Health Science Center Department of Oral and Maxillofacial Surgery
Houston
TX
USA

ISBN 978-3-319-93667-3 ISBN 978-3-319-93668-0 (eBook)
https://doi.org/10.1007/978-3-319-93668-0

Library of Congress Control Number: 2019931834

This Springer imprint is published by the registered company Springer Nature Switzerland AG
The registered company address is: Gewerbestrasse 11, 6330 Cham, Switzerland

Preface

The replacement of worn or damaged body parts has been a focus of medicine for centuries. As early as the sixth century BC, Indian surgeons described techniques for nasal reconstruction utilizing pedicled flaps from the arm. Since then, significant advances, especially in the field of cell biology, heralded a shift from flaps fashioned from dissimilar tissue to the transplantation of cells capable of restoring structural defects in perhaps an early iteration of regenerative medicine. Maxillofacial reconstruction has employed these techniques successfully since the 1800s grafting bone, cartilage, and skin cells. But limitations such as harvest site defects and the need for vascularized tissue beds continue to pose challenges, and multiple strategies are necessary for the optimal reconstruction of the maxillofacial defect. In addition, the special demands of craniofacial reconstruction produced by the prominent exposure of the face, lack of thick soft tissue camouflage, proximal bilateral structures, and the special senses housed within the region impose high fidelity standards on our reconstruction techniques.

Maxillofacial defects represent a heterogenous group of conditions requiring varied reconstructive approaches. The correction of bone and soft tissue deficiencies for dental implant placement employs different strategies from those used to restore skeletal continuity and even for similar indications, different techniques have been described. Some are time-tested and have been refined extensively to produce accurate, functional results while others are still undergoing development, attempting to take advantage of advances in molecular science and tissue engineering technology. This textbook contains the contributions of many centers of excellence, and different reconstructive options are described with practical suggestions based on the extensive experience of the different authors. The spectrum of techniques ranges from hard and soft tissue grafts to vascularized free tissue transfers. Tissue engineering techniques are covered along with adjunctive measures, which incorporate blood or marrow bioactive components and exogenous proteins to improve graft outcomes. A section on nerve reconstruction has also been added to address a difficult reconstructive challenge, which is only recently gaining attention from reconstructive surgeons. Finally, space has been provided in this textbook for a description of preclinical initiatives to develop novel regenerative medicine solutions to the limitations of existing reconstructive techniques.

Even though the editors for this project are based in one institution, we represent the combined influence of our distinguished mentors. These pioneers provided us with strong foundations in reconstructive surgery and regenerative medicine, inspired us to improve existing techniques, and encouraged us to share our knowledge for all who practice maxillofacial reconstructive surgery. This textbook reflects the commitment of all contributors to this noble cause and stands as a testimony to the trust and confidence placed in us by our patients, to employ every means possible to make them whole again.

Houston, TX, USA James C. Melville
Houston, TX, USA Jonathan W. Shum
Houston, TX, USA Simon Young
Houston, TX, USA Mark E. Wong

Acknowledgments

This book is dedicated to all those who inspired and given me support. To my parents Daryl and Julia for showing me the value of hard work, being patient and letting me find my career path. To my wife Dolfin and daughter Madelyn for their unconditional love and support through many long days and nights. You are the foundation that makes everything possible; you make me a better father, husband, and person.

My eternal gratitude to my friend and mentor Dr. Robert E. Marx for his guidance, mentorship, and inspiring me to set on a career path as an academic Oral and Maxillofacial surgeon and igniting my interest in tissue engineering, oncology, and reconstruction. To my friend, teacher, and colleague, Ramzey Tursun, thank you for teaching me the skill of microvascular surgery. It has been invaluable.

To my chairman and friend Mark E. Wong, I am inspired by your leadership and dedication to the specialty and our program every day; there is no more exceptional leader I can model my career after.

To my friends, co-editors, and colleagues Simon Young and Jon Shum who make work fun. Working with you both every day makes me a wiser and better surgeon.

To my colleagues, James Wilson, James Bertz, Issa Hanna, Joe Marchena, Kamal F. Busaidy, Nagi M. Demian, Jonathon S Jundt, Justin Seaman, Phil Freeman, Tim Woernley, Craig Pearl, Bob Debes, and Harry Gilbert, thank you for your camaraderie, knowledge, and making UTHealth OMFS one of best programs in the country.

To Sverre Klemp and Springer for making this book possible.

Lastly to the countless interns, residents, and fellows whom I have had the privilege to work with, thank you for your dedication and hard work; you make it possible to give our patients the highest treatment and care.—James C. Melville

To our chairman, Dr. Mark Wong, you have been a great mentor, teacher, and guide. Your support and advice have helped shape my professional career. To my friends and colleagues Dr. Jonathan Shum and Dr. James Melville, thank you for making my job fun—working with you both has been an honor, and I look forward to many more collaborations. To my mother, May, who has always been one of my staunchest supporters while holding me to the highest standards and my late father, Tony, who taught me simply to relax and take it all in. To my sons, Emerson and Paxton, who inspire me to be introspective and strive to be a better person and father every day. Lastly, to my incredible wife, Michelle, who brings endless joy and laughter to my life, I'm thankful for every day we've shared; you're the bedrock of our family and an endless source of love and understanding.—Simon Young

As luck would have it, I have been very fortunate to have had the opportunity to train under exceptional surgeons. Thank you RB MacIntosh for sparking a desire to learn, train, and to be better. I am grateful to my friends and mentors: JE Lubek, EJ Dierks, RA Ord, TG Bui, RB Bell, and D Dyalram. Thank you for the experiences, your expertise, and your time.

I am especially grateful to ME Wong for the incredible opportunity to be a part of the UTHSC-H faculty. Thank you to JC Melville and S Young for your friendship, support, and to our future collaborations!

To my parents, Annie and David Shum, brother Jeff, and my wife Jen for your love, guidance, and support.

To Gavin and Keaton, I hope to inspire your ambition.—Jonathan Shum

I am grateful for my gifted teachers and students for giving me my profession and my wife Marlene and children Denise, Lindsay, and Matthew for being an important part of the journey.—Mark E. Wong

Contents

Part I Regenerative Medicine and the Role of Tissue Engineering in Maxillomandibular Reconstruction

1 **Principles of Regenerative Medicine of the Maxillomandibular Region** 3
Mary C. Farach-Carson and F. Kurtis Kasper

2 **Traditional Use of Autogenous and Non-autogenous Grafts in Head and Neck Reconstruction: Principles of Conventional Bone Grafting** 13
Pooja Gangwani, Shahid R. Aziz, and Jose M. Marchena

3 **Reconstruction Considerations for the Posttraumatic, Benign Tumor, and Oncologic Patient** 25
Jeffrey D. Johnson and Tang Ho

4 **The Role of rhBMP-2 in Oral and Maxillofacial Reconstruction** 33
Alan Herford and Isaac Lowe

5 **Techniques of Obtaining BMAC, PRP, and PRF** 43
Hisham Marwan, Todd Sawisch, Avi Schetritt, and Ramzey Tursun

6 **Principles of Soft Tissue Engineering for Craniomaxillofacial Reconstruction** 53
Shiuhyang Kuo, Atsuko Miyazawa, and Stephen E. Feinberg

Part II Surgical Techniques Utilizing Biotechnology for Regeneration and Reconstruction

7 **In Situ Tissue Engineering** 73
Robert E. Marx

8 **Tissue Engineering Techniques for the Reconstruction of the Atrophic and Local Maxillary Defect** 87
Issa Hanna, Huy Tran, Jonathan Swope, and Michael Peleg

9 Tissue Engineering Solutions for Dental Implantology, Dentoalveolar Surgery, and Maxillary Sinus Floor Augmentation ... 109
R. Gilbert Triplett, Marianela Gonzalez, and Francisco Curiel

10 Alveolar Ridge Augmentation (Vertical and Horizontal) with Bioengineering ... 123
Tibebu Tsegga and Brian Rethman

11 Tissue Engineering for Transoral Reconstruction of Large Mandibular Continuity Defects ... 143
James C. Melville, Huy Q. Tran, Dhafer Alsuwied, Srinivasa R. Chandra, and Jonathan W. Shum

12 Contemporary Management of Mandibular Resection-Related Nerve Injury ... 155
Shadi Alzahrani, Arshad Kaleem, Hisham Marwan, and Ramzey Tursun

13 Use of Skin and Soft Tissue Substitutes in Maxillofacial Reconstruction ... 161
Jonathon Jundt and Nagi Demian

Part III State of the Art Free Vascular Transfer for Maxillomandibular Reconstruction

14 Reconstruction of Maxillary Defects ... 175
Baber Khatib, Ashish Patel, Eric J. Dierks, R. Bryan Bell, Melissa Amundson, and Allen Cheng

15 Principles of Mandibular Reconstruction with Vascularized Flaps ... 201
Nicholas Callahan, Joshua E. Lubek, and Donita Dyalram

16 Principles of Fibula Jaw in a Day Reconstruction: A Step-by-Step Approach ... 211
Mohammed Qaisi, James Murphy, and Henry Fung

17 Role of Microvascular Free Flaps Combined with Tissue Engineering ... 235
Waleed Zaid, Andrew Yampolsky, and Beomjune Kim

Part IV Future Trends for Maxillomandibular Regeneration and Reconstruction

18 3D Printing and Adenosine Receptor Activation for Craniomaxillofacial Regeneration 255
Christopher D. Lopez, Lukasz Witek, Roberto L. Flores, Andrea Torroni, Eduardo D. Rodriguez, Bruce N. Cronstein, and Paulo G. Coelho

19 The Future of Bioengineering for Head and Neck Reconstruction: The Customized Free Flap 269
Huy Tran, James C. Melville, Jonathan W. Shum, F. Kurtis Kasper, Mark E. Wong, and Simon Young

Part I

Regenerative Medicine and the Role of Tissue Engineering in Maxillomandibular Reconstruction

1 Principles of Regenerative Medicine of the Maxillomandibular Region

Mary C. Farach-Carson and F. Kurtis Kasper

1.1 Considerations in Maxillomandibular Reconstruction

Maxillomandibular defects pose unique challenges for reconstruction given the high visibility of the region and the specialized functions of the associated structures. Indeed, the maxilla and mandible contribute to important functions such as mastication, deglutition, speech, airway maintenance, and facial expression. In parallel, the visual prominence of the maxillomandibular region underscores the important role of cosmesis in quality of life after reconstruction. Defects to the area can arise from a variety of etiologies, including trauma, tumor extirpation, and congenital conditions, each of which presents unique reconstructive considerations. However, a defect from any cause disturbs the form and function of the affected structures. Accordingly, approaches to maxillomandibular reconstruction generally seek to restore function with improved esthetics insofar as is possible.

Primary considerations associated with maxillomandibular reconstruction include the location and size of the defect as well as the specific tissues requiring reconstruction. While the skeletal structures typically serve as the focus of maxillomandibular reconstruction, they are often not reconstructed in isolation. Traumatic and post-ablative defects each can involve multiple tissues, and they can present hostile environments for wound healing. For example, traumatic injuries may present contaminated wound beds, and local vascularization may be compromised as in cases of cancer patients who have received radiation therapy. The unique environments associated with the oral and nasal cavities also can complicate maxillomandibular reconstruction. Regardless of the complexity of the defect and the associated wound environment, the concept of replacing the missing or injured tissue with like tissue guides the reconstructive approach.

1.2 Tissue Transfer

Operative techniques and approaches for maxillomandibular reconstruction have evolved considerably over the years. Although an overview of the history of surgical technique is beyond the scope of the present discussion and can be found elsewhere [1], advances in tissue transfer have enabled improvement in outcomes in maxillomandibular reconstruction, especially in cases of large defects involving soft tissues. In simple

M. C. Farach-Carson
Department of Diagnostic and Biomedical Sciences, The University of Texas School of Dentistry at Houston, Houston, TX, USA
e-mail: Mary.C.FarachCarson@uth.tmc.edu

F. K. Kasper (✉)
Department of Orthodontics, The University of Texas School of Dentistry at Houston, Houston, TX, USA
e-mail: Fred.K.Kasper@uth.tmc.edu

J. C. Melville et al. (eds.), *Regenerative Strategies for Maxillary and Mandibular Reconstruction*, https://doi.org/10.1007/978-3-319-93668-0_1

terms, tissue transfer involves the relocation of tissue harvested from a donor site for reconstruction of a tissue defect. Grafting procedures invoke transfer of tissue harvested without an intact blood supply from a donor site to a recipient site. For example, bone grafting involves transfer of harvested bone tissue to fill a bony defect and relies on development of new vascularity to support its viability at the recipient site. While bone grafting techniques may be appropriate for reconstruction of small bony defects, larger defects typically require transfer of vascularized tissues to support continued viability of the tissue volume. Tissue flap techniques have been developed to meet this need and involve rotation or microsurgical transfer of a flap of tissue with an intact vascular supply to enable continued perfusion of the tissue at the recipient site.

A wide variety of tissue transfer techniques have been described for mandibular and maxillary reconstruction, with the fibula free flap being described as "the workhorse flap in most situations," owing in part to the length of bone that can be harvested, the pliability of the associated soft tissue, compatibility with osteotomies for shaping, and limited donor site morbidity [2]. Nevertheless, selection of an appropriate tissue for transfer from among the available options involves identification of the optimal combination of bone with associated soft tissue for use in maxillary or mandibular defect reconstruction with minimal morbidity at the donor and recipient sites [2]. Accordingly, the reconstructive approach should reflect consideration of the requirements of the defect (e.g., size, location, and affected tissues), any comorbidities that may be present, the availability and morbidity associated with potential donor sites, and the health and vascularity of the recipient site [2]. While reconstructive options include grafting and rotation or transfer of flaps, each requires a donor site morbidity, may not provide optimal cosmesis, and may not fully address other compromised or missing tissues. These limitations have motivated the development of tissue engineering and regenerative medicine technologies aimed at repair of skeletal and soft tissues in the maxillomandibular region (see Fig. 1.1).

1.3 Tissue Engineering and Regenerative Medicine

An expanding interest has arisen in recent decades toward the development of strategies to facilitate tissue regeneration or even to engineer replacement tissues to meet a wide variety of clinical needs, including maxillomandibular defects. Tissue engineering and regenerative medicine are complementary fields united in the objective of replacing or repairing missing or compromised tissues and organs. The terms "tissue engineering" and "regenerative medicine" often are used interchangeably, which reflects in part the similarities that exist between the fields, although distinctions between them have been the subject of

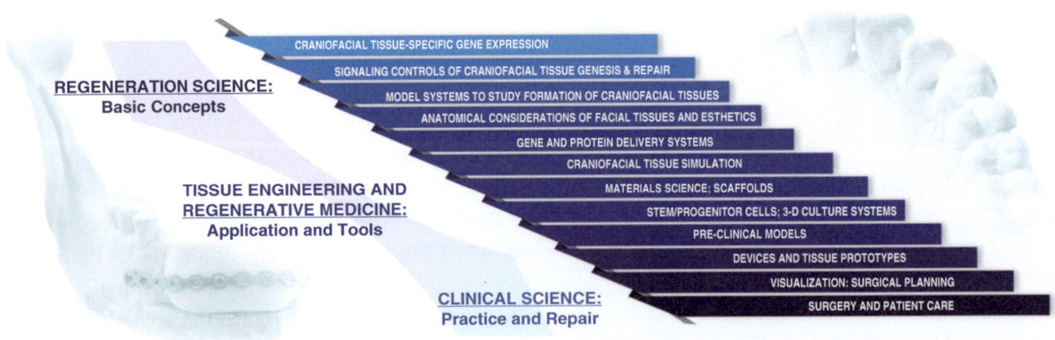

Fig. 1.1 A schematic representation of the continuum of concepts involved in maxillomandibular reconstruction, spanning from the basic principles of regeneration science to the clinical practice of surgical repair. Credit: Ms. Allison Haynie

considerable ongoing debate [3]. In broad strokes, the field of tissue engineering involves "the application of principles and methods of engineering and life sciences toward the fundamental understanding of structure-function relationships in normal and pathological mammalian tissues and the development of biological substitutes to restore, maintain, or improve tissue functions" [4]. The National Institute of Biomedical Imaging and Bioengineering describes regenerative medicine as "a broad field that includes tissue engineering but also incorporates research on self-healing – where the body uses its own systems, sometimes with the help of foreign biological material to rebuild tissues and organs" [5]. While a variety of additional definitions of these fields have been offered, they largely share the notion that tissue engineering presents a narrower focus than regenerative medicine, although synergies certainly exist between the fields. Both tissue engineering and regenerative approaches can utilize stem cells, including both adult-derived and embryonic cells, as well as induced pluripotent stem (iPS) cells [3, 6–8]. Nevertheless, when considered in the context of regulatory oversight and commercialization strategies for technologies aimed at repairing or replacing tissues and organs, semantics may present considerable importance [3]. Indeed, various approaches to tissue regeneration or repair may present unique regulatory considerations that should be measured at the onset of development; yet, the various approaches generally depend upon the action of scaffold materials, cells, and biologically active factors in various combinations to realize the common goal of restoring tissue function.

1.4 Biomaterial Scaffolds and Extracellular Matrix

Surgical approaches for maxillomandibular reconstruction often involve the placement of a graft, flap, or alloplastic material to fill the tissue deficit. In cases of tissue transfer, the graft of flap generally presents an extracellular matrix and associated biologically active factors to direct appropriate tissue remodeling in support of repair. Similarly, tissue engineering and regenerative medicine approaches commonly employ biomaterials as scaffolds to conduct, and in some cases promote, tissue repair or regeneration in a specific volume. Broadly defined a biomaterial is "a material intended to interface with biological systems to evaluate, treat, augment, or replace any tissue, organ or function of the body" [9]. In the context of tissue engineering and regenerative medicine, biomaterials can be applied as scaffolds to support tissue formation within a volume they define. A variety of design requirements guide selection of a suitable scaffold material for tissue engineering and regenerative medicine applications from the ever-expanding library of biomaterials.

A primary requirement for biomaterial scaffolds in tissue engineering and regenerative medicine is biocompatibility, or the ability to evoke an appropriate host response in a given application [9]. For instance, a biomaterial in a scaffold application should direct regeneration of the desired tissue over induction of chronic inflammation. Biomaterials applied as scaffolds also should present appropriate mechanical properties to meet the functional demands of the recipient site and to support formation of the desired tissue. Additionally, biomaterials in scaffold applications should be compatible with processing techniques to enable generation of the geometries required for appropriate function and cosmesis at the recipient site, and they should be compatible with clinically acceptable methods of sterilization.

A maxillomandibular defect ideally should be reconstructed without persistence of an alloplastic material. Accordingly, some tissue engineering and regenerative medicine approaches to maxillomandibular repair involve the application of a degradable biomaterial scaffold to support tissue formation transiently as the scaffold degrades. Such approaches might target complete degradation of the scaffold with time to preclude a secondary surgery for removal of the material. Degradable scaffolds are particularly attractive in strategies for bone tissue regeneration, as the mechanical load can be transferred with time from the degrading scaffold to the developing tissue to mitigate complications from potential stress shielding. However, appropriately tuning the scaffold degradation kinetics with the rate of tissue formation presents a considerable challenge.

In order to support tissue formation, a scaffold for tissue engineering and regenerative medicine applications should enable both cell migration and extracellular matrix deposition. In some cases, void volume in the form of interconnected pores supports cell migration and tissue formation throughout the scaffold, which is essential for tissue morphogenesis in soft tissue formation [10]. However, the mechanical strength of a scaffold generally decreases as the porosity increases, which presents particular limitations in load-bearing applications needed to restore hard tissues (reviewed in [11]). In other cases, scaffold materials are designed to incorporate enzymatically labile moieties to enable local generation of space for cell migration and extracellular matrix formation as directed in response to the actions of cells. Examples of this include use of cross-linking reagents that are designed to be degraded specifically by active matrix metalloproteinases [12].

The ultimate success of a scaffold in a tissue engineering and regenerative medicine application depends upon selection of an appropriate biomaterial. Metals present an extensive history of application in medicine and dentistry, and they have been investigated extensively in tissue engineering and regenerative medicine applications. While many biocompatible metallic biomaterials present mechanical strength that may be desirable for load-bearing applications in maxillomandibular reconstruction, most cannot degrade appropriately in the body to support tissue formation, which can limit application for both tissue engineering and regenerative medicine strategies. However, ongoing advances in the development of biodegradable metals may provide new opportunities in tissue engineering and regenerative medicine applications of metals [13].

Ceramic biomaterials also have been applied in tissue engineering and regenerative medicine approaches for maxillomandibular reconstruction. As ceramic materials can be tailored to mimic the inorganic component of mineralized tissues, ceramics can be leveraged as osteoconductive, and in some cases osteoinductive, materials for bone regeneration. Moreover, ceramics can be formulated to resorb in the body to enable development of degradable scaffolds to support regeneration of mineralized tissues. However, ceramics tend to be brittle, which can dampen their application for tissue regeneration in certain load-bearing sites where forces tend to be concentrated.

Polymeric biomaterials have been widely embraced for a variety of tissue engineering and regenerative medicine applications, owing in part to the range of chemical and physical properties that can be realized with these materials. Polymers can be synthesized from chemical precursors, or they can be derived from naturally occurring sources, such as proteins or carbohydrates. The chemical techniques applied in the generation of synthetically derived polymers generally enable consistent production in large quantities with properties tailored to the needs of the desired application. While a broad range of properties can be achieved with synthetic polymers, they typically lack biologically active domains to direct interactions with cells, unless modified specifically to incorporate such moieties. Alternatively, polymers derived from biological sources often retain biologically active domains, which can be harnessed to direct interactions with cells and tissues in tissue engineering and regenerative medicine applications. In some cases, tissue engineering and regenerative medicine approaches seek to leverage advantages of multiple biomaterials through fabrication of composite or hybrid constructs. For example, tissue engineering principles can be applied in the culture of osteogenic cell populations seeded onto a synthetic polymer scaffold to promote generation of a biologically active extracellular matrix material coating [14].

1.5 Cells

Although a scaffold may be applied to support tissue repair or regeneration, cells play an indispensable role in the process of tissue formation. Regenerative medicine approaches to tissue repair can seek to introduce cell populations to a tissue site alone or in combination with a carrier material to promote tissue formation. Similarly, tissue engineering approaches can involve application of cell populations in culture outside of the body to create a construct for tissue repair. Other approaches may seek to recruit host cells from tissues surrounding a defect or even from a distal site to facilitate tissue regeneration or repair [15]. In cases involving cell transplantation or cell transfer, the potential effects of the transplanted or transferred cells in contributing directly to repair, providing trophic effects, and interacting with cells at the recipient site should be carefully considered.

Cell-based approaches can supply autologous cells derived from the patient or allogenic cells from a human donor (or pool of donors). Xenogenic cells present another option in theory, although present concerns regarding immunocompatibility and potential pathogen transfer diminish enthusiasm for clinical application of animal-derived cells. In addition to the type of donor, the site from which cells are harvested presents another consideration. For example, tissue- or organ-specific differentiated cells, such as osteoblasts harvested from bone, can be applied in strategies aimed at repair of the respective tissue or organ. However, the number of differentiated cells that can be retrieved from a specific tissue or organ is typically limited, and cell harvest can present an associated donor site morbidity. While cell culture techniques could be applied to expand the isolated cells to numbers of therapeutic relevance, maintenance of tissue-specific phenotype in culture presents a challenge. Moreover, any manipulation of the harvested cells may complicate regulatory considerations governing their clinical application.

Stem and progenitor cells present another option for application in tissue engineering and regenerative medicine strategies. Stem and progenitor cells present the capacity to differentiate into various specialized cell types. For instance, mesenchymal stem cells are multipotent cells that can be harvested from the bone marrow, and they present the potential to differentiate down a variety of pathways to become one of multiple types of specialized cells, ranging from osteoblasts to myocytes [16]. Embryonic stem cells have received considerable interest, given their capacity to differentiate into cells derived from each of the three germinal layers. However, ethical considerations presently associated with the sourcing of embryonic stem cells have dampened enthusiasm for their use in research and clinical therapies. To circumvent this source limitation, various techniques have been developed and applied over the years toward genetic manipulation of cells, which potentially can be harnessed toward therapeutic ends. For example, methods for genetic manipulation of adult somatic cells to make them more embryonic have emerged in recent years to produce iPS cells. The use of iPS cells differentiated under controlled conditions into cells of various lineages has been touted as a compromise source solution; however to date their use has been limited primarily to in vitro applications [17].

Fortunately, various tissues throughout the body contain resident stem cells that can be harvested for application in tissue engineering and regenerative medicine strategies. Although the differentiation potential of adult-derived progenitor cells may be narrower than that of embryonic stem cells, the ethical burden associated with the harvest and use of adult-derived cells is generally lower. As previously mentioned, mesenchymal stem cells can be harvested from bone marrow and adipose tissue aspirates, and the multipotency can be leveraged in tissue engineering and regenerative medicine approaches for repair of numerous tissues of craniofacial relevance, including bone, cartilage, and muscle. Other sources of stem and progenitor cell populations can be found in the oral and maxillofacial regions and include dental pulp, the periodontal ligament, salivary glands [18], and oral epithelium [19, 20].

Although derived from adult cells, iPS cells can share the ability of embryonic stem cells to differentiate down various lineages into cell types from the three germinal layers [21, 22]. The unique pluripotency of this adult-derived cell population has inspired investigation of iPS cells for a variety of tissue engineering and regenerative medicine applications. However, the genetic manipulation associated with the derivation of iPS cells likely will complicate the regulatory considerations associated with their potential clinical application. Concerns also exist about their genetic stability, including the possibility that neoplasms may present after long-term implant [23, 17].

1.6 Biologically Active Factors and the Extracellular Environment

To effect repair or regeneration of the tissues of interest, cells applied in tissue engineering and regenerative medicine strategies must receive appropriate signals to direct their actions. The extracellular matrix in which cells normally reside in tissues presents a complex milieu of biologically active factors that provides signals to cells in a spatiotemporally controlled fashion in support of tissue function and maintenance. The functions of cells within tissues also may be influenced by additional factors, including mechanical forces transmitted through the extracellular matrix, oxygen tension, and osmotic pressure. Tissue transfer techniques generally leverage the tissue-specific biological activity of the extracellular matrix associated with the transferred tissue in combination with the action of viable cells that may be transferred with the tissue and cells at the recipient site to achieve tissue reconstruction, albeit often with an associated donor site morbidity.

Tissue engineering and regenerative medicine approaches for tissue reconstruction often leverage scaffolds to conduct tissue repair, but such materials should ideally mimic key properties associated with the extracellular matrix of the target tissue to direct formation of the appropriate tissue type. As previously discussed, scaffold materials provide a structure to support tissue formation in a specific volume, and they can be tailored in many cases to transmit mechanical forces to cells within them. In addition, various techniques have been developed to incorporate biologically active moieties within scaffolds to direct tissue repair. These methods are of particular interest for synthetically derived scaffold materials that may inherently lack biologically active cues. As growth factors serve a number of critical roles in the development and maintenance of tissues, a variety of tissue engineering and regenerative medicine approaches have sought to leverage select growth factors alone or in combination to inform tissue formation. Some regenerative medicine approaches, for example, seek to deliver the growth factor(s) of interest directly to the tissue defect either alone or in combination with a carrier material. For example, bone morphogenetic protein-2 (BMP-2) has been applied clinically in combination with collagen carriers to promote bone tissue regeneration in the craniofacial skeleton [24]. Our own laboratories have used recombinantly produced heparan sulfate proteoglycan domains derived from perlecan/HSPG2 for time-controlled release of growth factors including BMP-2, fibroblast growth factor 2 (FGF2), and transforming growth factor β [25, 26]. Many growth factors present a complex three-dimensional structure that must be retained for proper biological activity, yet the structure can be rapidly degraded in the extracellular environment. Accordingly, a variety of approaches have sought to employ biomaterial carriers for targeted and controlled delivery of biologically active factors. In some cases, the carrier shields the factors from the extracellular environment until release. However, the released growth factors can be rapidly cleared, which presents a considerable challenge in achieving and maintaining sufficient concentrations for therapeutic effect. In addition, delivery of growth factors and other molecular therapeutics in appropriate spatiotemporal patterns to promote the desired tissue regeneration outcome presents an exceptional challenge.

Alternative strategies seek to leverage specific peptide sequences, which may be derived from or inspired by growth factors or other components of the extracellular matrix, to direct formation of the desired tissue. As peptide sequences generally depend less on three-dimensional structure for biological activity than growth factors, they may better withstand degradation in the extracellular environment and may even present less potential immunogenicity [27]. In some tissue engineering and regenerative medicine approaches, specific peptide sequence can be tethered to scaffold materials to direct cell attachment and migration, and in certain cases differentiation of progenitor cells. These motifs can be presented either as pendant or "hanging" motifs or may be immobilized into the scaffold materials by cross-linking reactions [28].

Another approach for introduction of biologically active cues to inform tissue formation can be found in gene therapy and nucleic acid delivery technologies. Specific nucleic acid sequences can be targeted for delivery to promote or to inhibit production of desired proteins. The structural stability of nucleic acids relative to proteins presents a key advantage for delivery in tissue engineering and regenerative medicine approaches. Moreover, nucleic acids containing a specific gene of interest with an associated promoter can be delivered to cells to enable production of the factor of interest. Such an approach enables cells that receive the gene of interest to produce the encoded factor, in some cases in a sustained fashion, which circumvents some of the challenges associated with direct delivery of the factor of interest. An important consideration in gene therapy and nucleic acid delivery approaches is control of the expression of the therapeutic gene. Several methods have been explored to induce or repress expression of delivered genes by cells through chemical or physical triggers. However, delivery of nucleic acids to targeted cells presents a variety of challenges, especially considering natural mechanisms that have evolved to detect and remove exogenous nucleic acids from the body. To address these challenges, a range of strategies have emerged to leverage viral carriers for gene delivery with high efficiency. Although the efficiency of gene transfer with viral carriers can be quite high, a variety of ethical and regulatory considerations presently associated with their use dampen broad clinical application. Accordingly, non-viral gene delivery strategies remain an attractive area of investigation.

1.7 Current and Future Applications in Maxillofacial Reconstruction

Strategies aimed at reconstruction and regeneration of maxillomandibular structures require the combined actions of matrix, cells, and biologically active factors. Although tissue transfer, tissue engineering, and regenerative medicine approaches may adopt different methods for introducing and directing the actions of these players, they are united by an overall objective of restoring the function and cosmesis of missing or compromised tissues. Surgical approaches have long leveraged materials, cells, and biologically active factors resident within grafts and flaps, presented by clinical products, and persistent at the recipient site to effect tissue reconstruction. Indeed, in some respects, surgeons leveraged these fundamental principles of tissue engineering and regenerative medicine in maxillomandibular reconstruction well in advance of the emergence of these topics as specialized fields of investigation and regulation [3]. While the successes of existing surgical approaches to maxillomandibular reconstruction should not be discounted, limitations underscore the need for continued development of technologies and approaches to improve reconstructive outcomes while minimizing morbidity.

Tissue engineering and regenerative medicine tools present the potential to provide alternative sources of bone and other tissues for use in maxillomandibular reconstruction with little or minimal donor site morbidity. Beyond bone and vasculature, tissue engineering and regenerative medicine strategies are emerging to target a variety of additional tissues and structures of relevance to maxillomandibular reconstruction, including condylar cartilage, muscle, nerve, salivary glands, and dentition. A key concept emerging in the arenas of surgical repair, tissue engineering, and regenerative medicine for maxillomandibular reconstruction involves recognition of the importance of early integration of multiple tissues. For example, some current surgical approaches involve immediate place-

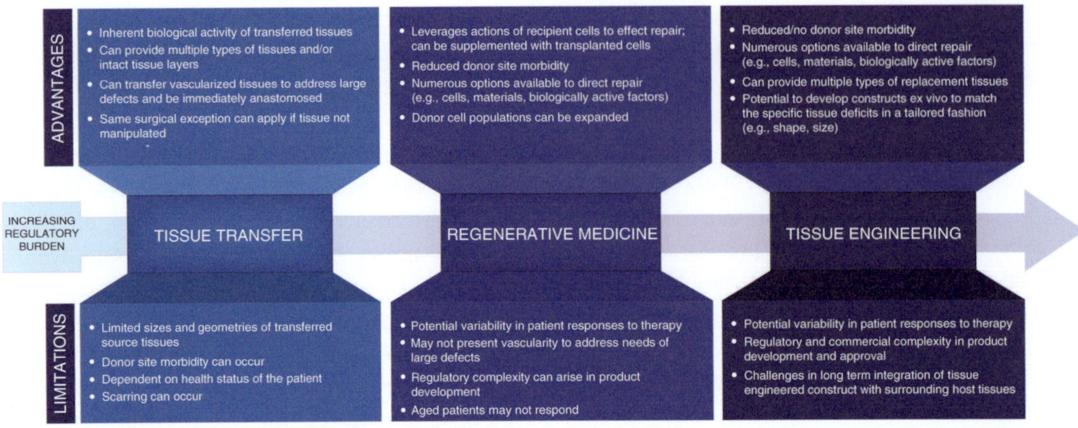

Fig. 1.2 An overview of the advantages and limitations of tissue transfer, regenerative medicine, and tissue engineering approaches to maxillomandibular repair, highlighting the increasing regulatory burden associated with each respective approach. Credit: Ms. Allison Haynie

ment of dental implants in reconstructed jaws to facilitate mechanical integration and return to masticatory function [29, 30]. In parallel, tissue engineering and regenerative medicine approaches are emerging that involve combined development of bone with integrated dentition for maxillomandibular reconstruction [31–33]. However, the complexity of any approach for maxillomandibular reconstruction should be thoughtfully considered, especially in terms of development of tissue engineering and regenerative medicine technologies for repair in the region. Indeed, the commercialization and regulatory burden associated with tissue engineering and regenerative medicine technologies generally scales with the complexity of the technology, which motivates consideration of simple solutions whenever appropriate [34] (see Fig. 1.2). Nevertheless, continued advances in maxillomandibular reconstruction remain on the horizon and will involve collaborative and multidisciplinary development of innovative approaches harnessing the fundamental principles of tissue engineering and regenerative medicine to meet the clinical need.

Acknowledgments The authors thank Ms. Allison Haynie for support in the graphic design of the figures in this chapter. The authors also acknowledge partial support for this work from a Peter Geistlich Research Award from the Osteo Science Foundation (FKK) and the National Institutes of Health grant AR054385 (MCFC).

References

1. Genden EM. Reconstruction of the mandible and the maxilla: the evolution of surgical technique. Arch Facial Plast Surg. 2010;12(2):87–90. https://doi.org/10.1001/archfacial.2010.18.
2. Neligan PC, Lipa JE. Reconstruction of mandible, maxilla, and skull base. In: Butler CE, Fine NA, editors. Principles of cancer reconstructive surgery. Boston: Springer; 2008. p. 117–40.
3. Katari R, Peloso A, Orlando G. Tissue engineering and regenerative medicine: semantic considerations for an evolving paradigm. Front Bioeng Biotechnol. 2014;2:57. https://doi.org/10.3389/fbioe.2014.00057.
4. Fox CF, Skalak R. Tissue engineering: proceedings of a workshop, held at Granlibakken, Lake Tahoe, California, February 26-29, 1988, UCLA symposia on molecular and cellular biology. New York: Liss; 1988.
5. National Institute of Biomedical Imaging and Bioengineering. Science education: glossary of terms. 2018. https://www.nibib.nih.gov/science-education/glossary. Accessed 2 Jan 2018.
6. Martin U. Therapeutic application of pluripotent stem cells: challenges and risks. Front Med. 2017;4:229. https://doi.org/10.3389/fmed.2017.00229.
7. Simkin J, Seifert AW. Concise review: translating regenerative biology into clinically relevant therapies: are we on the right path? Stem Cells Transl Med. 2018;7(2):220–31. https://doi.org/10.1002/sctm.17-0213.
8. Vining KH, Mooney DJ. Mechanical forces direct stem cell behaviour in development and regeneration. Nat Rev Mol Cell Biol. 2017;18(12):728–42. https://doi.org/10.1038/nrm.2017.108.
9. Williams DF. The Williams dictionary of biomaterials. Liverpool: Liverpool University Press; 1999.

10. Hawkins AM, Milbrandt TA, Puleo DA, Hilt JZ. Composite hydrogel scaffolds with controlled pore opening via biodegradable hydrogel porogen degradation. J Biomed Mater Res A. 2014;102(2):400–12. https://doi.org/10.1002/jbm.a.34697.
11. Vega SL, Kwon MY, Burdick JA. Recent advances in hydrogels for cartilage tissue engineering. Eur Cell Mater. 2017;33:59–75. https://doi.org/10.22203/eCM.v033a05.
12. Ali S, Saik JE, Gould DJ, Dickinson ME, West JL. Immobilization of cell-adhesive laminin peptides in degradable PEGDA hydrogels influences endothelial cell tubulogenesis. Biores Open Access. 2013;2(4):241–9. https://doi.org/10.1089/biores.2013.0021.
13. Seitz JM, Durisin M, Goldman J, Drelich JW. Recent advances in biodegradable metals for medical sutures: a critical review. Adv Healthc Mater. 2015;4(13):1915–36. https://doi.org/10.1002/adhm.201500189.
14. Thibault RA, Mikos AG, Kasper FK. Scaffold/extracellular matrix hybrid constructs for bone-tissue engineering. Adv Healthc Mater. 2013;2(1):13–24. https://doi.org/10.1002/adhm.201200209.
15. Mao JJ, Stosich MS, Moioli EK, Lee CH, Fu SY, Bastian B, et al. Facial reconstruction by biosurgery: cell transplantation versus cell homing. Tissue Eng B Rev. 2010;16(2):257–62.
16. Murphy MB, Moncivais K, Caplan AI. Mesenchymal stem cells: environmentally responsive therapeutics for regenerative medicine. Exp Mol Med. 2013;45:e54. https://doi.org/10.1038/emm.2013.94.
17. Ortmann D, Vallier L. Variability of human pluripotent stem cell lines. Curr Opin Genet Dev. 2017;46:179–85. https://doi.org/10.1016/j.gde.2017.07.004.
18. Srinivasan PP, Patel VN, Liu S, Harrington DA, Hoffman MP, Jia X, et al. Primary salivary human stem/progenitor cells undergo microenvironment-driven acinar-like differentiation in hyaluronate hydrogel culture. Stem Cells Transl Med. 2017;6(1):110–20. https://doi.org/10.5966/sctm.2016-0083.
19. Egusa H, Sonoyama W, Nishimura M, Atsuta I, Akiyama K. Stem cells in dentistry—part I: stem cell sources. J Prosthodont Res. 2012;56(3):151–65. https://doi.org/10.1016/j.jpor.2012.06.001.
20. Mao JJ, Giannobile WV, Helms JA, Hollister SJ, Krebsbach PH, Longaker MT, et al. Craniofacial tissue engineering by stem cells. J Dent Res. 2006;85(11):966–79.
21. Nakagawa M, Koyanagi M, Tanabe K, Takahashi K, Ichisaka T, Aoi T, et al. Generation of induced pluripotent stem cells without Myc from mouse and human fibroblasts. Nat Biotechnol. 2008;26(1):101–6. https://doi.org/10.1038/nbt1374.
22. Takahashi K, Tanabe K, Ohnuki M, Narita M, Ichisaka T, Tomoda K, et al. Induction of pluripotent stem cells from adult human fibroblasts by defined factors. Cell. 2007;131(5):861–72. https://doi.org/10.1016/j.cell.2007.11.019.
23. Martin U. Genome stability of programmed stem cell products. Adv Drug Deliv Rev. 2017;120:108–17. https://doi.org/10.1016/j.addr.2017.09.004.
24. Spagnoli D, Choi C. Extraction socket grafting and buccal wall regeneration with recombinant human bone morphogenetic protein-2 and acellular collagen sponge. Atlas Oral Maxillofac Surg Clin North Am. 2013;21(2):175–83. https://doi.org/10.1016/j.cxom.2013.05.003.
25. Srinivasan PP, McCoy SY, Jha AK, Yang W, Jia X, Farach-Carson MC, et al. Injectable perlecan domain 1-hyaluronan microgels potentiate the cartilage repair effect of BMP2 in a murine model of early osteoarthritis. Biomed Mater. 2012;7(2):024109. https://doi.org/10.1088/1748-6041/7/2/024109.
26. Yang WD, Gomes RR Jr, Alicknavitch M, Farach-Carson MC, Carson DD. Perlecan domain I promotes fibroblast growth factor 2 delivery in collagen I fibril scaffolds. Tissue Eng. 2005;11(1-2):76–89. https://doi.org/10.1089/ten.2005.11.76.
27. Kesireddy V, Kasper FK. Approaches for building bioactive elements into synthetic scaffolds for bone tissue engineering. J Mater Chem B. 2016;4(42):6773–86. https://doi.org/10.1039/C6TB00783J.
28. Liu S, Dicker KT, Jia X. Modular and orthogonal synthesis of hybrid polymers and networks. Chem Commun (Camb). 2015;51(25):5218–37. https://doi.org/10.1039/c4cc09568e.
29. Levine JP, Bae JS, Soares M, Brecht LE, Saadeh PB, Ceradini DJ, et al. Jaw in a day: total maxillofacial reconstruction using digital technology. Plast Reconstr Surg. 2013;131(6):1386–91. https://doi.org/10.1097/PRS.0b013e31828bd8d0.
30. Runyan CM, Sharma V, Staffenberg DA, Levine JP, Brecht LE, Wexler LH, et al. Jaw in a day: state of the art in maxillary reconstruction. J Craniofac Surg. 2016;27(8):2101–4. https://doi.org/10.1097/SCS.0000000000003139.
31. Abukawa H, Zhang W, Young CS, Asrican R, Vacanti JP, Kaban LB, et al. Reconstructing mandibular defects using autologous tissue-engineered tooth and bone constructs. J Oral Maxillofac Surg. 2009;67(2):335–47. https://doi.org/10.1016/j.joms.2008.09.002.
32. Young CS, Abukawa H, Asrican R, Ravens M, Troulis MJ, Kaban LB, et al. Tissue-engineered hybrid tooth and bone. Tissue Eng. 2005;11(9-10):1599–610. https://doi.org/10.1089/ten.2005.11.1599.
33. Zhang W, Abukawa H, Troulis MJ, Kaban LB, Vacanti JP, Yelick PC. Tissue engineered hybrid tooth-bone constructs. Methods. 2009;47(2):122–8. https://doi.org/10.1016/j.ymeth.2008.09.004.
34. Atala A, Kasper FK, Mikos AG. Engineering complex tissues. Sci Transl Med. 2012;4(160):160rv12. https://doi.org/10.1126/scitranslmed.3004890.

2 Traditional Use of Autogenous and Non-autogenous Grafts in Head and Neck Reconstruction: Principles of Conventional Bone Grafting

Pooja Gangwani, Shahid R. Aziz, and Jose M. Marchena

Maxillary and mandibular defects may occur secondary to infection, trauma, or tumor resection [1–3]. This may result into functional and esthetic compromise which can affect quality of life [1, 2, 4, 5]. Reconstruction of such defects remains challenging [5, 6], and may require vascularized or non-vascularized grafts [7, 8]. Contemporary vascularized grafts for jaw reconstruction include fibular, radial forearm, scapular, and deep circumflex iliac artery free flaps [9–14]. Traditional non-vascularized grafts are generally obtained from the anterior or posterior iliac crests [15–17] or from the chest as costochondral grafts when reconstruction of the ramus-condyle unit is necessary [18, 19]. Calvarial and tibial bone may be used as alternatives, either alone in smaller defects or as graft expanders [20, 21]. Finally, intraoral grafts from the lateral ramus and symphysis may be used for smaller dentoalveolar and implant surgery-related augmentation needs [22–24]. The purpose of this chapter is to review conventional non-vascularized grafts commonly used in oral and maxillofacial surgery. Accepted indications, advantages, limitations, and potential associated complications will be reviewed. Modern approaches utilizing free vascularized grafts, computer-assisted planning and surgery, or tissue engineering techniques will be discussed elsewhere in this book.

P. Gangwani
Saint Joseph's University Medical Center, Paterson, NJ, USA

Oral and Maxillofacial Surgery, University of Rochester, Rochester, NY, USA

S. R. Aziz
Oral and Maxillofacial Surgery, Rutgers School of Dental Medicine, Newark, NJ, USA

Division of Plastic and Reconstructive Surgery, Division of Surgery, Rutgers New Jersey Medical School, Newark, NJ, USA

East West Medical College/Update Dental School, Dhaka, Bangladesh

J. M. Marchena (✉)
East West Medical College/Update Dental School, Dhaka, Bangladesh

Oral and Maxillofacial Surgery, Ben Taub Hospital, Houston, TX, USA

Oral and Maxillofacial Surgery, University of Texas Health Science Center, Houston, TX, USA
e-mail: Jose.M.Marchena@uth.tmc.edu

2.1 Autogenous Versus Non-autogenous Bone Grafts

It is important for the surgeon to understand the difference between the healing of autogenous versus non-autogenous grafts, as this impacts surgical treatment planning and the strategy chosen for the particular problem at hand. Depending

J. C. Melville et al. (eds.), *Regenerative Strategies for Maxillary and Mandibular Reconstruction*,
https://doi.org/10.1007/978-3-319-93668-0_2

on the reconstructive needs, autogenous grafts may be harvested as block grafts containing both cancellous and cortical bone, cancellous bone only, or as pure cortical grafts. These autogenous bone grafts essentially function as transplants of viable bone and have the advantage of providing osteogenic, osteoinductive, and osteoconductive properties [25, 26]. The transplanted viable osteogenic cells allow for primary bone formation or osteogenesis. These grafts also allow for bone formation via osteoinduction which is the recruitment and induction of pluripotent mesenchymal stem cells by growth factors to differentiate into osteoblasts and chondroblasts and to ultimately form bone [25–27]. Finally, the bony architecture of the graft itself also provides a biological scaffold for osteoconduction which is the guided apposition and ingrowth of new bone from the recipient site [25–27]. These biological properties allow for more predictable graft healing and incorporation compared to nonviable non-autogenous grafts.

Non-autogenous grafts are available either as allogeneic or xenografts. Allogenic bone grafts are obtained from genetically dissimilar donors of the same species, whereas xenografts are obtained from a different species [27]. These grafts are available as mineralized, demineralized, fresh, frozen, or freeze-dried bone. The bone grafts are then processed to ensure sterility and to remove the antigenic components to minimize a host immune response and subsequent graft rejection [27]. Processing techniques include antibiotic washing, ultrasonic washing, physical debridement, treatment with ethanol, and gamma irradiation [27]. Mineralized allogenic bone grafts act predominantly as osteoconductive scaffolds, whereas demineralized allogenic bone grafts can possess both osteoconductive and osteoinductive properties [27]. Grafts can be obtained as preparations of particulate chips or as corticocancellous blocks.

Particulate allograft chips or powder grafts generally undergo more predictable revascularization compared to cortical or block allografts and are, therefore, incorporated faster at the recipient site [27]. Because of their particulate nature, they do not possess favorable mechanical properties and are more prone to resorption [25, 27]. As a result, their use as the sole graft component is generally limited to walled dentoalveolar defects (socket preservation technique) and periodontal bony deficiencies around dental implants. Block allografts are generally used for bridging of small bony gaps such as those resulting from inferior repositioning of the maxilla during orthognathic surgery. They can be rigidly fixated with screws or plates and provide stability at the bony gap while providing an osteoconductive bridge. More recent tissue engineering techniques using particulate allografts combined with autogenous bone marrow and recombinant bone morphogenic proteins for major segmental defects have shown good results [28]. Such strategies will be discussed later in this book.

2.2 Conventional Autogenous Non-vascularized Grafts

In maxillofacial surgery, conventional non-vascularized bone grafts are generally harvested from the anterior or posterior iliac crests, tibia, rib, calvarium, lateral ramus, or symphysis (chin). Grafts harvested from intraoral sites are used to correct intraoral defects only. Grafts harvested from extraoral sites may be used to correct major contour or segmental maxillofacial defects. This section will review conventional reconstructive strategies and techniques involving the most commonly used non-vascularized autogenous bone grafts. Case examples will be shown to highlight the conventional principles and techniques discussed.

2.2.1 Anterior and Posterior Iliac Crest

The anterior and posterior iliac crests are commonly harvested as particulate cancellous grafts, as block grafts containing only cortex, or as blocks containing both cortical and cancellous bone. The advantage of including cortical bone is that it allows for screw fixation of the graft. This will prevent mobility thus reducing the risk of extrusion, resorption, infection, or lack of incorporation of the graft. Furthermore, if used to span a continuity defect, fixation of the graft across the defect provides immediate stability of the graft itself and the recipient site segments. Both sites are popular donor sites as they are readily accessible and provide generous amounts of required autogenous bone stock.

The anterior ilium is more commonly harvested and can provide both cancellous marrow and corticocancellous blocks [29]. Approximately 50 cc of uncompressed cancellous bone or a corticocancellous block of 5 × 3 cm may be obtained from the anterior ilium [16, 25, 29]. Reconstruction of defects up to 5 cm in length can be managed with a single anterior ilium [25, 29]. If a corticocancellous block is required, a medial approach whereby the iliacus muscle is reflected off the medial cortex up to a depth of 5–7 cm is preferred [25, 29]. In children, the harvesting procedure is fundamentally the same as in adults with the exception of preservation of the cartilaginous cap at the crest since this area functions as a potential growth and ossification center prior to the completion of growth [30].

Harvested cancellous grafts may be used for fracture non-unions, alveolar clefts, sinus lifting, or mandibular reconstruction in conjunction with titanium mesh or cadaveric cribs. Block grafts may be used for contour augmentation or segmental defects. Since block grafts can be fixated in place with screws, they may be placed via a trans-oral approach provided that the recipient site is not compromised and that there is sufficient soft tissue to allow for primary closure over the graft (Fig. 2.1). A similar trans-oral reconstruction of the maxilla can be carried out using block grafts (Fig. 2.2). Complications related to anterior iliac crest bone graft harvesting have been reported to range from 1 to 25% and may include hematoma and seroma formation, nerve injuries, cosmetic deformity of the hip, abdominal hernia, pelvic instability, ileus, infection, and persistent pain [31]. In a more recent review of 372 adult patients undergoing anterior iliac crest harvesting, Almaiman et al. reported seroma formation in one patient, fractures in two patients, hematoma formation in one patient, and temporary nerve injury in three patients. None of the patients developed an infection or sustained permanent nerve injuries [31].

The posterior ilium is harvested when a larger amount of bone is needed and may provide up to 100 cc of uncompressed cancellous bone or a corticocancellous block of greater than 5 × 5 cm [15, 16, 25]. The majority of the bone reservoir is found below the insertion of the gluteus maximus muscle [15, 16, 25]. The surgical approach involves marking a 6–8 cm curvilinear incision between superior (superiorly) and medial cluneal nerves (inferiorly) [15, 16, 25]. The incision is placed 3 cm lateral to the midline, centered over the bony prominence to which the gluteus maximus muscle is inserted [15, 16, 25]. The skin, subcutaneous tissue, thoracodorsal fascia, and periosteum are sharply incised. The gluteus maximus muscle is then elevated off the tubercle, and a block of 5 × 5 cm corticocancellous bone may be harvested [15, 16, 25]. Cancellous bone can then be harvested from the resulting defect with bone curettes and bone gouges [15, 16, 25]. The block graft may be augmented with particulate cancellous bone when reconstructing segmental defects of the mandible (Figs. 2.3 and 2.4). Harvesting of posterior iliac crest bone has been associated with lower complication rates compared to the anterior ilium [32]. The main disadvantage associated with harvesting bone from the posterior ilium is that the patient needs to be placed in the prone position. This adds significant time to the operation, as well as risk of accidental extubation when repositioning the patient back to the supine position. Because a slight "jackknife" position is preferred to optimize surgical access, care must be taken to properly pad all pressure points. In male patients, attention must be directed to ensure the genitalia are not compressed in the prone position.

2.2.2 Costochondral Graft

The costochondral graft is generally considered to be a workhorse for the reconstruction of the mandibular ramus, condyle, and temporoman-

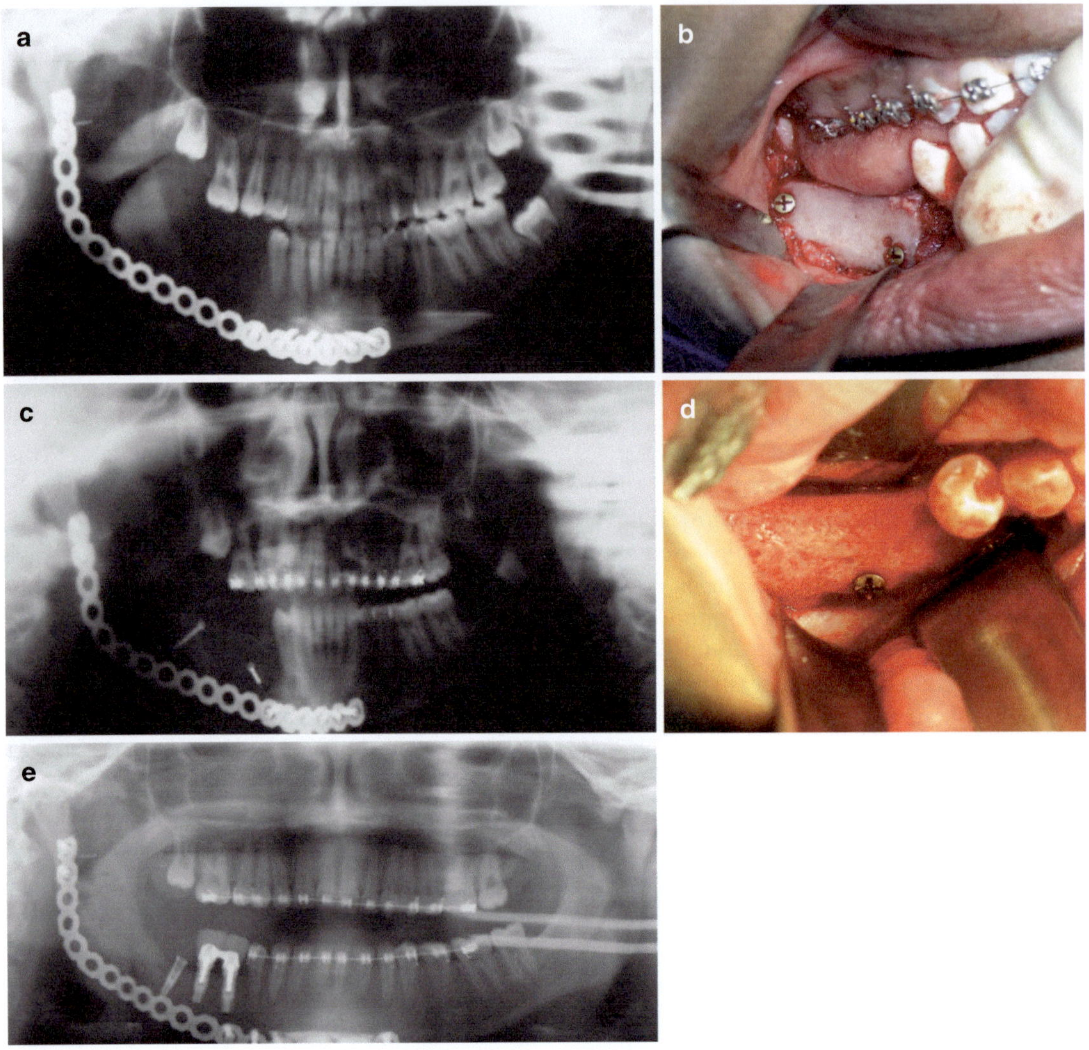

Fig. 2.1 Reconstruction of a mandibular tumor resection defect with a trans-oral anterior iliac crest bone graft. (**a**) Preoperative panoramic image showing right the mandibular defect. (**b**) Intraoperative photograph showing the iliac crest block graft fixated to the recipient bone with screws. (**c**) Immediate postoperative radiograph showing the position and screw fixation of the graft. (**d**) Intraoperative photograph prior to implant placement showing excellent graft incorporation and volume retention. (**e**) Panoramic image taken at long-term follow-up showing excellent graft incorporation and retention of vertical height

dibular joint [19, 33, 34]. However, in a recent review of the literature, Kumar et al. concluded that there is significant unpredictability with costochondral grafts with many grafts showing overgrowth, undergrowth, lateral overgrowth, and fractures [35]. Typically, the 5th, 6th, or 7th contralateral rib is chosen to reconstruct the ramus-condyle unit [19, 34, 36]. Approximately 9 cm of the rib can be harvested in children and about 15 cm in adults [19, 34]. Alternate ribs are obtained when more than one rib is required in order to minimize postoperative morbidity [19, 36]. Reconstruction of the ramus-condyle unit in pediatric patients require the use of 1–2 mm of cartilaginous cap along with the rib graft [18, 19]. Since the costochondral is rigid similar to a block graft, it can be fixated onto a pre-bent reconstruction plate or native mandible. This allows for such graft to be placed immediately at the time of tumor resection (Fig. 2.4).

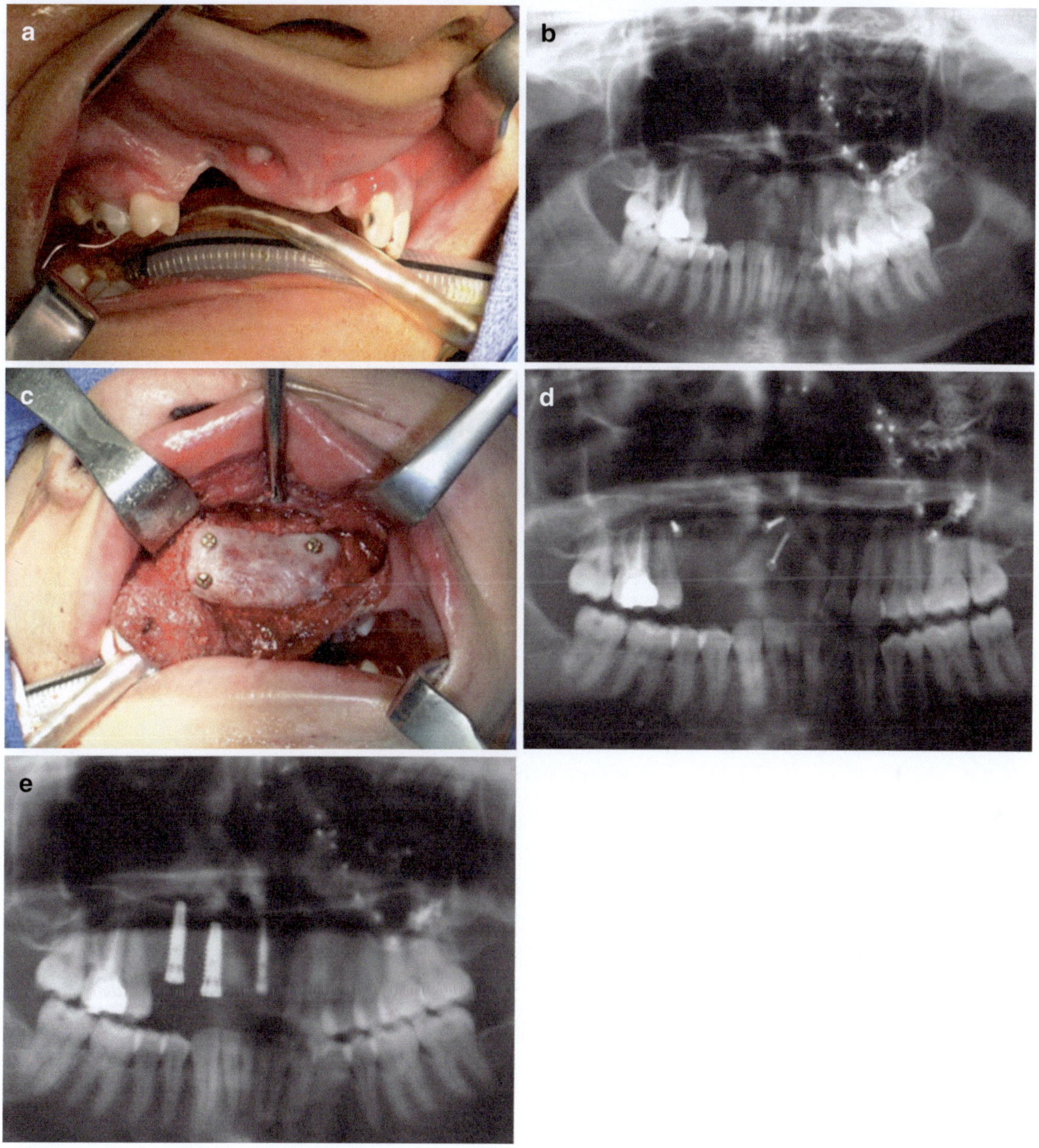

Fig. 2.2 Reconstruction of a post-traumatic anterior maxillary defect with an anterior iliac crest block graft. (**a**) Preoperative clinical picture showing the anterior maxillary defect. (**b**) Preoperative radiograph of the same defect. (**c**) Intraoperative photograph showing the anterior iliac crest block graft fixated in place with screws. (**d**) Immediate postoperative radiograph showing the position and screw fixation of the graft. (**e**) Panoramic radiograph taken after placement of dental implants showing excellent graft healing and graft incorporation

2.2.3 Tibial Metaphysis Graft

Bone from the proximal tibia can be obtained by either a medial or lateral approach [37]. In children, a medial approach is preferred [38]. These procedures are relatively easy to perform and are associated with a low complication rate of 1.3% [20, 39, 40]. Approximately 25 cc of cancellous bone can be harvested from a single site [20, 39]. If more than 25 cc of bone is needed, bilateral tibial grafts or additional autogenous harvest sites can be considered [20]. Cancellous grafts harvested from the tibia may be used for sinus lift procedures, nonunions, or small continuity defects. In addition, tibial cancellous bone is also commonly used for grafting of alveolar clefts (Fig. 2.5) [20, 39]. Postoperative complications like hematoma, infection, and proximal tibial fracture have been reported but occur infrequently [16, 20, 39, 40].

2.2.4 Calvarial Graft

A calvarial bone graft remains a feasible option for reconstruction of the maxilla and mandible in situations of complex maxillofacial trauma when a coronal flap is utilized for access [16, 21]. Such grafts are typically harvested as a monocortical block grafts [16, 41]. Typically, 1.5 × 3 cm cortical strips of the outer table of the thicker portion of the parietal bone are obtained [21]. Calvarial grafts are generally used for reconstruction of traumatic midface injuries (Fig. 2.6). Complications associated with harvesting of calvarial bone include alopecia along the incision line, bleeding, and dural tear [16, 21, 41].

2.2.5 Oral Bone Grafts

Common intraoral bone harvest sites include the mandibular ramus and symphysis [22–24]. Because they are inherently contaminated with

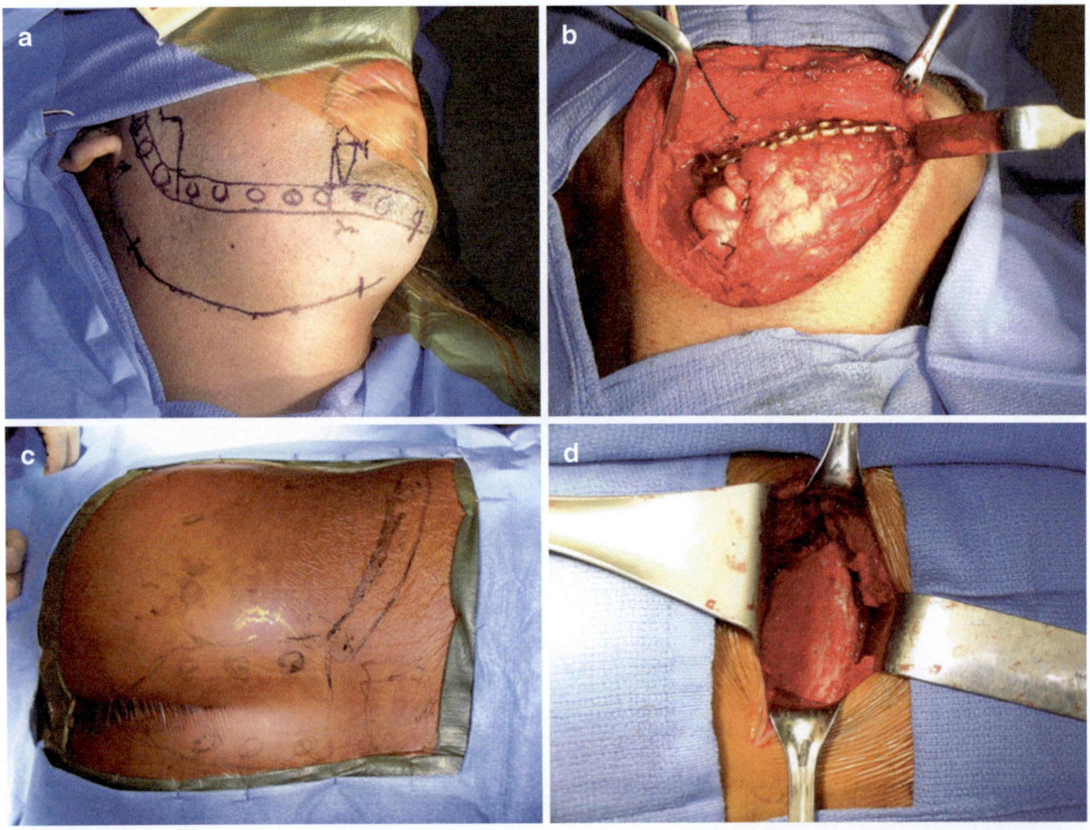

Fig. 2.3 Secondary reconstruction of a tumor defect of the posterior mandible with a block graft from the posterior ilium. (**a**) Intraoperative markings of the defect and planned incision. (**b**) Intraoperative photograph showing the extent of the defect. (**c**) Intraoperative markings of the posterior iliac crest donor site. (**d**) Block graft is harvested. (**e**) Graft is divided to create buccal and lingual cortices. (**f**) Grafts are fixated in place as buccal and lingual cortices. (**g**) Cancellous bone is packed around the block grafts. (**h**) Radiograph taken at 1-year follow-up after placement of dental implant fixtures

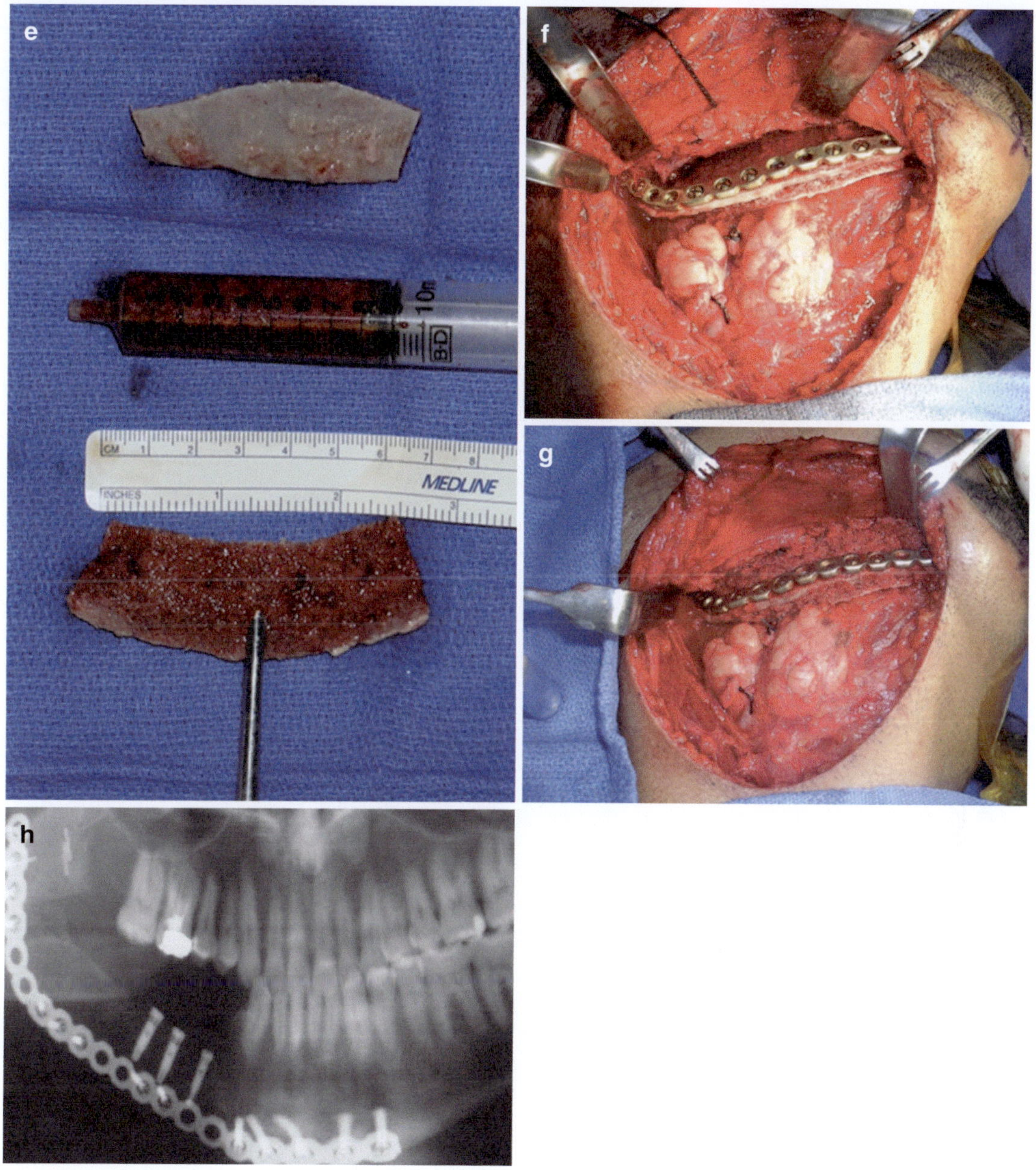

Fig. 2.3 (continued)

oral bacteria, these grafts are limited to intraoral recipient sites. They are generally used for the augmentation of alveolar defects and for grafting around dental implant fixtures [22–24]. Often non-autogenous (allogenic) bone graft products may be used to augment these smaller grafts.

Symphysis grafts primarily consist of cortical bone [22–24]. Two corticocancellous block grafts of up to 1.5 × 3 cm in dimension can be obtained from each side of the midline [23]. A bone mill can be used when a particulate graft is needed. About 7 cc of particulate graft can be obtained from the symphysis region [23]. The surgical procedure involves placing a superior osteotomy 5 mm inferior to the apices of the anterior mandibular teeth and an inferior osteotomy 10 mm

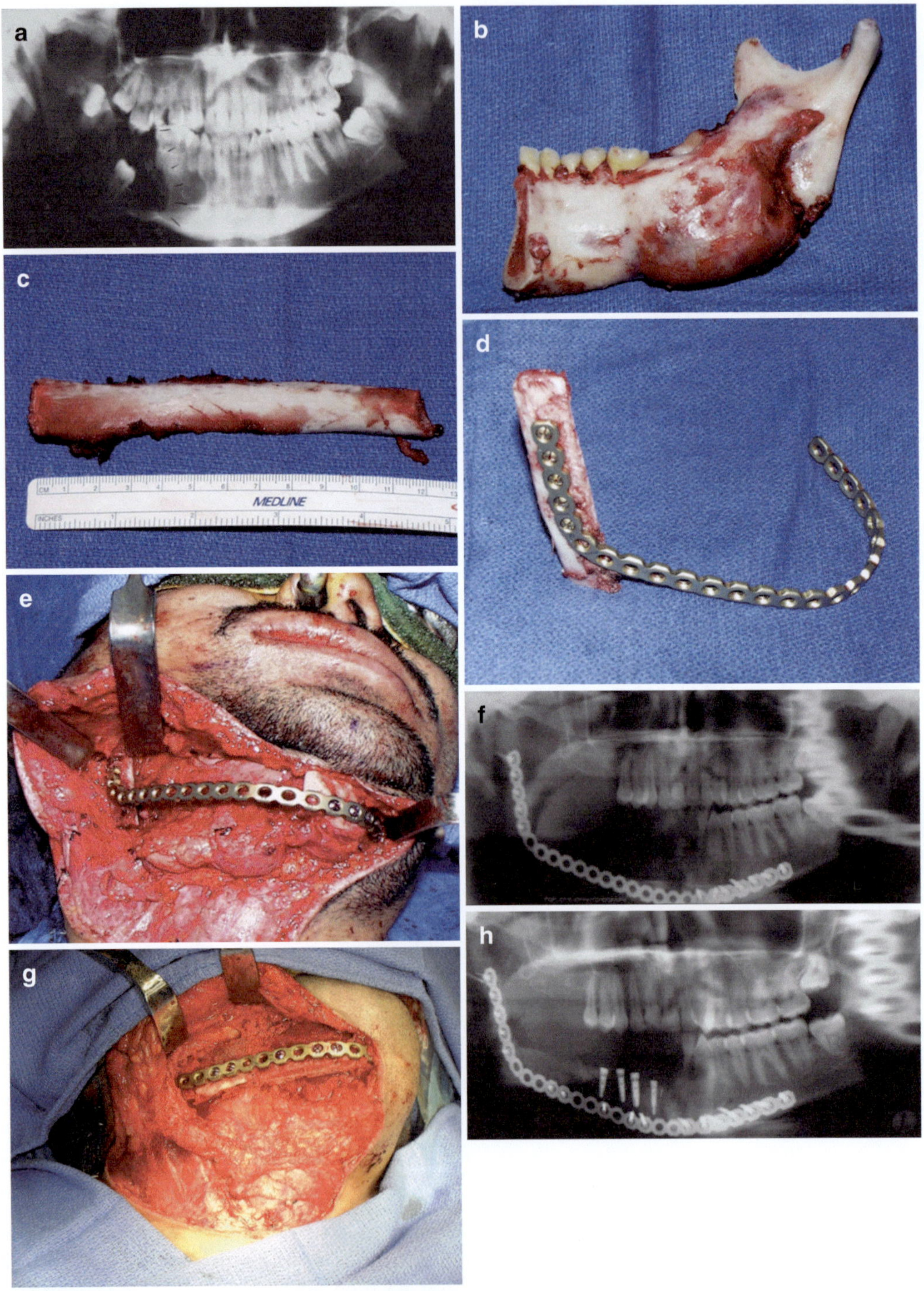

Fig. 2.4 Immediate reconstruction of the ramus-condyle unit and secondary reconstruction of the posterior mandible after resection of the right hemi-mandible with disarticulation. (**a**) Preoperative radiograph showing a large biopsy-proven ameloblastoma of the posterior mandible. (**b**) Resected specimen. (**c**) Harvested costochondral graft at the time of resection. (**d**) The costochondral graft is secured to a pre-bent reconstruction plate. (**e**) Intraoperative image showing the costochondral graft in place. (**f**) Postoperative image showing costochondral graft in in place. (**g**) Second-stage reconstruction of the body of the mandible with a posterior iliac crest block graft 5 months after the initial resection and immediate reconstruction of the ramus and condyle. (**h**) Radiograph taken at long-term follow-up after consolidation of the grafts and placement of dental implants

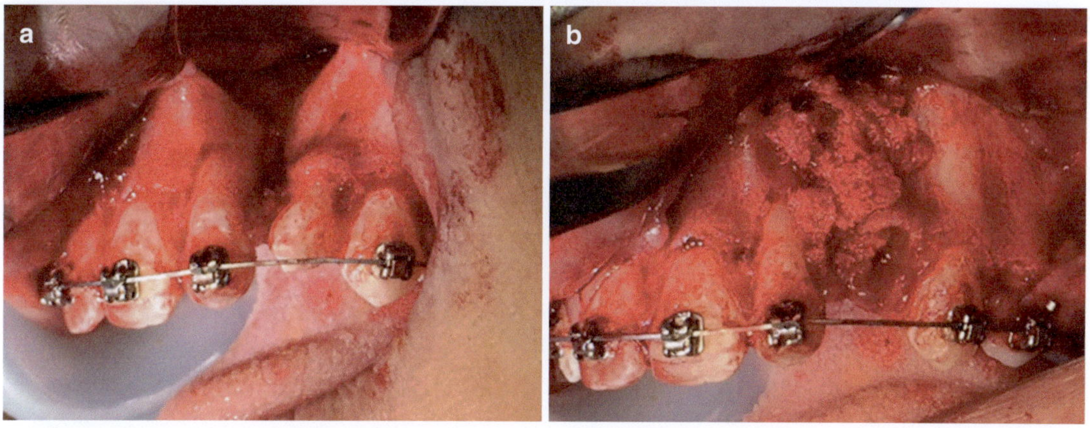

Fig. 2.5 Grafting of an alveolar cleft with autogenous cancellous bone. (**a**) Intraoperative photograph taken after exposure of the alveolar cleft. (**b**) Photograph showing cancellous graft in place

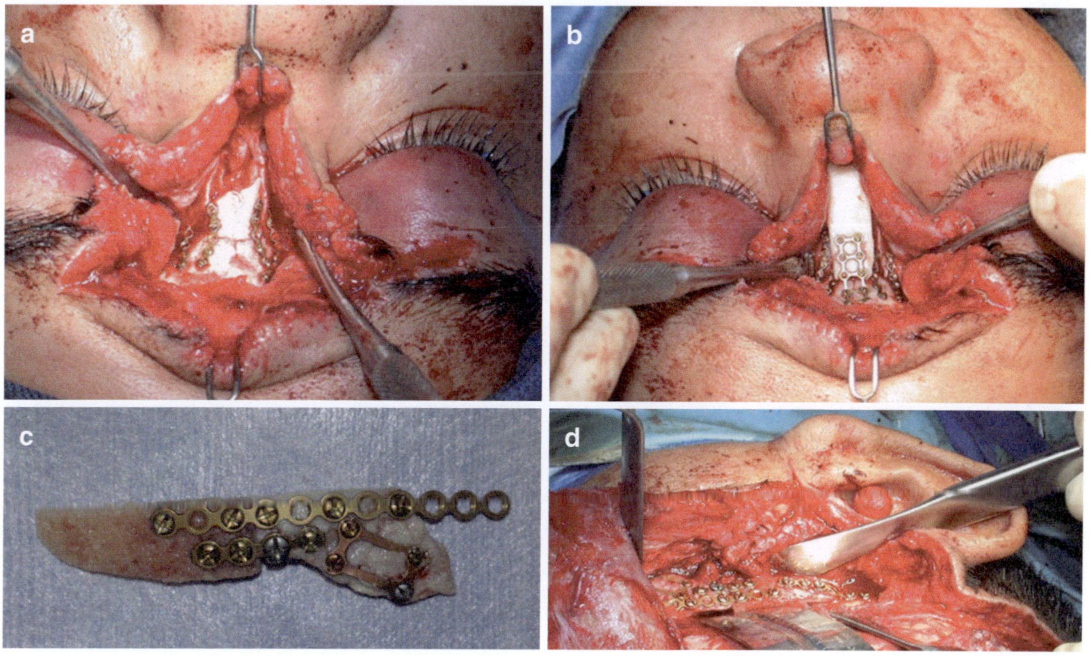

Fig. 2.6 Calvarial grafts used in maxillofacial trauma reconstruction. (**a**) Intraoperative photograph showing comminuted nasal fractures causing flattening of the nasal bridge. (**b**) Photograph showing the calvarial graft in place. (**c**) Calvarial graft harvested for reconstruction of a comminuted fractured zygomatic arch. (**d**) Photograph showing the calvarial graft in place to reestablish lateral facial projection and stability of the zygoma

above the inferior border of the mandible [22, 23]. Lateral osteotomies are made 5 mm anterior to the mental foramen bilaterally to connect the superior and inferior osteotomies [22, 23]. The authors prefer to leave a midline strut of intact bone in order to preserve chin morphology. Although harvested blocks may be morselized into particulate grafts, they are typically used as blocks and fixated in place (Fig. 2.7). Complications associated with symphysis graft harvesting include chin ptosis, paresthesia in the mental nerve distribution, wound dehiscence, and rarely mandibular fracture [22, 23, 42]. Occurrence of a mandibular fracture may be prevented by keeping the inferior osteotomy 8–10 mm superior to the inferior border of the mandible [22, 23, 42].

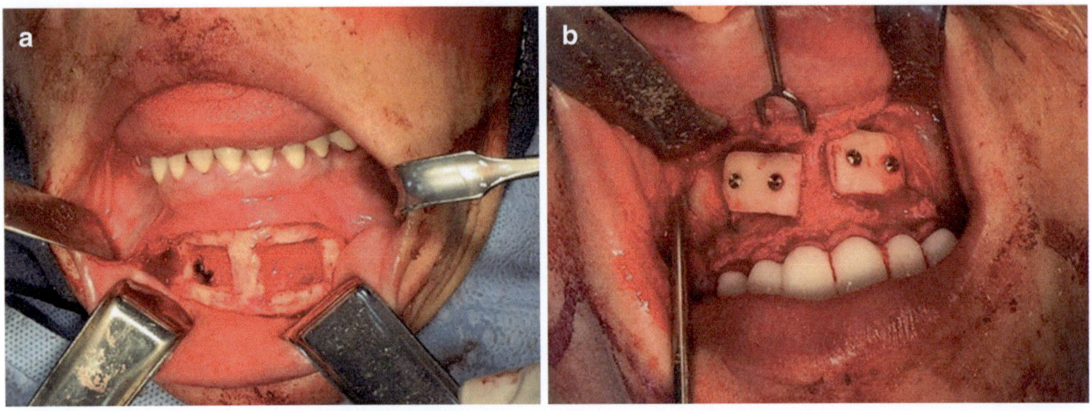

Fig. 2.7 Anterior maxillary augmentation using bilateral symphysis block grafts. (**a**) Intraoperative photograph showing the donor sites after harvesting of bilateral block grafts. Notice that a vertical midline strut is preserved to maintain chin morphology. (**b**) Photograph showing typical screw fixation of these grafts

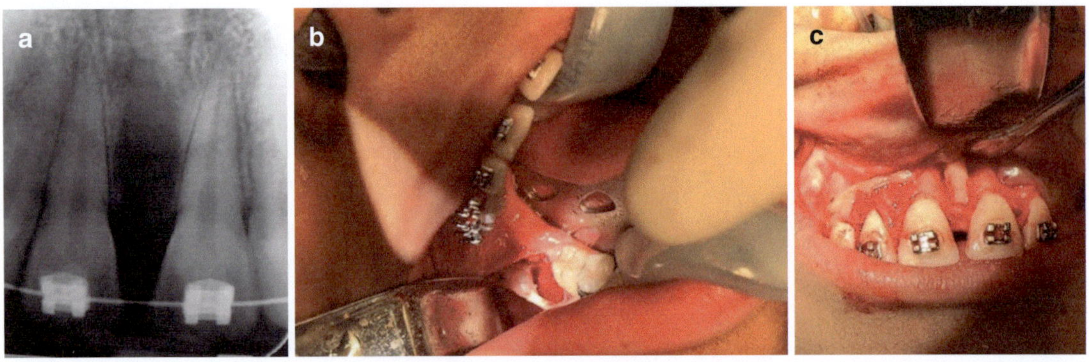

Fig. 2.8 Ramus cortical graft to a non-healing bony defect resulting from a surgically assisted palatal expansion. (**a**) Preoperative radiograph showing the midline bony defect. (**b**) Photograph showing lateral ramus donor site. (**c**) Intraoperative image showing the graft wedged in place which bridges the bony gap and provides stability to the segments

Ramus grafts are also harvested as cortical blocks. The blocks tend to be thinner compared to those harvested from the symphysis; however, a single block of up to 3 × 5 cm in size may be harvested [22, 24, 43]. The indications for a ramus graft generally overlap with those for a symphysis graft. They can be used as contour augmentation grafts or for grafting of bony gaps resulting from osteotomies (Fig. 2.8). If indicated, harvesting of a ramus graft can be carried out in conjunction with the removal of a problematic lower wisdom tooth as the surgical approach and exposure are similar. Injury to the inferior alveolar nerve is a possible complication when harvesting a ramus graft [22, 24, 43]. Such injury can be prevented by careful planning with a detailed analysis of relevant images and by placing the inferior osteotomy 5 mm inferior to the mandibular canal [24].

2.3 Conclusion

Conventional approaches with non-vascularized autogenous bone grafts provide predictable results in common oral and maxillofacial surgical practice provided the recipient sites have healthy vascularity and adequate soft tissue for would closure. Autogenous bone grafts are essentially transplants of viable bone and tend to undergo more predictable healing and graft incorporation compared to allografts or xenografts which only allow for osteoinduction and osteoconduction. The amount of bone and degree of stability needed will dictate whether the surgeon chooses block or particulate grafts. Particulate grafts are generally used for grafting of non-unions, alveolar clefts, maxillary sinus lifting, or small segmental defects. Because of their particulate nature, they become readily vascularized. Disadvantages of particulate grafts include risk of graft mobility, extrusion, and poor resistance to compressive forces. Block grafts, in contrast, can be fixated in place with screws thus providing graft stability as well as immediate stability across bony gaps. Similar to block grafts, costochondral grafts are rigid and can also be fixated onto a reconstruction plate or native mandible when reconstructing the ramus-condyle unit. Modern bone graft reconstruction approaches applying tissue engineering techniques, free vascularized grafts, and computer-assisted surgery and planning will be discussed elsewhere in this book.

References

1. Pogrel MA, Podlesh S, Anthony JP, Alexander J. A comparison of vascularized and nonvascularized bone grafts for reconstruction of mandibular continuity defects. J Oral Maxillofac Surg. 1997;55(11):1200–6.
2. Moura LB, Carvalho PH, Xavier CB, et al. Autogenous non-vascularized bone graft in segmental mandibular reconstruction: a systematic review. Int J Oral Maxillofac Surg. 2016;45(11):1388–94.
3. Hayden RE, Mullin DP, Patel AK. Reconstruction of the segmental mandibular defect: current state of the art. Curr Opin Otolaryngol Head Neck Surg. 2012;20(4):231–6.
4. Schlieve T, Hull W, Miloro M, Kolokythas A. Is immediate reconstruction of the mandible with nonvascularized bone graft following resection of benign pathology a viable treatment option? J Oral Maxillofac Surg. 2015;73(3):541–9.
5. Zou D, Huang W, Wang F, et al. Autologous ilium grafts: long-term results on immediate or staged functional rehabilitation of mandibular segmental defects using dental implants after tumor resection. Clin Implant Dent Relat Res. 2015;17(4):779–89.
6. Iino M, Fukuda M, Nagai H, et al. Evaluation of 15 mandibular reconstructions with Dumbach Titan Mesh-System and particulate cancellous bone and marrow harvested from bilateral posterior ilia. Oral Surg Oral Med Oral Pathol Oral Radiol Endod. 2009;107(4):e1–8.
7. Van Gemert JT, van Es RJ, Van Cann EM, Koole R. Nonvascularized bone grafts for segmental reconstruction of the mandible—a reappraisal. J Oral Maxillofac Surg. 2009;67(7):1446–52.
8. Bai XF, Wushou A, Zheng J, Li G. An alternative approach for mandible reconstruction. J Craniofac Surg. 2013;24(2):e195–8.
9. Chen YB, Chen HC, Hahn LH. Major mandibular reconstruction with vascularized bone grafts: indications and selection of donor tissue. Microsurgery. 1994;15(4):227–37.
10. Wei FC, Seah CS, Tsai YC, Liu SJ, Tsai MS. Fibula osteoseptocutaneous flap for reconstruction of composite mandibular defects. Plast Reconstr Surg. 1994;93(2):294–304; discussion 305-296.
11. Wang YY, Fan S, Zhang HQ, Lin ZY, Ye JT, Li JS. Virtual surgical planning in precise maxillary reconstruction with vascularized fibular graft after tumor ablation. J Oral Maxillofac Surg. 2016;74(6):1255–64.
12. Baker SR, Sullivan MJ. Osteocutaneous free scapular flap for one-stage mandibular reconstruction. Arch Otolaryngol Head Neck Surg. 1988;114(3):267–77.
13. Villaret DB, Futran NA. The indications and outcomes in the use of osteocutaneous radial forearm free flap. Head Neck. 2003;25(6):475–81.
14. Kimata Y, Uchiyama K, Sakuraba M, et al. Deep circumflex iliac perforator flap with iliac crest for mandibular reconstruction. Br J Plast Surg. 2001;54(6):487–90.
15. Aziz SR. Posterior iliac crest bone grafting. In: Bagheri SC, Bell RB, Khan HA, editors. Atlas of oral and maxillofacial surgery. St. Louis: Elsevier; 2016. p. 1270–4.
16. Zouhary KJ. Bone graft harvesting from distant sites: concepts and techniques. Oral Maxillofac Surg Clin North Am. 2010;22(3):301–16.
17. Marx R, Stevens M. Posterior ileum. In: Marx R, Stevens M, editors. Atlas of oral and extraoral bone harvesting. Chicago: Quintessence; 2010. p. 19–38.
18. Kaban LB, Bouchard C, Troulis MJ. A protocol for management of temporomandibular joint ankylosis in children. J Oral Maxillofac Surg. 2009;67(9):1966–78.
19. Krishnan DG. Costochondral graft. In: Bagheri SC, Bell RB, Khan HA, editors. Atlas of oral and maxillofacial surgery. St. Louis: Elsevier; 2016. p. 1282–8.
20. Kushner GM, Alpert B. Tibial bone graft. In: Bagheri SC, Bell RB, Khan HA, editors. Atlas of oral and maxillofacial surgery. St. Louis: Elsevier; 2016. p. 1275–81.
21. Curtis WJ, Golden B, Jaskolka MS. Calvarial bone graft. In: Bagheri SC, Bell RB, Khan HA, editors. Atlas of oral and maxillofacial surgery. St. Louis: Elsevier; 2016. p. 1289–99.
22. Yates D, Brockhoff HC, Finn R. Intraoral bone harvesting techniques. In: Bagheri SC, Bell RB, Khan HA, editors. Atlas of oral and maxillofacial surgery. St. Louis: Elsevier; 2016. p. 1300–4.

23. Marx R, Stevens M. Chin. In: Marx R, Stevens M, editors. Atlas of oral and extraoral bone harvesting. Chicago: Quintessence; 2010. p. 101–10.
24. Marx R, Stevens M. Mandibular ramus. In: Marx R, Stevens M, editors. Atlas of oral and extraoral bone harvesting. Chicago: Quintessence; 2010. p. 111–23.
25. Peleg M, Sawatari Y, Marx RN, et al. Use of corticocancellous allogeneic bone blocks for augmentation of alveolar bone defects. Int J Oral Maxillofac Implants. 2010;25(1):153–62.
26. Myeroff C, Archdeacon M. Autogenous bone graft: donor sites and techniques. J Bone Joint Surg Am. 2011;93(23):2227–36.
27. Khan SN, Cammisa FP Jr, Sandhu HS, Diwan AD, Girardi FP, Lane JM. The biology of bone grafting. J Am Acad Orthop Surg. 2005;13(1):77–86.
28. Melville JC, Marx RE, Tursun R, Moody MS, Hew D, Schacht SM, Starlley E, Peleg M, Sawatari J, Armentano L, Gomez J. The utilization of allogeneic bone, bone morphogenetic protein and bonem arrow aspirate concentrate for immediate reconstruction of benign tumor continuity defects. J Oral Maxillofac Surg. 2014;72(9 Suppl):e204–5.
29. Agarwal R, Obeid G. Anterior iliac crest bone grafting (AICBG). In: Kademani DTP, editor. Atlas of oral and maxillofacial surgery. St. Louis: Elsevier; 2016. p. 1261–9.
30. Potter J. Mandibular reconstruction. In: Bagheri SC, Bell RB, Khan HA, editors. Current therapy in oral and maxillofacial surgery. St. Louis: Elsevier; 2012. p. 483–96.
31. Almaiman M, Al-Bargi HH, Manson P. Complication of anterior iliac bone graft harvesting in 372 adult patients from may 2006 to may 2011 and a literature review. Craniomaxillofac Trauma Reconstr. 2013;6(4):257–66.
32. Ahlmann E, Patzakis M, Roidis N, Shepherd L, Holtom P. Comparison of anterior and posterior iliac crest bone grafts in terms of harvest-site morbidity and functional outcomes. J Bone Joint Surg Am. 2002;84-A(5):716–20.
33. Ahmad I, Choudhary R. Wide surgical excision with split rib graft reconstruction of mandible for ameloblastoma; our 10 year experience. Indian J Otolaryngol Head Neck Surg. 2013;65(1):40–3.
34. Marx R, Stevens M. Rib. In: Marx R, Stevens M, editors. Atlas of oral and extraoral bone harvesting. Chicago: Quitessence; 2010. p. 61–28.
35. Kumar P, Rattan V, Rai S. Do costochondral grafts have any growth potential in temporomandibular joint surgery? A systematic review. J Oral Biol Craniofac Res. 2015;5(3):198–202.
36. El-Sheikh MM, Zeitoun IM, Medra AM. The split rib bundle graft in mandibular reconstruction. J Craniomaxillofac Surg. 1992;20(8):326–32.
37. Benninger B, Ross A, Delamarter T. Approaches to proximal tibial bone harvest techniques. J Oral Maxillofac Res. 2012;3(2):e2.
38. Corre PK, Laure B, Bonnet R, Khonsari HR, Cordova LA, Mercier J. Medial proximal tibia donor site: contribution to alveolar cleft repair in children. Rev Stomatol Chir Maxillofac. 2011;112:280–5.
39. Kushner GM. Tibia bone graft harvest technique. Atlas Oral Maxillofac Surg Clin North Am. 2005;13(2):119–26.
40. O'Keefe RM, Reimer BL, Butterfield SL. Harvesting of autogenous cancellous bone graft from the proximal tibial metaphysis: a review of 230 cases. J Orthop Trauma. 1991;5:469–74.
41. Marx R, Stevens M. Cranial bone. In: Marx R, Stevens M, editors. Atlas of oral and extra oral bone harvesting. Chicago: Quintessence; 2010. p. 79–100.
42. Weibull L, Widmark G, Ivanoff CJ, Borg E, Rasmusson L. Morbidity after chin bone harvesting—a retrospective long-term follow-up study. Clin Implant Dent Relat Res. 2009;11(2):149–57.
43. Gungormus M, Yavuz MS. The ascending ramus of the mandible as a donor site in maxillofacial bone grafting. J Oral Maxillofac Surg. 2002;60(11):1316–8.

Reconstruction Considerations for the Posttraumatic, Benign Tumor, and Oncologic Patient

3

Jeffrey D. Johnson and Tang Ho

3.1 Introduction

Patients undergoing reconstructive surgery are subjected to a complicated and often arduous process and must therefore be carefully selected to ensure the most desired outcomes. In planning for the functional and aesthetic reconstruction of the midface and mandible, there are common pre-surgical considerations which must be addressed, regardless of the patient's primary disease process. The most important of these include type and extent of defect, overall patient health, comorbidities, and patient preference.

Overall patient health is a broad term that includes many of modifiable and non-modifiable factors. Of these, age has been highly studied and has not been shown to affect flap outcome or perioperative mortality rate. It should therefore not be a limiting factor in deciding which type of reconstruction to pursue [1]. Biological age, rather than chronological age, should be the greatest concern. No significant differences in disease-specific or overall survival were shown when comparing free flap patients less than or greater than 70. However, medical comorbidities have been found to be significant predictors of medical complications. In the same study, disease stage was the only significant predictor of recipient site complications [2]. Looking at flap survival specifically, preoperative comorbidity and not age was associated with poorer outcomes. It is important to note, however, that older patients are statistically more likely to have a higher presence of medical comorbidities [3], though the overwhelming body of literature demonstrates that it is the preexisting comorbidities, and not age, that portend poorer outcomes.

3.2 Reconstructive Options

Once a thorough preoperative risk assessment has taken place, it is important to explore all surgical and nonsurgical options available to give a patient the best possible functional and aesthetic results while minimizing morbidity. Furthermore, a thorough evaluation of all available options will allow the surgeon to appropriately counsel the patient in making an informed treatment decision. While free tissue transfer occupies the peak of the reconstructive ladder and is considered by many to be the gold standard of maxillofacial reconstruction, patients may also benefit from a combination of local or regional flaps, nonvascularized grafts, alloplastic materials, or the use of regenerative medicine techniques to restore form and function. Additionally, there is a role for the use of prosthodontics and obturators, especially

J. D. Johnson · T. Ho (✉)
Division of Facial Plastic and Reconstructive Surgery, Department of Otolaryngology – Head and Neck Surgery, University of Texas Health Science Center at Houston, Houston, TX, USA
e-mail: Tang.Ho@uth.tmc.edu

J. C. Melville et al. (eds.), *Regenerative Strategies for Maxillary and Mandibular Reconstruction*,
https://doi.org/10.1007/978-3-319-93668-0_3

for those patients who are poor surgical candidates or at significant risk for decreased healing. Often, these may also be the ideal options simply due to patient preference.

Obturators can be used alone or in conjunction with other reconstructive surgeries to improve functional outcomes. Factors that must be considered in choosing to use an obturator include donor site morbidity, operative time, the possibility of recurrence, and willingness of a patient to undergo surgical reconstruction. Oral function has been shown to be equal in patients who received implant-supported obturators compared to implant-supported fixed prostheses in free vascularized flaps, although there was lower patient-reported quality of life for patients with obturators [4]. Furthermore, an obturator can provide a valuable bridging option until a more definitive reconstruction can be performed.

Another option to consider in patients who are not good candidates for a prolonged or staged reconstructive surgery is the use of alloplastic materials, such as an immediate reconstruction with titanium bridging plates. While the use of alloplastic materials is a common aesthetic technique, their role in functional reconstruction is less clear. The little evidence available shows that the benefits afforded by a faster reconstruction and avoidance of donor site morbidity can be offset by relatively high failure (30.8%) and complication (40.1%) rates when used for mandibular reconstruction [5]. Failure rates with this approach increase in correlation with larger defect size, when the defect includes the mandibular symphysis, and in smokers [6]. Alloplastic reconstruction may have a role in patients with small defects or as a temporary measure, but the replacement of resected bone is usually necessary for successful reconstruction in the long term.

The functional success of a bony reconstruction primarily depends on bony union and viability of endosseous implants. Nonvascularized bone grafts (NVBG) are more often used after resection of benign lesions and vascularized bone grafts (VBG) after malignancies due to the size of the defect. Despite larger defect size, older age, presence of malignancy, and prior irradiation, patients with VBG showed higher incidence of bony union and implant success rate compared to NVBG [7]. Although there is theoretically decreased operative time and blood loss with NVBG, they have shown a higher complication and failure rate, with infection being the most common complication [8]. NVBG require soft tissue coverage often in the form of adjacent tissue transfer, which may or may not be a viable option depending on the extent of the initial resection and viability of the surrounding tissue. Local or regional flaps may be used in patients who are poor candidates for microvascular anastomosis and may be performed in conjunction with NVBG, as the use of NVBG often requires additional soft tissue coverage.

Despite the myriad of proposed options for VBG, the modern-day workhorse for maxillo-mandibular bony reconstruction is the fibular free flap (FFF), with high reported success rates and relatively low donor site morbidity [9]. While there is a high success rate for the FFF, certain considerations must be taken regarding patient selection including patient mobility and the evidence of peripheral vascular disease. Decreased patient mobility may negatively affect rehabilitation after a FFF. In cases where donor site morbidity from a FFF is a concern, the scapular free flap may be a more viable option [10]. Even with expected functional loss and postoperative symptoms experienced by FFF recipients, overall quality of life is not significantly lower than reference populations [11].

The use of regenerative medicine techniques may be implemented in conjunction with any of the aforementioned reconstructive strategies and will be discussed in greater detail in the remaining chapters. All the available reconstructive options and their associated risks and benefits should be discussed with the patient to allow the patient to make an informed decision. Patient preference should be taken into consideration, and sufficient preoperative counseling should be provided given that more complex surgeries will entail a much longer time for recovery and rehabilitation. Patient compliance can play an important role in surgical outcomes especially in staged reconstructions.

After considering all variables related to patient health, comorbidities, and patient preference, we can broadly classify patients requiring maxillofacial reconstruction into three main groups: posttraumatic, oncologic, and benign tumor. Each of these clinical scenarios merits special considerations that address the characteristics unique to their basic pathophysiology.

3.3 Posttraumatic

Maxillofacial trauma secondary to avulsion or severe ballistic injuries can create complex composite defects. A common factor in the management of these traumatic injuries is timing of intervention and specifically how it relates to the body's inflammatory response to said traumatic event. In the case of graft placement, one such concern is the viability of the tissue at the recipient site and how this may affect overall graft survival.

The value for a staged approach to reconstruction of severe facial trauma has been previously delineated [12]. Initial assessment of complex facial wounds involves patient stabilization, the evaluation of multi-system trauma, and consultation of neurosurgical and ophthalmological teams as indicated. Preoperative imaging, specifically CT angiogram, is often indicated as the incidence of major vascular injury in facial gunshot wounds (GSWs) is 10–50% depending on the entry point.

Early surgical reconstruction of the underlying bony framework is important in preventing soft tissue contracture [13]. This can be accomplished with the use of bone grafting and locking reconstruction plates depending on availability of soft tissue coverage. Thorough investigation of remaining viable bone and soft tissue is important in determining the extent of reconstruction necessary [12]. Waiting an adequate amount of time for the nonviable tissue to declare itself, while logical, should be weighed against the risks of progression of the inflammatory cascade and eventual scar formation when determining timing of definitive repair [14]. The amount of time needed for demarcation of tissue viability differs depending on the mechanism but often will become clear within 24–72 h.

Unlike blunt facial trauma, avulsive or projectile injuries are more apt to progressive tissue loss secondary to necrosis or infection seen after the time of initial injury and may therefore require more time to determine the extent of tissue loss. Projectile injuries are generally classified as low or high energy, relating to the amount of damage inflicted on surrounding tissue. Resistance of energy transfer from the projectile is greater in dense tissues such as bone and causes compressive waves which damage surrounding tissues [15]. When evaluating blast injuries, damage is directly proportional to the distance from the blast, and damage may be much more severe than what is seen superficially.

In addition to tissue necrosis, there is a significantly increased risk of infection in penetrating facial trauma, with a reported 39% wound healing complication if a gunshot passes through the oral cavity [16] and as high as 100% after close range shotgun blast [17]. Infection can occur either from bacteria carried into a wound by a projectile or contamination of the wound after initial injury. Animal models have shown that the volume of necrotic tissue needing debridement is decreased by earlier debridement and use of antibiotics [18]. There is limited data, however, demonstrating the benefit of prophylactic antibiotics in facial trauma. It has been shown to be beneficial in preoperative mandibular fractures, but the advantage is less certain for midfacial fractures or in the postoperative setting [19]. Therefore, while early surgical debridement may be important, the role of prophylactic antibiotics is less clear.

Over time, there has been a shift from delayed repairs to more immediate definitive reconstruction in an attempt to decrease scar contracture of the face [20]. Bringing well-vascularized bone and soft tissue into a seemingly unhospitable wound in the form of free tissue transfer has allowed for reconstruction of the bony skeleton and overlying soft tissue shortly after extensive facial injury without compromising aesthetic or functional outcomes [21]. The use of free tissue transfer in posttraumatic reconstruction is beneficial due to the improved vascularity, decreased contracture and scarring, ability to reduce infection, and replacement of composite tissues that this method affords. Its use is sometimes even necessary in settings such as violation of the anterior skull base and active cerebrospinal fluid leak, where a vascular tissue seal is required. There are no clear guidelines on the exact timing of the reconstructive process, and further prospective investigations are still needed, but current trends favor more immediate reconstruction as it may help to avoid soft tissue contracture and lead to improved long-term functional results.

3.4 Malignant Tumors

Head and neck cancer patients can present a unique set of challenges for the reconstructive surgeon. Perioperative care, usually by an interdisciplinary team, is important for optimal recovery from surgery. A systematic review by the Enhanced Recovery After Surgery Society examined 17 different areas of perioperative care for head and neck cancer patients, recommending a focus on nutritional status; antibiotic, antiemetic, and thromboembolic prophylaxis; prevention of hypothermia; early mobilization and decannulation; opioid-sparing analgesia; and pulmonary physical therapy [22]. Ensuring a successful reconstruction outcome starts long before the actual surgery, with a focus on one very important and controllable risk factor: nutrition.

Up to 80% of head and neck cancer patients are malnourished, often as a direct result of their disease process. Unfortunately, this is often not addressed at the time of a reconstructive surgery. Decreased BMI has been shown to be a negative prognostic factor independent of tumor stage, with a 5-year survival of 27.1% (underweight) vs 59.9% (normal weight) patients undergoing chemoradiotherapy [23]. Nutrition, especially on the day of surgery, is also important as traditional preoperative fasting has been associated with delayed recovery and poorer outcomes compared to a 2-hour fasting window after clear liquids [24]. Malnourished patients are also not surprisingly at a greater risk for infection and poor wound healing.

Infection after reconstructive surgery is associated with longer hospital stays, increased healthcare cost, and poorer functional and aesthetic outcomes. Postoperative wound infection in clean-contaminated head and neck surgeries has been reported as high as 80% [25]. Infection rate is higher in patients of the male sex and with previous alcohol and tobacco use and tumors of the base of the tongue or mandibular gingiva and those who underwent segmental mandibular resection or had nasogastric tubes at the time of surgery [26]. Given the importance of perioperative nutrition and possible infection risk related to nasogastric tubes, an emphasis on prophylactic gastrostomy placement may also improve patient outcomes. Randomized controlled trials of perioperative antibiotic prophylaxis continued for at least 24 h after surgery in clean-contaminated procedures consistently show decreased wound infection [27]. Longer courses of antibiotics, however, do not show any additional benefit compared to 24 h, even in patients undergoing free flap reconstruction [28]. Given the existing data, all patients undergoing reconstruction for a defect secondary to resection of a malignancy would likely benefit from a short course of perioperative antibiotics.

In addition to the risks of postoperative wound infection, all cancer patients are at an increased risk for venous thromboembolism (VTE), including both deep vein thrombosis (DVT) and pulmonary embolus (PE). The risk of VTE is also increased in patients undergoing reconstructive surgery given long operative times and postoperative immobility secondary to donor site morbidity. The use of pharmacologic prophylaxis in head and neck cancer patients undergoing free flap reconstruction has been shown to effectively protect against VTE [29]. Despite this benefit, the use of antithrombotics has not been shown to definitively decrease the rate of flap thrombosis or failure, but their use does increase the risk of postoperative hematoma [30].

One main reason that head and neck cancer patients differ from other reconstructive candidates is the high likelihood that they will receive or have already received adjuvant radiation and/or chemotherapy. The introduction of radiation, in particular, into the healing reconstructive field further complicates wound healing. Any maxillofacial reconstruction in a cancer patient must be able to withstand the tissue-damaging effects of these adjuvant therapies.

Patients who have already undergone radiation therapy present a unique set of challenges. Flap failure is believed to be more common after radiation, and a recent meta-analysis showed a statistically increased risk of flap failure (RR 1.48, $P < 0.004$), complications (RR 1.84, $P < 0.001$), reoperation (RR 2.06, $P < 0.001$), and fistula (RR 2.05, $P < 0.001$) in previously irradiated patients undergoing microvascular reconstruction [31]. Animal studies, however, have shown no effect of prior radiation therapy on the patency of anastomosed vessels despite observations of significant damage to the surrounding tissue [32]. In addition, irradiated tissue can become more difficult to work with due to scarring. Also, operative complications tend to increase as the time between radiotherapy and surgery increases [32]. Despite this, the overall dose of radiation has not shown to be an independent predictor of outcomes. Patients who had T3 or T4 cancers and those who continue to smoke, however, have been shown to be poor candidates for salvage surgery with microvascular reconstruction as they have an increased rate of recurrence [33]. In addition, many patients undergoing head and neck oncologic surgery are former and current tobacco users. The continued use of tobacco is clearly detrimental to the success of reconstructive patients, and abstinence from smoking has been shown to improve wound healing [34], with cessation at least 3 weeks prior to surgery showing the greatest benefit [35]. All potential reconstructive patients should be counseled and given resources for smoking cessation prior to surgery.

The need for reconstruction in head and neck cancer patients is not only important after tumor extirpation but also as a result of osteoradionecrosis (ORN) from radiation therapy. Although irradiated tissue is inherently damaged and therefore at risk for poor healing, patients who undergo free flap surgery for ORN do not have greater risks of 90-day perioperative complications or differences in free flap viability compared with patients who undergo microvascular reconstruction for other causes [36]. This is believed to be secondary to the introduction of vascularized free tissue from a non-irradiated donor site into the reconstructive surgical field.

The use of regenerative medicine in cancer patients presents a unique challenge. One of the most commonly used regenerative substances in maxillofacial reconstruction is recombinant human bone morphogenetic protein 2 (rhBMP-2). However, bone morphogenetic proteins (BMPs) are involved in both tumorigenesis and regulation of cancer progression. Regarding tumor growth, they have been used as biomarkers for the prognosis of certain cancers. BMP-2 level, specifically, is an independent negative predictor of prognosis in patients with non-small cell lung cancer. This is important given that many head and neck cancer patients have an extensive history of smoking and are therefore at increased risk of developing lung cancer as well. In direct opposition to this are the findings that BMPs may also play a role in tumor suppression [37]. Hence, the exact role BMPs might play in head and neck cancer has not yet been fully elucidated.

Despite the potential risks of increased tumor growth or recurrence, the use of BMPs has been attempted as an adjunct to reconstruction of head and neck cancer patients. A case series looking at the use of rhBMP-2 at the time of free flap reconstruction for osteoradionecrosis of the mandible demonstrated no increase in the rate of cancer recurrence but also failed to show any improvement in healing or complications compared to the control group [38]. Regenerative medicine may yet to have an important role to play in reconstruction of cancer patients, but more research is needed looking at the interaction of BMPs and head and neck cancers.

3.5 Benign Tumors

In certain aspects, benign tumors may arguably be seen as a less complicated group of patients requiring reconstruction. There is often limited damage to surrounding tissues, unlike trauma or postradiation, and patients are less likely to have the myriad of other perioperative concerns often present in patients with malignancies as discussed above. Two common types of benign maxillofacial tumors include ameloblastoma and keratocystic odontogenic tumor (KOT). Clinical presentation of the most common ameloblastoma subtype, solid or multicystic, can include malocclusion, asymptomatic swelling, or mucosal ulceration. Controversy still exists regarding the management of this aggressive benign tumor between conservative and radical treatments. Conservative treatment through enucleation and curettage has a reported 48.7% recurrence rate [39], while primary resection generally yields recurrence rate <15%. However, "recurrence" is believed by some to be in fact residual disease, with gross pathologic tumor extension measured at a mean of 4.5 mm beyond radiographic margins [40]. Given this, adequate margin should be taken at initial resection if further reconstruction is planned, as this tumor has the ability to extend into and through bone. The same can be said for the KOT, which has been classified as a neoplasm because of its aggressive behavior and tendency for recurrence. Therefore, the classification of benign tumor should not be interpreted as a provision that allows for less meticulous or thorough evaluation and management.

Attention to the type of resection performed and risk of recurrence for benign maxillofacial neoplasms is essential to ensuring a lasting reconstruction. The use of regenerative medicine in the form of rhBMP-2 has shown excellent regeneration of large mandibular defects after resection of ameloblastoma, obviating the need for autogenous bone graft [41]. Additionally, a case series of mandibular reconstructions after removal of benign lesions or trauma demonstrated only one of six patients with a postoperative complication of infection [42].

3.6 Conclusion

Despite the course that leads a patient to requiring reconstructive maxillofacial surgery, care must be taken to investigate all patient comorbidities and barriers to surgery. Biological rather than chronological age should be used to help guide the decision-making process. All available surgical options should be discussed fully with the patient as they are indubitably an important stakeholder in the reconstructive process. Patients undergoing reconstruction after severe facial trauma should have all devitalized tissue debrided and receive reconstruction surgery as soon as possible to minimize scarring that may limit reconstructive options and outcomes. Head and neck cancer patients present a unique set of circumstances which may hinder successful reconstruction. More focus is needed on perioperative and preoperative health status including nutrition and smoking cessation. Prior radiation therapy complicates the reconstructive process and portends to potentially poorer overall outcomes. Benign tumors of the maxilla and mandible often have a high recurrence rate, and proper attention must be given to the method of tumor extirpation prior to reconstruction. Reconstructive surgery of the midface and mandible, regardless of the primary disease process, is a complex endeavor. Careful preoperative assessment, prudent surgical intervention, and thorough perioperative management are necessary for optimizing treatment outcomes in these patients.

References

1. Grammatica A, Piazza C, Paderno A, Taglietti V, Marengoni A, Nicolai P. Free flaps in head and neck reconstruction after oncologic surgery. Otolaryngol Head Neck Surg. 2015;152(5):796–802.
2. Peters TT, Post SF, Van Dijk BA, et al. Free flap reconstruction for head and neck cancer can be safely performed in both young and elderly patients after careful patient selection. Eur Arch Otorhinolaryngol. 2015;272(10):2999–3005.
3. Hwang K, Lee JP, Yoo SY, Kim H. Relationships of comorbidities and old age with postoperative complications of head and neck free flaps: a review. J Plast Reconstr Aesthet Surg. 2016;69(12):1627–35.
4. Wang F, Huang W, Zhang C, Sun J, Qu X, Wu Y. Functional outcome and quality of life after a maxillectomy: a comparison between an implant supported obturator and implant supported fixed prostheses in a free vascularized flap. Clin Oral Implants Res. 2017;28(2):137–43.

5. Sadr-Eshkevari P, Rashad A, Vahdati SA, Garajei A, Bohluli B, Maurer P. Alloplastic mandibular reconstruction: a systematic review and meta-analysis of the current century case series. Plast Reconstr Surg. 2013;132(3):413e–27e.
6. Kämmerer PW, Klein MO, Moergel M, Gemmel M, Draenert GF. Local and systemic risk factors influencing the long-term success of angular stable alloplastic reconstruction plates of the mandible. J Craniomaxillofac Surg. 2014;42(5):e271–6.
7. Foster RD, Anthony JP, Sharma A, Pogrel MA. Vascularized bone flaps versus nonvascularized bone grafts for mandibular reconstruction: an outcome analysis of primary bony union and endosseous implant success. Head Neck. 1999;21(1):66–71.
8. Akinbami BO. Reconstruction of continuity defects of the mandible with non-vascularized bone grafts. Systematic literature review. Craniomaxillofac Trauma Reconstr. 2016;9(3):195–205.
9. Hayden RE, Mullin DP, Patel AK. Reconstruction of the segmental mandibular defect: current state of the art. Curr Opin Otolaryngol Head Neck Surg. 2012;20(4):231–6.
10. Dowthwaite SA, Theurer J, Belzile M, et al. Comparison of fibular and scapular osseous free flaps for oromandibular reconstruction: a patient-centered approach to flap selection. JAMA Otolaryngol Head Neck Surg. 2013;139(3):285–92.
11. Löfstrand J, Nyberg M, Karlsson T, et al. Quality of life after free fibula flap reconstruction of segmental mandibular defects. J Reconstr Microsurg. 2017;34:108.
12. Futran ND. Maxillofacial trauma reconstruction. Facial Plast Surg Clin North Am. 2009;17(2):239–51.
13. Gruss JS, Antonyshyn O, Phillips JH. Early definitive bone and soft-tissue reconstruction of major gunshot wounds of the face. Plast Reconstr Surg. 1991;87(3):436–50.
14. Clark N, Birely B, Manson PN, et al. High energy ballistic and avulsive facial injuries: classification, patterns, and an algorithm for primary reconstruction. Plast Reconstr Surg. 1996;98(4):583–601.
15. Rozen N, Dudkiewicz I. In: Lerner A, Soudary M, editors. Wound ballistics and tissue damage in armed conflict injuries to the extremities. Berlin: Springer; 2011. p. 21–33.
16. Kihtir T, Ivatury RR, Simon RJ, Nassoura Z, Leban S. Early management of civilian gunshot wounds to the face. J Trauma. 35(4):569–77.
17. Suominen E, Tukiainen E. Close-range shotgun and rifle injuries to the face. Clin Plast Surg. 2001;28(2):323–37.
18. Dahlgren B, Almskog B, Berlin R, et al. LOcal effects of antibacterial therapy (benzyl-penicillin) on missile wound infection rate and tissue devitalization when debridement is delayed for twelve hours. Acta Chir Scand Suppl. 1982;508:271–9.
19. Morris LM, Kellman RM. Are prophylactic antibiotics useful in the management of facial fractures? Laryngoscope. 2014;124(6):1282–4.
20. Olding M, Winski FV, Aulisi E. Emergency free flap reconstruction of a facial gunshot wound. Ann Plast Surg. 1993;31(1):82–6.
21. Kaufman Y, Cole P, Hollier L. Facial gunshot wounds: trends in management. Craniomaxillofac Trauma Reconstr. 2009;2(2):85–90.
22. Dort JC, Farwell DG, Findlay M, et al. Optimal perioperative care in major head and neck cancer surgery with free flap reconstruction: a consensus review and recommendations from the enhanced recovery after surgery society. JAMA Otolaryngol Head Neck Surg. 2017;143:292.
23. Takenaka Y, Takemoto N, Nakahara S, et al. Prognostic significance of body mass index before treatment for head and neck cancer. Head Neck. 2015;37(10):1518–23.
24. Lambert E, Carey S. Practice guideline recommendations on perioperative fasting: a systematic review. JPEN J Parenter Enteral Nutr. 2016;40(8):1158–65.
25. Skitarelić N, Morović M, Manestar D. Antibiotic prophylaxis in clean-contaminated head and neck oncological surgery. J Craniomaxillofac Surg. 2007;35(1):15–20.
26. Belusic-gobic M, Zubovic A, Cerovic R, Dekanic A, Marzic D, Zamolo G. Multivariate analysis of risk factors for postoperative wound infection following oral and oropharyngeal cancer surgery. J Craniomaxillofac Surg. 2018;46(1):135–41.
27. Rodrigo JP, Alvarez JC, Gómez JR, Suárez C, Fernández JA, Martínez JA. Comparison of three prophylactic antibiotic regimens in clean-contaminated head and neck surgery. Head Neck. 1997;19(3):188–93.
28. Carroll WR, Rosenstiel D, Fix JR, et al. Three-dose vs extended-course clindamycin prophylaxis for free-flap reconstruction of the head and neck. Arch Otolaryngol Head Neck Surg. 2003;129(7):771–4.
29. Shuman AG, Hu HM, Pannucci CJ, Jackson CR, Bradford CR, Bahl V. Stratifying the risk of venous thromboembolism in otolaryngology. Otolaryngol Head Neck Surg. 2012;146(5):719–24.
30. Lee KT, Mun GH. The efficacy of postoperative antithrombotics in free flap surgery: a systematic review and meta-analysis. Plast Reconstr Surg. 2015;135(4):1124–39.
31. Herle P, Shukla L, Morrison WA, Shayan R. Preoperative radiation and free flap outcomes for head and neck reconstruction: a systematic review and meta-analysis. ANZ J Surg. 2015;85(3):121–7.
32. Thankappan K. Microvascular free tissue transfer after prior radiotherapy in head and neck reconstruction - a review. Surg Oncol. 2010;19(4):227–34.
33. Kim AJ, Suh JD, Sercarz JA, et al. Salvage surgery with free flap reconstruction: factors affecting outcome after treatment of recurrent head and neck squamous carcinoma. Laryngoscope. 2007;117(6):1019–23.
34. Sorensen LT, Karlsmark T, Gottrup F. Abstinence from smoking reduces incisional wound infection: a randomized controlled trial. Ann Surg. 2003;238(1):1–5.

35. Kuri M, Nakagawa M, Tanaka H, Hasuo S, Kishi Y. Determination of the duration of preoperative smoking cessation to improve wound healing after head and neck surgery. Anesthesiology. 2005;102(5):892–6.
36. Zaghi S, Danesh J, Hendizadeh L, Nabili V, Blackwell KE. Changing indications for maxillomandibular reconstruction with osseous free flaps: a 17-year experience with 620 consecutive cases at UCLA and the impact of osteoradionecrosis. Laryngoscope. 2014;124(6):1329–35.
37. Zhang L, Ye Y, Long X, Xiao P, Ren X, Yu J. BMP signaling and its paradoxical effects in tumorigenesis and dissemination. Oncotarget. 2016;7(47):78206–18.
38. Sweeny L, Lancaster WP, Dean NR, Magnuson JS, Carroll WR, Louis PJ, Rosenthal EL. Use of recombinant bone morphogenetic protein 2 in free flap reconstruction for osteonecrosis of the mandible. J Oral Maxillofac Surg. 2012;70(8):1991–6.
39. Hasegawa T, Imai Y, Takeda D, et al. Retrospective study of ameloblastoma: the possibility of conservative treatment. Kobe J Med Sci. 2013;59(4):E112–21.
40. Carlson ER, Marx RE. The ameloblastoma: primary, curative surgical management. J Oral Maxillofac Surg. 2006;64(3):484–94.
41. Cicciù M, Herford AS, Cicciù D, et al. Recombinant human bone morphogenetic protein-2 promote and stabilize hard and soft tissue healing for large mandibular new bone reconstruction defects. J Craniofac Surg. 2014;25(3):860–2.
42. Desai SC, Sclaroff A, Nussenbaum B. Use of recombinant human bone morphogenetic protein 2 for mandible reconstruction. JAMA Facial Plast Surg. 2013;15(3):204.

4 The Role of rhBMP-2 in Oral and Maxillofacial Reconstruction

Alan Herford and Isaac Lowe

4.1 Introduction

For many maxillofacial reconstructions, the preferred method for reconstruction has incorporated autografts harvested from various parts of the body. This includes microvascular free flap as well as other types of tissue transfers. Recently, this standard has been challenged by the rapid increased research into osteogenic factors in an attempt to eliminate the need for a donor site and associated morbidities such as longer operating time, secondary surgical site, higher amounts of blood loss, donor site pain, ambulation, infection, impaired wound healing, hematoma, and donor site fracture [1–3]. Donor site pain is the most common postoperative complication in regard to ICBG harvest, and current research suggests a long term complication rate ranging from 25 to 46.5% [4–6]. This means, for ICBG cases alone, we can expect an average of one patient in four experiencing a complication which may possibly last long term [4–6]. Currently research is focusing on numerous osteogenic factors (BMP-2, PRP, PDGF, TFG-β, and IGF) with recombinant human bone morphogenetic protein-2 (rhBMP-2) showing the most promise with regard to bone formation [7].

A. Herford (✉) · I. Lowe
Department of Oral and Maxillofacial Surgery, Loma Linda University, Loma Linda, CA, USA
e-mail: aherford@llu.edu

4.2 What Are They?

BMPs are a family of cytokines, in the TGF-β family, that are involved in angiogenesis, cell apoptosis, inflammation, and the development and maintenance of many processes throughout of the body [8]. BMP's main role, for the limits of the discussion of maxillofacial reconstruction, is the mediation of bone and cartilage formation and maintenance [8, 9]. BMP-2 in particular emerged as a major cytokine of interest, in regard to its potential for clinical application, due to its specific osteogenic properties [3, 8, 9]. Bone grafting in the maxillofacial region is primarily completed via autografting from the iliac crest because the use of other grafting materials (xenografts and allografts) is not considered osteogenic and the body takes longer to replace the grafting material because osteogenic and osteolytic cells need to be recruited to the grafting site and differentiate into mature osteogenic cells before bone turnover can begin. In some cases, a decalcifying stage may be needed before the material can be turned over if the grafting material is calcified adding to the turnover time.

For larger defects such as segmental defects of the jaw, non-autogenous grafts have not been shown to have a high success rate, and thus autografts are indicated. The combination of allografts or xenografts with BMP-2 has shown promise for reconstrution by the addition of a growth factor

J. C. Melville et al. (eds.), *Regenerative Strategies for Maxillary and Mandibular Reconstruction*,
https://doi.org/10.1007/978-3-319-93668-0_4

to non-autogenous grafts, granting osteoconductive properties closer to autogenous grafts.

4.3 How Do They Work?

BMP-2 is uniquely osteogenic, when compared to other growth factors because BMP-2 induces undifferentiated mesenchymal cells to differentiate into osteoblasts and exhibits chemotactic effects toward human osteoblasts and osteoblast progenitor cells in the human marrow [4, 8]. BMP-2 binds to the surface proteins of undifferentiated mesenchymal cells and activates transcription factors SMAD-1, SMAD-5, and SMAD-8 which complexes with SMAD-4 in the cytosol [3, 8, 9]. The complex then travels to the cell's nucleus and activates transcription factors, Runx-2 and TCF/TLEF, to trigger osteoblastic differentiation, maturation, and osteogenic effects (increased production of extracellular matrix, alkaline phosphate, and collagen-1) [3, 8, 9] (Fig. 4.1). BMP-2 also indirectly induces osteoclastic effects through the RANK/RANKL and TRAP mechanisms which facilitates the replacement of grafting material eventually resulting in new bone formation [4, 3, 8, 9]. Because autografts feature no risk of graft rejection and the graft itself intrinsically has osteogenic properties, it's easy to understand why autografting has remained the gold standard for so long. The fact that BMP-2 is osteoinductive is critical because using it with allografting and xenografting materials (which are not inherently

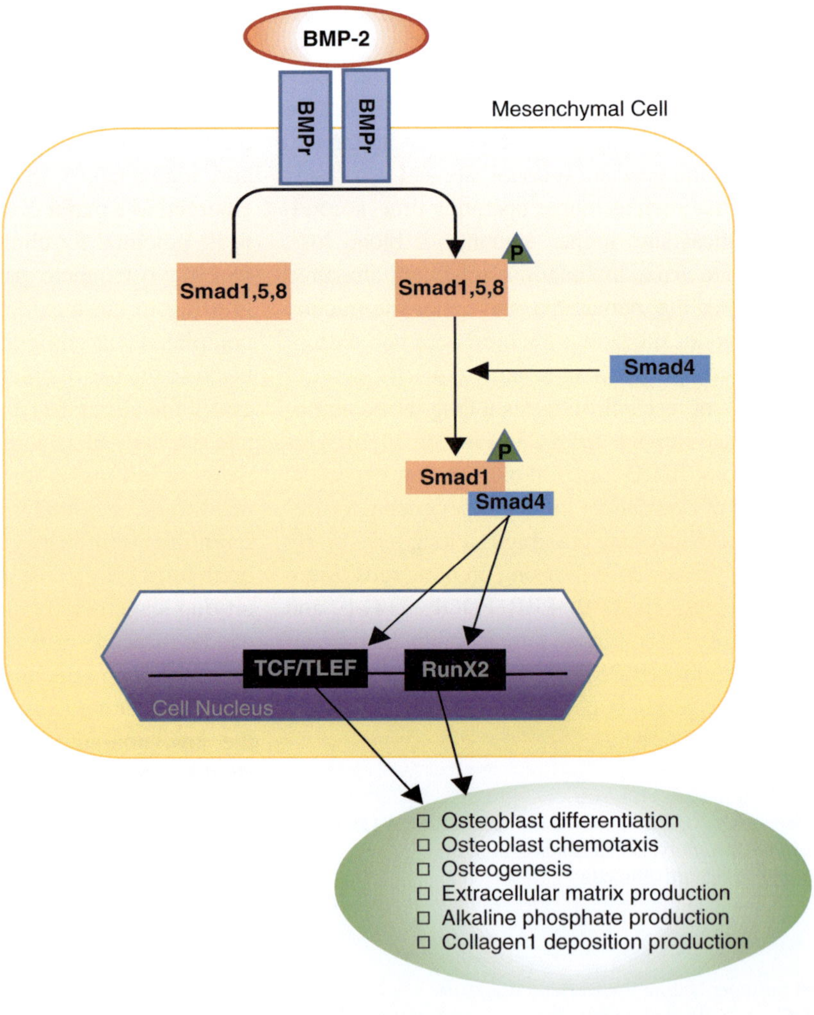

Fig. 4.1 Cartoon describing BMP-2's osteogenic pathway

osteoinductive) potentially accelerates the substitution of the material with natural "self" bone shortening the gap between these materials and autografting [4].

4.4 Clinically What Does It Look Like and How Do I Use It?

For maxillofacial reconstruction cases, rhBMP-2 kits commonly present as a vial of rhBMP-2 and a vial of sterile saline and several ACSs that are easily manipulated to fit into a variety of shapes and sizes. The vials of rhBMP-2 come in concentrations of 1.5 mg/cc which has been found to be the optimum concentration for osteoinductive properties while minimizing postoperative complications in the maxillofacial region [3, 10]. The size and location of the osseous defect and restoration plan typically dictate the amount of rhBMP-2 used for the procedure. The ACS is soaked in the indicated amount of rhBMP-2 for 15 min and placed into the grafting site within 2 hours of opening the kit. The collagen sponge is then cut into small pieces and combined with a "graft extender." This may be either autograft, non-autograft, or both. It is important to mix the sponge throughout the graft materials in order to incorporate the slow release properties of the BMP/ACS graft. When ready, the rhBMP-2/ACS graft material is placed into the defect. In most cases, a titanium mesh and membrane are used to maintain the graft material in the defect and to prevent soft tissue invasion into the grafting site that would normally result in grafting failure or not obtaining an adequate bone volume [4]. Fixation techniques vary in the literature depending on the grafting site [2, 3, 7, 8, 11, 12].

4.5 Does It Work? What Can I Use It For?

Traditionally, research has shown that autogenous grafts are associated with features the greatest success rates, in general for maxillofacial reconstruction, when compared to allografting and xenografting procedures [13, 14]. Research reflects this trend for dental and prosthetic implants in that implants placed in allografted and xenografted bone are slightly less successful than in autografted sites, but depending on the situation, autografting may not always be possible [13, 14]. Autogenous grafting requires more surgical time, more blood loss, and the aforementioned complications [6, 10]. Also, in some situations, there may not be enough available tissue, in terms of volume and quality, to harvest if the defect is too extensive [1–6]. Therefore, even though autografting techniques are slightly more successful, allografting and xenografting can lead to safe and successful outcomes for maxillofacial reconstruction.

Extensive animal studies [4, 7, 8, 10, 11, 15–17] show that rhBMP-2 (INFUSE®) is a successful, safe, and reliable material to use for critical-sized maxillary and mandibular bone defects, maxillary cleft reconstruction [13], and long-term dental implant support and retention [17]. Some animal studies even reported superior results with rhBMP-2 when compared with ICBG for certain procedures [16].

The first human trial using BMPs was conducted in 2001 by Moghaden et al. and showed that BMPs can safely and successfully be used for xenografting in the mandible after healing in the trapezius muscle [18]. A few years later, rhBMP-2 with ACS was also shown to be safe and successful in repairing continuity defects in the mandible [3]. Thus far, the FDA approved rhBMP-2 with ACS use for maxillofacial sinus augmentation in 2007 (INFUSE®) but has been effective in meeting the needs for a variety of osseous maxillofacial reconstruction situations, as an off-label use [2, 3, 7, 8, 11, 12, 19].

More recent research shows rhBMP-2 with ACS and other allografting or xenografting materials features comparable and sometimes superior results compared to the standard ICBG modalities for alveolar cleft repair [2] (Fig. 4.2), localized alveolar bone defects [10] (Fig. 4.3), critical maxillary and mandibular bone defects [3, 4, 12], autogenous sinus augmentations [10], dental endosseous implant placement with restoration [20], xenografting [18], alveolar socket

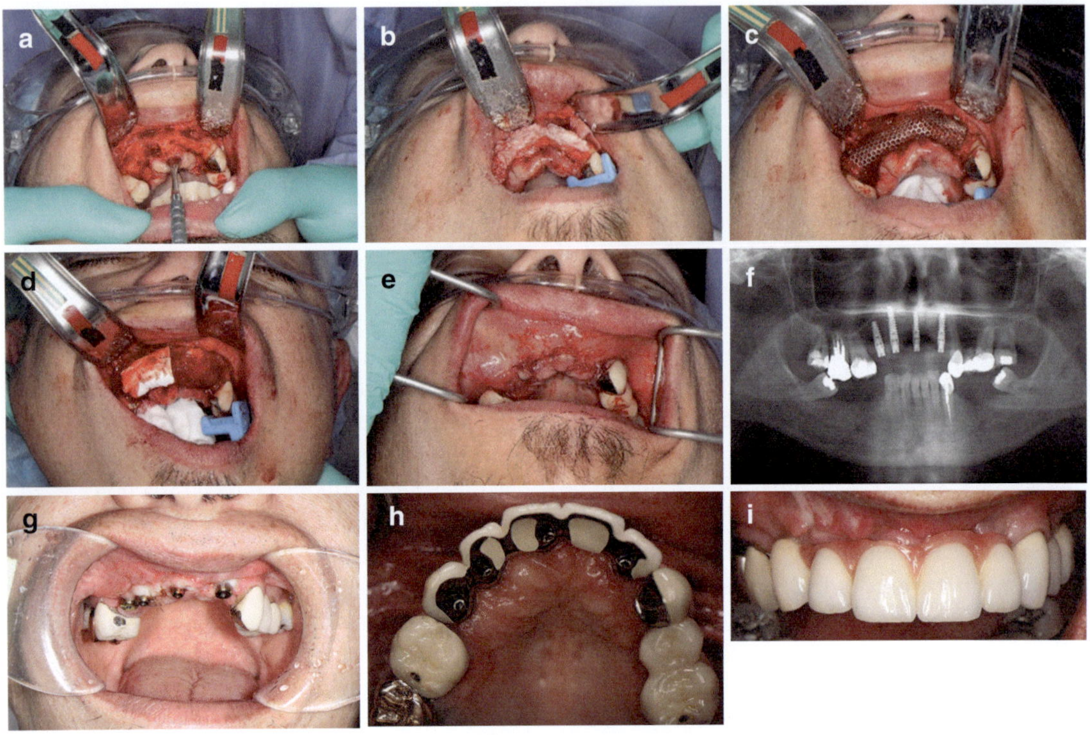

Fig. 4.2 (**a**) Critical maxillary anterior osseous defect spanning tooth #3–11. (**b**) With anterior and posterior flaps reflected, the collagen sponge with rhBMP-2 was placed on the maxillary alveolar ridge. (**c**) Prefabricated titanium mesh placed to retain the graft. (**d**) Resorbable membrane placed over the titanium mesh to prevent soft tissue invasion. (**e**) Primary closure via 3.0 chromic suture. (**f**) One-year follow-up panoramic radiograph after implant placement and osseointegration

preservation [10], and craniofacial defects in general [7, 21]. When rhBMP-2 with ACS is used for alveolar socket preservation (Fig. 4.4), results showed double the amount of bone regeneration compared to empty control sites [10].

Alveolar cleft defect closure with rhBMP-2 and ACS featured no statistically significant difference in success rates, failure rates, and needs for secondary surgery to facilitate hard tissue closure when compared to standard ICBG therapy [2]. For all other aforementioned procedures, case reports and reviews indicate similar success rates, failure rates, and needs for secondary operations when compared to the standard ICBG therapy [2, 3, 7, 8, 11, 12, 19]. Furthermore, there is no evidence to show that rhBMP-2 and ACS use in bone grafting for maxillofacial reconstruction prevents tooth eruption or orthodontic tooth movement [2]. On average, patients were evaluated on a 6- and 18-month basis after surgery was completed and/or after functional loading at the grafting site [2, 3, 7, 8, 11, 12, 19].

Overall, articles report that rhBMP-2 with allo-/xenografting material features excellent grafting volume retention [2, 3, 7, 8, 11, 12, 19, 21]. Histologic evaluation most commonly reports the rhBMP-2 sites being indistinguishable from the "normal bone" [2, 3, 7, 8, 11, 12, 19, 21]. Integration results at grafting sites feature the same if not better results when compared to autografting sites radiographically and histologically [2, 3, 7, 8, 11, 12, 19, 21]. Measurement modalities included histologic sampling (for trabecular pattern, trabecular thickness, and hard tissue volume) and radiologic evaluation (dental radiographs, CBCT scans, panoramics) [2, 3, 7, 8, 11, 12, 19, 21].

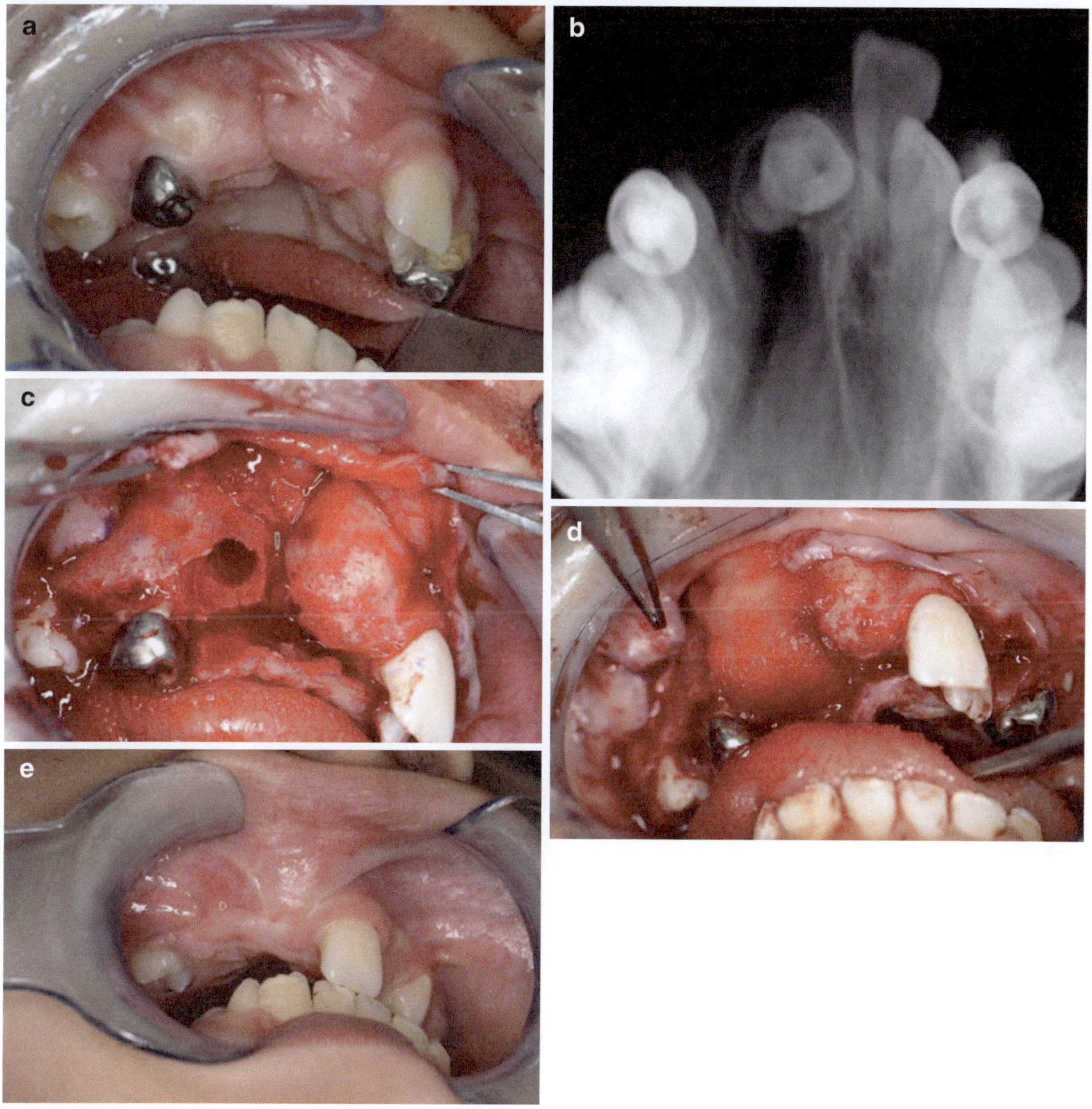

Fig. 4.3 (**a**) Panoramic radiograph showing alveolar cleft defect between teeth #6 and #7. (**b**) Occlusal radiograph showing alveolar cleft defect between teeth #6 and #7. (**c**) Occlusal radiograph after cleft restoration with rhBMP-2/ACS and orthodontic positioning. (**d**) Clinical view of cleft repair after rhBMP-2/ACS and orthodontic correction

4.6 Combination with Autografts

In some cases, it may be preferable to combine autogenous grafts with a growth factor. One example would be those cases which require reconstruction of the mandibular condyle as part of the primary reconstruction [22]. A costochondral graft can be used to reconstruct the joint and limit the risk of ankylosis. An alteration would be to use a prosthesis to replace the joint. Another benefit of adding BMP to an autograft is that it reduces the size of graft harvest [22].

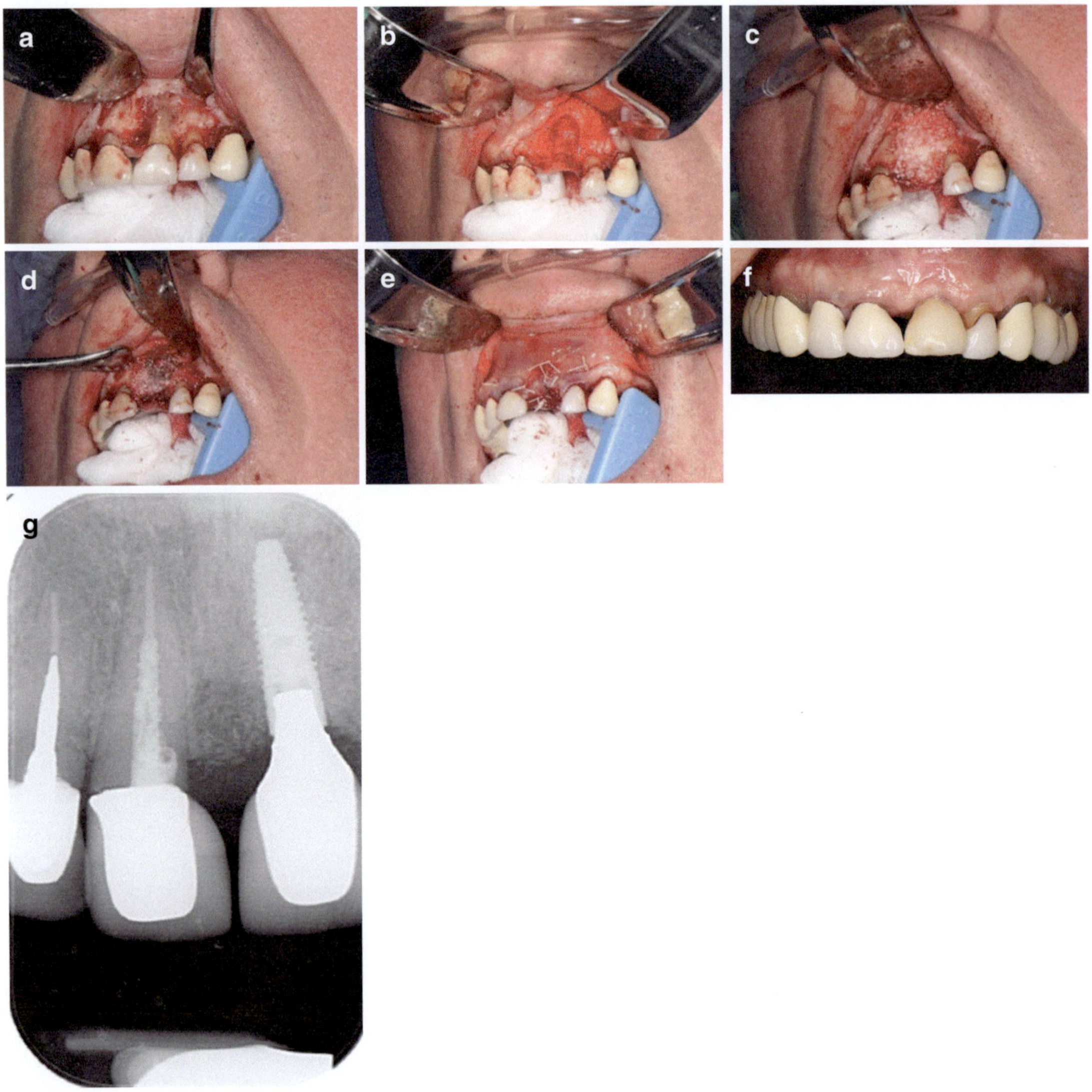

Fig. 4.4 (**a**) Maxillary buccal flap reflected to appreciate the three-walled osseous defect at tooth #9. (**b**) Surgical site after atraumatic extraction of tooth. (**c**) Bone graft with rhBMP-2 placed in the alveolus. (**d**) Adapted titanium mesh fixated. (**e**) Primary closure with 3.0 chromic gut

4.7 Limits, Adverse Effects, and Contraindications with rhBMP-2/ACS

Despite the debate regarding rhBMP-2's success and adverse events rates in animal and human trials, the vast majority of articles show that most of the more serious postoperative complications are primarily seen with spinal fusion cases and not maxillofacial reconstruction [15, 23]. These most common complications for spinal fusion cases with rhBMP-2 are retrograde ejaculation, dysphagia, heterotopic bone formation, hematoma, seroma, osteolysis, and wound complications that all occur in differing rates [15, 23]. For maxillofacial cases, certain postoperative complications, like postoperative infection and wound dehiscence, did not occur at a significantly different rate compared to autografting complication rates [2]. Other adverse events specifically correlated with the use of rhBMP-2 in maxillofacial reconstruction are edema, erythema, and dysphagia which are consistent with BMP-2's inherent role in the inflammatory process [2, 5, 12, 20, 24]. The complication rate for dysphagia ranges from 3 to 85%, with 25% being the most commonly reported. These complications most often are self-resolving, but about 10% require steroid therapy but no other interventions [2, 5, 24]. The rates for significant erythema and edema are much more difficult to quantify.

While the majority of edema and erythema complications self-resolve without the need for intervention and proceed onto successful reconstruction, these complications should be taken seriously because they can be profound [4, 25–27]. Reviews and case reports indicate that the maximum amount of postoperative edema and erythema occurs 3 to 4 days post-op with rhBMP-2 and usually lasts up to 8 days which is not outside the normal range for maxillofacial reconstructions in general anyway [25]. It is difficult to definitively identify the rate of significant rhBMP-2-related postoperative edema and erythema because these are regularly occurring complications when preforming surgery in general. The vast majority of these cases do self-resolve with only a hand full of cases requiring postoperative steroids and did not require any further intervention [2, 3, 5, 7, 8, 11, 12, 19, 21].

The vast majority of serious complications requiring immediate intervention are not related to rhBMP-2. One reported case featured a patient requiring immediate return to the operating room due to postoperative hematoma formation [5]. This complication was attributed to the patient's von Willebrand factor disease and found no relation to the rhBMP-2 used [5]. There was also one case of prolonged intubation attributed to patient's history of obstructive sleep apnea also not related to rhBMP-2 use [2]. That being said, life-threatening situations do occur when using rhBMP-2 that are related to the profound edema discussed earlier. Care should be taken to manage the airway for mandibular reconstruction when utilizing rhBMP-2.

Finally, one article reported a risk but no cases of seroma development as a possible adverse event associated with localized elevated levels of rhBMP-2 [28]. Akeel et al. showed that 10–50 ng/mL rhBMP-2 triggers cell cultures to elevate levels of angiogenic and inflammatory factors (IL-6 and VEGF) via a ROS-dependent mechanism [5]. While the article stated this mechanism might contribute to a risk of seroma formation, it also stated that triggering these angiogenic effects may help potentiate beneficial effects [29].

The contraindications for using INFUSE® are if the operation site is proximal to tumors recently treated and for patients that have a history of allergic reaction, currently pregnant, that have an active infection in the operating site, currently being treated for malignancies [31]. Despite there being a narrow FDA approval for surgical procedures, INFUSE is successfully used in a number of procedures in the head and neck area for a variety of surgical procedures.

There have been FDA warnings that highlight the increased safety risks children face when they undergo spinal surgery with INFUSE® [30]. These warnings are focused on spinal surgeries because there is less space between the spinal cord and the bones surrounding. This warning and increased risk do not extend to its use in pediatric facial surgery patients. However, due to the nature of pediatric surgical cases, complications can turn develop more rapidly compared to adults patients. Subsequently, the safety of BMP in this patient population requires further study [30–33]. BMP is shown to be successful in a variety of procedures for this age group: craniocervical reconstruction, tibial pseudarthrosis, tibial nonunion repair, femoral nonunion repair, and even talonavicular joint repair [32].

Complications for this patient age range resemble those seen for adult patients, primarily associated with profound inflammatory response associated to the use of BMP [32, 33]. The reported overall complication rate with BMP ranges from 20 to 39% with only 8% requiring intervention [32, 33]. The rare cases of intervention for complications are most commonly due to postoperative infection, compartment syndrome, and hematoma with all documented patients recovering successfully after [32, 33]. Retreatment with BMP or repeated exposure is not currently associated with an elevated risk of complications in this age group and is not associ-

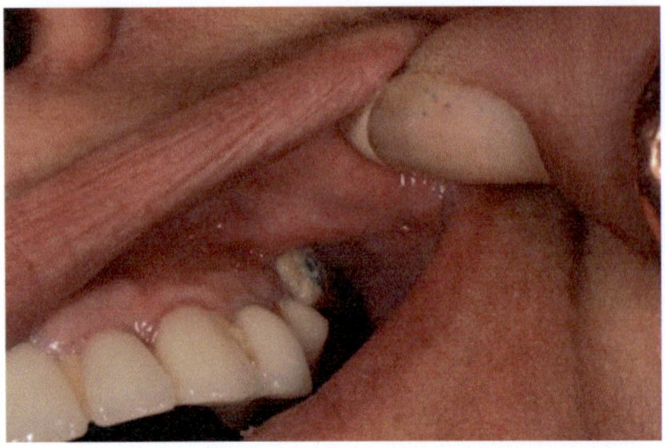

Fig. 4.5 S/p maxillary posterior bilateral rhBMP-2 bone grafting, maxillary left posterior wound dehiscence on 2-week follow-up due to history of smoking and inability to stop during healing period despite informed and understood risks by the patient. Retracted left lip and vestibule to better visualize complication site

ated with the development of cancer [32, 33]. There have been documented cases of patients with neurofibromatosis featuring enlargement of pre-existing intracranial gliomas, but no direct association with the use of BMP has been established [32, 33]. While rhBMP with ACS is a safe alternative to autografting in pediatric patients, caution and the patient's airway should be secured for facial surgery due to the inherent nature of pediatric complications turning severe rapidly.

Overall success and failure rates of bone grafting with rhBMP-2 do not differ at a statistically significant rate from ICBG in the presence of patient risk factors (smoking, alcohol use, hypothyroidism, osteoporosis, diabetes, radiation, and bisphosphonate therapy), but should a patient feature any of these risk factors, they must be informed of the risk of graft failure, postoperative infection, wound dehiscence (Fig. 4.5), delayed healing, and even the need for subsequent grafting [1, 4, 15].

4.8 Conclusion

A variety of maxillofacial defects have been shown to be amenable to reconstruction with BMP-2. RhBMP-2 with allografting or xenografting material offers the benefits of having osseous defects successfully and safely replaced in the human body by the patient's own tissues.

Even though autografting procedures are the current standard for maxillofacial reconstruction that requires bone grafting, we need to continually strive to develop new methods and techniques to reduce the morbidity and burden for our patients. Out of all the current osteogenic mediators being researched, rhBMP-2 with ACS features the most promise and clinically offers comparable results compared to the standard ICBG procedures for a variety of maxillofacial reconstruction cases. While rhBMP-2 with ACS holds great promise as an alternative to ICBG, we need more research into the long-term viability of bone grafting sites accomplished with rhBMP-2 with a variety of alloplastic materials, post-op complication rates, and success rates with greater patient pools in a greater variety of maxillofacial reconstruction defects. For now, rhBMP-2 appears to be an excellent alternative to ICBG, but with more research and time, we may even be able to replace many of the gold standard bone grafting treatment options for maxillofacial reconstruction and eliminate the need for a donor site all together.

References

1. Buser Z, Brodke DS, Youssef JA, Meisel H-J, Myhre SL, Hashimoto R, Park J-B, Tim Yoon S, Wang JC. Synthetic bone graft versus autograft or allograft for spinal fusion: a systematic review. J Neurosurg Spine. 2016;25(4):509.
2. Hammoudeh JA, Fahradyan A, Gould DD, Liang F, Imahiyerobo T, Urbinelli L, Nguyen JT, Magee W 3rd, Yen S, Urata MM. A comparative analysis of rhBMP-2/DBM vs. ICBG for secondary alveolar bone grafts in patients with cleft lip and palate: review of 501 cases. Plast Reconstr Surg. 2017;140:318e.
3. Herford AS, Boyne PJ. Reconstruction of mandibular continuity defects with bone morphogenetic protein-2 (rhBMP-2). J Oral Maxillofac Surg. 2008;66:616.
4. Herford AS, Dean J. Recombinant human bone morphogenic protein-2 (rhBMP-2) used in maxillofacial surgery. J Oral Maxillofac Surg. 2008;16:1.
5. Riederman BD, Butler BA, Lawton CD, Rosenthal BD, Balderama ES, Bernstein AJ. Recombinant human bone morphogenetic protein-2 versus iliac crest bone graft in anterior cervical discectomy and fusion: dysphagia and dysphonia rates in the early

postoperative period with review of the literature. J Clin Neurosci. 2017;44:180.
6. Zhang H, Wang F, Ding L, Zhang Z, Sun D, Feng X, An J, Zhu Y. A meta analysis of lumbar spinal fusion surgery using bone morphogenetic proteins and autologous iliac crest bone graft. PLoS One. 2014;9:e97049.
7. Schilephake H. Bone growth factors in maxillofacial skeletal reconstruction. Int J Oral Maxillofac Surg. 2002;31:469.
8. Reddi AH. Bone morphogenetic proteins and skeletal development: the kidney-bone connection. Pediatr Nephrol. 2000;14:598.
9. Termaat MF, Den Boer FC, Bakker FC, Patka P, Haarman HJTM. Bone morphogenetic proteins. J Bone Joint Surg Am. 2005;87:1367.
10. McKay WF, Peckham SM, Badura JM. A comprehensive clinical review of recombinant human bone morphogenetic protein-2 (INFUSE® Bone Graft). Int Orthop. 2007;31:729.
11. Boyne PJ. Application of bone morphogenetic proteins in the treatment of clinical oral and maxillofacial osseous defects. J Bone Joint Surg Am. 2001;83-A(Suppl 1):S146.
12. Cicciù M, Herford AS, Stoffella E, Cervino G, Cicciù D. Protein-signaled guided bone regeneration using titanium mesh and Rh-BMP2 in oral surgery: a case report involving left mandibular reconstruction after tumor resection. Open Dent J. 2012;6:51.
13. Boyne PJ, Nath R, Nakamura A. Human recombinant BMP-2 in osseous reconstruction of simulated cleft palate defects. Br J Oral Maxillofac Surg. 1998;36:84.
14. Motamedian SR, Khojaste M, Khojasteh A. Success rate of implants placed in autogenous bone blocks versus allogenic bone blocks: a systematic literature review. Ann Maxillofac Surg. 2016;6(1):78–90. https://doi.org/10.4103/2231-0746.186143.
15. Hustedt JW, Blizzard DJ. The controversy surrounding bone morphogenetic proteins in the spine: a review of current research. Yale J Biol Med. 2014;87:549.
16. Toriumi DM, Kotler HS, Luxenberg DP, Holtrop ME, Wang EA. Mandibular reconstruction with a recombinant bone-inducing factor. Functional, histologic, and biomechanical evaluation. Arch Otolaryngol Head Neck Surg. 1991;117:1101.
17. Wikesjo UM, Sorensen RG, Wozney JM. Augmentation of alveolar bone and dental implant osseointegration: clinical implications of studies with rhBMP-2. J Bone Joint Surg Am. 2001;83-A(Suppl 1):S136.
18. Smolka W, Eggensperger N, Kollar A, Iizuka T. Midfacial reconstruction using calvarial split bone grafts. Arch Otolaryngol Head Neck Surg. 2005;131:131.
19. Jung RE, Windisch SI, Eggenschwiler AM, Thoma DS, Weber FE, Hämmerle CHF. A randomized-controlled clinical trial evaluating clinical and radiological outcomes after 3 and 5 years of dental implants placed in bone regenerated by means of GBR techniques with or without the addition of BMP-2. Clin Oral Implants Res. 2009;20:660.
20. Boyne PJ. Animal studies of the application of rhBMP-2 in maxillofacial reconstruction. Bone. 1996;19:83.
21. Davies SD, Ochs MW. Bone morphogenetic proteins in craniomaxillofacial surgery. Oral Maxillofac Surg Clin North Am. 2010;22:17.
22. Bachelet J-T, Bourlet J, Chateau J, Jacquemart M, Dufour C, Mojallal A, Gleizal A. Costal grafting in mandibular reconstruction. Plast Reconstr Surg Glob Open. 2015;3:e565.
23. Skovrlj B, Marquez-Lara A, Guzman JZ, Qureshi SA. A review of the current published spinal literature regarding bone morphogenetic protein-2: an insight into potential bias. Curr Rev Musculoskelet Med. 2014;7:182.
24. Shah MM, Smyth MD, Woo AS. Adverse facial edema associated with off-label use of recombinant human bone morphogenetic protein–2 in cranial reconstruction for craniosynostosis. J Neurosurg Pediatr. 2008;1:255.
25. Balaji SM. Use of recombinant human bone morphogenetic protein (rhBMP-2) in reconstruction of maxillary alveolar clefts. J Maxillofac Oral Surg. 2009;8:211.
26. Leal CT, Calvo AM, de Souza Faco RA, de Cunha Basto Júnior JC, Yaedú RY, da Silva Dalben G, Carvalho RM. Evolution of postoperative edema in alveolar graft performed with bone morphogenetic protein (rhBMP-2). Cleft Palate Craniofac J. 2015;52(5):e168–75. https://doi.org/10.1597/14-169.
27. Ribeiro Filho SA, Francischone CE, De Oliveira JC, Ribeiro LZ, Do Prado FZX, Sotto-Maior BS. Bone augmentation of the atrophic anterior maxilla for dental implants using rhBMP-2 and titanium mesh: histological and tomographic analysis. Int J Oral Maxillofac Surg. 2015;44:1492.
28. Moghadam HG, Urist MR, Sandor GKB, Clokie CML. Successful mandibular reconstruction using a BMP bioimplant. J Craniofac Surg. 2001;12:119.
29. Boyne PJ, Nakamura A, Shabahang S. Evaluation of the long-term effect of function on rhBMP-2 regenerated hemimandibulectomy defects. Br J Oral Maxillofac Surg. 1999;37:344.
30. Oral Maxillofacial Surgery, Medtronic INFUSE® Bone Graft. http://www.infusebonegraft.com/healthcare-providers/oral-maxillofacial-surgery/index.htm. Accessed 17 Jan 2018.
31. FDA Warning for Infuse Bone Graft Risks in Kids. https://www.schmidtlaw.com/fda-warning-for-infuse-bone-graft-risks-in-kids/. Accessed 17 Jan 2018.
32. Stiel N, Hissnauer TN, Rupprecht M, Babin K, Schlickewei CW, Rueger JM, Stuecker R, Spiro AS. Evaluation of complications associated with off-label use of recombinant human bone morphogenetic protein-2 (rhBMP-2) in pediatric orthopaedics. J Mater Sci Mater Med. 2016;27:184.
33. Oetgen ME, Richards BS. Complications associated with the use of bone morphogenetic protein in pediatric patients. J Pediatr Orthop. 2010;30:192.

Techniques of Obtaining BMAC, PRP, and PRF

5

Hisham Marwan, Todd Sawisch, Avi Schetritt, and Ramzey Tursun

The popularity of bioactive surgical additives in oral surgery procedures has seen an increase in recent years. These bioactive additives are used to regulate inflammation and increase the speed of the healing process. These agents augment the wound healing process through anabolic bone formation, angiogenesis, osteoblast differentiation, mitosis, chemotaxis, and other processes that improve the healing environment.

While the entire hard and soft tissue healing process is not completely understood, platelets play a crucial role not only in hemostasis but in wound healing as well. The first mention of platelets' regenerative potential was by Ross et al. in 1974. They described a growth factor from platelets which stimulated the mitogenic response in the bone periosteum during bone repair wound healing. Specific growth factors contained in the a-granules of platelets (such as platelet-derived growth factor, transforming growth factor-b1, epithelial growth factor, vascular endothelial growth factor, insulin-like growth factor-I, basic fibroblast growth factor, and hepatocyte growth factor) promote the bone regeneration of oral and maxillofacial bone defects. Platelets also play a role in the host defense mechanism at the wound site by delivering signaling peptides that attract macrophage cells. In addition, platelet concentrates contain leukocytes that synthesize interleukins involved in the non-specific immune reaction. All of which has led to an increase interest in platelet concentrates as bioactive additives. In this chapter we will discuss the three most common regenerative techniques utilizing the platelets as a bioactive additive, the platelet-rich plasma (PRP), the bone marrow aspirate concentrate (BMAC), and the platelet-rich fibrin (PRF).

Bone marrow aspirate and bone marrow aspirate concentrate (BMAC) is a generous source of mesenchymal stem cells and osteoprogenitor cells. Once harvested, the cells can be added to any type of bone graft in order to facilitate osteogenesis. It has been proven that mixing the stem cells from the bone marrow with a biocompatible bone graft will enhance the differentiation and proliferation of stem cells into mature bone. Furthermore, the utilization of allogenic graft acting as a scaffold with the BMAC will avoid the morbidity associated with harvesting autogenous bone and will yield to bone formation.

H. Marwan
Oral/Head and Neck Oncology and Microvascular Surgery, University of Miami/Jackson Memorial Hospital, Miami, FL, USA

T. Sawisch · A. Schetritt
Oral and Maxillofacial Surgery, Fort Lauderdale, FL, USA

R. Tursun (✉)
Clinical Surgery, Head and Neck Oncologic and Microvascular Reconstructive Surgery, Division of Oral and Maxillofacial Surgery, University of Miami - Miller School of Medicine, Miami, FL, USA
e-mail: r.tursun@med.miami.edu

J. C. Melville et al. (eds.), *Regenerative Strategies for Maxillary and Mandibular Reconstruction*,
https://doi.org/10.1007/978-3-319-93668-0_5

The preferred harvest site is either from the anterior or posterior ilium (Fig. 5.1). Bone marrow aspiration from the anterior iliac crest is virtually free from complications. The procedure can be done as an outpatient with local anesthesia, with or without sedation. On the other hand, aspiration from the posterior ilium is usually accomplished under general anesthesia for patients' comfort.

To harvest a sufficient amount of mesenchymal cells, two puncture sites are recommended to aspirate 60–120 ml of bone marrow aspirate. A 2 mm incision is made at the puncture site; blunt dissection is preceded down to the cortex (Fig. 5.2). The sharp trocar punch, with heparin coating, is pressed through the outer cortex into the marrow space using a twisting motion. Once in the marrow, the trocar is removed leaving the hollow aspiration sleeve inside (Fig. 5.3). Also, a bone core can be harvested using the sharp trocar (Fig. 5.4), and this bone core can be used as an autogenous source of bone graft to the final bone mixture. We use a 20 ml syringe with 0.5 ml of heparin. The 0.5 ml of heparin is injected into the marrow space before the plunger is drawn back to aspirate the bone marrow. The harvested bone marrow must remain anticoagulated. A 4 ml of anticoagulant citrate dextrose-A (ACD-A) is placed in a blood collection bag. Subsequently, each syringe with bone marrow is injected into the bag (Fig. 5.5). Finally, the bone marrow aspirate is drawn from the anticoagulated blood col-

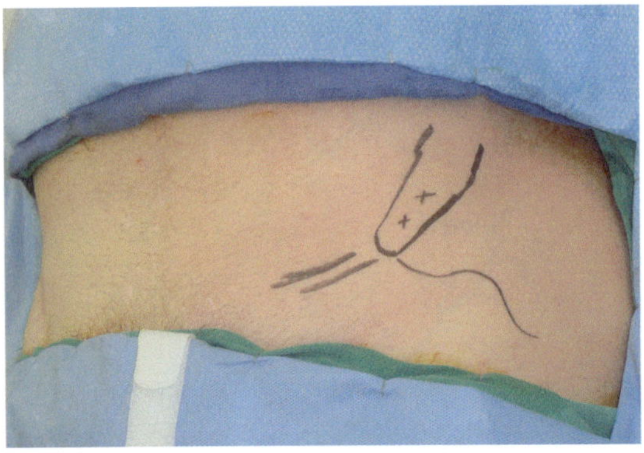

Fig. 5.1 Preparation of anterior ilium for BMAC harvest

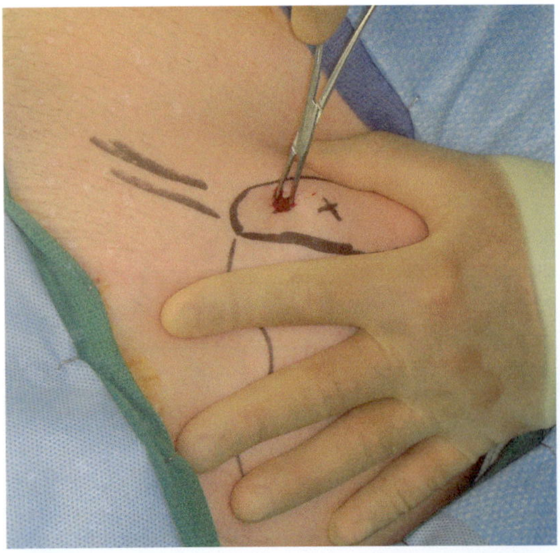

Fig. 5.2 Blunt dissection of the puncture site

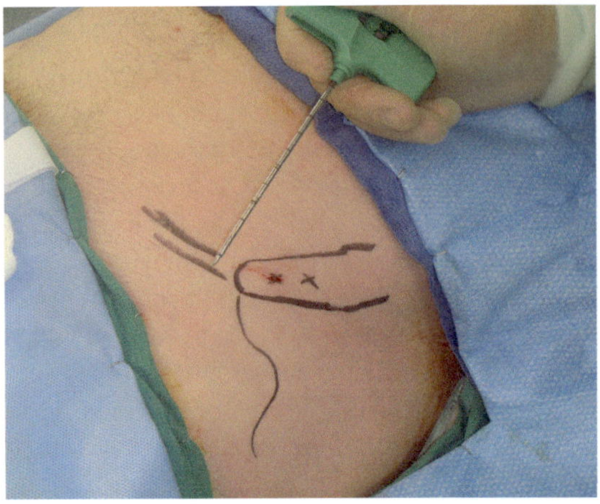

Fig. 5.3 Sharp trocar before insertion into the anterior iliac puncture site

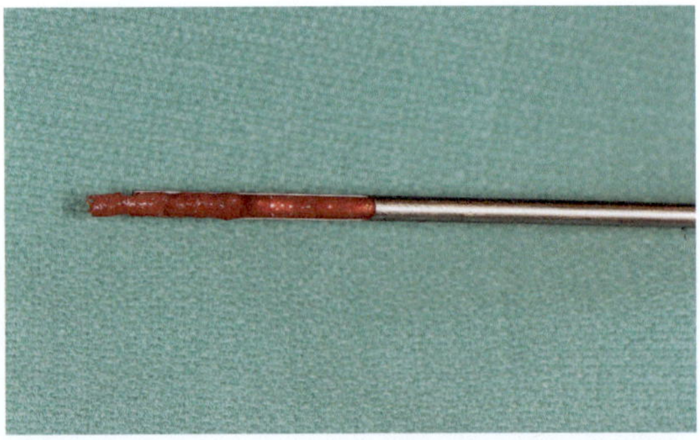

Fig. 5.4 Obtaining of the bone core to be mixed with the BMAC

lection bag and placed into the chamber for centrifuging process (Fig. 5.6). The skin incision is then closed with 5-0 fast resorbing gut suture and no dressing is required.

After centrifuging, two fractions are formed into the chamber. First one is the plasma fraction, which contains the adhesion molecules, fibronectin and fibrinogen. The other fraction is the cellular fraction, which contains the mesenchymal stem cells and the osteoprogenitor cells (Fig. 5.7). The cellular fraction can be added to the bone graft to enhance the bone formation (Fig. 5.8). Finally, the plasma fraction will be placed on top of the bone graft to act as a scaffold and improve the osteoconductivity of the graft.

The second technique is the platelet-rich plasma (PRP). This technique is characterized by obtaining the growth factors from a simple blood drawing obviating the need for any autogenous bone procurement with minimal morbidity. The authors usually combine the PRP with bone grafting materials (autogenous or allogenic) to achieve more predictable outcomes. It was reported in the literature that the PRP will help the bone graft by providing the required tissue adhesion molecules, namely, the fibrin, fibronectin, and vitronectin. In addition, it will amplify the signaling effect of rhBMP-2/ACS.

The technique of harvesting PRP starts with aseptic phlebotomy using a 18-gauge needle

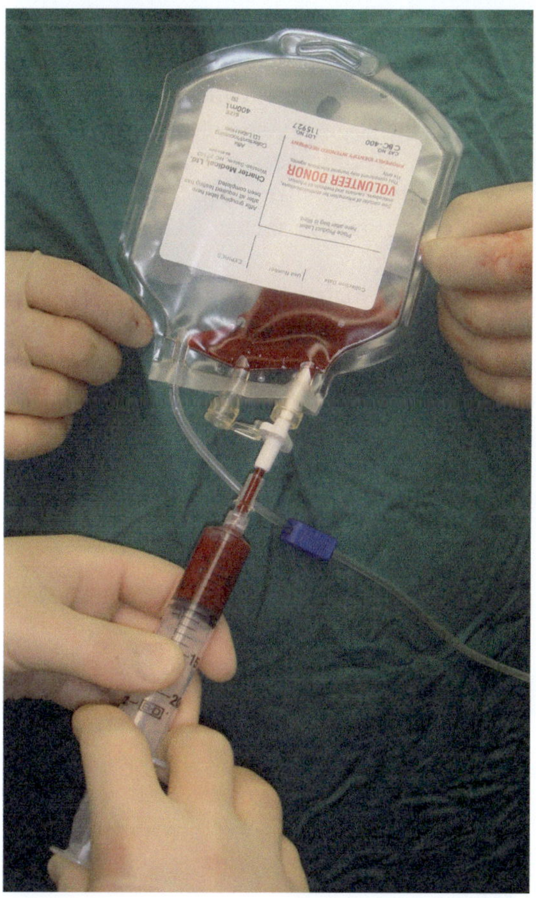

Fig. 5.5 Placement of harvested bone marrow in the blood-collecting bag

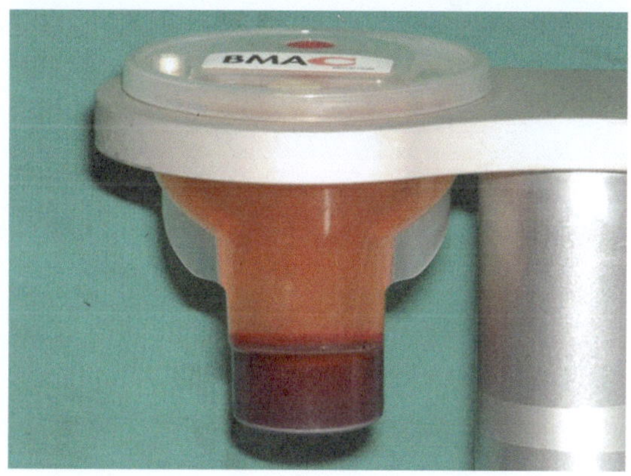

Fig. 5.6 Bone marrow aspirate in the chamber after centrifuging

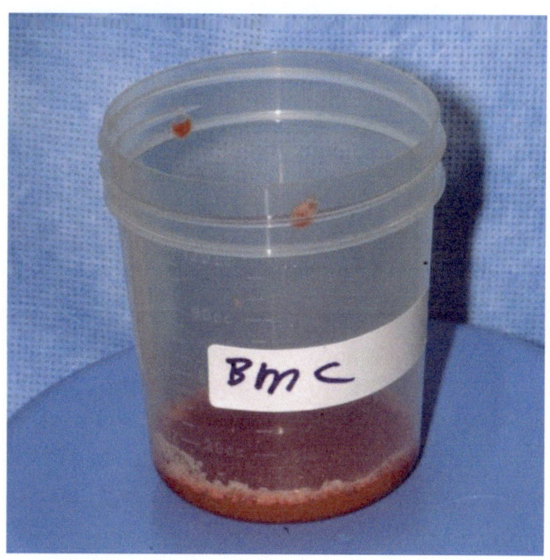

Fig. 5.7 The cellular fraction component of the BMAC

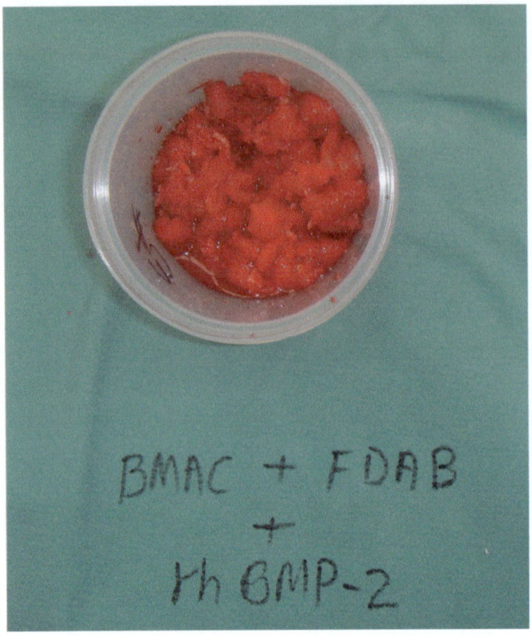

Fig. 5.8 Mixture of the bone graft with the bone marrow aspirate before placement in the recipient bed

to prevent platelets disruption during the aspiration technique. We usually prefer using the antecubital vein because of the ease of use and its large diameter. The syringe should contain Anticoagulant Citrate Dextrose A (ACD-A). According to the manufacturer recommendations, when a 20 ml syringe is used, 2 ml of ACD-A should be placed in the syringe prior to the withdrawal. On the other hand, when 60 ml of autogenous blood is drawn, 7 ml of ACD-A should be used. Once the whole venous blood drawing is completed, we use the SmartPrep device by Harvest Technologies for the centrifuging process. The total time of processing using the SmartPrep device is 14 min.

Once the platelets are separated and concentrated, a distinct layer of the plasma is formed, and this is the platelet-poor plasma (PPP). This layer is drawn separately and can be used as a separate layer to achieve cohesiveness of the grafted material. The other layer contains the anticoagulated PRP that can be stored up to 8 h. Activation of the PRP is required and this is accomplished by adding calcium to the PRP. The authors' preferred method to add the calcium is by adding 5 ml of 10% calcium chloride to 5000 unit of topical bovine thrombin. Usually, 2–4 drops of this mixture will suffice to activate the clotting cascade of the PRP.

The last technique is the platelet-rich fibrin (PRF), renamed leukocyte PRF (L-PRF) due to its higher leukocyte content, which represents a second generation of platelet concentrates, and its development can be attributed to Choukroun et al. in 2001 [1]. The major advantages of PRF include having completely immune-compatible growth factors collected at relatively no costs without anticoagulants or bovine thrombin or any other gelling agent.

PRF consists of a fibrin matrix of platelets, leukocytes, cytokines, and circulating stem cells. The PRF clot is yielded by a natural polymerization process during centrifugation, and its natural fibrin architecture is responsible for the slow release of growth factors and matrix glycoproteins for 7 or more days. The growth factors released from PRF are transforming growth factor-b (TGF-b), platelet-derived growth factor-AB (PDGF-AB), vascular endothelial growth factor (VEGF), and glycoproteins (such as thrombospondin-1, fibronectin, and vitronectin). Leukocytes in PRF have been shown to be highly important immune cells capable of directing and recruiting various cell types during the wound healing process. Additionally, PRF has a strong induction effect on proliferation of various cells, especially osteoblasts in vitro.

Advanced PRF (A-PRF) is a modification of the L-PRF protocol, whereby the centrifugation G-force is reduced. An increase in total leukocyte numbers within PRF matrix scaffolds was observed in A-PRF. It has also been shown that the release of several growth factors, including PDGF, TGF-b1, VEGF, epidermal growth factor (EGF), and insulin-like growth factor (IGF), was significantly higher in A-PRF compared with L-PRF.

5.1 PRF Preparation

5.1.1 PRF Membrane Fabrication

The PRF process requires the following steps:

1. Collection of whole venous blood in a tube (glass tubes or glass-coated plastic tubes) without anticoagulant. These are red-top tubes.
2. Tube(s) are then centrifuged at
   (a) 2700 RPM for 12 min for standard L-PRF.
   (b) 1300 RPM for 8 min for advanced A-PRF.
3. Once centrifuged, three layers are formed (Fig. 5.9): a red blood cell (RBC) base at the bottom, acellular plasma (platelet-poor plasma [PPP]) as a supernatant, and a PRF clot in the middle.
4. PRF clot is removed from collection tube (Fig. 5.10), and the RBCs are gently scraped off the clot (Fig. 5.11).
5. PRF clots are placed on compression box (Fig. 5.12) to express plasma.
6. Compressed PRF clots (Fig. 5.13) are comprised of a strong fibrin membrane containing leukocytes, platelets, cytokines, and circulating stem cells.

5.1.2 PRF Bone Putty

PRF can be used to enhance particulate bone graft material healing qualities and handling characteristics by creating a putty-like consis-

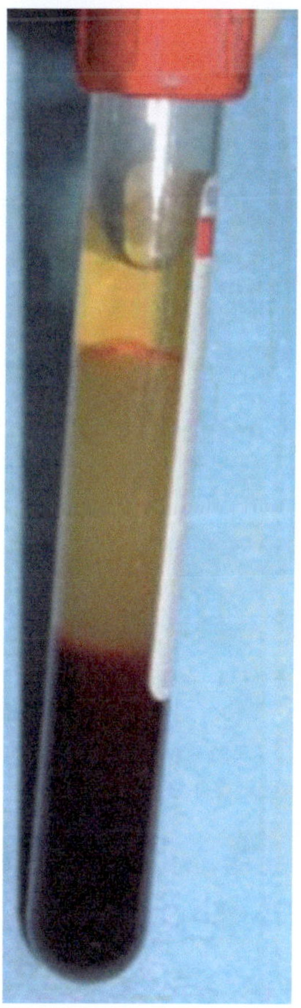

Fig. 5.9 PRF clot in tube post-centrifuge

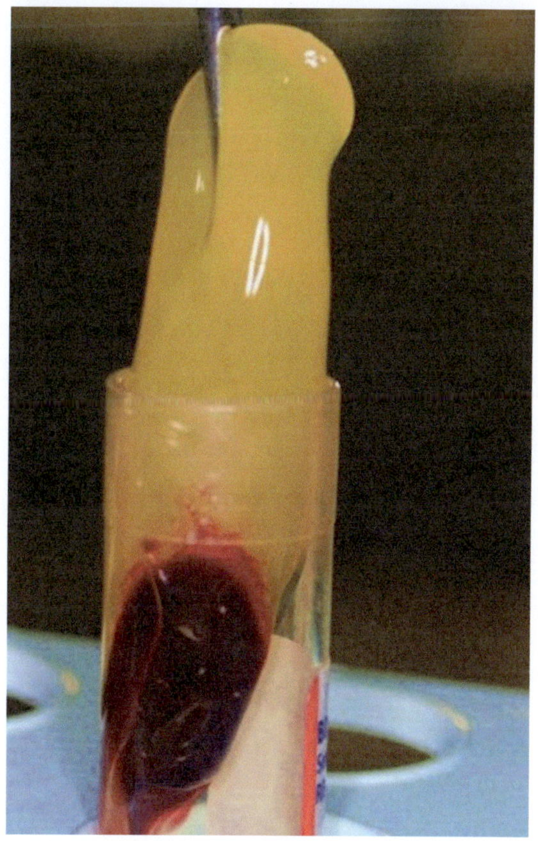

Fig. 5.10 PRF clot removed from tube

Fig. 5.11 Scraping RBC off PRF clot

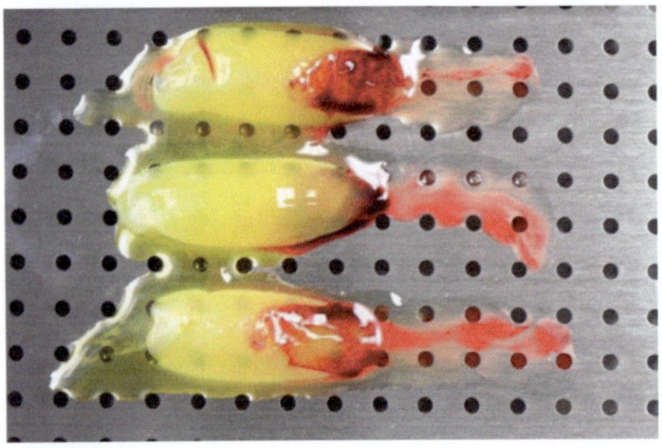

Fig. 5.12 PRF clots on compression box

tency. The process to create a "bone putty" is as follows:

1. Collection of whole venous blood in a plastic tube (white top) without anticoagulant for hydration of the particulate bone.

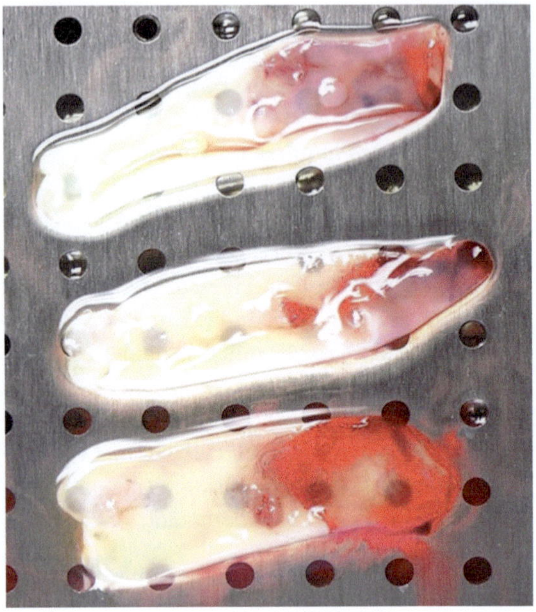

Fig. 5.13 Compressed PRF clots

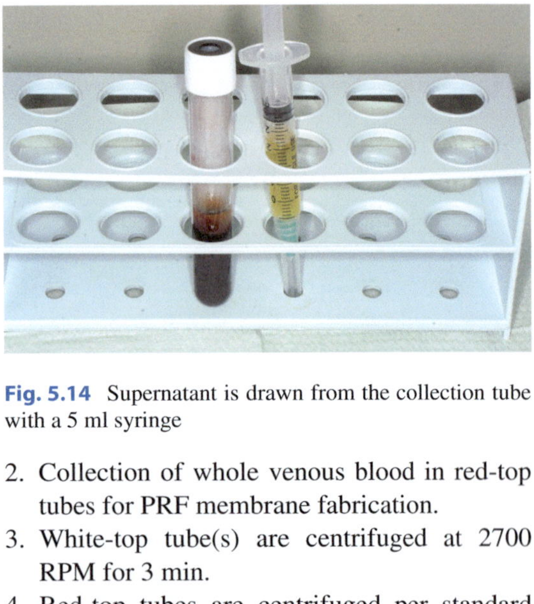

Fig. 5.14 Supernatant is drawn from the collection tube with a 5 ml syringe

2. Collection of whole venous blood in red-top tubes for PRF membrane fabrication.
3. White-top tube(s) are centrifuged at 2700 RPM for 3 min.
4. Red-top tubes are centrifuged per standard PRF protocol.
5. Supernatant is drawn from the white-top collection tube with a 5 ml syringe (Fig. 5.14).
6. The particulate bone graft is hydrated with the supernatant (Fig. 5.15a, b).
7. Once PRF membranes are formed, the particulate graft is further hydrated with the PRF membrane supernatant (Fig. 5.16a, b).

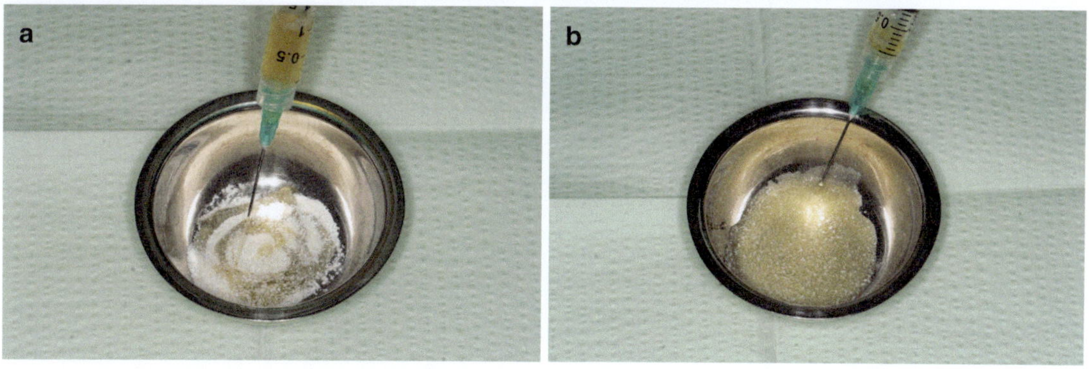

Fig. 5.15 (**a**, **b**) Hydrating the particulate bone graft with the supernatant

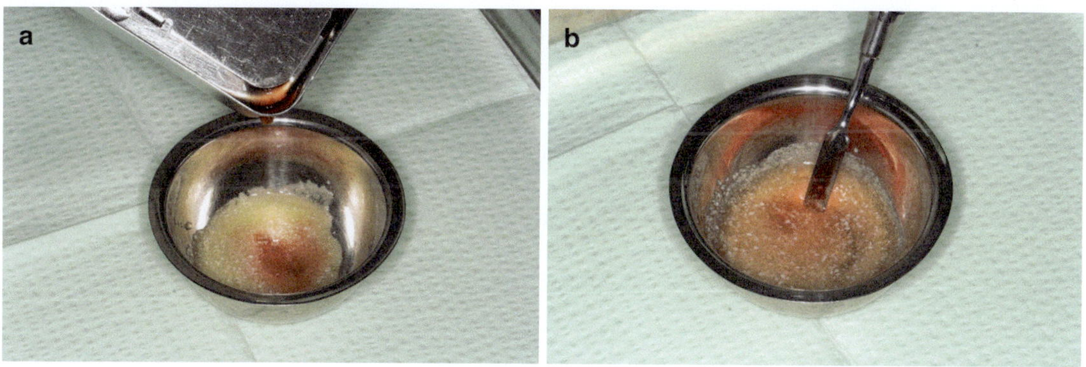

Fig. 5.16 (**a**, **b**) Hydrating the particulate bone graft with the PRF supernatant

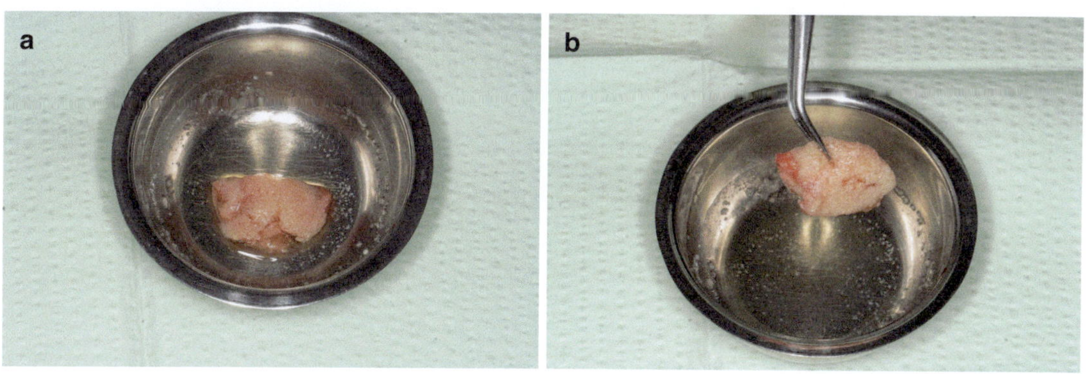

Fig. 5.17 (**a**, **b**) Resulting particulate graft putty

This will initiate coagulation and result in a particulate graft putty (Fig. 5.17a, b). Small fragments of PRF membrane can be added to the mix to accelerate the formation of the putty.

5.2 PRF Applications

The compressed PRF clot containing platelets, leukocytes, and growth factors in a strong fibrin membrane will release growth factors for up to 7 days. This biological PRF membrane has multiple oral surgery applications.

5.2.1 PRF in Particulate Grafting Procedures

The formation of a PRF bone putty enhances the handling characteristics of the particulate bone graft and incorporates the enhanced wound healing benefits of PRF. It is of interest to note that PRF placed into extraction sockets has been shown to decrease the rate of complications and infection. This is likely due to the leukocytes present in PRF as leukocytes are the cell type responsible for preventing infiltration by pathogens.

The PRF bone putty can be used in socket graft applications (Fig. 5.18a, b) and in lateral ridge augmentations (Fig. 5.19a, b) where its graft containment characteristics are particularly beneficial. The PRF membrane is also beneficial in these procedures where its physical characteristics are used to create compartmentalized wound healing spaces (Fig. 5.19c).

5.2.2 PRF Membrane as a Palatal Bandage for Free Gingival Graft (FGG) Donor Site

Donor site bleeding, pain, and discomfort are the common problems encountered during FGG procedures. PRF promotes microvascularization and epithelial cell migration to its surface and can be used as a bandage for FGG door sites (Figs. 5.20 and 5.21). This technique reduces postoperative morbidity and complications associated with donor sites by accelerating the healing process.

Platelet-rich fibrin is a second-generation autologous platelet concentration that is simple and inexpensive to process. PRF contains high amounts of bioactive growth factors to enhance wound healing through increased chemotaxis, proliferation, differentiation, and angiogenesis. It can be used in a variety of oral surgery and implant-related procedures to increase the speed of wound healing and decrease morbidity.

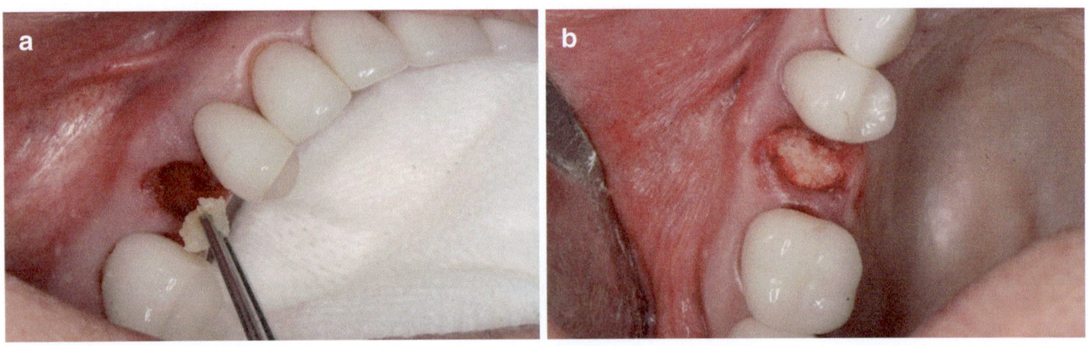

Fig. 5.18 (**a**, **b**) Particulate bone graft PRF putty placed in extraction socket

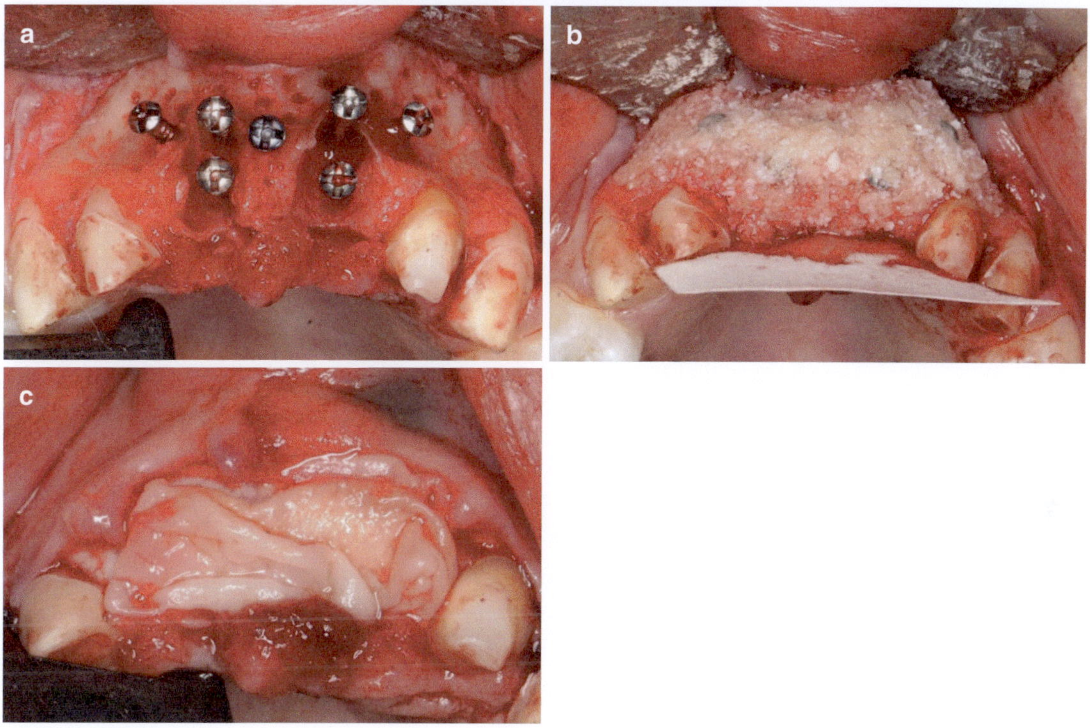

Fig. 5.19 (**a**, **b**) Particulate bone graft PRF putty placed in lateral ridge augmentation procedure. (**c**) PRF membrane placed over bone graft putty to compartmentalize wound healing space

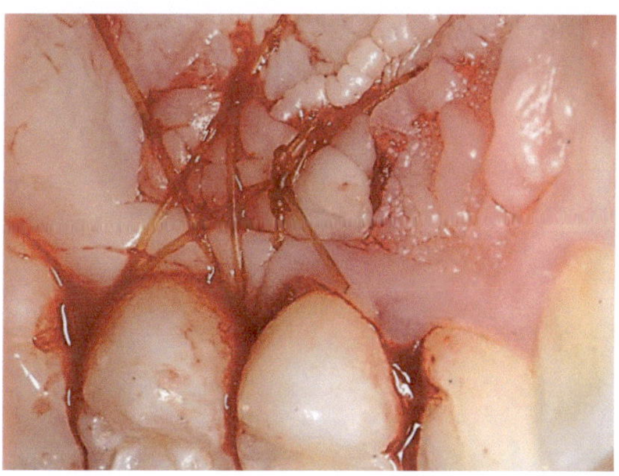

Fig. 5.20 PRF membrane sutured into FGG donor site

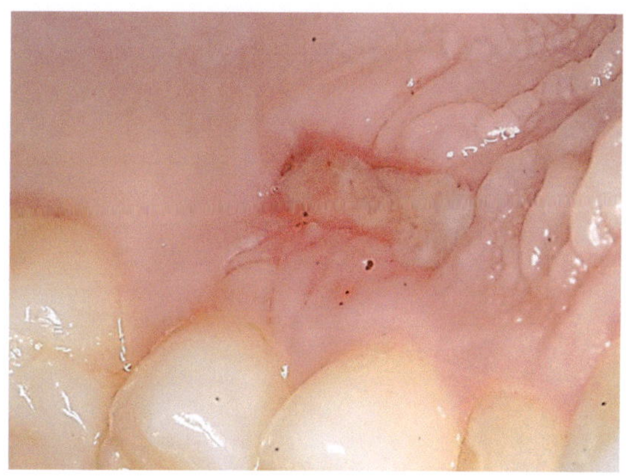

Fig. 5.21 1-week healing of FGG donor site

Reference

1. Dohan DM, Choukroun J, Diss A, Dohan SL, Dohan AJ, Mouhyi J, et al. Platelet-rich fibrin (PRF): a second-generation platelet concentrate. Part I: technological concepts and evolution. Oral Surg Oral Med Oral Pathol Oral Radiol Endod. 2006;101:e37–44.

6 Principles of Soft Tissue Engineering for Craniomaxillofacial Reconstruction

Shiuhyang Kuo, Atsuko Miyazawa, and Stephen E. Feinberg

6.1 Introduction

A tissue-engineered device is needed when the supply of autogenous native tissue is limited or scarce to repair defects due to severe wounds, tumor removal, chronic or acute degenerative changes, or congenital symptoms. Tissue engineering creates new tissue by combining cell and developmental biology with material principles. Eventually, a tissue-engineered product has to find its way into a host. The autogenous cells are used to eliminate the immune responses of the host to the tissue-engineered product. The supporting materials used in manufacturing tissue-engineered devices should cause no immunogenicity reactions and easily be integrated into a host and biodegradable and possess good mechanical properties [1].

The manufacture of tissue-engineered devices must meet current Good Manufacturing Practice (cGMP) standards which are enforced by the US Food and Drug Administration under Title 21 of the Code of Federal Regulations [2, 3]. To meet these standards, our laboratory cultured keratinocytes and constructed autogenous tissue-engineered devices in a serum-free chemically defined medium without a feeder layer. Our laboratory routinely uses a tissue-engineered device termed ex vivo produced oral mucosa equivalent (EVPOME) [4–6], which is composed of oral keratinocyte cellular layers and a supporting scaffold made from human decellularized cadaver skin, as a study model or to repair the loss of oral soft tissue in our clinical trials [7]. We cultured oral keratinocyte stem/progenitor cells to manufacture EVPOMEs. In oral mucosa tissue, oral keratinocyte stem cells reside in the basal cellular layer, and progenitor cells can residue in the basal cellular layer and the inner part of the spinous layer [8]. The percentage of oral keratinocyte stem cells is less than 10% [9, 10]. The expressions of Ki67, a proliferation marker, and integrin α6 and CD71, two frequently used oral keratinocyte stem cell markers, on native gingiva are shown in situ in Fig. 6.1. The majority of oral basal keratinocytes lack the expression of Ki67 and CD71. However, Ki67 was detected on the cellular layer on top of basal cellular layer. The expression of integrin α6 was detected along the contact area between basal membrane and basal keratinocytes and is important for cell adhesion to the underlying basement membrane. The negative expression of Ki67 suggests that the cells within the basal layer are oral keratinocytes in a quiescent stage within the native tissue. However, these oral keratinocyte stem cells will turn into highly proliferative cells when placed into an in vitro culture which simulates a

Author contributed equally with all other contributors. Shiuhyang Kuo and Atsuko Miyazawa

S. Kuo · A. Miyazawa · S. E. Feinberg (✉)
Department of Oral and Maxillofacial Surgery, University of Michigan Health System, Ann Arbor, MI, USA
e-mail: skuo@med.umich.edu; sefein@umich.edu

J. C. Melville et al. (eds.), *Regenerative Strategies for Maxillary and Mandibular Reconstruction*,
https://doi.org/10.1007/978-3-319-93668-0_6

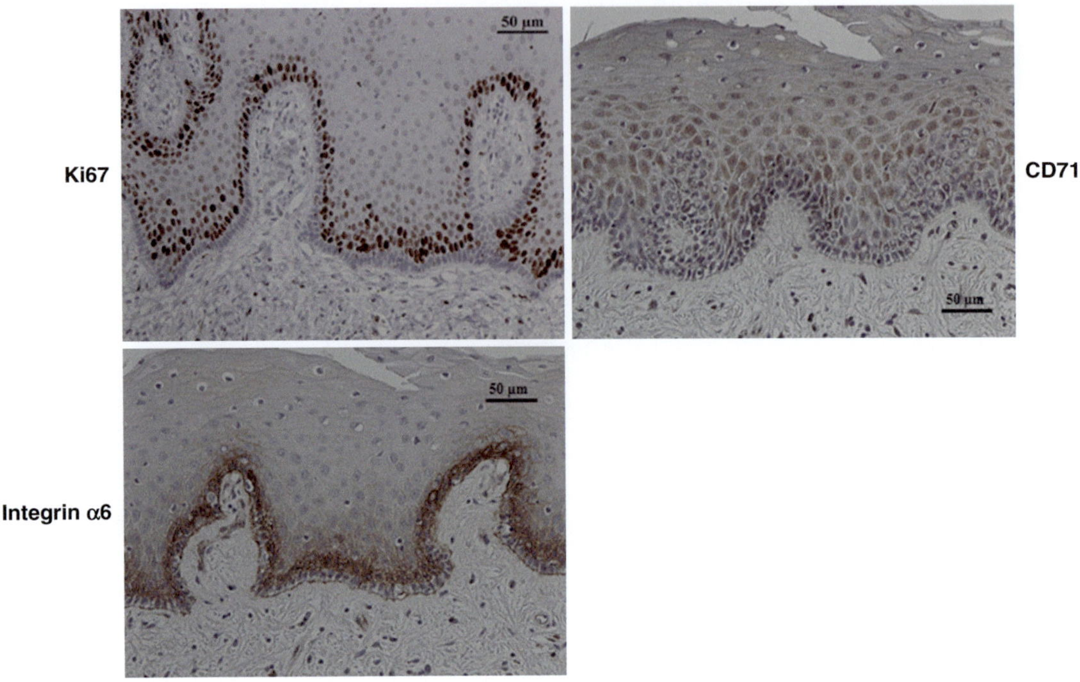

Fig. 6.1 Ki67, integrin α6, and CD71 expression on gingiva. Majority of basal keratinocytes do not express Ki67 and CD71. Integrin α6 was detected along the contact area between basal membrane and basal keratinocytes. Each scale bar represents 50 μm

wounding environment [9, 11]. It is this property that allows us to amplify and obtain a high number of oral keratinocytes possessing differentiation capacity used in manufacturing tissue-engineered devices, such as EVPOMEs.

Mammalian skin is composed of two distinct layers, epidermis and the underlying dermis. The epidermis' basal layer is composed of two keratinocyte populations, keratinocyte stem cells (KSCs) and transitory amplifying (TA) cells. Epidermal stem cells have been found to express markers which are typically expressed by stem cells in other epithelial tissues, i.e., CD34, K15, alpha6 integrin, and Lgr5 [12–14]. However, these phenotypic markers cannot be used to differentiate between the two epidermal cell populations, KSCs and TA cells, because of their lack of expression on their cell surfaces [15].

EVPOMEs manufactured by culturing a single cell type, such as oral or skin keratinocytes, can be used to repair oral mucosa or skin defects caused by diseases or trauma. There are areas within our bodies where the mucosa and skin meet and create a special structure of tissue termed mucocutaneous junctions, such as the lip, eyelid, anus, or vagina. To repair these mucocutaneous junctions, a tissue-engineered device was developed and manufactured by co-culturing oral and skin keratinocytes on either side of a cell-free zone created by a barrier [6, 16–18].

In this chapter, we present the concepts of clinical applications using tissue-engineered devices with detailed protocols that we used to culture primary human keratinocytes, skin and oral mucosa, to manufacture tissue-engineered devices such as EVPOMEs and oral and skin co-cultures that can be used to create mucocutaneous junctions. We report a unique culture system that allows cells to grow beyond contact inhibition, such that cells are produced in excess of available growth surface. The newly produced cells called "pop-up" cells float up into the culture medium and when passed into a new growth vessel reattach to the growth surface and continue to grow as adherent monolayer cells that upon confluence can begin to continuously produce additional floating "pop-up" cells [19–21]. In addition to the traditional method of using attached monolayer keratinocytes to culture keratinocytes and manufacture EVPOMEs, we also report the method of manufacturing EVPOMEs using floating "pop up" oral keratinocytes.

6.2 Proposed Clinical Applications for Reconstruction of the Lips

The lips form a dynamic complex facial structure that is difficult to reconstruct after an avulsion injury. The lips are a composite tissue consisting of the mucosa, skin, and innervated muscles. Significant loss of the lips is a functional and esthetic concern because the neuromuscular control of normal lip structures is required for eating, drinking, talking and social gesturing. Avulsion of the lips is a survivable injury; but without functional lip reconstruction, life for injured individuals is burdened by drooling, food spillage while eating, unintelligible speech, and social rejection. Functional reconstruction of the lips is compelling; when greater than 50% of the lips are avulsed, face transplantation under lifetime immunosuppression is the contemporary option.

Regeneration of oral mucosal tissue has not been previously addressed in craniomaxillofacial (CMF) soft tissue injuries. The lack of available TE/RM (tissue-engineered/regenerative medicine)-produced oral mucosa has limited surgeons to reconstruct the oral cavity and other functional units that contain an M/C junction, i.e., the lips, anal sphincter, vagina, and eyelids.

The in vitro development of a human oral mucosa is fundamental. Oral mucosa forms the inner aspect of the lip. The oral mucosa extends onto the vermillion border to unite with the skin of the face forming an M/C junction. Though skin equivalents have been developed for treatment of burns and chronic wounds, development of an *ex vivo-p*roduced *o*ral *m*ucosal *e*quivalent (EVPOME) has lagged behind that for the skin. EVPOME for lip reconstruction will be a transforming esthetic-functional option.

The TE/RM approach faces several barriers preventing its transition into the clinical arena. These barriers are (1) inability to create composite tissue structures especially in the area of soft tissue reconstruction and (2) the difficulty in developing an in vivo perfusion system (blood vessels) to supply nutrition for large segments of tissue created in vitro. Though TE/RM in vitro culture systems have been designed to mimic the natural environment, artificial in vitro systems only allow viability of uncomplicated pieces of tissue based on oxygen perfusion of 100–200 microns. Vascularization and nutrition remain a problem limiting the size and complexity of tissue-engineered constructs that can be successfully transplanted after they are developed in vitro [22].

Most surgeons have addressed the limitations stated above by creating prefabricated flaps with their own blood supply. In this method, a vascular carrier is implanted to a new skin territory. Following a period of maturation and neovascularization, the prefabricated flap can be transferred, based on the implanted pedicle. Prefabricated flaps have been utilized in surgery for reconstruction of individual esthetic units such as the nose, ear, cheek, lip, and neck [23–26]. With flap prefabrication, a distally ligated vascular pedicle is implanted underneath the desired donor tissue, and after a period of 8 weeks of neovascularization, this donor tissue can then be transferred based on the pre-implanted vascular pedicle as its axial blood supply [27]. We propose to use a parallel but distinct concept called flap prelamination. Lamination refers to the process of bonding or integration of several different tissue layers. Prelamination designates a reconstructive process where a three-dimensional structure is built at a remote site by laminating several different tissue layers together as composite grafts into a reliable existing axial vascular bed. The structure matures 2–3 weeks before transferring the unit en bloc to the defect based on its native axial blood supply. The technique of prelamination allows reconstruction to begin at a remote site. This is important as the recipient site being reconstructed may lack either the blood supply or healthy tissue substratum necessary to support construction of a sophisticated three-dimensional construct at the defect site. Furthermore, remote reconstruction in an unscarred vascular bed offers the best chance for the composite grafts to mature [24, 25]. The technique of prelamination is often used in reconstructing structures with multiple functional layers, i.e., full-thickness reconstruction of the nose, lip, cheek, ear, maxilla, mandible, and trachea [28, 29].

Prefabricated innervated prevascularized prelaminated (PIPP) flaps increase the options available to reconstructive surgeons. Present lip reconstructions are limited to harvesting existing anatomical units. For example, the radial forearm free flap with the palmaris longus tendon is a standard option for reconstructing total lower lip defects and offers an excellent pedicle, appropriate tissue thickness, and a method to suspend the lip with the incorporated tendon rather than muscle [30]. However, existing data indicates that most injured combatants sustain multiple injuries to the extremities along with the head and neck regions [31]. The difficulty then becomes obvious if we entertain a total lip reconstruction in a battle-injured warrior by using upper extremity parts. Flap prefabrication allows the reconstructive surgeon to choose an uninjured wound bed with adequate tissue characteristics, a set of vessels, a motor nerve as well as a place to fabricate the construct.

6.2.1 Advantages of Our TE/RM Approach

The tissue-engineered human mucosa and skin keratinocytes on the same piece of dermal matrix (M/C constructs) are cultured in vitro, transferred to an existing muscle, such as the latissimus dorsi muscle (LDM), which can be harvested as an IPP flap. In a defined period of time, the M/C constructs mature, integrate, and develop a micro-capillary system with the underlying muscle flap. The PIPP flap is then transferred to the area of the lips. Vascular and nerve pedicles are microvascularly anastomosed to existing pedicles in the face (Fig. 6.2). The advantages of a regenerative medicine approach to reconstruct craniofacial soft tissue injuries are:

1. There is no or minimal donor site morbidity. We use the protected torso and avoid the use of injured extremities as a tissue source thus shortening the rehabilitation phase. This results in a decrease in patient morbidity.
2. Restoration of function is accelerated; esthetic and functional outcome are better than contemporary options.
   (a) A decrease in the number of surgeries needed with a concomitant decrease in operating room time

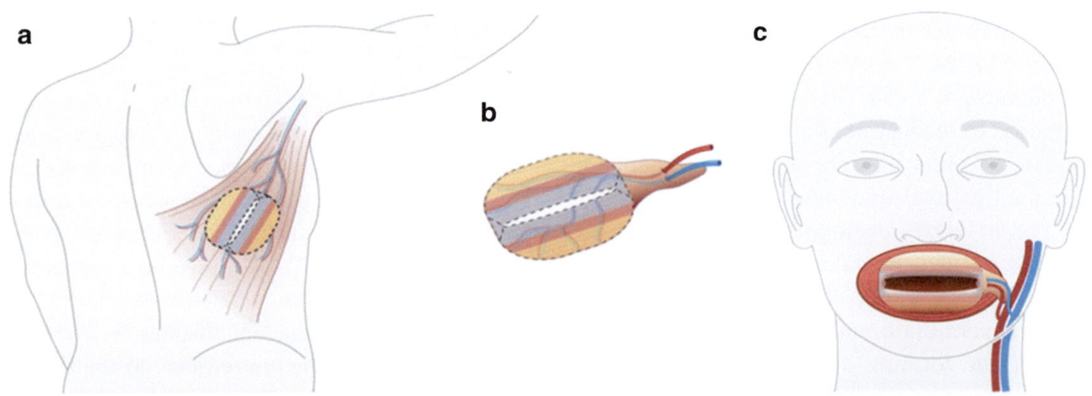

Fig. 6.2 (**a**) M/C construct integrated onto the LDM with a surgically created stoma. (**b**) The LDM with integrated M/C surgically removed to create a prevascularized prelaminated flap with thoracodorsal artery and vein. We will also harvest the thoracodorsal nerve. (**c**) The LDM prevascularized prelaminated flap inserted into the face to form lips with microanastomosis of the thoracodorsal artery and vein to the external carotid artery and external jugular vein, respectively. The thoracodorsal nerve will be anastomosed to the facial nerve for motor function (from 18 with permission)

(b) An enhancement in the quality, shape, and function of soft tissue regenerated

3. The source of the cells to develop the soft tissue constructs (lips) will come from the patient. One centimeter punch biopsies will be excised from the cheek for oral mucosa and the posterior auricular region for skin (optimal color match to face), thus making the construct autochthonous (self to self).

6.2.2 Proposed Study Design

We propose a multicenter, prospective, interventional non-randomized open label single arm safety and efficacy in subjects with equal to or greater than 50% loss of the lips who will be treated with a prefabricated innervated prevascularized prelaminated (PIPP) designer microvascular free flap for functional lip reconstruction. We could not ethically have a control population for randomization that was treated with contemporary surgical techniques to functionally reconstruct greater than 50% of the lips as these techniques have proven to be less than adequate. A functional lip will be reconstructed by the following steps:

1. Two 1.0 cm circular punch biopsies, the posterior auricular area for autogenous skin keratinocytes and the buccal (cheek) mucosa for autogenous oral keratinocytes, will be taken.
2. A mucocutaneous (M/C) construct will be manufactured in vitro in a cGMP facility with the autogenous skin and oral keratinocytes using an appropriate template for M/C fabrication that would address the size and geometry of the defect to be reconstructed (Fig. 6.3).
3. An ex vivo fabricated M/C construct will be grafted onto the latissimus dorsi muscle (LDM) to create a prefabricated innervated prevascularized prelaminated (PIPP) flap. The

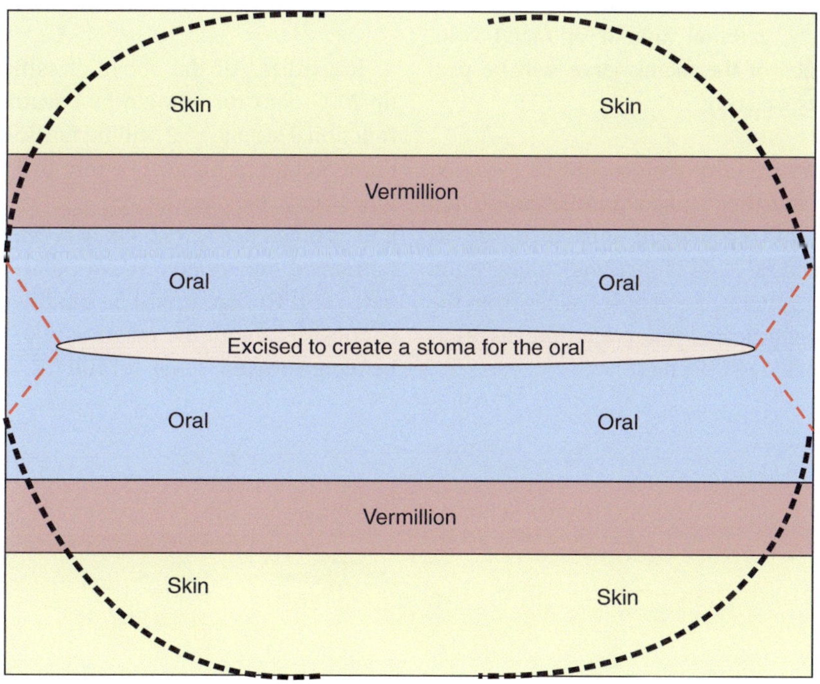

Fig. 6.3 Template for making either an upper or lower lip (50% of above template) or a complete "set" of lips. Oral = oral keratinocytes. Skin = epidermal (skin) keratinocytes. Black filled circle = excised skin tissue from the periphery of the template. Red filled circle = cuts in oral mucosa to form a commissure of the lip which will be "turned in" (from 18 with permission)

implantation of the M/C construct will be placed parallel to the muscle fibers of the LDM as well as parallel to the motor nerve and blood supply based on the thoracodorsal vessels. The parallel position of the neurovascular bundle and M/C construct will facilitate the surgical inset of the flap and the functional dilation and contraction of the muscle fibers. The location of the M/C construct will be just below the angle of the scapula in order to allow for an appropriate length of the neurovascular pedicle.

4. A stoma within the M/C construct will be created at the time of grafting onto the LDM such that the long axis of the opening is parallel to the muscle fibers to simulate the orbicularis oris muscle. The stoma or opening will be maintained with an inert obturator fabricated from a 3D printer.
5. The designer prelaminated flap at 2–3 weeks postimplantation will be harvested.
6. A microvascular free transfer to the recipient site containing the thoracodorsal artery, vein, and motor nerve to the external carotid or facial artery, external jugular or facial vein, and a branch of the facial nerve will be performed, respectively.

We can successfully restore muscle volume and function and support neovascularization and reinnervation of regenerated muscle tissue as well as structural and functional integration between the regenerated and host tissues from the transposed prefabricated innervated prevascularized prelaminated (PIPP) flap.

6.2.3 Challenges and Potential Problems

One potential problem is contraction of the devices during the manufacturing phase due to the large size of tissue-engineered mucocutaneous construct. To circumvent this untoward scenario, devices 15–20% larger is suggested to counteract the contraction based on our experience. Another challenge is that each subject will present "diverse" and "unique" types of lip defects. To address the lip diversity issue, we will group them into specific and suitable categories based on defect size, location, depth, and involvement of lip. The specific categories are:

1. Total lip reconstruction involving both commissures (Fig. 6.4a)
2. Upper lip: 50% or more of lip to be reconstructed (Fig. 6.4b)
3. Lower lip: 50% or more of lip to be reconstructed (Fig. 6.4b)
4. Involvement of 50% or more of the lower and upper lip involving the commissure (Fig. 6.4c)

Regardless of the above classification, a full lip M/C construct should be constructed, with a functional stoma, that will be fabricated based on the template in Fig. 6.3 within the prelaminated innervated prevascularized prefabricated (IPP) flap as noted in Fig. 6.2. On harvesting of the prefabricated innervated prevascularized prelaminated (PIPP) flap, it will be modified such that it is consistent with the magnitude of the defect to be reconstructed. Thus, for full lower or upper lip

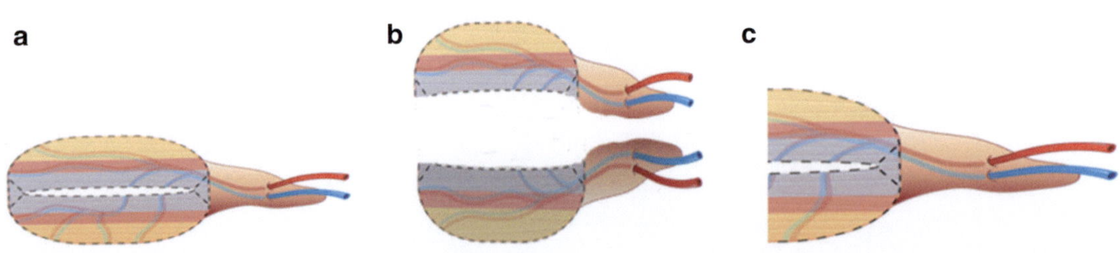

Fig. 6.4 Types of mucocutaneous constructs to be used for lip reconstruction. (**a**) M/C construct that will be used for all defects and for total upper and lower lip reconstruction, (**b**) M/C that could be used for total upper or lower lip reconstruction, and (**c**) M/C construct to be used for partial upper and lower lip reconstruction

reconstruction, only one half of the template on the LDM will be used for restoration of the lip (Fig. 6.4b), or if the area to be reconstructed involves only half of the upper and lower lip, only half of the template on the LDM, to include one commissure, will be harvested for lip reconstruction (Fig. 6.4c).

In the following sections, we will detail the protocol of cell culture and manufacturing EVPOMEs and mucocutaneous (M/C) constructs.

6.3 Materials and Reagents

6.3.1 Materials

1. Surgical scalpels: blade no. 15 (Fisher Scientific, catalogue no. 08-927-5C), blade no. 21 (Fisher Scientific, catalogue no. 08-927-5D). Scalpels no. 15 are used on small size of oral mucosa with the epidermal surface dimension of 5 × 5 mm or smaller.
2. Forceps: dissecting tissue forceps (Fisher Scientific, catalogue no. 13-812-41), tissue forceps with teeth (Roboz Surgical Instrument Co., catalogue no. RS8162), used for oral tissue manipulation; #5 Dumont Forceps (Roboz Surgical Instrument Co., catalogue no. RS-4955), used for dermal equivalent scaffolds manipulation.
3. Operating scissors: used for cutting dermal equivalent scaffolds, lengths, and types of scissors depending on individual preference.
4. Sterilized round petri dish: 100 × 15 mm petri dish (Fisher Scientific, catalogue no. 08-757-100D) used for oral tissue washing and dermal equivalent scaffolds rehydration; 35 × 10 mm (Fisher Scientific, catalogue no. 08-772A) used for trypsinization treatment on oral tissue.
5. Sterilized round cell culture dish: 100 × 20 mm cell culture dish (Fisher Scientific, catalogue no. 08-772-22). 150 × 25 mm cell culture dish (Fisher Scientific, catalogue no. 08-772-24).
6. Cell strainers: 100 μm pores (Fisher Scientific, catalogue no. 08-771-19).
7. Cell culture flasks: T25 (Fisher Scientific, catalogue no. 10-126-28) used for initial cell seeding when total cell number dissociated from oral tissue is equal or lower than 5M, T75 (Fisher Scientific, catalogue no. 07-202-000) used for initial cell seeding when total cell number dissociated from oral tissue is higher than 5M and for cell subculture.
8. 48-Well flat bottom cell culture plates (Fisher Scientific, catalogue no. 07-200-86): used for single cell type tissue engineering device culture.
9. Transwell multiple 6-well plate with 3 μm pore size inserts (Fisher Scientific, catalogue no. 07-200-166) and Transwell insert and dish with 75 mm diameter insert of 3 μm pore size inserts (Fisher Scientific, catalogue no. 07-200-173): used for creating air-liquid phase cell culture.
10. Sterilized rectangular 128 × 86 mm 8-well dishes (Fisher Scientific, catalogue no. 12-565-497) used for two cell types co-culture.
11. Transwell inserts and dishes: 75 mm in diameter with 3 μm pore size inserts used for two cell type tissue engineering device co-culture (Fisher Scientific, catalogue no. 07-200-173).
12. Dermal equivalent scaffolds: our laboratory uses two types of dermal equivalent scaffolds, 14-20/1000 in. thickness AlloDerm® (LifeCell Corporation, Branchburg, NJ, USA, catalogue no. 101020), and thin (0.4–0.7 mm thickness) and dehydrated Allopatch (Musculoskeletal Transplant Foundation, Edison, NJ, USA, catalogue no. 370408). Our laboratory does not use fibroblast feeder layer to culture oral keratinocytes because FDA regulations require the serum-free cell culture medium and to avoid cross contamination from feeder cellular layer.
13. 200 μm nylon mesh filter (BioDesign Inc of New York, NY, catalogue no. N200S).
14. 250 ml filter system (Fisher Scientific, catalogue no. 09-761-100).

6.3.2 Reagents

1. 1X DPBS (Dulbecco's phosphate-buffered saline) without calcium chloride and magnesium chloride (ThermoFisher Scientific, catalogue no. 14190144). 10X PBS, pH 7.4 (ThermoFisher Scientific, catalogue no. 70011069).
2. Trypsin solution for cell dissociation from oral tissue: Final concentration of 0.04% trypsin (Sigma-Aldrich, catalogue no. T-0303-10G) with 6 mM glucose (Sigma-Aldrich, catalogue no. G7021-1KG) in 1X DPBS, filter to sterilize by vacuum filter system (Fisher Scientific, catalogue no. 09-761-1), aliquot into 5 or 10 ml/tube, and stored at −20 °C.
3. Trypsin solution for cell dissociation from skin tissue: Final concentration of 0.125% trypsin in 1X DPBS and sterilize with 250 ml filter system.
4. Cell culture medium: EpiLife™ and calcium-free (ThermoFisher Scientific, catalogue no. MEPICF-500) that comes with 0.5 ml of 0.06M calcium chloride (ThermoFisher Scientific, catalogue no. 509703), stored at 4 °C.
5. Cell culture medium for skin keratinocytes: EpiLife™ with 10% FBS, and EpiLife™ with 2% FBS.
6. EpiLife™ defined growth supplement (EDGS): ThermoFisher Scientific, catalogue no. S-012-5, stored at −20 °C. EDGS will be thawed right before the addition of 5 ml of EDGS into 500 ml EpiLife™ base medium to prepare full cell culture medium.
7. Trypsin/EDTA solution for cell subculture: ThermoFisher Scientific, catalogue no. R001100. Trypsin/EDTA solutions are aliquot into 10 ml/tube and stored at −20 °C for long period of time. The solution is thawed right before the application.
8. Defined Trypsin Inhibitor (DTI, ThermoFisher Scientific, catalogue no. R007100). DTI can be stored at −20 °C for long period of time. DTI solution can be thawed and stored at 4 °C for short period of time.
9. Gentamicin/Fungizone solution for washing and sterilizing oral and skin tissue: Gentamicin (50 mg/ml, ThermoFisher Scientific, catalogue no. 15750060) and amphotericin B (250 μg/mL, ThermoFisher Scientific, catalogue no. 15290018) – Adding 12.5 ml of gentamicin into 20 ml of amphotericin B Fungizone, mix well, aliquot into 3.25 ml/vial, and store at −20 °C. Thaw and add 3.25 ml of gentamicin/Fungizone solution into 500 ml 1X DPBS to make final concentrations of 125 μg/ml gentamicin and 1 μg/ml amphotericin B.
10. Gentamicin/Fungizone solution for cell culture: Gentamicin (50 mg/ml, ThermoFisher Scientific, catalogue no. 15750060) and amphotericin B (250 μg/ml, ThermoFisher Scientific, catalogue no. 15290018) – Adding 6.7 ml of gentamicin into 20 ml of amphotericin B Fungizone, mix well, aliquot into 1 ml/vial, and store at −20 °C. Thaw and add 1 ml of gentamicin/Fungizone solution into 500 ml EpiLife™ Full Medium to make final concentrations of 25 μg/ml gentamicin and 0.375 μg/ml amphotericin B.
11. 1M calcium chloride solution (Sigma-Aldrich, catalogue no. 21115): Sterilization by filter with 0.22 μm pore size and aliquot into 1 ml/tube.
12. Human type IV collagen (Sigma-Aldrich, catalogue no. C8374): Use filter-sterilized 0.5M acetic acid to prepare 1 mg/ml human type IV collage, aliquot into 500 μl/tube, and store at −20 °C.
13. SYLGARD® 184 Silicone Elastomer Base and Curing Agent (Fisher Scientific, catalogue no. NC9644388).
14. Formaldehyde solution, 37% by weight (Fisher Scientific, catalogue no. BP531-500): Add 100 ml 10X PBS and 100 ml 37% formaldehyde into 500 ml ddH_2O, then add ddH_2O to 1 l to make 3.7% formalin solution. 3.7% formalin solution can be stored in a bottle wrapped with aluminum foil at 4 °C.
15. Antibodies: Anti-cytokeratin 10/13 antibody (Abcam, catalogue no. ab9383), anti-CD44 antibody (Cell Signaling Technology, catalogue no. 3570), anti-K2 antibody (PROGEN

Biotechnik, Heidelberg, Deutschland, catalogue no. 65191), anti-small proline-rich protein 3 antibody (Sigma-Aldrich, catalogue no. HPA044467).

6.4 Methods

6.4.1 Keratinocytes Culture

6.4.1.1 Oral Keratinocytes Culture

All materials and reagents should be sterile.

1. All procurement of human tissues needs to be approved by experimenters' institutes.
2. Freshly dissected oral mucosa tissues can be transported in washing solution containing 125 μg/ml gentamicin/1 μg/ml amphotericin B (Fungizone) and stored at 4 °C for short period of time before processing tissues. Oral tissues can also be stored in full medium before tissue process procedures begun to maintain the viability of cells inside tissue. However, it can also increase the chance of contamination.
3. All procedures should be performed inside the hood of a biological safety cabinet from this point on.
4. Using forceps to transfer oral tissues into a 100 × 15 mm petri dish containing 15 ml washing solution. Blood on tissues and unwanted parts of tissues can be removed by a scalpel during this step.
5. Transfer oral tissues into a new 100 × 15 mm petri dish containing 15 ml washing solution. The tissues are held by forceps, scraped gently by a scalpel, submerged in washing solution, and incubated for 20–30 min. The purpose of this step is to sterilize the tissues.
6. Transfer tissues into a new petri dish and repeated step 5.
7. Repeat step 6.
8. Transfer tissues into a 35 × 10 mm dish containing 5 ml of 0.04% Trypsin with 6 mM glucose, and incubate for 16–17 h at room temperature inside the hood of the biological safety cabinet. Depending on the size of tissue, the size of dish and the volume of trypsin/glucose solution should be adjusted.
9. Transfer tissues into a 100 × 15 mm petri dish containing 5 ml (equal volume) of DTI solution.
10. A loose epidermal part of tissue should be observed at this point. Using forceps to hold down the tissue and a scalpel to scrape epidermal part of tissue, where the keratinocyte basal cells reside, away from the rest of tissue, which is discarded.
11. Pipette the DTI solution containing cells dissociated from oral tissue, and filter them through a 100 μm cell strainer placed on top of a 50 ml tube. Apply 2–3 ml of fresh DTI solution or full cell culture medium in the dish to wash the remaining cells and filter them through the same cell strainer.
12. Count viable oral keratinocytes' number.
13. Seed oral keratinocytes onto up to 5M cells/T25 cell culture flask. If cell number is higher than 5M, cells can be divided and seeded into multiple T25s or up to 15M cells/T75 cell culture flask.
14. Incubated cells inside a 5% CO_2 incubator at 37 °C for 48 h without disturbing cells.
15. Change medium every 2 days and subculture oral keratinocytes when cells reach appropriate confluence.

6.4.1.2 Skin Keratinocyte Culture

1. For preparation of isolation of primary adult human skin keratinocytes, in laminar flow hood, clean blood and fat off of tissues with washing solution. If the specimen is not going to be used immediately, it can be put on ice or alternatively left at 4 °C and submersed in DPBS.
2. Hold the sample with the forceps with teeth, and scape the sample with scalpel to remove excess fat and blood residue with washing solution in the petri dish.
3. Remove spent washing solution and rinse sample with fresh washing solution then transfer to new petri dish.
4. Repeat three times this process for each 30 min. If rinsing solution comes out too red,

repeat washing in order to eliminate blood cells.

5. After the last washing, slice full-thickness epithelium into 5-mm-wide strips fringe with scalpel. The thinner the strips, the better the trypsinization will be.
6. Add 0.125% trypsin solution in petri dish to be exposed in maximum area to trypsin.
7. Leave in laminar hood for 16 h for trypsinization. At this time, epidermal side is up and dermis side is down.
8. After 16 h, aspirate off most of the trypsin solution and add the same volume of EpiLife™ with 10% FBS. The serum can help to stop the action of trypsin.
9. Gently scrape off the epidermal layer down to the basal layer by a #21 scalpel with holding one end of the tissue with sterile forceps. Discard dermis afterward.
10. Filter the cell suspension through 200 μm nylon mesh (see Note 1) into 50 ml tubes using a sterile serological 10 ml pipette.
11. Rinse petri dish with fresh EpiLife™ with 10% FBS and filter into cell suspension 50 ml centrifuge tubes.
12. Count cells and centrifuge cells for 5 min.
13. Aspirate supernatant and resuspend cell pellet in warmed EpiLife™ with 2% FBS.
14. Plate 2–3 × 10^6 cells/ml.
15. Incubate at 37 °C, 5% CO_2 for 24 h for cells to attach.
16. Change medium to regular EpiLife™ medium without serum.
17. Feed regular EpiLife™ every other day and subculture cells when 70–80% confluence.

6.4.2 3D Organotypic Culture Using EVPOMEs (Ex Vivo Produced Oral Mucosa Equivalents) to Manufacture Soft Tissue Engineering Devices

There is a need to restore damaged craniomaxillofacial soft tissue either intraorally and/or extraorally. The tissue-engineered device manufactured by autogenous cells became a solution to meet the requirements with minimal morbidity to the patient. The tissue-engineered device has to integrate into host tissue easily and in as short a period of time as possible without causing any immunogenic reaction within the recipient. The tissue-engineered device has to facilitate neovascularization such that the major vascular supply to the construct is based on the major vessels of the implant site, i.e., muscle. EVPOMEs were proven to have such properties [5, 7, 32]. The manufacture of EVPOMEs is basically composed of two phases, liquid phase and an air-liquid phase. The liquid phase establishes the seeding and proliferation of basal cellular layer on top of scaffold. The air-liquid phase promotes the stratification of keratinocytes and differentiation, i.e., keratinization.

6.4.2.1 EVPOME Using Single Cell Type, Oral Keratinocytes, Culture

1. Decellularized cadaver skin used as a scaffold is cut into 1 cm in diameter by scissors or by an 8 mm in diameter punch. The size of cut scaffolds is determined according to the experimental designs. In our laboratory, we use AlloDerm® or Allopatch as scaffolds. The ideal thickness of scaffolds should be around 0.5 mm. A thick scaffold will impede cell growth and development on scaffolds; however, it will be difficult to physically manipulate a too thin scaffold.
2. Rehydrate scaffolds in 100 mm dish containing 15 ml 1X DPBS for 30 min. Use forceps to gently tap the scaffolds to make sure air is pressed out of the scaffolds.
3. Aspirate old DPBS and replace with fresh DPBS in the same dish. Tap the scaffolds, and let the scaffolds submerge in DPBS for 30 min.
4. Repeat step 3.
5. Transfer scaffolds into wells of 48-well plate with epidermal side up. Epidermal side is the cell seeding side. Epidermal side is the side DPBS won't stay on the surface which appear rougher than the dermal side. Dermal side is the side where DPBS will stay on the surface and look shinier than the epidermal side. There is a notch on one corner of AlloPatch scaffolds to assist the identification of epidermal side.

6. Add 100 μl 1X DPBS and 5 μl of 1 mg/ml human type IV collagen onto each scaffold. Rock 48-well plate back and forth a few times, and incubate the plate at 4 °C overnight. The volumes of 1X DPBS and 1 mg/ml collagen should be proportionally adjusted according to the culture surface area of the well.
7. On the next day, place the 48-well plate inside the hood of the biological safety cabinet to allow the temperature warm up to the room temperature.
8. Add 1 ml full medium into each well with scaffold to test pH value. The color of the medium should be pink. If color of the medium is yellowish orange or yellow, then aspirate out the solution, and replenish with fresh 1 ml full medium.
9. Add trypsin/EDTA solution to just cover cell surface to dissociate cells from culture flasks, and incubate at 37 °C inside 5% CO_2 incubator for approximately 5 min.
10. Add equal volume of DTI solution into culture flask to stop trypsin activity. Rock culture flask gently a few times to allow DTI and trypsin/EDTA to mix homogenously.
11. Collect cell suspension into a 50 ml tube.
12. Count viable cells.
13. Centrifuge cells at 100 g for 5 min.
14. From this step on, full medium with 1.2 mM calcium (0.6 ml 1M calcium chloride in 500 ml full medium) should be used to culture cells.
15. Aspirate trypsin/EDTA/DTI solution completely, and resuspend cell pellet in full medium to make cell density 100 k cells/100 μl.
16. Seed 200 k cells onto each scaffold, add additional full medium drop by drop by up to the top of each well, and incubate cells at 37 °C inside a 5% CO_2 incubator.
17. Change medium every day for 4 days.
18. Transfer each scaffold onto an insert in 6-well plate to create an air-liquid phase to promote cellular stratification. Add 1.5 ml full medium in each well where the insert sits or the appropriate volume of full medium to allow medium to meet the bottom of the insert if different setup is used.
19. Change medium every 2 days for 7 days.
20. Evaluate EVPOMEs by histology and immunohistochemistry at the end of 11 culture days (Fig. 6.5) (See note 2).

6.4.2.2 EVPOME Using Floating Keratinocytes Culture

Huang et al. [33] showed that trypsin can damage cellular membranes and cellular proteome structures and upregulate p53 and p21 Waf1/Cip1 involving cellular senescence process. It becomes obvious that an important advantage of using floating cells is to eliminate the insult of trypsinization, which is a required step to dissociate attached monolayer cells.

1. The processes of scaffolds are the same as in steps 1–7 under Sect. 6.4.2.1.
2. Use 1 ml full medium to test the pH value of collagen-coated scaffold, and then aspirate the solution right before cell seeding.
3. Collect floating cells in spend medium and count viable cells. Centrifuge cells to concentrate viable floating cells if cell density is lower than 150 k cells/ml (See note 3).
4. Add full medium containing 0.06 mM into the spent medium to make the floating cell density equal to 150 k cells/ml.
5. Seed floating drop by drop onto the scaffolds.
6. Add full medium containing 0.06 mM calcium drop by drop into the well until medium reaches the top of the well.
7. Incubate inside a CO_2 incubator at 37 °C for 24 h.
8. Change medium with fresh full medium containing 1.2 mM calcium. Follow steps 17–20 under Sect. 6.4.2.1.
9. Evaluate EVPOMEs by histology and immunohistochemistry at the end of 12 culture days (Fig. 6.6).

6.4.2.3 EVPOME Using Two Cell Types Co-cultures

The mucocutaneous junction, an area where mucosa and skin tissue meet, is a complex structure. The restoration of such structure by a tissue-engineered device requires a co-culture of oral and skin keratinocytes separated by cell-free

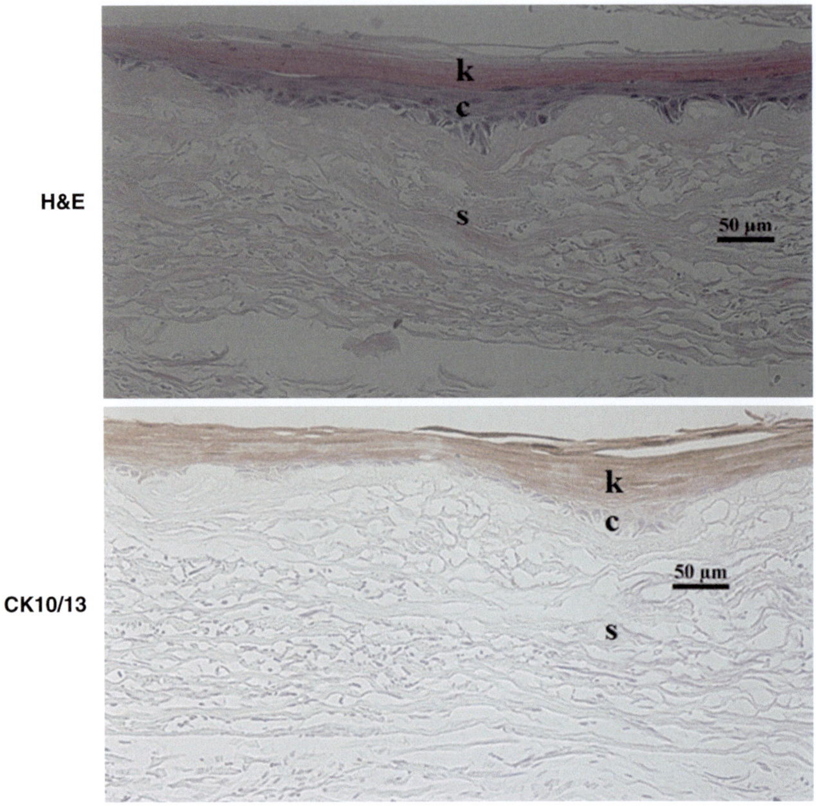

Fig. 6.5 Histology of oral keratinocyte EVPOME. H&E histology was shown on the top and immunohistochemistry of CK10/13 with dilution of 1:300 on the bottom. K represents keratin layer, c cellular layers, and S scaffold. Each scar bar represents 50 µm

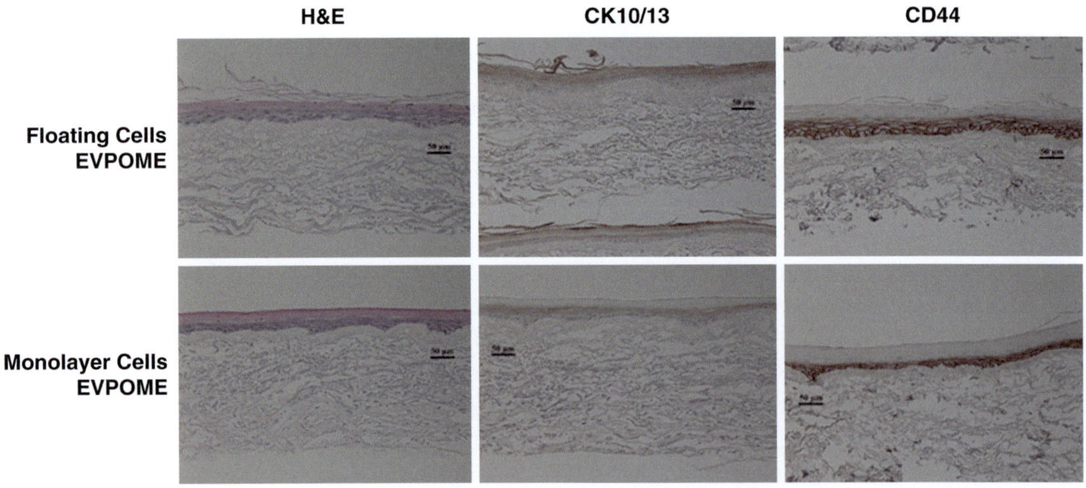

Fig. 6.6 Histology of floating and monolayer oral keratinocyte EVPOMEs. Floating and monolayer oral keratinocytes were collected from the same cell culture flask. H&E, CK10/13 with dilution of 1:300, and CD44 with dilution of 1:50. Each scar bar represents 50 µm

zone that allows both types of cells to gradually merge and migrate toward one another to create the mucocutaneous junction. In the case of lip restoration, the tissue-engineered mucocutaneous junction construct will be implanted into the latissimus dorsi muscle, with the stoma or opening in the device paralleling the muscle fibers. Once the device has been integrated into the muscle bed, it will be harvested along with the thoracodorsal neurovascular bundle to create a prefabricated prelaminated microvascular free flap mimicking lip muscular structure [18]. The following are protocols for manufacturing an in vitro tissue-engineered mucocutaneous junction construct.

Making of Barrier, Corral, and Waffle

1. Corral, waffle, and 2.8 mm wide barrier frames were made out of ABS*plus* plastic by a Dimension Elite 3D Printer (Stratasys, Eden Prairie, MN, USA). The designs of corral and barrier based on dimensions of the well of 128 × 86 mm 8-well dishes and waffle based on 150 × 25 mm dish were shown in Fig. 6.7. The designs of barrier, corral, and waffle should be based on the culture dishes used in experiments.
2. Place each barrier inside a well of rectangular 128 × 86 mm 8-well dishes, and make sure the barrier sits tightly inside the well.
3. Mix SYLGARD® 184 with a ratio of base to curing agent ranging from 10 to 11 by weight.

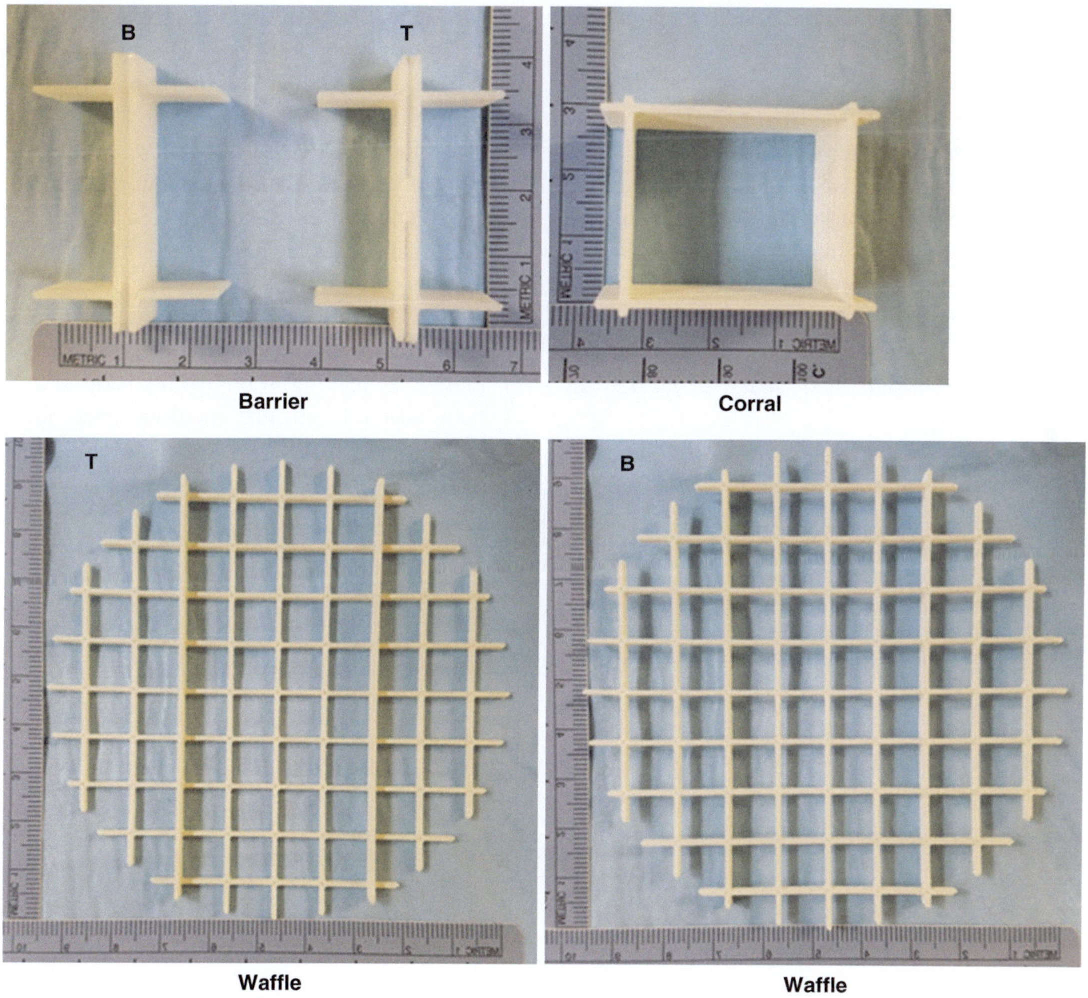

Fig. 6.7 Barrier, corral, and waffle used for oral and skin keratinocytes co-cultured EVPOMEs. Barrier is used to create a cell-free zone for oral and skin keratinocytes co-culture, corral is used during liquid phase culture, and waffle is used during air-liquid phase culture. T is the top side facing up. B is the bottom side facing up. Barrier bottom side is the side contacting the scaffold. Waffle top side is the side where the insert sits

4. Fill the central space of the frame with thoroughly mixed SYLGARD® 184 [PDMS, poly(dimethylsiloxane)] and cured for 24 h at room temperature followed by an additional 4 h at 65 °C to complete curing.
5. Remove PDMS-filled barrier from the well and trim extra PDMS on the surface of barrier.
6. Place the barrier in a new well to examine whether there is a leak by adding 1 ml of dyed solution in one side of barrier and wait for a few minutes. The dyed solution will flow into the other side of barrier if the barrier cannot make a tight seal.
7. Sterilize corrals and barriers with ethanol and/or UV.

Manufacturing EVPOME Using Oral and Skin Keratinocytes Co-cultures

1. Cut scaffolds into 3 × 3 cm squares. The size of scaffolds depends on the choice of culture dishes and experimental designs. Cut two notches on two corners.
2. Rehydrate the scaffolds in 15 ml 1X DPBS inside a 100 mm dish. Tap and press air out of the scaffolds and soak in DPBS for 20–30 min.
3. Aspirate DPBS and replenish with fresh DPBS and repeat step 2 two more times.
4. Let the scaffolds soak in fresh DPBS overnight at room temperature inside the hood of the biological safety cabinet.
5. On the next day, prepare 2 ml 0.05 mg/ml human type IV collagen by adding 100 μl 1 mg/ml collagen into 1900 μl 1X DPBS.
6. Transfer DPBS-soaked scaffolds from dish into the well of rectangular 128 × 86 mm 8-well dish with epidermal side up and two notched corners on the same side, either right or left.
7. Add 2 ml 0.05 mg/ml human type IV collagen into the well and incubate at 4 °C overnight.
8. Aspirate collagen solution completely from the well.
9. Insert the sterilized barrier into well with forceps, and place it firmly on top of scaffold along the central line of scaffold to make a tight seal and create a cell-free zone between the two cellular areas (Fig. 6.8).
10. Add 0.5 ml full medium containing 0.06 mM calcium into one side of barrier to make sure there is no leaking. If no leaking is observed, then add 0.5 ml full medium containing

Fig. 6.8 Barrier setup for oral and skin keratinocytes co-culture cell seeding. Barrier is firmly pressed on top of the scaffold whose two notched corners were placed on the same either right or left side of the barrier. Oral (O) and skin (S) keratinocytes will be seeded on either side of the barrier, respectively

0.06 mM calcium into the other side of barrier.

11. Prepare oral and skin keratinocytes for cell seeding. Seed 500 k cells/cm^2 in 0.25 ml medium containing 0.06 mM calcium of oral and skin keratinocytes onto each side of the scaffold, respectively (See note 4). Make sure cells are evenly disturbed over the scaffold surface, and allow cells to sit for 30 min without disturbance.
12. Add 2.75 ml medium containing 0.06 mM calcium into each side of the barrier. The total volume of the medium of each side will be 3.5 ml.
13. Incubate cells inside a 5% CO_2 incubator at 37 °C for 24 h (See note 5).
14. Aspirate medium completely first, remove the barrier, transfer scaffold to a 100 × 20 mm dish, place a corral on top of the scaffold to pin down the scaffold and prevent its movement, and add 30 ml medium containing 1.2 mM calcium to continue liquid phase culture (Fig. 6.9).
15. Change medium containing 1.2 mM calcium every 2 days for a total of 4 days.
16. Transfer scaffold onto a 75 mm diameter insert sitting on top of a waffle inside a 150 × 25 mm cell culture dish, and add 80 ml medium containing 1.2 mM calcium to create an air-liquid phase (Fig. 6.10). The volume of the medium applied should depend on the height of the waffle and the size of culture dish.
17. Change medium every 2 days for 10 days (See note 6).
18. Evaluate by histology and immunohistochemistry at the end of 15 culture days (Fig. 6.11).

There is a continued need for reconstruction of complex soft tissue defects of the head and neck tissue. There have been advances in reconstruction through regenerative medicine, tissue engineering, and advanced surgical reconstruction. However, each of these endeavors has its positive and negative aspects. Throughout our past and present studies, the issue of immunosuppression and graft rejection has been addressed by harvesting the patient's own cells. We have also addressed the junctions involved in mucocutaneous transition, by creating a tissue-engineered mucocutaneous junction. We have addressed vascularity of engineered tissue by the prefabrication and prelamination and transfer via free tissue

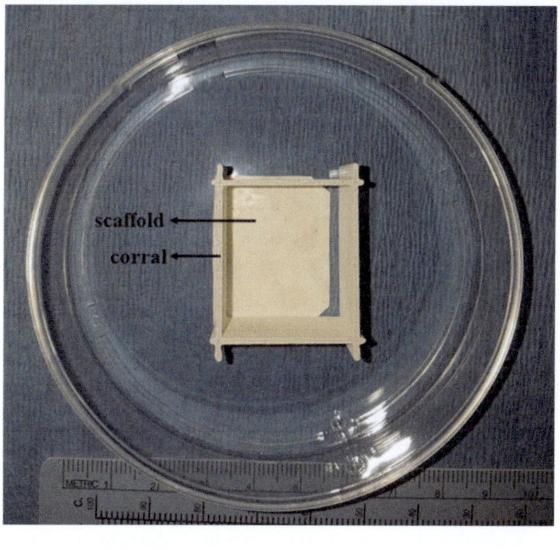

Fig. 6.9 Liquid phase setup for oral and skin keratinocytes co-cultured EVPOMEs. The corral is placed on top of scaffold to prevent floating around of scaffold and ensure the stability of cell growth

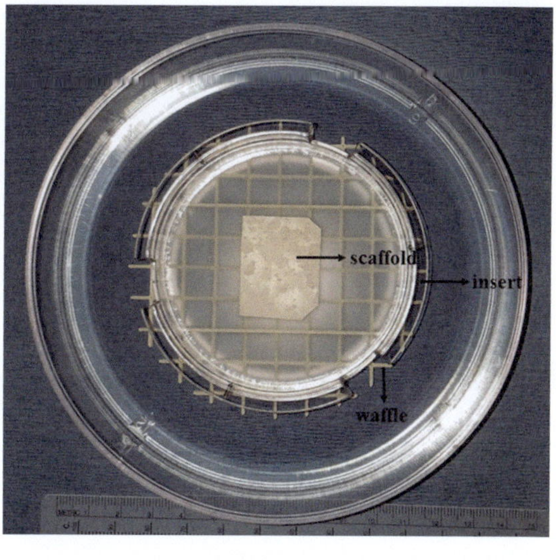

Fig. 6.10 Air-liquid phase setup for oral and skin keratinocytes co-cultured EVPOMEs. Insert sits on top of the waffle, which functions as a raise in order to create an air-liquid phase culture. Scaffold is placed on the insert

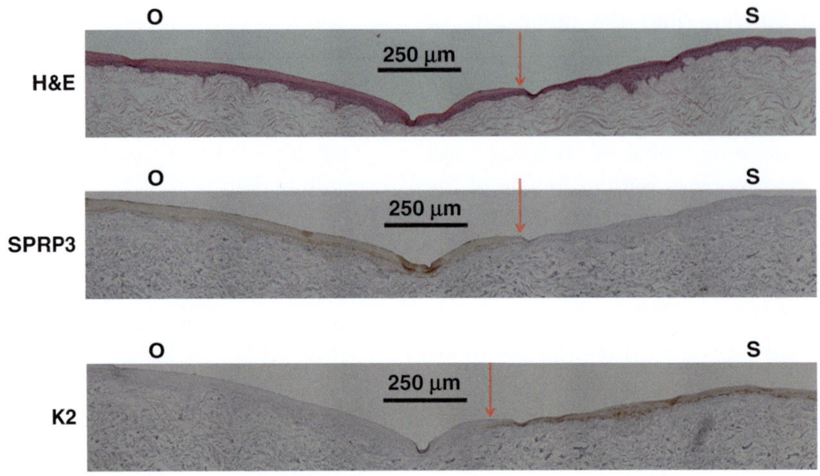

Fig. 6.11 Histology of oral and skin keratinocyte co-culture EVPOME. Cells were culture in medium with 0.06 mM Ca^{+2} for 1 day, at liquid phase in medium with 1.2 mM Ca^{+2} for 4 days, and at air-liquid phase in medium with 1.2 mM Ca^{+2} for 10 days. O is oral keratinocytes culture side and S skin keratinocytes culture side. Red arrows represent where oral and skin keratinocytes met and merged. Anti-K2 antibody at a dilution of 1:250 and anti-small proline-rich protein 3 antibody at a dilution of 1:2000 were used to distinguish skin and oral keratinocytes, respectively

transfer. Further studies are needed to address function and sensation as well as multicenter clinical trials to evaluate the viability and effectiveness of development of this type of specialized therapy. In addition, long-term studies are needed to assess color, function, and durability of patients with prelaminated/prefabricated rotational and free flaps.

6.5 Notes

1. Prepare a funnel-shaped nylon net filter by folding the round-shaped filter twice and then stapling at the edge of the overlapping portion. Soak several filters in 95% ethanol overnight for sterilization.
2. Evaluation of EVPOMEs by histology and immunohistochemistry:

   At day 11 post-seeding, EVPOME samples were collected and fixed in 5 ml of 3.7% phosphate-buffered formalin in wells of 6-well plate at 4 °C overnight. The fixing time can be evaluated empirically to get the best signals of specific biomarkers examined. EVPOMEs are washed in 1X PBS solution, cut into 1 mm wide strips, and then stored in 70% ethanol for paraffin embedding.
3. Centrifuge floating cells at 10 g for 2.54 cm radius, 4 °C for 5 min. Aspirate spent medium and resuspend cell pellets in appropriate volume of full medium containing 1.2 mM calcium.
4. The larger the cell seeding surface, the higher the cell seeding number/cm^2 is required to obtain a well-developed EVPOME. However, the relationship between cell seeding number/cm^2 and cell growth area is not lineally proportional. The optimal cell seeding number/cm^2 needs to be determined empirically based on the surface area of dish where the scaffold sits not the surface area of scaffold itself.
5. The number of cell culture days in medium containing 0.06 mM calcium can be changed based on the experimental designs.
6. The number of air-liquid phase culture days should be determined empirically based on the width of cell-free zone.

Acknowledgment We would like to thank our colleagues, Dr. Cynthia Marcelo, Dr. Hiroko Kato, Eve Bingham, and James Washington, for their contributions to our projects presented in this manuscript. Dr. Feinberg's laboratory was supported by NIH grants R01 DE 019431 and R01 DE 013417 and the Department of Defense grants W81XWH-08-2-0034, W81XWH-12-2-0057, and W81XWH-13-2-0052.

References

1. Obrien FJ. Biomaterials & scaffolds for tissue engineering. Mater Today. 2011;14:88–95.
2. Steel MP, Roessler BJ. Compliance with good manufacturing practices for facilities engaged in vector production, cell isolation, and genetic manipulations. Curr Opin Biotechnol. 1999;10:295–7.
3. Izumi K, Song J, Feinberg SE. Development of a tissue-engineered human oral mucosa: from the bench to the bed side. Cells Tissues Organs. 2004;176:134–52.
4. Izumi K, Takacs G, Terashi H, Feinberg SE. Ex vivo development of a composite human oral mucosal equivalent. J Oral Maxillofac Surg. 1999;57:571–7.
5. Kuo S, Zhou Y, Kim HM, Kato H, Kim RY, Bayar GR, Marcelo CL, Kennedy RT, Feinberg SE. Biochemical indicators of implantation success of tissue-engineered oral mucosa. J Dent Res. 2015;94:78–84.
6. Kuo S, Kim HM, Wang Z, Bingham EL, Miyazawa A, Marcelo CL, Feinberg SE. Comparison of two decellularized dermal equivalents. J Tissue Eng Regen Med. 2017;12:983. https://doi.org/10.1002/term.2530.
7. Izumi K, Neiva RF, Feinberg SE. Intraoral grafting of tissue-engineered human oral mucosa. Int J Oral Maxillofac Implants. 2013;28:e295–303.
8. Squier CA, Kremer MJ. Biology of oral mucosa and esophagus. J Natl Cancer Inst Monogr. 2001;2001(29):7–15.
9. Calenic B, Ishkitiev N, Yaegaki K, Imai T, Kumazawa Y, Nasu M, Hirata T. Magnetic separation and characterization of keratinocyte stem cells from human gingiva. J Periodontal Res. 2010;45:703–8.
10. Calenic B, Greabu M, Caruntu C, Tanase C, Battino M. Oral keratinocyte stem/progenitor cells: specific markers, molecular signaling pathways and potential uses. Periodontol. 2015;69:68–82.
11. Izumi K, Tobita T, Feinberg SE. Isolation of human oral keratinocyte progenitor/stem cells. J Dent Res. 2007;86:341–6.
12. Trempus CS, Morris RJ, Bortner CJ, Cotsarelis G, Faircloth RS, Reece JM, Tennant RW. Enrichment for living murine keratinocytes from the hair follicle bulge with the cell surface marker CD34. J Invest Dermatol. 2003;120:501–11.
13. Jaks V, Barker N, Kasper M, van Es JH, Snippert HJ, Clevers H, Toftgård R. Lgr5 marks cycling, yet long-lived, hair follicle stem cells. Nat Genet. 2008;40:1291–9.
14. Webb A, Li A, Kaur P. Location and phenotype of human adult keratinocyte stem cells of the skin. Differentiation. 2004;72:387–95.
15. Metral E, Bechetoille N, Demarne F, Rachidi W, Damour O. α6 Integrin (α6 high)/transferrin receptor (CD71) low keratinocyte stem cells are more potent for generating reconstructed skin epidermis than rapid adherent cells. Int J Mol Sci. 2017;18:282. https://doi.org/10.3390/ijms18020282.
16. Peramo A, Marcelo CL, Feinberg SE. Tissue engineering of lips and muco-cutaneous junctions: in vitro development of tissue engineered constructs of oral mucosa and skin for lip reconstruction. Tissue Eng Part C Methods. 2012;18:273–82.
17. Bayar GR, Kuo S, Marcelo CL, Feinberg SE. In vitro development of a mucocutaneous junction for lip reconstruction. J Oral Maxillofac Surg. 2016;74:2317–26.
18. Kim RY, Bae SS, Feinberg SE. Soft tissue engineering. Oral Maxillofac Surg Clin North Am. 2017;29:89–104.
19. Chaffer CL, Brueckmann I, Scheel C, Kaestli AJ, Wiggins PA, Rodrigues LO, Brooks M, Reinhardt F, Su Y, Polyak K, Arendt LM, Kuperwasser C, Bierie B, Weinberg RA. Normal and neoplastic nonstem cells can spontaneously convert to a stem-like state. PNAS. 2011;108:7950–5.
20. Marcelo CL, Peramo A, Ambati A, Feinberg SE. Characterization of a unique technique for culturing primary adult human epithelial progenitor/“stem cells”. BMC Dermatol. 2012;12:8. https://doi.org/10.1186/1471-5945-12-8.
21. Peramo A, Feinberg SF, Marcelo CL. Characterization of cultured epithelial cells using a novel technique not requiring enzymatic digestion for subculturing. Cell Tissue Bank. 2013;14:423–35.
22. Spector M. Tissue engineering. In: Basic principles of tissue engineering, chap. 1. Chicago: Quintessence; 1999.
23. Evans DM. Facial reconstruction after a burn injury using two circumferential radial forearm flaps, and a dorsalis pedis flap for the nose. Br J Plast Surg. 1995;48:471–6.
24. Pribaz JJ, Fine N, Orgill DP. Flap prefabrication in the head and neck: a 10-year experience. Plast Reconstr Surg. 1999;103:808–20.
25. Pribaz JJ, Weiss DD, Mulliken JB, Eriksson E. Prelaminated free flap reconstruction of complex central facial defects. Plast Reconstr Surg. 1999;104:357–365; discussion 366-367.
26. Teot L, Cherenfant E, Otman S, Giovannini UM. Prefabricated vascularized supraclavicular flaps for face resurfacing after postburns scarring. Lancet. 2000;355:1695–6.
27. Guo L, Pribaz JJ. Clinical flap prefabrication. Plast Reconstr Surg. 2009;124:e340.
28. Vranckx JJ, Delaere P, Vanderpoorten V. Prefabrication and prelamination procedures for larynx and tracheal reconstruction. Paper presented at Fourth Congress of the World Society for Reconstructive Microsurgery, Buenos Aires, Argentina, 23–25 October 2005. 2005.
29. Vranckx JJ, Delaere P, Vanderpoorten V. Prefabrication and prelamination procedures for larynx and tracheal reconstruction. Paper presented at 17th Congress of the European Association of Plastic Surgeons, Istanbul, Turkey, 25–28 May 2006. 2006.
30. Faria JC, Scopel GP, Alonso N, Ferreira MC. Muscle transplants for facial reanimation: rationale and results of insertion technique using the palmaris longus tendon. Ann Plast Surg. 2009;63:148–52.
31. Lew TA, Walker JA, Wenke JC, Blackbourne LH, Hale RG. Characterization of craniomaxillofacial battle

injuries sustained by United States service members in the current conflicts of Iraq and Afghanistan. J Oral Maxillofac Surg. 2010;68:3–7.
32. Izumi K, Feinberg SE, Terashi H, Marcelo CL. Evaluation of transplanted tissue-engineered oral mucosa equivalents in severe combined immunodeficient mice. Tissue Eng. 2003;9:163–74.
33. Huang H-L, Hsing H-W, Lai T-C, Chen Y-W, Lee T-R, Chan H-T, Lyu P-C, Wu C-L, Lu Y-C, Lin S-T, Lin C-W, Lai C-H, Chang H-T, Chou H-C, Chan H-L. Trypsin-induced proteome alteration during cell subculture in mammalian cells. J Biomed Sci. 2010;17:36. https://doi.org/10.1186/1423-0127-17-36.

Part II

Surgical Techniques Utilizing Biotechnology for Regeneration and Reconstruction

In Situ Tissue Engineering

7

Robert E. Marx

7.1 In Situ Bone Tissue Engineering

In situ tissue engineering refers to the regeneration of tissue, in this case bone, within the volume space of its intended functional site and at the point of care. It has the advantage of eliminating the donor site with all of its inherent morbidities, reducing operating time, reducing overall costs, and returning patients back to their workplace and families much sooner.

The biologic basis of in situ tissue engineering is the same as that for any tissue regeneration and follows the sequence of events noted in embryology. That is, the development or regeneration of tissue requires *cells* capable of differentiating into the target tissue or that have already partially differentiated to do so, i.e., stem cells and/or progenitor cells; a *signal* to differentiate, proliferate, and begin the tissue synthesis process; and a *matrix* (*scaffold*) upon which the tissue regeneration can occur. This is referred to as the "tissue engineering triangle," the legs of which are cells, signal, and matrix. However, more recently this tissue engineering triangle has been recognized as requiring a "bioreactor" which for the laboratory scientist represents a tissue growth medium and a temperature- and pH-controlled nutrient environment. For the clinician the "bioreactor" is the natural wound space of the recipient site and why the term in situ tissue engineering is used (Fig. 7.1)

Stem Cells The term "stem cells" is often used to apply to a varied group of cells with indeterminate capabilities. For the purpose of this chapter the following practical definitions apply.

Omnipotent Stem Cells Primitive cells akin to a blastomere capable of regenerating a complex set of interdependent tissues and organs thus forming an entire complete organism (cloning)

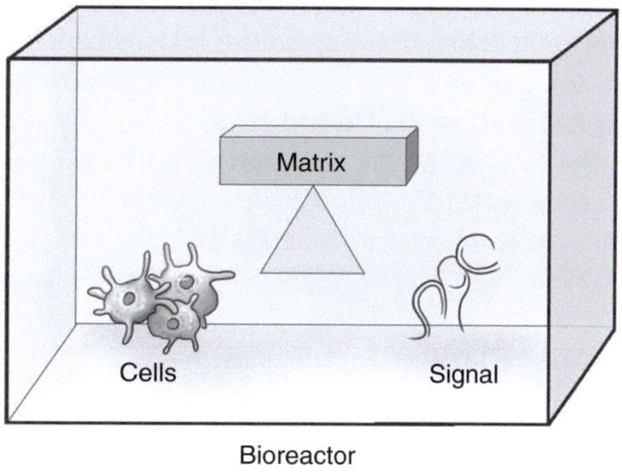

Fig. 7.1 The tissue engineering cube

R. E. Marx (✉)
Division of Oral and Maxillofacial Surgery, University of Miami, Miller School of Medicine, Miami, FL, USA
e-mail: rmarx@med.miami.edu; mruiz3@med.miami.edu

J. C. Melville et al. (eds.), *Regenerative Strategies for Maxillary and Mandibular Reconstruction*, https://doi.org/10.1007/978-3-319-93668-0_7

Pluripotent Stem Cells Cells capable of regenerating a complete organ, i.e., the liver, kidney, eye, tooth, etc.

Multipotent Stem Cells Cells capable of forming more than one of the several tissues, i.e., the bone, cartilage, blood vessels, etc.

Progenitor Cells Cells capable of forming only one tissue, i.e., bone, cartilage, blood vessels, etc.

7.1.1 How Can the Clinician Access Stem Cells?

Although allogeneic stem cells have recently become available as an off-the-shelf FDA-approved product, they have yet no proven clinical track record. Additionally, the theoretical concern of these cells which bear cell surface markers, i.e., CD105+, etc., may be antigenic despite corporate reassurances giving caution to their widespread use at this time. Therefore, autogenous stem cells and progenitor cells which today already have a track record of success without antigenicity are preferred and are emphasized in this chapter.

The practical source of stem cells for in situ tissue engineering is from platelet-rich plasma (PRP) or bone marrow aspirates (BMA) and bone marrow aspirate concentrates (BMAC). The authors' testing studies have shown PRP to contain 250–300 × 10^3 CD34+ cells/ml producing only 600 colony-forming units (CFUs) which is a measure of their bone-forming capabilities. Tests of BMAC have shown BMAC to yield 2.3 × 10^6 stem cells/ml producing 1200–1800 CFUs and the newer BMA needles/trocars to yield 2.9 × 10^6 stem cells/ml producing 3200 CFUs.

7.1.2 Platelet-Rich Plasma

Platelet-rich plasma remains a centrifugation process of whole blood. Today, several approved devices are available that can produce the required 1 × 10^6 platelets/ul or five times baseline to qualify as regenerative PRP. The value of platelet-rich plasma is its high concentration of growth factors (PDGFaa, PDGFab, PDGFbb, TGfb1, TGfb2, VEGF, EDGF, and SDAF-1 alpha) as well as its cell adhesion molecules (fibrin, fibronectin, vitronectin) and the nutrient-rich plasma fraction environment [1].

7.1.3 Bone Marrow Aspirate Concentrates

Bone marrow aspirate concentrates (BMAC) are a point of care yield of mesenchymal stem cells (MSC) and osteoprogenitor cells (OPC) as well as the bone marrow plasma environment. The technique uses four needle/trocar punctures, two in each side of either the anterior or posterior ilium. A trocar that is coated with a heparin solution of 2000 units/ml is worked into the iliac crest to a depth of 3.0 cm (Fig. 7.2). A 15 ml syringe where the inner barrel has been also coated with the heparin solution is attached. The bone marrow is then aspirated 5 ml at a time while rotating the aspiration cannula one full circle (360°) (Fig. 7.3). This is repeated for each 5 ml draw until either 15 ml is withdrawn from each puncture site or 30 ml is drawn from each puncture site (Fig. 7.4). The four 15/ml site draws will yield 60 ml of BMA which after centrifugation yields 10 ml of BMAC useful for most oral and maxillofacial surgery applications. The four 30 ml/site draws will yield 120 ml of BMA which after centrifugation results in 20 ml of BMAC useful for larger defects and bone regeneration needs in compromised tissues (i.e., radiated tissue).

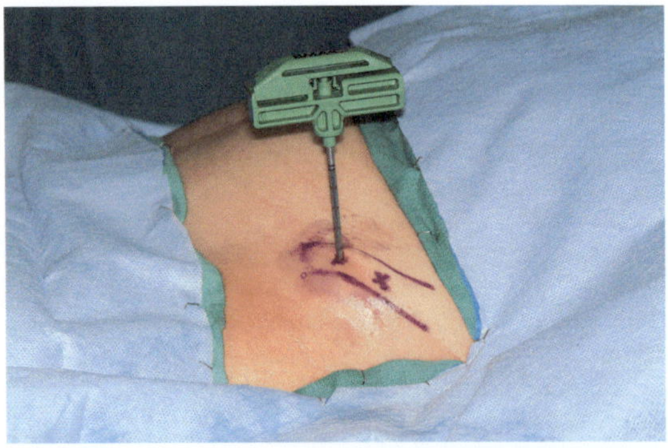

Fig. 7.2 Bone marrow trocar placed 3 cm into the anterior ilium between the cortices

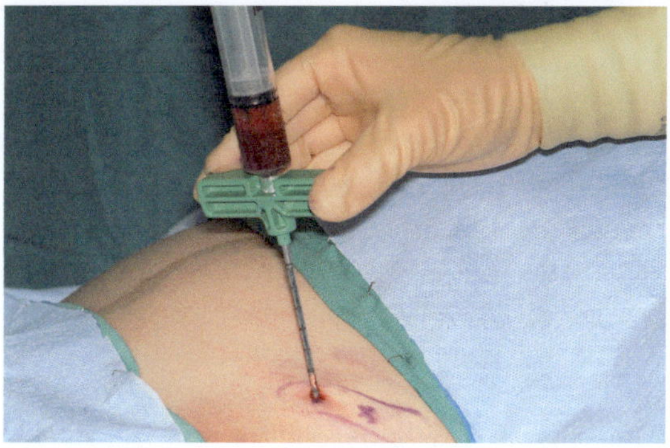

Fig. 7.3 Aspiration of bone marrow while turning the trocar 360° improves stem cell yield

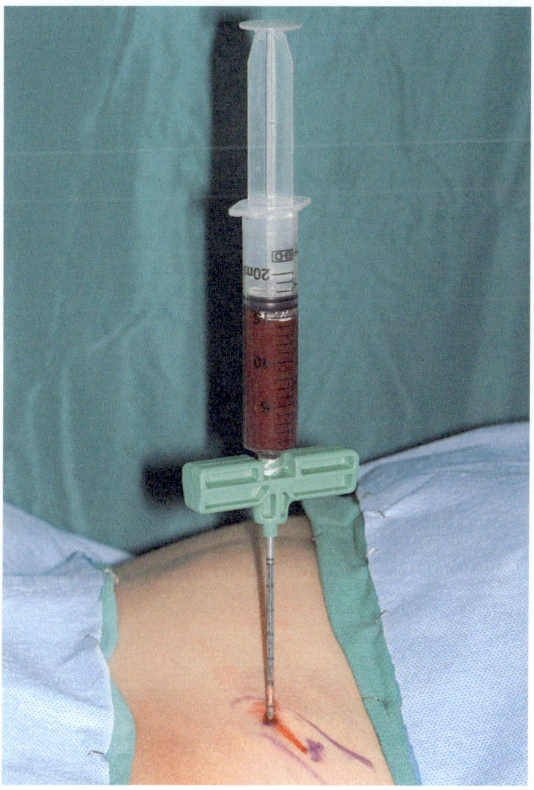

Fig. 7.4 Fifteen ml of bone marrow is aspirated from each puncture site

This BMA is placed into the centrifuge with the appropriate counter balance in place (Fig. 7.5). A programed 14 min double spin cycle separates and concentrates stem cells and osteoprogenitor cells as well as bone marrow platelets and megakaryocytes (Fig. 7.6). Stem cells and osteoprogenitor cells proven to be contained in BMA and BMAC are CD34+, CD44+, CD90+, CD105+, and CD271+.

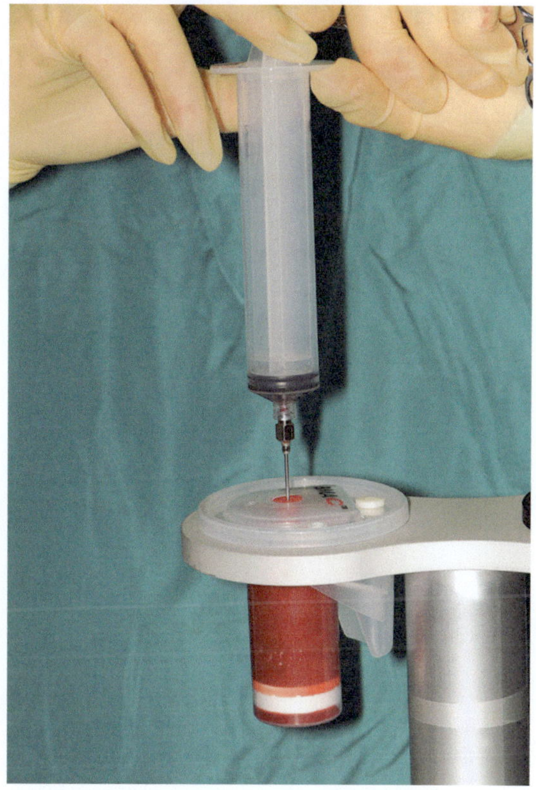

Fig. 7.5 The bone marrow aspirate is placed into the centrifugation canister that has a floating shelf that separates and concentrates marrow stem cells and progenitor cells

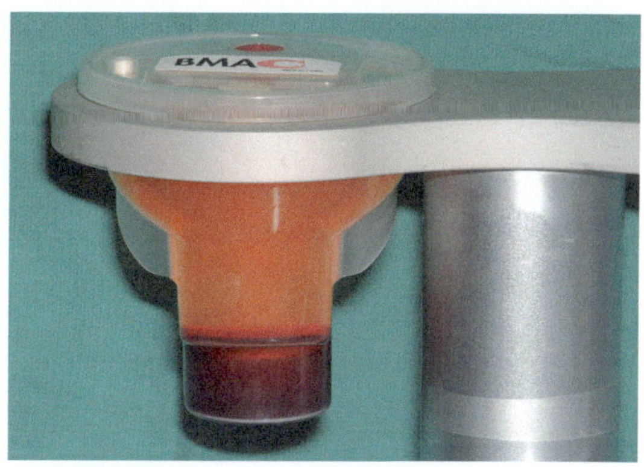

Fig. 7.6 The stem cell-/progenitor cell-rich layer is in the area of the buffy coat

New Design Bone Marrow Aspirates New design bone marrow aspirators are specifically engineered to exceed the stem cell/progenitor cell yield of BMAC while eliminating the centrifugation and thus the centrifuge. They take advantage of the knowledge that stem cells are adherent cells that mostly reside on the trabecular surface of cancellous bone and on the inner cortical wall of all bones [2]. The ilium puncture is identical to that for BMAC, but the screw design harvests 1 ml of BMA at a time (Figs. 7.7 and 7.8). The screw device in the system turns one turn to reposition the cannula 3 mm so that it is in a new adjacent untapped space in the bone marrow (Figs. 7.9, 7.10, and 7.11). This reduces the dilution of the aspirate with bone marrow blood. Each of several new positions of the cannula yielding 1 ml of BMA toward the entry site yields 10 ml of BMA with more stem cells/progenitor cells than centrifugation of 60 ml of BMA diluted with bone marrow blood (Fig. 7.12).

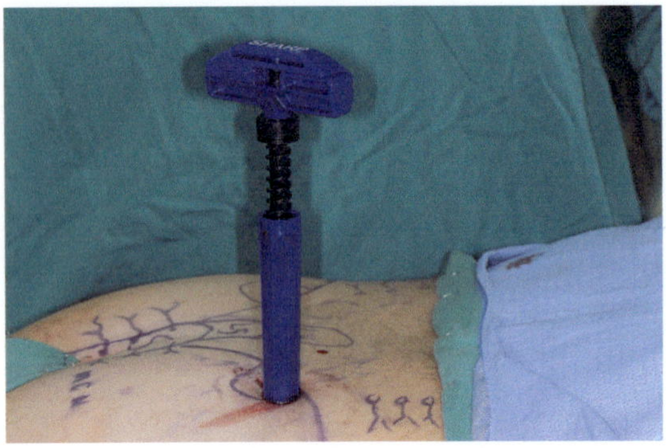

Fig. 7.7 This marrow aspirator is placed into the marrow space but has a threaded turning assembly to sequentially draw marrow from different locations

The Signal The clinically proven signal for bone regeneration is recombinant human bone morphogenetic protein-2/acellular collagen sponges [rhBMP-2/ACS (Medtronic, Memphis TN)] (Fig. 7.13). Studies showing efficacy gained approval in 2003 [3] for orthopedic lumbar spinal fusion and in 2005 [4] for tibia fractures [5]. Additional studies led to FDA approval for the oral and maxillofacial indications of ridge preservation [5] and sinus augmentation in 2007 [6, 7]. However, its more common usages have been the off-label applications of ridge augmentation, cystic defects, and continuity defects.

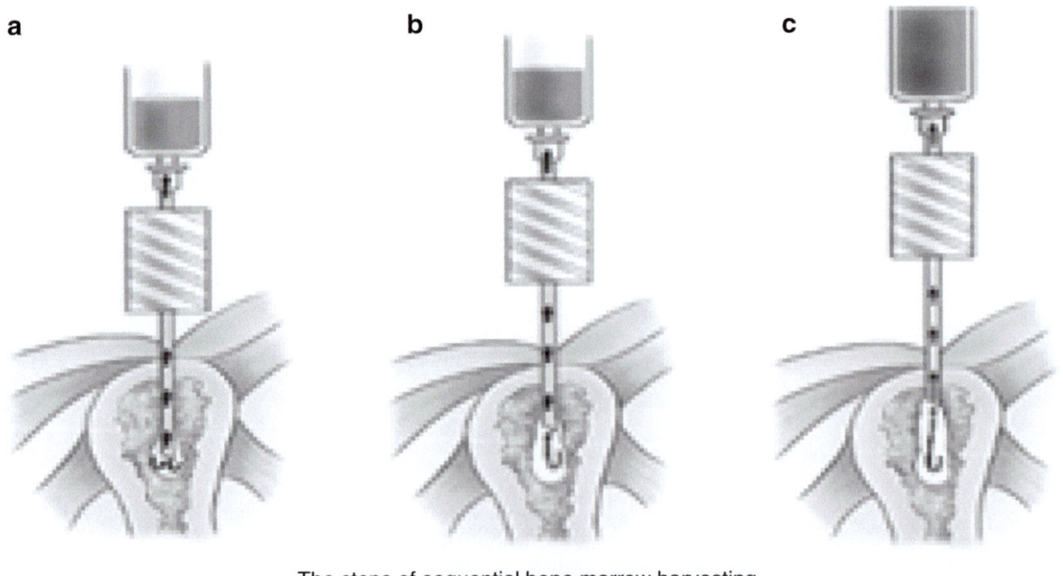

Fig. 7.8 Sequential positioning stem cell harvester. This advanced bone marrow harvester is capable via a screw design to reposition the aspirating tip 0.5–1 cm per turn of the handle, thereby aspirating a greater number of stem cell/osteoprogenitor cell via each new location and limiting dilution by marrow blood. (**a**) Sequential marrow aspirating device flush with skin with the aspirating port 3–4 cm in the marrow space. (**b**) Sequential marrow aspirating device now in a more superficial and new position in the marrow space. (**c**) Sequential marrow aspirating device repositioned again to be in a more superficial and new position in the marrow space

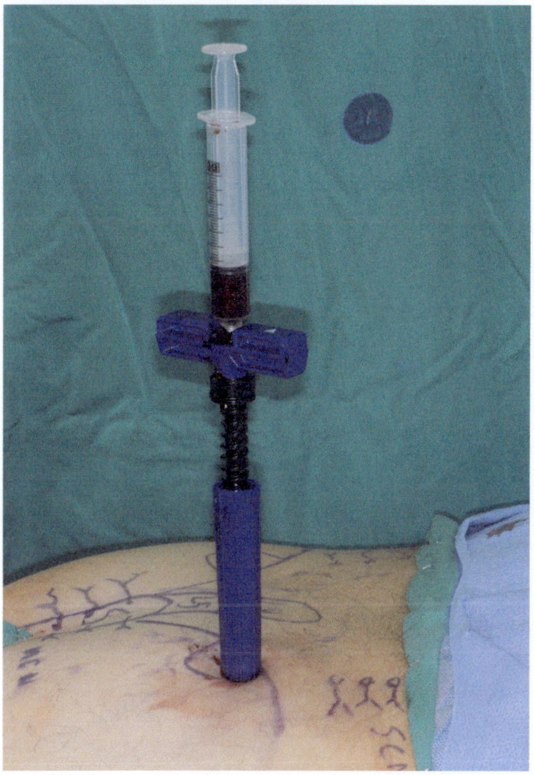

Fig. 7.9 Sequential marrow aspirating device now in a more superficial and new position in the marrow space

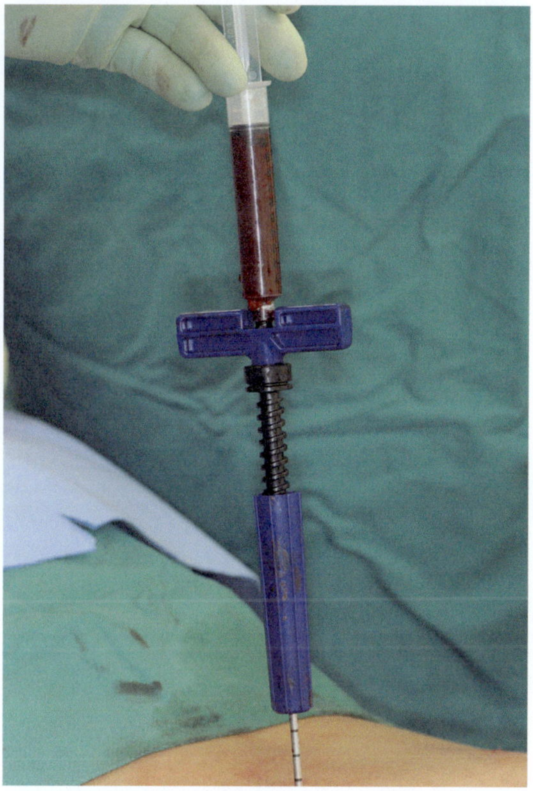

Fig. 7.10 Sequential marrow aspirating device after several 1 cc marrow volume turns of the assembly

In nature BMPs direct and regenerate the entire skeleton. Initially they form a soft osteoid bone or "baby's bone." This is recapitulated in in situ tissue engineering which produces an initial soft but normal patient-specific bone capable of osseointegrating dental implants, growing like normal bone, and maturing quickly with function.

The signal of rhBMP-2 is released out of the acellular collagen sponge over 21 days [8]. This small protein dimer binds to the stem cell and progenitor cell membranes inducing a proliferation, differentiation, and osteoid production [9, 10].

7.1.4 The Matrix

The matrix of in situ tissue engineering is the large surface area of cancellous allogeneic bone and the cell adhesion molecules in the plasma fraction of PRP, BMA, or BMAC. Cancellous allogeneic bone binds the cell adhesion molecules of fibrin and fibronectin from the plasma as

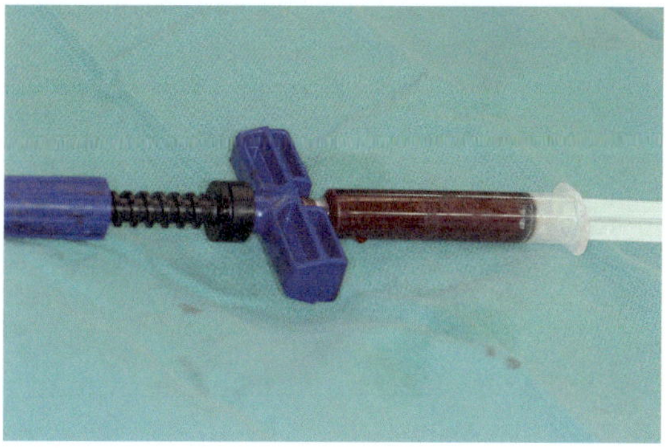

Fig. 7.11 The sequential marrow aspirator is turned 180° for each 1 cc marrow volume until the device becomes displaced from the marrow space

well as vitronectin and stromal-derived activation factor-1 alpha from platelets and the stem cells themselves. These strands of protein molecules and the surface area of the cancellous allogeneic bone act as an excellent scaffold upon which osteoid will be deposited. The new osteoid will remodel over 6 months removing the cell adhe-

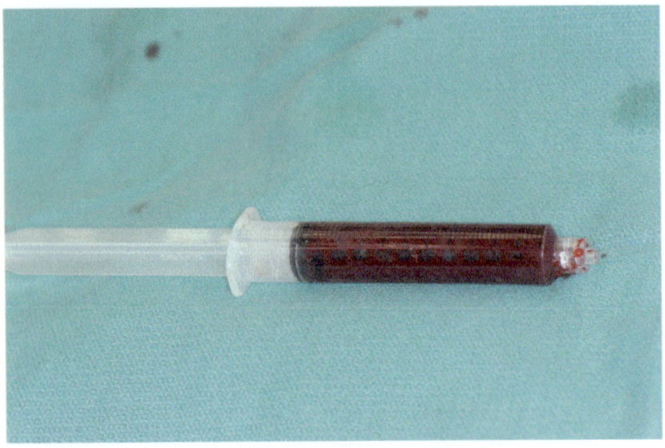

Fig. 7.12 The sequential marrow aspirator will yield 10–20 ml of a stem cell-/progenitor cell-rich marrow without the need for centrifugations

Fig. 7.13 rhBMp-2 is allowed to bind to the absorbable collagen sponge over 15 min before use

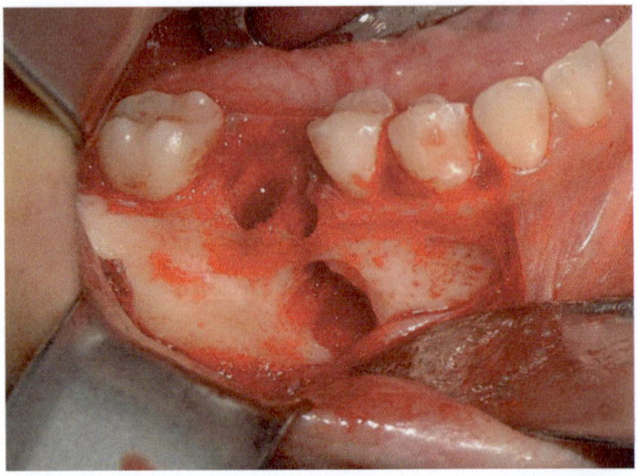

Fig. 7.14 Large socket defect and buccal wall defect after a molar extraction

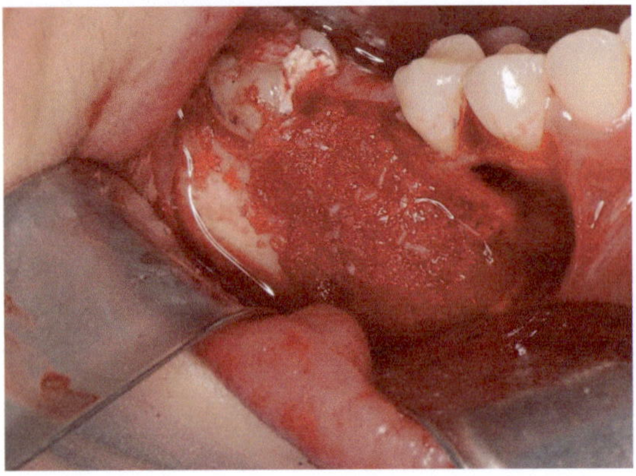

Fig. 7.15 In situ tissue-engineered graft in place

sion molecules and allogeneic bone by macrophage digestion and osteoclastic resorption resulting in a totally viable patient-specific bone.

7.1.4.1 The Bioreactor

In in situ tissue engineering, the bioreactor is represented by the environment of the recipient tissue space. While laboratory tissue engineering attempts to duplicate or enhance this cellular niche, the body's natural wound space has been perfected over millions of year of evolution and remains the best environment. The natural bioreactor is a low pH, low O_2 tension environment with the cells initially surviving from the nutrient-rich plasma fraction of the graft and responding to the growth factors secreted by the cellular fraction. By day 3, capillary ingrowth begins and is completed by day 17 at which time osteoid formation has already begun.

7.2 Clinical Applications of In Situ Tissue Engineering

A. Ridge Preservation: The use of in situ tissue engineering for ridge preservation is only indicated in tooth sockets, devoid of all but one or two walls. In such cases, 3 ml of PRP is mixed together with just enough cancellous allogeneic bone to fill the socket and 1.05 mg of rhBMP-2. The socket can be closed primarily, or if not, a collagen plug or a membrane applied over the surface will allow undisturbed bone regeneration. Mature bone will be apparent in 6 months (Figs. 7.14, 7.15, and 7.16).

B. Vertical Ridge Augmentation: Most patients who lose teeth also lose bone. In many cases the loss of vertical height contraindicates dental implant placement or forces the clinician

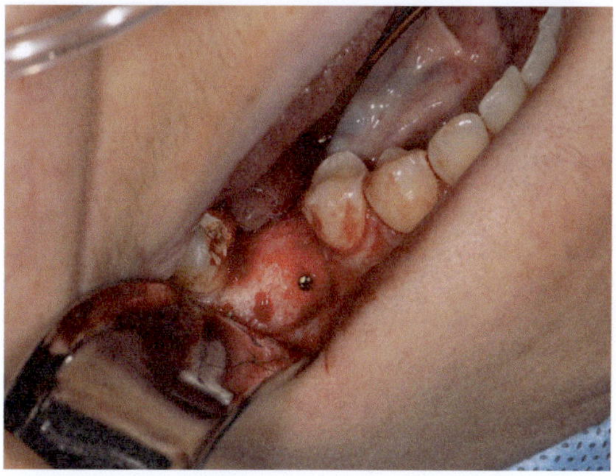

Fig. 7.16 Reentry documents complete bone regeneration

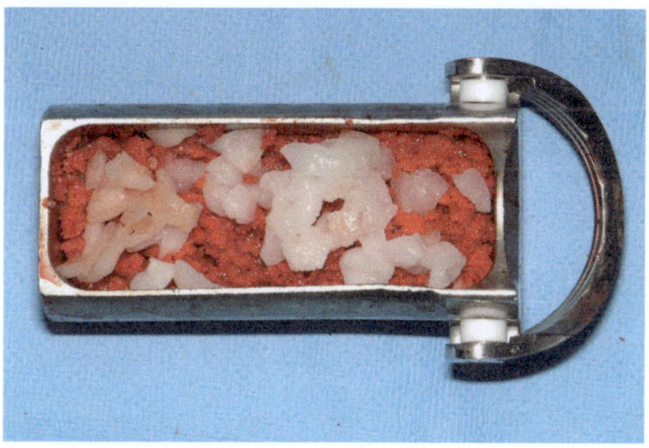

Fig. 7.17 The rhBMP-2/absorbable collagen sponge must be cut up and distributed through the volume of the graft for best results

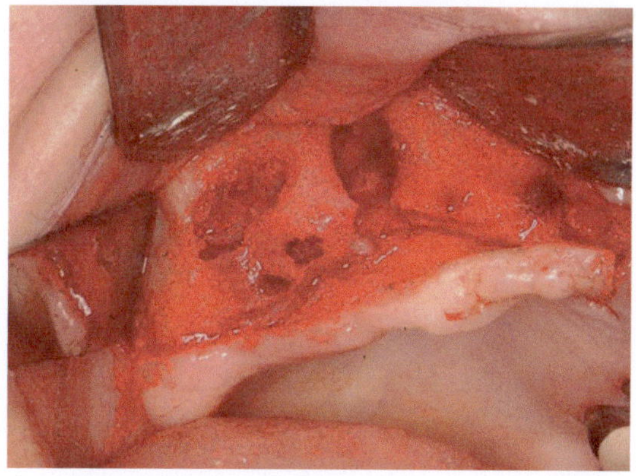

Fig. 7.18 Large vertical ridge defect requiring grafting

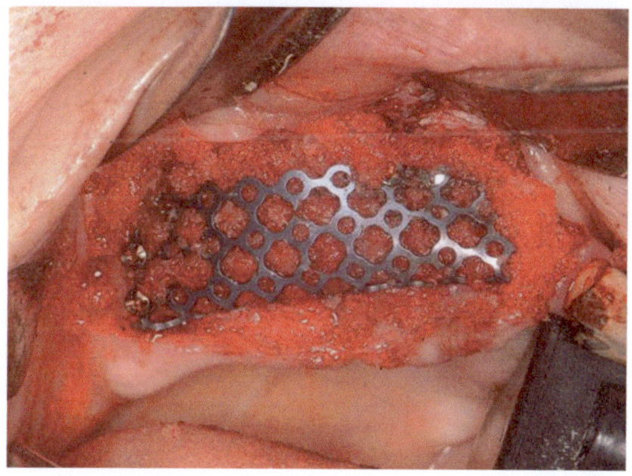

Fig. 7.19 Titanium mesh containing in situ tissue-engineered graft

into a compromised treatment plan where the implant is not completely within the bone.

Successful vertical ridge augmentation using in situ tissue engineering is accomplished with 7 ml of PRP, 1 cc of cancellous allogeneic bone per tooth area lost, and 0.5 mg of rhBMP-2/ACS per tooth area lost. Combined together and mixed thoroughly so that the cutup rhBMP-2/acellular collagen sponge is uniform throughout the graft will insure multiple foci of bone induction within the graft (Fig. 7.17).

This graft composite requires containment to maintain the space for bone regeneration. This is best accomplished with a titanium mesh or a reinforced membrane (Figs. 7.18 and 7.19). In the author's experience, resorbable polylactic acid/polygalactic acid polymers of any ratio have caused excessive early and late fragmentation resulting in swelling and pain for the patient. Reentry into the area has shown a significant inflammatory response and a foreign body giant cell reaction that inhibits bone regeneration (Figs. 7.20 and 7.21). Therefore, resorbable mesh covers of this material cannot be recommended. Bone regeneration under a firm immobile crib containment matures predictably to an implant capable bone in 6 months (Fig. 7.22).

C. Sinus Augmentation: Closed sinus lift augmentations and localized areas requiring sinus augmentation usually do not require the enhancement of growth factors or of in situ tissue engineering. However, the large hyper-pneumatized sinus with less than 3 mm of residual sinus floor should strongly be considered for in situ tissue engineering. In such cases the same surgical techniques of sinus

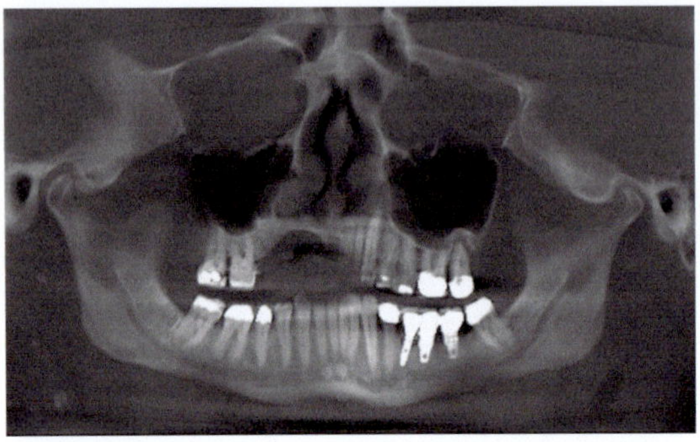

Fig. 7.20 Inadequate bone regeneration results in a graft contained by a PLA/PGLA polymer resorbable mesh

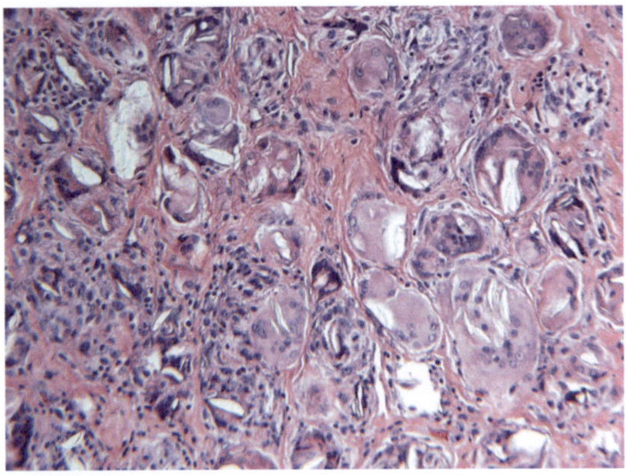

Fig. 7.21 Severe foreign body giant cell reaction to the particles of a resorbing PLA/PGLA mesh

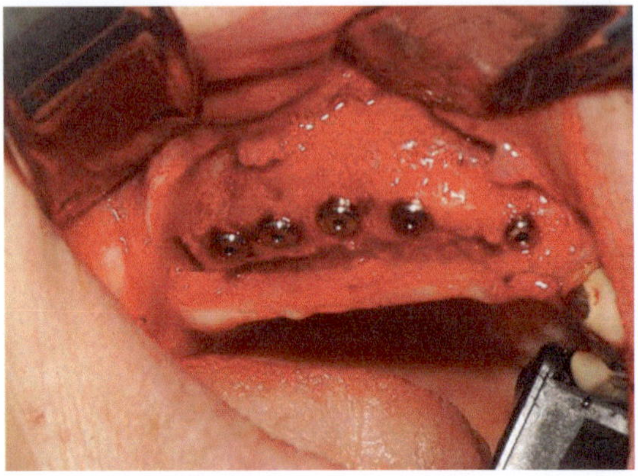

Fig. 7.22 Contained by a ridged titanium mesh, the excellent bone regeneration of the graft in Fig. 7.19 is seen housing five dental implants

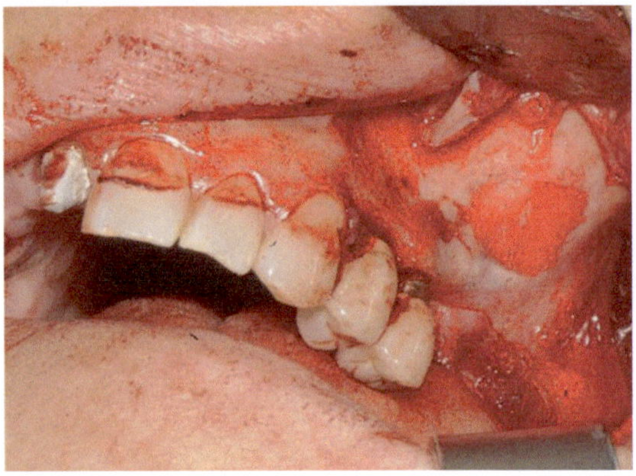

Fig. 7.23 In situ tissue-engineered sinus graft in place

membrane elevation and/or covering of any sinus membrane perforation are accomplished. However, instead of placing autogenous or allogenic bone particulates into the created space, 2.1 mg of rhBMP-2/ACS together with 4–7 cc of cancellous allogeneic bone and 7 ml of platelet-rich plasma are used by the author (Fig. 7.23). This will result in a complete resorption of the allogeneic bone and its complete replacement with the patient own viable bone capable of receiving dental implants in 6 months (Fig. 7.24).

D. Cystic Defects: Most intrabony defects, even after a large cyst is removed, will regenerate bone in the defect due to internal endosteal cells, marrow stem cells, and periosteal osteoblasts that surround the defect and therefore do not require bone grafting. However, in very large cystic defects that threaten a pathologic fracture or in areas where dental implants are later planned, in situ tissue engineering can regenerate bone faster, thus resisting fracture, and regenerate a more complete bone to support dental placements later (Figs. 7.25, 7.26, and 7.27). In such cases, the author uses 1 mg of rhBMP-2/ACS per each 1 cm diameter of the defect, 7 ml of platelet-rich plasma, and 1 cc of cancellous allogeneic bone per each 1 cm diameter of the defect.

In such cases, as well as continuity defects of the mandible, the maturity of the graft can be followed clinically and radiographically. Clinically the graft will transition from a particulate feel on palpation to firm within 2 weeks and bone hard within 1 month. Radiographically, the particulate quality of the allogeneic bone will appear to coalesce and consolidate within

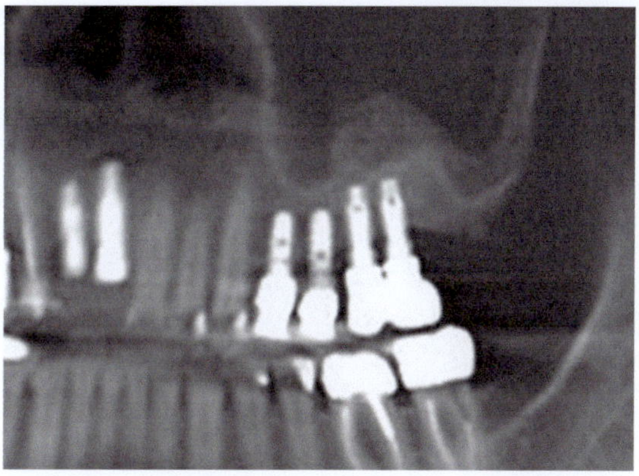

Fig. 7.24 Bone regenerated in the maxillary sinus

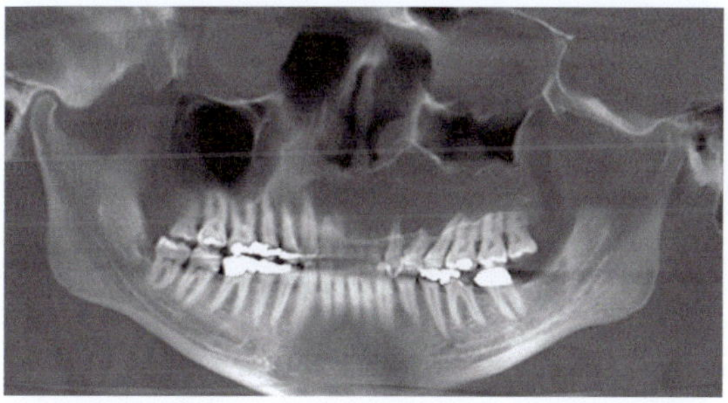

Fig. 7.25 Large cystic defect in the maxilla

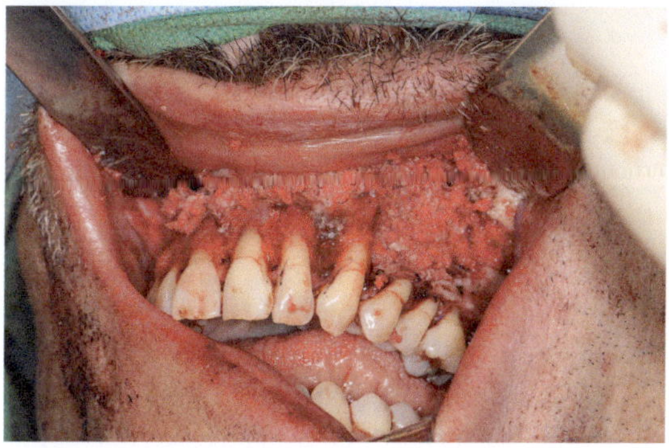

Fig. 7.26 After removal of a large odontogenic keratocyst, an in situ tissue-engineered graft was placed

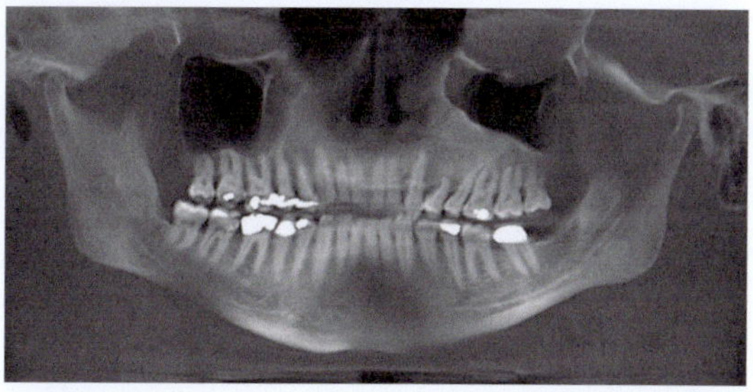

Fig. 7.27 Bone regeneration developed and was durable after 6 years, teeth mobility reduced from 2+ to non-mobile and all teeth regained vitality to pulp testing

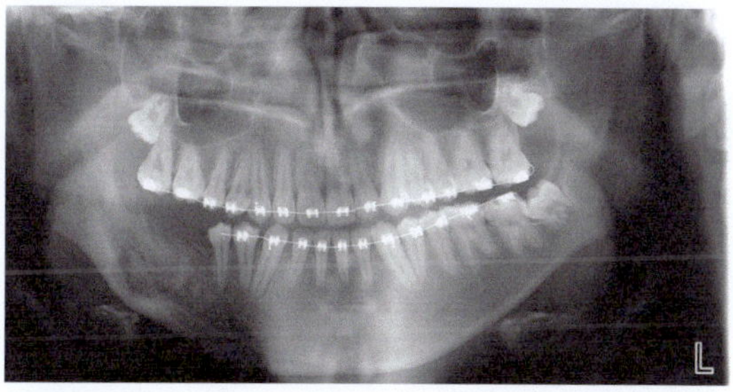

Fig. 7.28 Another large cystic defect shows the particulate nature of an early in situ tissue-engineered graft

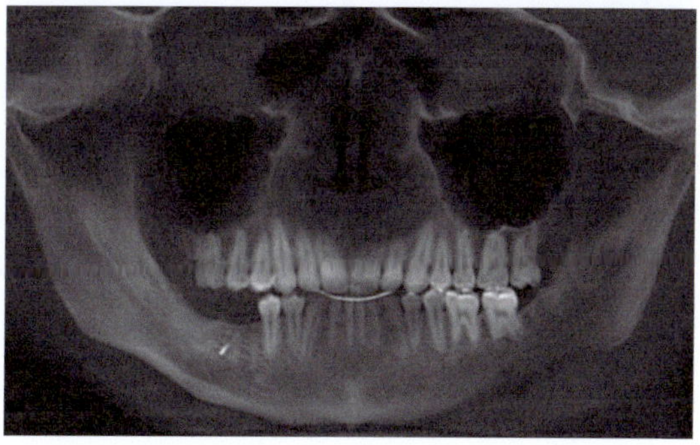

Fig. 7.29 After 3 months the graft in Fig. 7.28 can be seen to coalesce and consolidate

the first 3 months (Figs. 7.28, 7.29, and 7.30). An increase in mineral density and the emergence of a trabecular bone pattern at 6 months are an indication of mature bone capable of dental implant placement.

E. Continuity Defects: Until the advance of in situ tissue engineering, continuity defect reconstruction has required either open autogenous bone harvests of non-vascularized bone blocks and cancellous cellular marrow [11] or an open bone and soft tissue harvest of a vascularized bone pedicle, the most common today being that of a fibula [12]. With these approaches, added operating time, intensive care unit (ICU) management, in-hospital time, hospital wound care, as well as patient pain and morbidity must be considered. In situ tissue engineering can eliminate

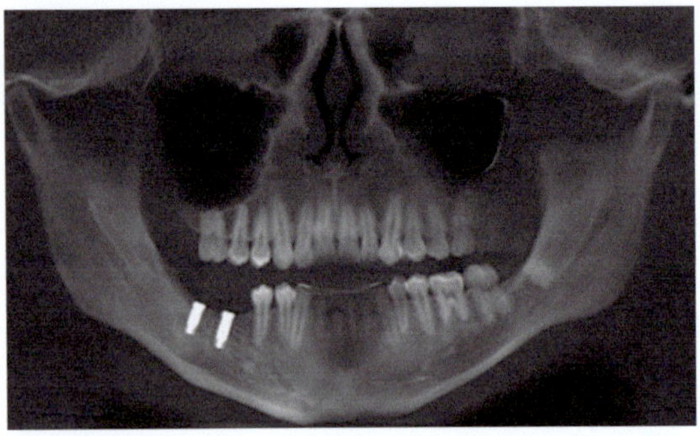

Fig. 7.30 A mature graft at 6 months will show a trabecular bone pattern indicating acceptability for implant placement

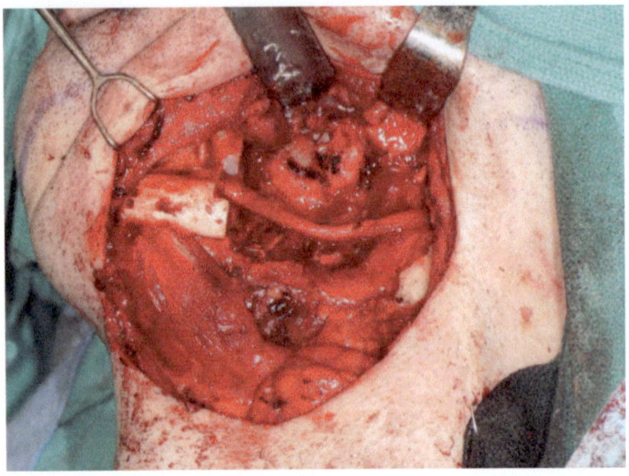

Fig. 7.31 Large continuity defect of the mandible after a resection for an ameloblastoma

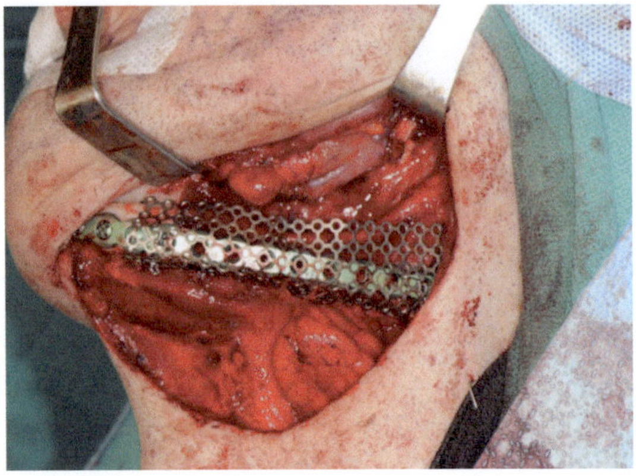

Fig. 7.32 A titanium reconstruction plate is used for internal stabilization, and a titanium mesh is used to contain the in situ tissue-engineered graft

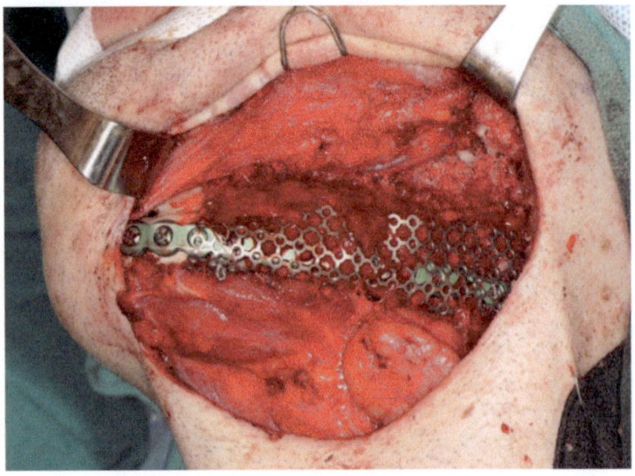

Fig. 7.33 In situ tissue-engineered graft in place

or reduce each of these negative factors. In continuity defects using in situ tissue engineering requires the use of BMAC or the newer BMA devices that yield greater numbers of osteoprogenitor cells/mesenchymal stem cells (cell solutions Ranfac or EmCyte) that are existent in the bone marrow. The graft composite for continuity defects requires 1 mg of rhBMP-2/ACS for each 1 cm linear length of the defect, 10–20 ml of BMAC or the same volume of new device BMA, and 1 cc of cancellous allogeneic bone for each 1 cm linear length of the defect.

The surgical technique must also insure graft stability and position the remaining mandible in the correct centric dental occlusion in the partially dentate patient and the appropriate arch alignment in the edentulous patient. This is accomplished by a 2.7 mm mandibular titanium reconstruction plate fixated with bicortical locking screws and maxillomandibular fixation via arch bars or bone screws (Figs. 7.31 and 7.32). It should be noted that only 3 weeks of maxillomandibular fixation is required in such cases due to the rapid bone regeneration from in situ tissue engineering. The author uses either a titanium mesh or an allogenic rib strut at the inferior border and sutures the neck fascia to these or to the titanium plate so as to condense and contain the graft (see Fig. 7.33).

Maxillomandibular fixation is released in 3 weeks with instructions for a soft diet (fish-pasta-ground beef consistency) for the following 1 month. The graft will clinically

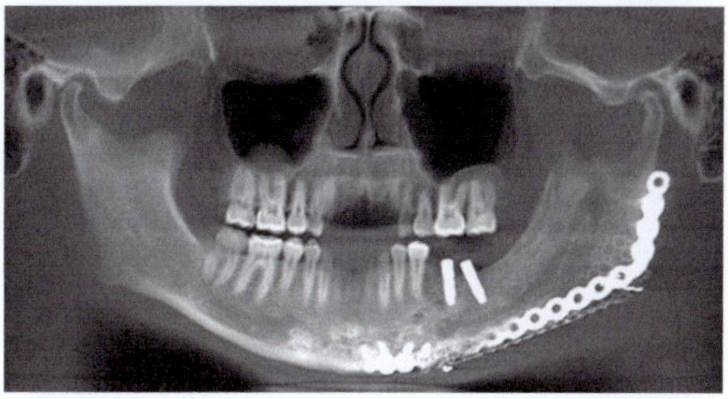

Fig. 7.34 Excellent regeneration of the size, form, and structure of the mandible as seen with a mandibular canal outline and dental implants

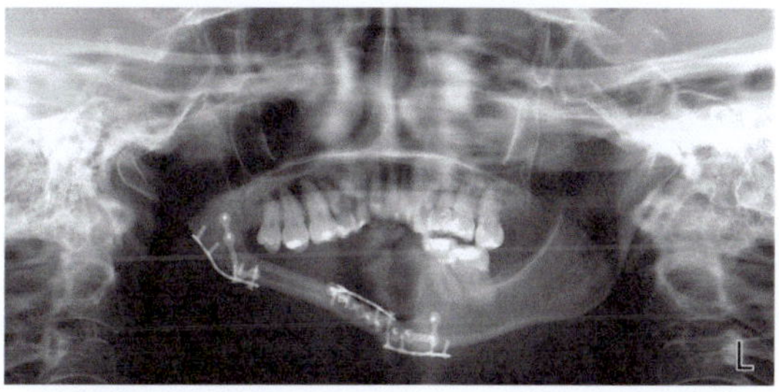

Fig. 7.35 Volume-deficient free vascular fibula incapable of dental rehabilitation

and radiographically mature in the appropriate arch coordination to the maxilla in 6 months. This arch coordination together with near-normal bone height and width allows for the best positioning of dental implants for a balanced occlusion (Fig. 7.34).

F. Rescuing Deficient Grafts: At times mandibular continuity grafts are insufficient in height or width to accept dental implants for complete restoration of function (Fig. 7.35). This is particularly noted in microvascular fibula grafts which are only 12 mm in height maximally, a straight bone, and one that is triangular in shape. In such cases, in situ tissue engineering can be 3sed to augment volume-deficient grafts and gain a better fusion to adjacent bone. A titanium mesh or dental implants may be required to maintain the submucosal volume space for the graft to mature (Fig. 7.36). The exact amount of rhBMP-2/ACS must be empirically assessed by the surgeon, but at least 10–20 ml of BMAC or new device BMA should be used together with sufficient cancellous allogeneic bone to fill in voids and gain the desired width and height for optimal dental implant placements in the alveolar of ridge or create a sufficient ridge for a non-implant supported prosthesis (Figs. 7.37 and 7.38).

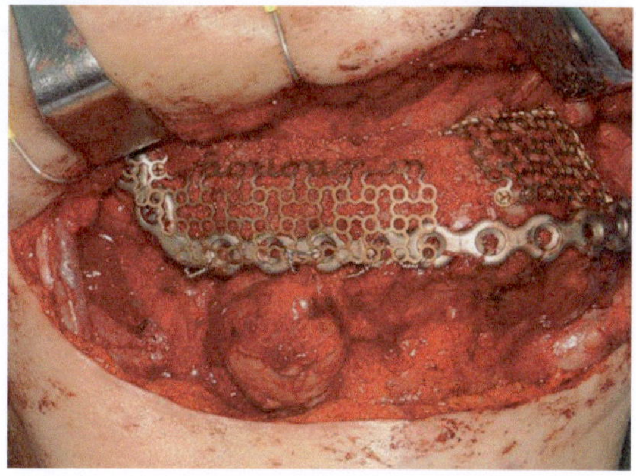

Fig. 7.36 Fibula graft removed and replaced by an in situ tissue-engineered graft contained by a titanium mesh

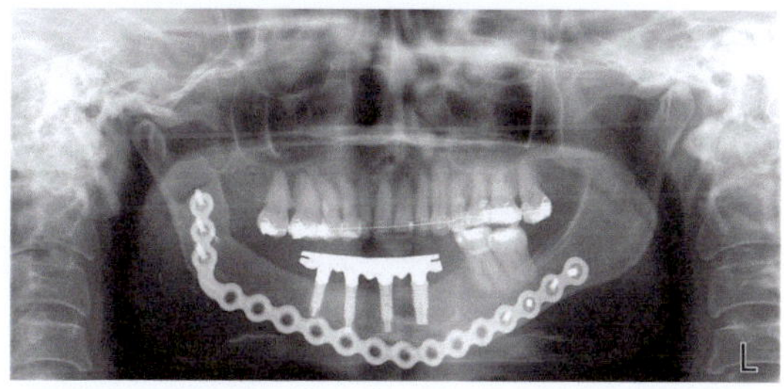

Fig. 7.37 The in situ tissue-engineered graft rescued the deficient fibula and fully reconstructed the mandible with a dental implant-retained prosthesis

G. Proof of Performance: While the forgoing previous six case applications detailed dosages of the tissue engineering triangle and the technical aspects of placement, it understates its significant advantages of bone regeneration without harvesting autogenous bone from a donor site. The author accomplished two studies that address the advantages of in situ tissue engineering:

1. Vertical Ridge Augmentation: Table 7.1

In this study 80 patients in each arm comparing in situ tissue engineering grafts to autogenous grafts

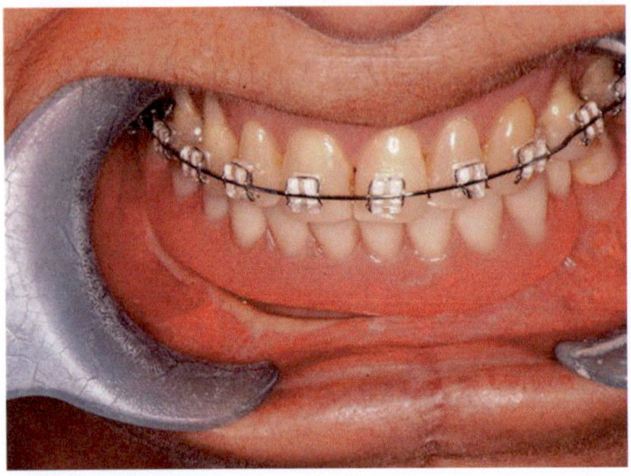

Fig. 7.38 Patient wears a denture with good retention and function

from the tibia or symphysis in a clinical setting were analyzed as to outcome (see Table 7.1).

The results show nearly equal outcomes of bone regeneration frequency and bone density as measured by trabecular bone volume. They also showed an equal ability of the bone regenerate to osseointegrate dental implants. However, the strong advantage of in situ tissue engineering is apparent in the lack of a donor site and the donor site morbidity in each patient as well as the time saving of a shorter procedure where bone is not harvested. The time saving also relates to a nearly equal cost factor overcoming the cost of the rhBMP-2/ACS. This study did not address the time and costs involved in the postoperative management of the donor site in cases of autogenous bone harvest which are not required with in situ tissue engineering.

Table 7.1 Maxillary vertical ridge augmentation

	In situ TE	Autograft	*P* value
N	80	80	N/A
Bone regeneration	74 (92.5%)	73 (91.5%)	0.94
TBA[a]	(71%)	(68%)	0.90
Implant osseointegration	176/190 (92.6%)	179/203 (88.2%)	0.88
Donor site	None	80/80	–
Average cost	$5900	$5300	0.89
Average time of procedure	48 min	81 min	0.05

[a]*TBA* trabecular bone area from bone cores

Mandibular continuity defect			
	TE	ACMG	MV fibula
N	100	100	100
Bone graft regeneration	94 (94%)	91 (91%)	90 (90%)
Donor site pain	0	100 (6/10)	100 (5/10)
Dental implant function	78 (78%)	79 (79%)	23 (23%)

Mandibular continuity defect			
	TE	ACMG	MV fibula
Mean hospital stay	1.7 days	4.2 days	10.4 days
ICU	0 days	0 days	2 days
Mean procedure time	3.7 h	5.7 h	9.8 h
Hospital cost factor	$19,200	$41,300	$87,200

2. Continuity Defects: In this study, 100 consecutive size-matched continuity defects indicated for elective bone graft surgery were compared between in situ tissue-engineered graft (ITE), autogenous cancellous marrow grafts (ACM), and microvascular fibular grafts (MVF); each patient outcome parameters are shown in Table. Time and cost parameters are shown in Table.

   The patient outcome parameters indicate a nearly equal bone regeneration across the three grafts. However, ITE and ACMG produced a greater height and width of bone than did the MVF grafts and thus resulted in a greater outcome related to implant placement and useful function (78%/79% to 23%). ITE grafts created no donor site pain, whereas both ACMG and MVF grafts produced a variable degree of postoperative pain in the 5/10 to 6/10 subjective score range.

   The time and cost parameter data shows a significant reduction in hospital care time between ITE, ACMF, and MVF grafts with the MVF

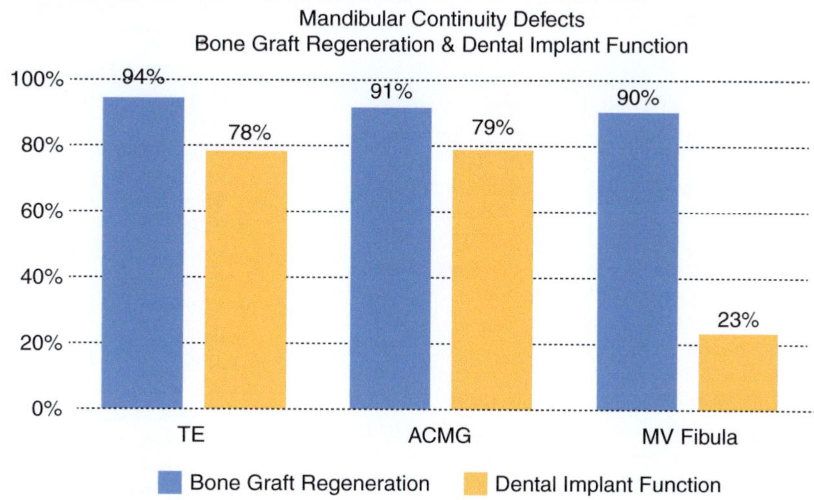

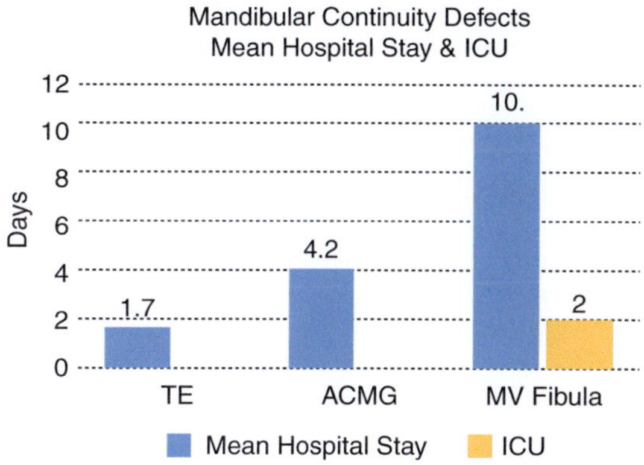

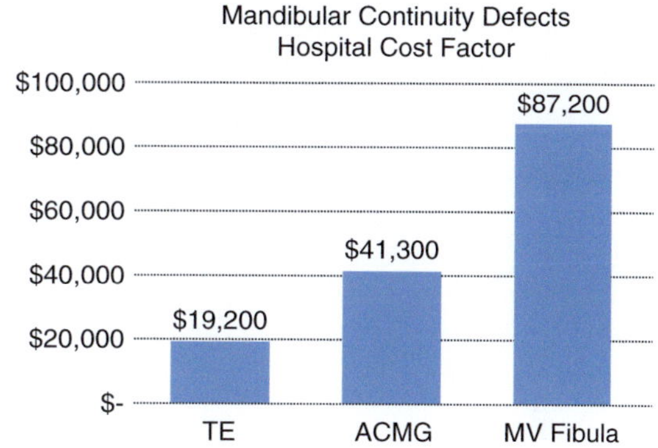

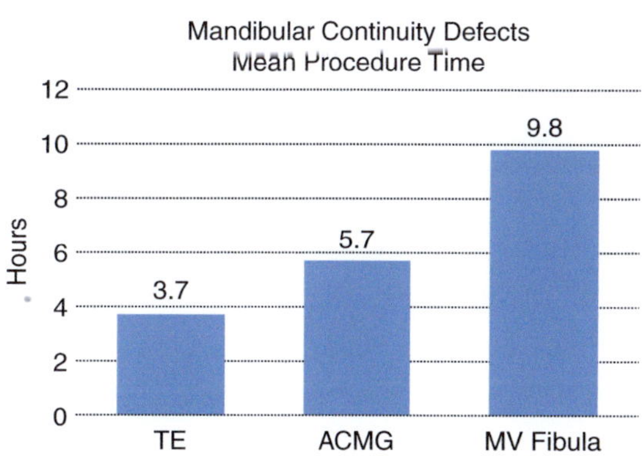

grafts requiring 2 days of ICU management during that time. Moreover, another significant difference between the three graft techniques is seen in operating time. This together with the length of hospital stay shows a very significant difference in hospital charges: ITE $19,200, ACMG $41,300, and MVF $87,200.

Such data strongly supports in situ tissue engineering as one of the several standards of rare options for surgeons aiming to replace a patient's lost or missing bone today. Moreover, it establishes ITE as a means consistent with today's trends in value analysis, tissue/organ preservation, and cost savings.

References

1. Marx RE. Platelet rich plasma: evidence to support its use. J Oral Maxillofac Surg. 2004;62:489–93.
2. Prockop DJ. Marrow stromal cells for nonhematopoietic tissues. Science. 1997;276:71–4.
3. Schimandle JH, Boden SD, Hutton WC. Experimental spinal fusion with recombinant human bone morphogenetic protein-2. Spine. 1995;20:1326–34.
4. Swiontkowski M, Goulet J, Paiement G, Bucholz R, Jones A, Fitzpatrick M, Valentin-Opran A. Safety and feasibility of implanting recombinant human BMP-2/absorbable collagen sponge in patients with open tibia fractures. Presented at the Annual Meeting of the Orthopedic Trauma Association. 1997.
5. Fiorellini JP, Howell TH, Cochran D, Malmquist J, Lilly LC, Spagnoli D, Tolzanic J, Jones A, Nevins MR. Randomized study evaluating recombinant human bone morphogenetic protein-2 for extraction socket documentation. J periodontol. 2005;76(4):605–13.
6. Boyne PJ, Marx RE, Nevins M, Triplett G, Lazaro E, Lilly L, Alder M, Nummikoksi P. A feasibility study evaluating rhBMP-2/absorbable collagen sponge for maxillary sinus floor augmentation. Int J Periodontics Restorative Dent. 1997;17:11–25.
7. Boyne PJ, Lilly LC, Marx RE, Moy PK, Nevins M, Spagnoli DB. De novo bone induction by recombinant human bone morphogenetic protein-2 (rhBMP-2) in maxillary sinus floor augmentations. J Oral Maxillofac Surg. 2005;63(12):1693–707.
8. rhBMP-2 and ACS. In: McKay WF, Peckman SM, Marotta JS, editors. The science of rhBMP-2, chap. 4. St Louis: Quality Medical Publishing; 2006. p. 67–85.
9. Wozney JM. The bone morphogenetic protein family and osteogenesis (review). Mol Reprod Dev. 1992;32:160–7.
10. Wang EA, Rosen V, D'alessandro JS, Baudy M, Cordes P, Harads J. Recombinant human bone morphogenetic protein induces bone formation. Proc Natl Acad Sci U S A. 1990;87:2220–9.
11. Marx RE. Mandibular reconstruction. J Oral Maxillofac Surg. 1993;51:466–79.
12. Hidalgo DA. Fibula free flap. A new method of mandible reconstruction. Plast Reconstr Surg. 1989;84:71–9.

Tissue Engineering Techniques for the Reconstruction of the Atrophic and Local Maxillary Defect

8

Issa Hanna, Huy Tran, Jonathan Swope, and Michael Peleg

8.1 Introduction

As extensively studied in paleontology literature, the evolution of human jaws and mastication system is crucial in the quest to trace mankind lineage and origin. The present morphologies of human dentition, maxilla, and mandibles are the result of million years of adaptation to changes in diet and complex speech [1]. While this discovery is important from an anthropologic standpoint, it also serves to underscore the roles of the maxilla and mandible in physiologic processes such as mastication and respiration. Furthermore, these structures contribute to facial esthetic profile that can considerably cultivate one's self-image and social-interpersonal relationships. As life expectancy is projected to increase globally, maxillofacial surgeons and dentists will continue to face a rising number of edentulism cases, as well as the enormous task of restoring quality of life for the prospective patient.

The difficulty of reconstructing the atrophic maxilla lies in the advance bone resorption upon patient presentation, the lack of soft tissue volume, and the associated risks of bone grafting that increase with age. In long-standing edentulism, bone deficits are multidimensional, and the quality of the residual bone is frequently poor due to decreased vascular density and bone remodeling. In addition, the patient's health history, compliance, and expectations can further complicate the reconstructive picture. It must be emphasized that a successful reconstructive outcome is not only measured by the quantity of bone regenerated but also by the final esthetic form and functional capability after prosthesis placement and/or implantation. Accordingly, prosthetic planning must dictate surgical development, and it is prudent that the relationship between the surgeon and restorative dentist remains close throughout the reconstructive process.

A variety of materials and techniques can be used in the reconstruction of the atrophic maxilla. Traditionally, autologous bone is preferred due to its superior regenerative potential. It is the only graft capable of all three bone-healing properties (osteoinduction, osteoconduction, and osteogenesis) that allogenic graft or bone substitutes have yet to achieve to date. There is a variety of autologous graft available for use in maxillofacial reconstruction, including cancellous, cortical, and bone marrow aspirates. Compared to cancellous graft, cortical graft provides good initial mechanical stability but less surface area, which

I. Hanna (✉) · H. Tran · J. Swope
Oral and Maxillofacial Surgery, The University of Texas Health Science Center at Houston, Houston, TX, USA
e-mail: Issa.A.Hanna@uth.tmc.edu

M. Peleg
Oral and Maxillofacial Surgery, University of Miami, School of Medicine, Miami, FL, USA
e-mail: mpeleg@med.miami.edu

J. C. Melville et al. (eds.), *Regenerative Strategies for Maxillary and Mandibular Reconstruction*,
https://doi.org/10.1007/978-3-319-93668-0_8

in turns provides fewer amount of osteocompetent cells, growth factors, and slower rate of graft revascularization and integration [2]. On the other hand, a combination of cortical and cancellous bone can compromise for excellent primary stability and bone generation potential, which makes mixed autogenous corticocancellous bone the ideal graft for maxillofacial reconstruction. While autologous graft provides excellent and predictable outcomes, it is often not available in sufficient quantity and always requires additional surgical site that may not be appropriate for selected patients. The advent of tissue engineering has provided alternatives for the maxillofacial surgeons with results comparable to that of autologous bone. Bone marrow aspirate concentrate (BMAC) and recombinant human bone morphogenetic protein (rhBMP-2) are two examples that have been discussed in the previous chapter. They will be further explored here in conjunction with the current reconstruction modality of the atrophic maxilla.

8.1.1 Principles of Guided Bone Regeneration and Barrier Membranes

In order to secure a deep understanding of tissue engineering principle, it is important to first examine the concept of guided bone regeneration (GBR) in relation to maxillofacial reconstruction. The basis of GBR is to provide the graft site an environment free of migrating connective and fibrous tissues in order to maximize the osteogenesis potential of the placed bone graft. A semipermeable, biocompatible membrane is essential to attain this objective, which serves to prevent migration of local fibroblastic cells yet allow adequate vascular ingrowth, oxygen and nutrient diffusion, and recruitment of osteocompetent cells to the site of the defect. Additionally, the membrane must be rigid enough to prevent prolapse of the surrounding soft tissue, effectively preserving the defect volume during the initial bone filling period. Barrier membrane has been extensively researched in the orthopedic literature since 1957 and has subsequently led to the first published application in oral and maxillofacial surgery in 1964. In that time span, Boyne and his team successfully demonstrated bone regeneration with the use of cellulose acetate filter in canine mandibles [3, 4].

There is a variety of membranes available in today's market, and they can generally be classified into one of the two categories: resorbable and non-resorbable. Non-resorbable membranes were the first of the two to be introduced with the advent of polytetrafluoroethylene (PTFE) membrane, and subsequently the expanded PTFE (e-PTFE) and dense PTFE with or without titanium reinforcement (d-PTFE or ti-d-PTFE) followed suit. The increased porous size of e-PTFE led to higher rate of bacterial colonization and premature intraoral exposure, necessitating tight mucosal closure over the membrane. Furthermore, e-PTFE creates a high rate of fibrous tissue ingrowth which made removal of the membrane traumatic. As a response to these issues, d-PTFE was created as a microporous biomaterial that is impermeable to bacteria but still allows for adequate oxygen and nutrient diffusion. This is especially advantageous in defects with insufficient soft tissue such as recent extraction socket because primary mucosal closure is not required. Removal of d-PTFE is also much easier due to a low degree of tissue ingrowth into the membrane surface. Titanium-reinforced d-PTFE increases the stiffness of the material and can be used in defects that necessitate a higher degree of space maintenance [5]. On the other hand, when dealing with a larger, more generalized alveolar ridge defect, non-resorbable titanium mesh is the material of choice due to its high strength and rigidity. Titanium mesh is also highly malleable, which allows excellent geometrical conformation to the natural shape of the alveolar ridge. The authors' experience shows an additional advantage to titanium mesh in that if the mesh is exposed to the oral cavity 3 weeks after bone graft procedure, bone consolidation has already been initiated, and so it is unnecessary to remove the mesh until graft consolidation is complete.

To eliminate the need for a second surgery for membrane or mesh removal, bioresorbable membranes were introduced as a second-generation material derived from either natural or synthetic material. The most common types of resorbable membranes are collagen and aliphatic polyesters, which are usually polyglycolide (PLG) or polylactide (PLA). These materials are degraded by hydrolysis or other physiologic processes associated with normal tissue turnover [6]. While resorbable membranes are appealing to both the clinicians and patients from comfort and cost-benefit perspective, studies have demonstrated an erratic rate of membrane resorption which may lead to overlying soft tissue collapse and poor final graft outcome. Furthermore, these membranes are extremely dependent on adjacent tissue health because of the increased rate of resorption induced by exposure to tissue inflammatory reactions. The authors have used resorbable mesh for generalized maxillary alveolar reconstruction cases in the past. Our experience has been yields of unpredictable and less-than-optimal bone volumes due to increased bone resorption and postoperative inflammatory response. We have subsequently reverted back to a more predictable titanium mesh containment system. If the goal is to reconstruct a large, generalized defect where a prolonged period of membrane stability is required, a non-resorbable membrane is a more reliable option.

8.1.2 Role of Tissue Engineering in Maxillofacial Reconstruction

Historically, large dentoalveolar reconstruction necessitated the harvest of autogenous bone from large-volume sites such as the anterior or posterior ilium. In addition to surgical morbidity, this option may be contraindicated in certain patients such as those with bone metabolic disorders or a history of hip trauma. With the advent of tissue engineering techniques, the era of autogenous harvesting is becoming less of a necessity, and predictable outcomes utilizing tissue engineering principles have become a reality. Although it is not the authors' intention to discuss in detail the biomechanism of tissue engineering components such as PRP, BMAC, or BMP (these are addressed in detail in separate chapters within this text), it is important to discuss what options are available to assist the surgeon in achieving predictable bone grafting outcomes. For a bone graft to maximize its bone-forming potential, all three osteogenic principles must be appreciated (Fig. 8.1). Osteogenesis is satisfied by the use of a concentrated solution of the patient's own mesenchymal cells (the authors utilize bone marrow aspirate concentrate) that is placed in an osteoconductive scaffolding network provided by allogeneic particulate or block graft. To further enhance this intricate relationship, bone-migrating and bone-differentiating signals are added through the use of recombinant human bone morphogenetic protein (rhBMP). A smaller and more localized dentoalveolar defect allows for the substitution of platelet-rich plasma (PRP) in place of BMAC because an adequate amount of osteoprogenitor cells is readily available at the native site. These cells can be recruited and activated by physiologic mechanisms which can be further enhanced by the application of rhBMP signaling. In these localized cases, the use of PRP does not in itself provide sufficient number of mesenchymal cells but offers excellent quantity of

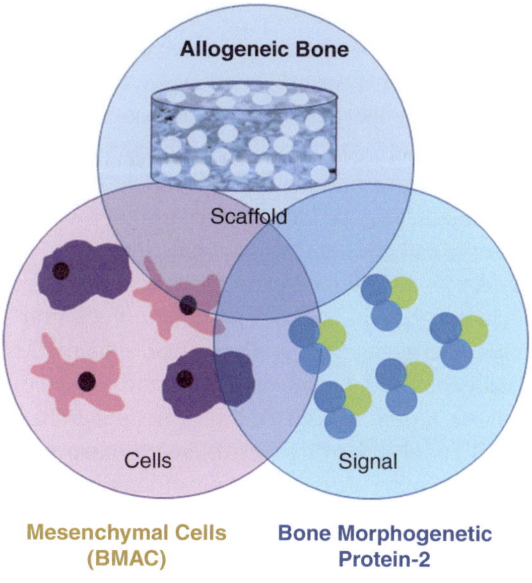

Fig. 8.1 Tissue engineering principle for osteogenesis is comprised of three critical components. Osteoprogenitor cells (BMAC) placed in an osteoconductive matrix (bone scaffold) and the addition of a recombinant signal (rhBMP)

growth factors (i.e., PDGF, TGF-b, VEGF, and EGF), aiding in accelerated bone graft maturation and soft tissue healing [7–9]. The engineered graft must be protected from contaminants of the oral cavity by primary mucosal coverage with a barrier membrane as previously discussed in the context of guided bone regeneration (GBR). If particulate allogenic graft is elected for a large, generalized maxillofacial defect, titanium mesh must be used due to its superior rigidity and plasticity properties, providing a containment system capable of conforming to the natural geometrical shape of dentoalveolar ridge. On the other hand, the intrinsic rigidity of block graft concomitantly serves as a space maintainer to prevent invasion of fibrous tissue and loss of defect volume. Close follow-up appointments leading to dental implant placement must be made to assess for graft viability and integration. Complications such as wound dehiscence without infection can have a quick and simple resolution if caught early.

8.2 Reconstruction of the Atrophic Maxilla

As previously mentioned, the surgical approach to reconstructing the atrophic maxilla is dependent on the final restorative plan agreed upon by the surgeon, the dentist, and the patient. If the defect(s) is localized, grafting with autogenous or allogenic bone prior to multiple single-unit dental implants may be the preferred option for satisfactory esthetic and functional outcomes. A variety of techniques exist for localized defect including guided bone regeneration, block grafting, interpositional grafting, and distraction osteogenesis. If the patient has been fully edentulous for a prolonged period of time, generalized maxillary alveolar ridge insufficiency exists from disuse atrophy. A combination of surgical techniques will be discussed to reconstruct the generalized atrophic maxilla. The authors will also provide a case example of localized maxillary reconstruction, though the majority of principles and techniques of generalized reconstruction can be easily extrapolated to the reconstruction of the localized defect.

8.2.1 Resorption Pattern

The atrophic maxilla resorbs centripetally and in a superior and palatal direction, reducing both the anterior-posterior (AP) and transverse dimensions. As a result, the maxilla becomes narrower overtime with characteristic knife-edge residual alveolar ridge. Additionally, resorption post-tooth extraction tends to favor the facial/buccal bone, furthering the distance between the atrophic maxillary and mandibular alveolar ridges. The most bone resorption occurs in the first month after tooth extraction and can reach 3–5 mm by the 6-month mark [10]. In a meta-analyses study by Hämmerle et al., it was demonstrated that the mean alveolar ridge reduction can reach 3.8 mm in width and 1.24 mm in height in the first year after tooth extraction without ridge preservation therapy [11]. The Cawood and Howell classification of the edentulous jaw was developed in 1988 based on a randomized, cross-sectional study of 300 dried skulls and demonstrated the temporal progression of the edentulous jaws (Fig. 8.2) [12]. In most cases of atrophic maxilla, patients often present with class IV, V, or VI classification.

8.2.2 Anatomical Consideration

The maxilla has a rich vascular network that is resilient against surgical manipulations. The nasopalatine and infraorbital arteries provide blood supply to the anterior maxilla and dentition, as the posterior superior alveolar (PSA), greater and lesser palatine arteries do for the posterior region. Due to this extensive anastomotic network, surgical reflection of the maxillary mucoperiosteal tissue is generally more forgiving in comparison to that of the mandible.

The maxillary sinuses, nasopalatine foramen, and nasal cavity are anatomical structures that must be considered in maxillary reconstruction. After posterior upper tooth extraction, the ipsilateral maxillary sinus is pneumatized inferiorly, further reducing the alveolar ridge vertical height. The extent of pneumatization increases

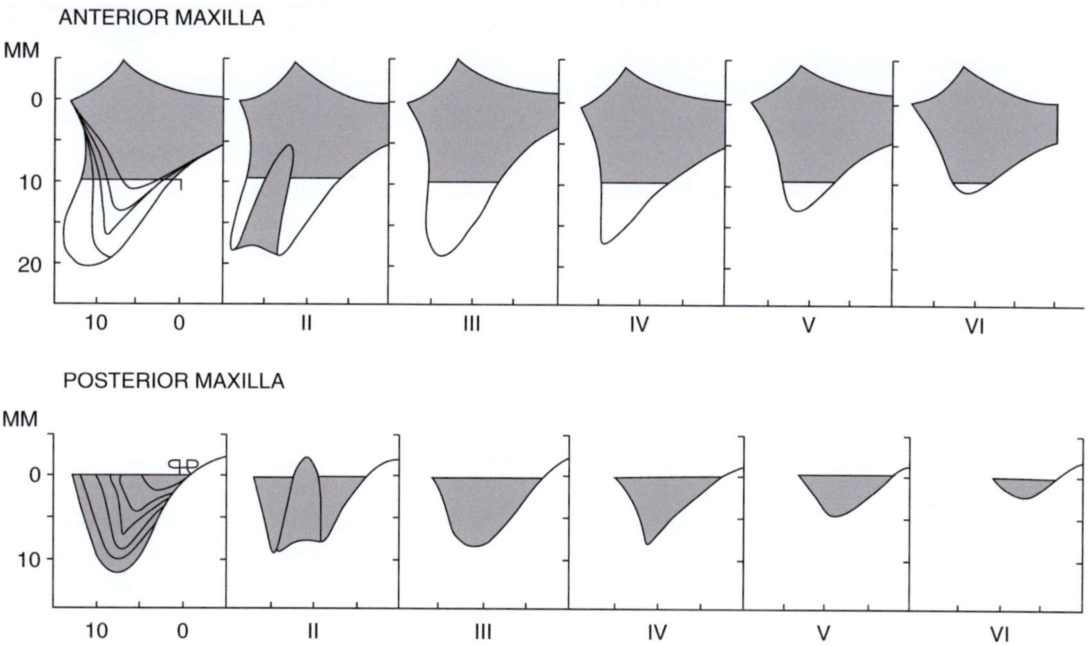

Fig. 8.2 Classification of the anterior and posterior maxilla. Adapted from Cawood and Howell [12]

based on several factors, such as extraction of teeth enveloped by a superiorly curving sinus floor, extraction of multiple adjacent teeth, and extraction of second molars [13]. Ridge preservation therapy must be considered if the patient has one or more factors while exhibiting desire for future dental implant. The maxillary sinus is supplied by the infraorbital artery and a vessel branch of the PSA, the alveolar antral artery. This anastomosis relationship supplies the Schneiderian membrane and anterolateral sinus wall and is a major source of hemorrhagic complication in sinus lift surgeries [14]. Care must be taken based on preoperative imaging assessment to avoid damage to the PSA branch, which can be intraosseous (64.3%), intra-sinus (29.1%), or superficial to the lateral wall of the maxillary sinus (6.7%) [15].

In most cases of the atrophic maxilla, the distance between the nasal cavity and alveolar crest of the anterior maxilla is severely reduced, particularly in patients suffering from combination syndrome (a long-term consequence of complete maxillary denture opposing partial mandibular denture and preserved anterior dentition). Additionally, a neurovascular bundle (nasopalatine nerve and terminal branch of nasopalatine artery) runs through the nasopalatine canal in the midline of the premaxillary segment, posterior to the alveolar ridge. Studies have shown that this canal increases in size with ridge atrophy [16, 17]. Though care should be taken to avoid disruption of the neurovascular bundle, which can lead to bothersome intraoperative bleeding and postoperative sensory disturbance, the authors find that generous reflection of the palatal tissue leads to sacrifice of this neurovascular bundle with no clinically relevant postoperative complications. It is worth mentioning that several studies in the literature have reported the use of nasopalatine canal as an anatomic buttress for dental implant in the rehabilitation of severely atrophic maxilla. Follow-up results showed good implant stability, and few cases demonstrated short-term paresthesia with eventual return of normal sensation [18–20].

The authors will discuss reconstruction of maxillary defects with the intent of dental implant rehabilitation. It is critical to discuss and understand the nuances of the surgical techniques described as they become an important component in successful outcomes.

8.2.2.1 Case 1

Tissue-engineered reconstruction of the atrophic maxilla utilizing corticocancellous allogeneic block graft and bone morphogenetic protein (rhBMP-2).

62-year-old male presented with complete maxillary and partial mandibular edentulism. Patient was unable to tolerate a trial of maxillary denture due to discomfort, which led him to inquire his general dentist to the possibility of implant-supported maxillary prosthesis. Additionally, patient still retained mandibular anterior teeth that presented a risk of *combination syndrome* with prolonged use of removable complete denture in the maxilla (Figs. 8.3 and 8.4). After discussion of treatment options, the patient elected for tissue engineering reconstruction using corticocancellous allogeneic block graft and recombinant human bone morphogenetic protein (rhBMP-2) followed by dental implant placement.

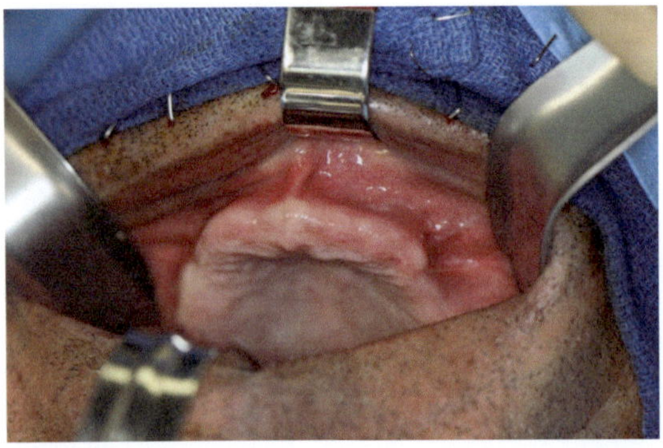

Fig. 8.3 A 62-year-old male presented with maxillary edentulism. Patient elected for reconstruction of maxilla with tissue engineering

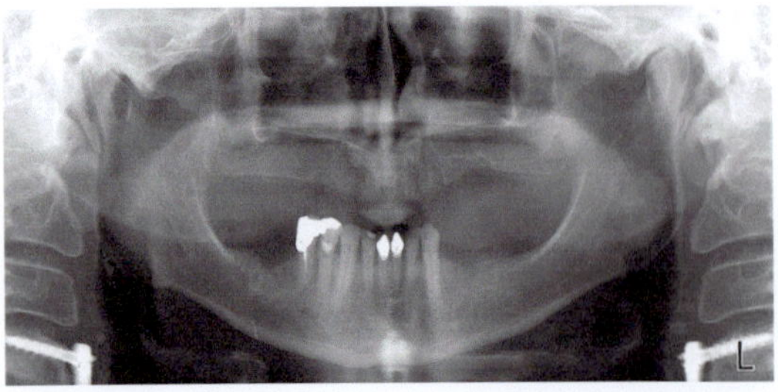

Fig. 8.4 Preoperative panoramic radiograph showed asymmetrical resorption of the maxilla. Left maxillary sinus lift was offered, but patient declined due to financial constraint

Armamentarium

1. Large rhBMP-2 (INFUSE) kit (total graft volume 8 cc, total rhBMP 12 mg)
2. Allogeneic anterior iliac crest sourced from a major tissue bank (the authors have access to this bone from either the Musculoskeletal Tissue Foundation or the University of Miami Tissue Bank)

Protocol

1. The patient is screened and medically optimized for tissue engineering surgery. Relative and absolute contraindications to rhBMP-2 use are assessed and ruled out. Additionally, off-label use to BMP outside of FDA-approved procedures, i.e., sinus augmentation and extraction socket preservation, must also be informed and consented by the patient. The surgeon must also advise the patient of postoperative sequelae such as moderate swelling and need for meticulous oral hygiene.
2. Peridex 0.12% is poured into the mouth and left for a minute prior to suctioning in order to decrease intraoral bacterial load. 2 g of ampicillin or Ancef can be administered in the perioperative period. If the patient has a penicillin allergy, 900 mg of IV clindamycin can be used.
3. Prior to incision, the surgeon has an understanding of the generalized area of reconstruction based on the restorative plan. The authors find it helpful to mark the bilateral extent of this reconstruction on the mucosa using a marking pen. This allows the surgeon to plan the distal extent of the incision and the associated posterior releasing incisions. It is important to keep the distal releasing incisions as far from planned surgical site as possible to protect the graft construct and assist in recruitment of tissue for end-of-case closure. A mid-crestal incision is made distal from each end of the planned reconstruction region as marked, with adequate distobuccal releasing incisions. Full-thickness mucoperiosteal tis-

sue is reflected off the residual ridge in both buccal and palatal directions, taking care not to tear or perforate the mucosa (Fig. 8.5). The palatal flap can be sutured together using a silk suture in a horizontal mattress fashion, keeping the flap away from the surgical field for improved visualization and ease of grafting. In block graft cases, this step is usually not needed.

4. The exposed recipient site is evaluated for anatomical shape in preparation of contouring the allogeneic blocks. In addition, host site preparation must allow sufficient access to the native trabecular bone blood vessels and osteogenic cells, which are critical components for bone incorporation). In the maxilla, perforation of the cortical bone is rarely needed, though this can be incorporated if the surgeon prefers. If perforation of the host cortical bone is elected, a 701 bur should be used judicially, and care should be taken to conserve and avoid overheating healthy, native bone for later application of fixation screws in the graft.
5. The authors prefer to use a bicortical allogeneic iliac crest. The graft is prepared by first removing one of the cortical plates using a reciprocating saw or 1.2 mm fissure bur, leaving a block with a thick cancellous marrow and remaining cortical bone (Fig. 8.6). However, if the graft has sufficient cancellous marrow width, it is more cost-effective to split the graft in half to be used in two different sites. The graft is trimmed to conform to the buccal dimension of the dentoalveolar ridge. This is then repeated for each graft until there is coverage for the planned reconstruction site. It is important to round the sharp edges of the segments with a surgical oval bur to avoid postoperative wound dehiscence and increased inflammatory reactions due to irritation of the overlying soft tissue (Fig. 8.6). This also takes the physiologic load off of the body to react and round these unnaturally sharp edges during the healing phase. The graft should be contoured to adapt perfectly to the host site, with absolutely no rocking movement. The maintenance of a robust cancellous marrow on the graft also helps facilitate stability by allowing for compression and adaption of the graft to the host site. Each block graft needed is prepared in the same fashion and then hydrated in normal saline. If the surgeon is utilizing BMAC or PRP, the graft should instead be hydrated with this cellular-rich fluid.
6. The inner aspect of the trimmed block graft is layered with the rhBMP and placed to the buccal side of the alveolar ridge (Fig. 8.7). Titanium screws are used to secure the segments to the natural ridge. The authors prefer a 1.5 mm screw of the appropriate length. If a smaller 1.2 mm screw is used, care must be taken upon screw retrieval to ensure that the screw head does not shear off, as the authors have historically experienced. At least two screws must be used per segment to avoid any rotational or sliding movements (Fig. 8.7). After the evaluation for satisfactory graft stability, additional BMP is then packed into the defective site surrounding the fixated block graft. The authors also recommend using a final strip of rhBMP on top of the grafted site (Fig. 8.8 is a representative photo and is not of the case documented here).
7. The reflected buccal and labial flap can be submucosally dissected in order to gain additional tissue length for primary coverage (Fig. 8.9). The surgeon must also anticipate moderate

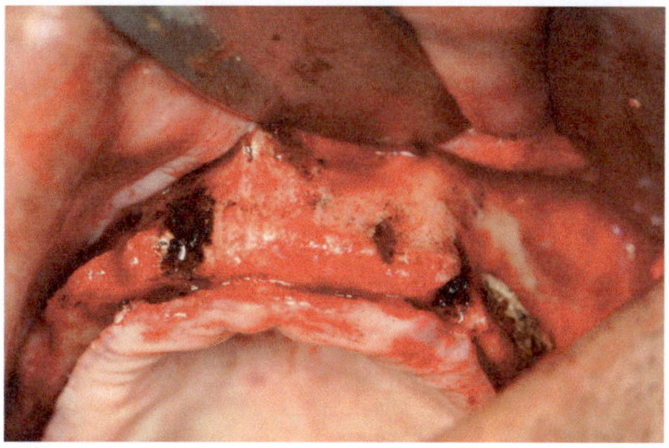

Fig. 8.5 Mid-crestal with distobuccal releasing incision was made from one side of the maxilla to the other. Full-thickness mucoperiosteal was reflected labially, revealing moderate horizontal ridge deficiency. Left maxillary posterior ridge also exhibits severe vertical bone loss

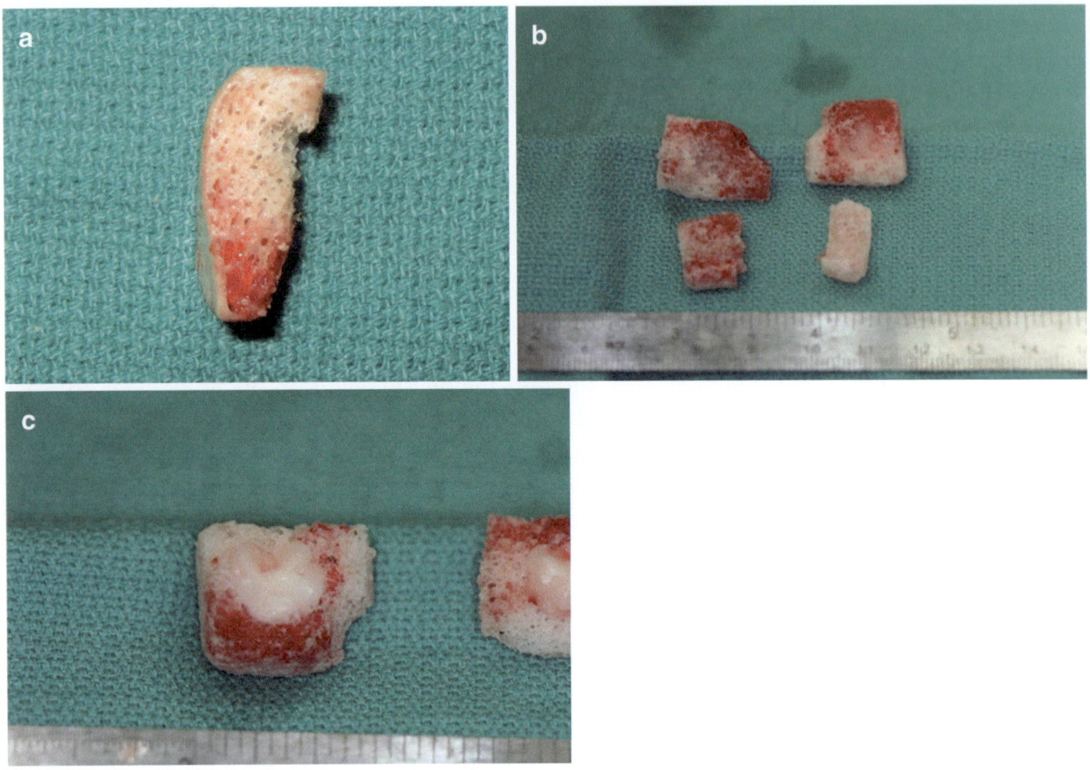

Fig. 8.6 (**a**) Cross section of allogeneic iliac crest block graft. Note that cortical bone was removed from one side of the graft, leaving thick cancellous marrow with one layer of cortical shell. (**b**) Allogenic block graft trimmed and conformed to the buccal aspect of the alveolar ridge. Note that smaller grafts are in a J pattern and will augment the vertical as well as horizontal dimension. (**c**) Bone morphogenetic protein is layered in the inner aspect of the block graft prior to fixation onto the host site

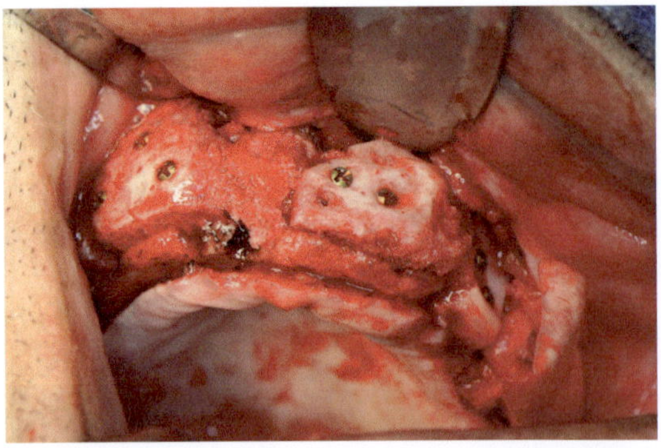

Fig. 8.7 Titanium screws are used to fixate the graft to the alveolar ridge. At least two screws must be used to achieve rigid fixation without rotational movement

postoperative edema due to the use of rhBMP, which can place additional stress on the overlying tissue. As such, it is critical to close the graft construct in a tension-free manner. Using forceps to hold traction on the flap superiorly, a pair of Metzenbaum scissors is used to carefully dissect the mucosal layer from the underlying periosteal tissue. Care must be taken to stay within the dissection plane to avoid tissue perforation. 4-0 vicryl suture is used to repair any perforation made during the dissection, though absolute care should be taken not to have any perforations. Adequate flap mobilization is confirmed by pulling the flap with a tissue forceps over the graft palatally. The flap should easily reach well into the palatal region. Prior to closure, the surgeon may elect to use a slowly resorbing membrane to cover the entire graft bed. The use of resorbing membrane is acceptable here due to the intrinsic space maintenance property of a block graft, thus preventing soft tissue prolapse. Though this is not an absolute need as noted in our case

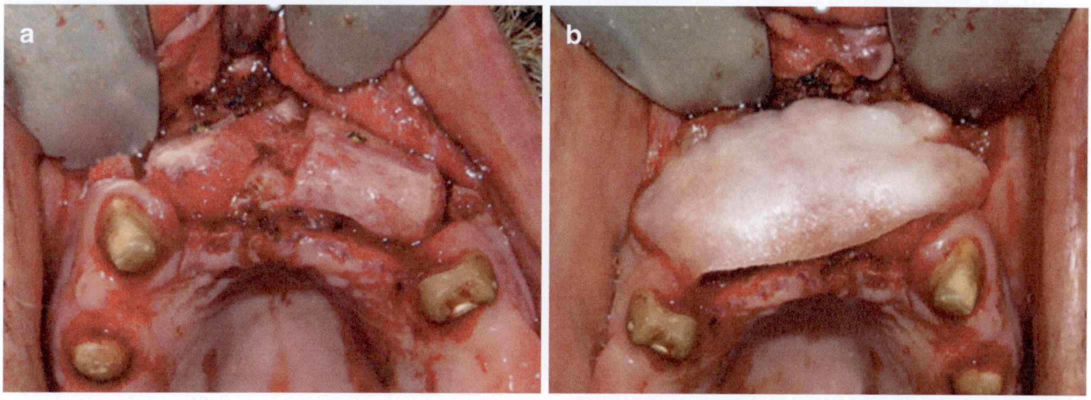

Fig. 8.8 (**a**) Allogeneic block graft fixated by titanium screws. BMP is lined in the inner aspect of the graft. (**b**) A strip of rhBMP was also placed on the outer aspect of the entire graft prior to closure. Note: these images are derived from a separate case and not representative of the case documented here

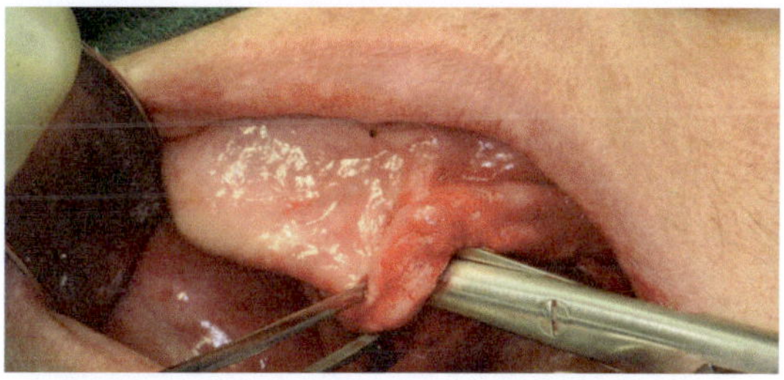

Fig. 8.9 Submucosal dissection was used in the buccal/labial flap in order to achieve primary, tension-free closure of the tissue-engineered graft. Forceps were used to hold traction of the flap, while Metzenbaum scissors were used to dissect the mucosal layer away from the underlying periosteum. Care must be taken to avoid mucosal perforation

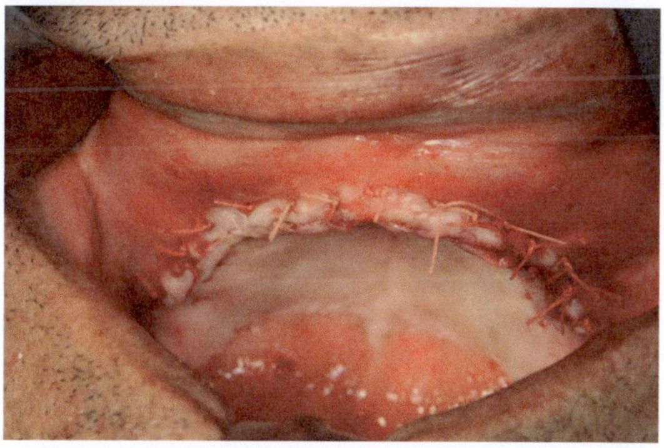

Fig. 8.10 Watertight, tension-free coverage of the graft using 3-0 vicryl suture utilizing interrupted horizontal mattress technique

example, it is preferred and does follow sound principles of guided bone regeneration as outlined in previous discussion. The surgeon should use a membrane he or she is most comfortable using. The authors have successfully used allogeneic dura and type I cross-linked porcine or bovine collagen membranes.

8. After confirming adequate tension-free closure, the defect is closed primarily with a slowly resorbing or permanent suture in a running locking or interrupted horizontal mattress fashion (Fig. 8.10). The surgeon may opt for a vicryl suture, yet the authors now find the use of a nylon suture to be preferable. Oversuturing with a 3-0 chromic gut on a tapered needle is performed as needed. A watertight closure is absolutely critical. The use of tissue sealants (i.e., TISSEEL, Dermabond) is not usually needed in maxillary reconstruction yet can be used depending on the surgeon's preference. Medications at discharge should include analgesics, antibiotics, and Peridex 0.12% oral rinse. The surgeon should also obtain a postoperative radiographic image (Fig. 8.11).
9. After a period of osteointegration (the authors prefer 6 months), the reconstruction site is uncovered (Fig. 8.12), fixation screws are removed, and the planned dental implants are placed (Fig. 8.13). As noted in Figs. 8.12 and 8.13, the grafted bone has been well integrated into the host site and has provided more than sufficient bone for dental implant placement and rehabilitation.

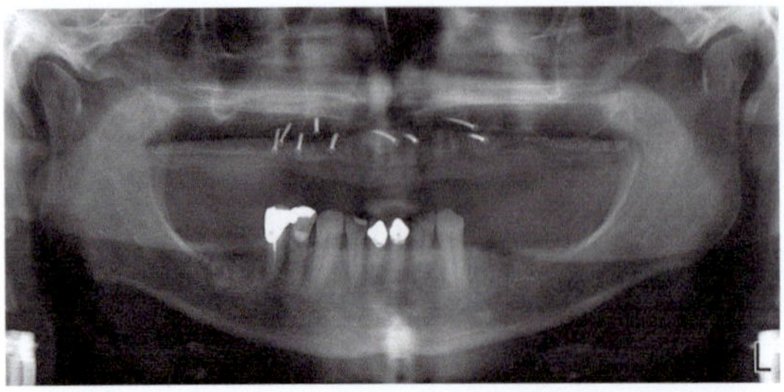

Fig. 8.11 Immediate postoperative panoramic radiograph

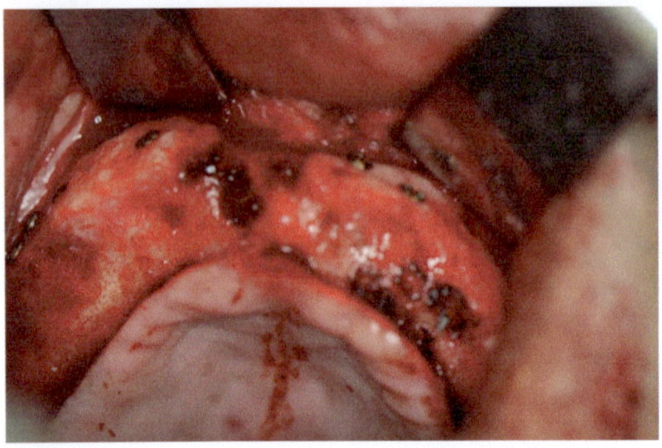

Fig. 8.12 Exposure of graft site 6-month postoperatively showed well-vascularized osseointegrated bone graft and a successful horizontal augmentation

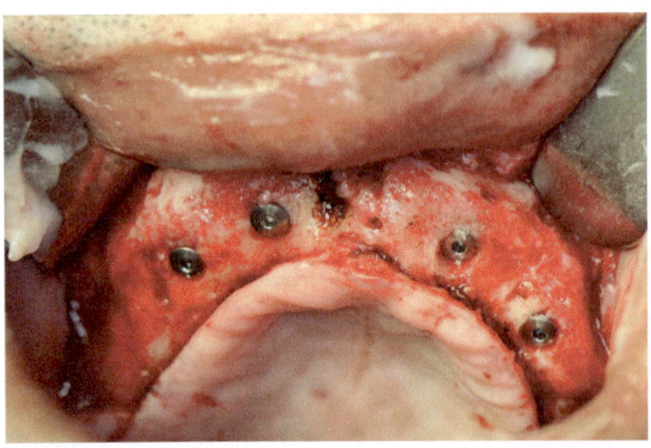

Fig. 8.13 Four dental implants were placed in the maxilla 6 months postoperatively. Primary stability was achieved

8.2.2.2 Case 2

Tissue-engineered reconstruction of the atrophic maxilla utilizing allogenic cancellous bone in particulate form, bone morphogenetic protein (rhBMP-2), bone marrow aspirate concentrate (BMAC), a titanium crib, and allogeneic dura mater membrane.

70-year-old female presented with *combination syndrome* (severely atrophic maxilla resulting from maxillary complete denture functioning against natural mandibular anterior dentition) (Fig. 8.14a, b). Block graft was deemed insufficient for the amount of bone needed and the type of defect (generalized horizontal and vertical insufficiency) present. The surgeon must keep in mind that the posterior atrophic maxilla usually has minimal vertical bone stock to the maxillary sinus; thus it is inadvisable to perform only horizontal ridge augmentation. The authors' goals were to augment both the horizontal and vertical dimensions of the anterior and posterior maxillary alveolus, as well as vertical augmentation of the bilateral maxillary sinuses. This type of defect lends itself well to full-arch reconstruction using a titanium mesh crib in an arch form and regenerative tissue engineering construct.

Very large grafts that are planned to be treated with allogeneic particulate grafting and mesh reconstruction benefit from introduction of osteoprogenitor cells via the use BMAC in the graft construct, as well as the use of rhBMP for bone signaling. The authors prefer to use

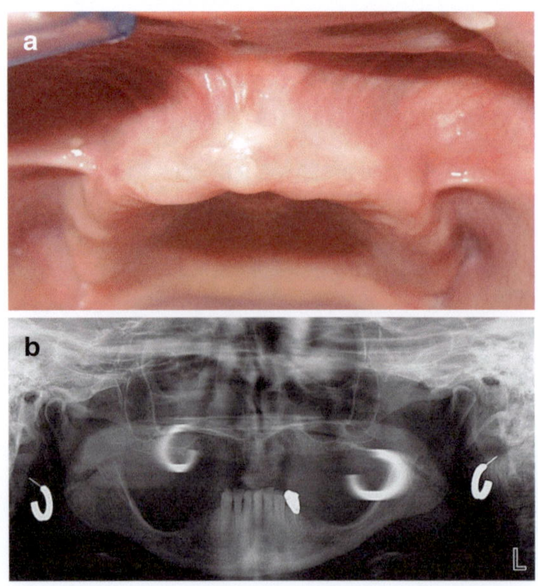

Fig. 8.14 (**a**) Generalized maxillary atrophy. (**b**) Panoramic radiograph demonstrating bilaterally pneumatized maxillary sinuses

milled allogeneic cancellous cubes from a bone bank. The authors prefer 12 mg of rhBMP and 120 cc of BMA. 120 cc of BMA, usually harvested from unilateral or bilateral anterior iliac crest, is spun down to approximately 10 cc of BMAC during the centrifugation process. In addition, the authors utilize a traditional lateral approach for bilateral sinus lifts. Sinus lift grafting can also utilize the tissue engineering construct.

Armamentarium

1. Large rhBMP-2 (INFUSE) kit (total graft volume 8 cc, total rhBMP 12 mg) (Fig. 8.15).
2. 30 cc unit of cancellous bone cubes from a major tissue bank (the authors have access to this bone from either the Musculoskeletal Tissue Foundation or the University of Miami Tissue Bank). The cancellous cubes are milled down to 1.0–2.0 mm particulate graft (Fig. 8.15).
3. 120 cc Harvest® Bone Marrow Aspirate Concentrate (BMAC®) System (Fig. 8.15).
4. Titanium mesh 0.4 mm malleable is preferred by the authors.
5. Dura mater (medium size) synthetic membrane.

Protocol

1. The patient is screened and medically optimized for tissue engineering surgery. Relative and absolute contraindications to BMAC and rhBMP-2 use are assessed and ruled out. Additionally, off-label use to BMP outside of FDA-approved procedures, i.e., sinus augmentation and extraction socket preservation, must also be informed and consented by the patient. The surgeon must also advise the patient of postoperative sequelae such as moderate swelling and need for meticulous oral hygiene.
2. Peridex 0.12% is poured into the mouth and left for a minute prior to suctioning in order to decrease intraoral bacterial load. 2 g of ampicillin or Ancef can be administered in the perioperative period. If the patient has a penicillin allergy, 900 mg of IV clindamycin can be used.
3. Prior to incision, the surgeon has an understanding of the generalized area of reconstruction based on the restorative plan. A full-arch reconstruction usually requires extension into the second molar position. It is important to keep the releasing incisions as far distal from planned surgical site as possible to protect the graft construct and assist in recruitment of tissue for end-of-case closure.
4. In the documented case, a Le Fort incision was chosen and is made approximately from the mesial second molar region from one side of the maxilla to the other (Fig. 8.16). Of note, the authors seldom use the Le Fort incision and now instead elect for mid-crestal incision with bilateral posterior releasing incisions, due to a consistent need for buccal tissue advancement for primary graft coverage. After the chosen incision is completed, a full-thickness mucoperiosteal tissue flap is reflected off the residual ridge in both buccal and palatal directions, taking care not to tear or perforate the mucosa. A 4-0 vicryl suture is used to repair any perforation made during the dissection, though absolute care should be taken not to have any perforations. The palatal flap can be sutured together with a silk suture in horizontal mattress fashion, keeping the flap away from the surgical field for improved visualization and ease of reconstruction (Fig. 8.17).

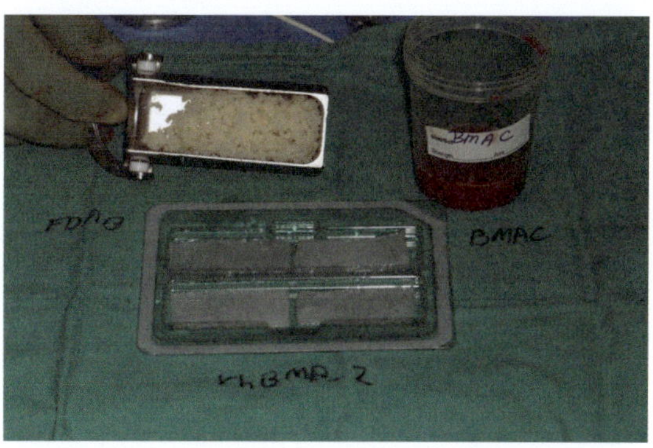

Fig. 8.15 Freeze-dried particulate allogenic bone, rhBMP-2, and BMAC (harvested from bilateral anterior iliac crests)

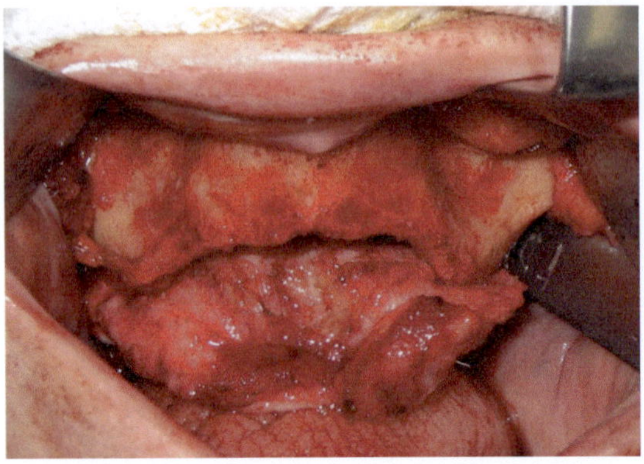

Fig. 8.16 Subperiosteal reflection of maxillary mucosa revealed irregular, knife-edge alveolar ridge. A traditional Le Fort incision was used in this case with preservation of much of the facial tissue in the palatal flap. Although this approach avoided an incision placed directly over the graft, the authors now seldom use this technique as it limits the advancement of buccal tissue in surgical closure, especially in large-type grafts

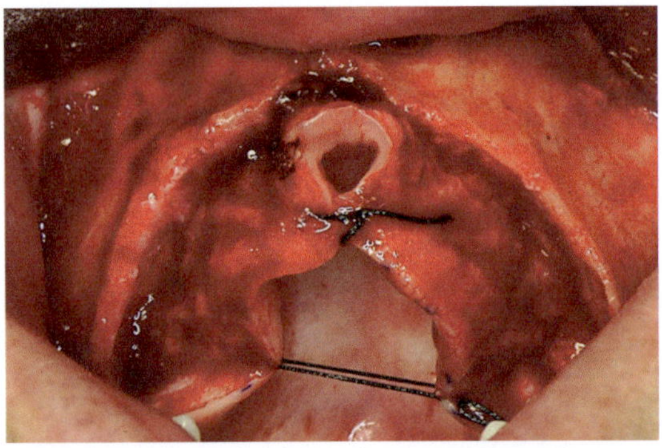

Fig. 8.17 The palatal tissues are sutured using a 3-0 silk suture in a horizontal mattress fashion to provide effective retraction and improved visualization of the surgical field. Of note, this picture significantly demonstrates the knife-edge alveolar ridge

5. Freeze-dried particulated allogenic cancellous graft is first hydrated with a portion of the BMAC. Any unabsorbed BMAC is recollected for future use. The reconstituted rhBMP-2 is cut into small segments using a tissue pickup and tissue scissor. The allogeneic bone, soaked with BMAC, and the segmented pieces of rhBMP are then combined into a tissue engineering construct and kept to the side until ready for use. The construct can be kept non-implanted and at room temperature for up to 4 h and still maintain viability (Fig. 8.18).

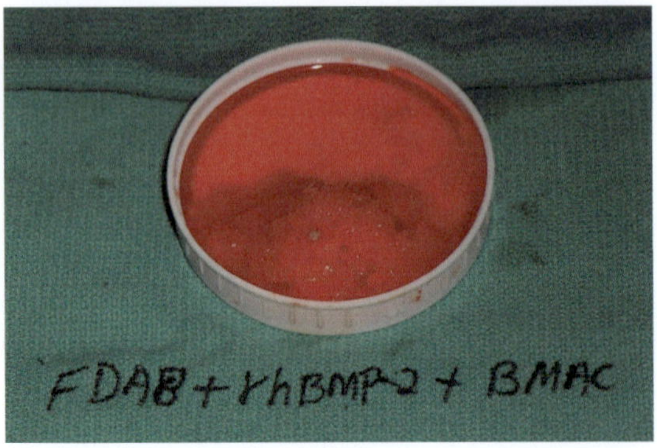

Fig. 8.18 Combination of freeze-dried allogenic bone (particulate), rhBMP-2, and BMAC mixed together. This tissue-engineered construct can be kept non-implanted up to 4 h at room temperature

6. Maxillary sinuses were exposed using a traditional lateral window approach (Fig. 8.19). The tissue-engineered construct was then densely packed from this access, carefully avoiding sinus membrane perforation (Fig. 8.20). If a perforation is noted, a collagen membrane can be used to cover the perforation in order to isolate the graft site from the sinus cavity. If the perforation is too large, suture repair should be done prior to placement of a collagen membrane.
7. Titanium mesh was trimmed and molded to the desired geometrical shape of an augmented alveolar ridge in both horizontal and vertical dimensions. The mesh must have stable points of contact on the palate as well as superiorly on the facial bone to facilitate mesh stabilization. At this point, the surgeon should locate areas that are acceptable for screw fixation. It is important to remove any sharp edges from the mesh to prevent tissue trauma and graft exposure in the postoperative period. This can be facilitated using a sharp wire cutter, an oval bur, and bovie scratch pad. Synthetic allogeneic dura mater membrane was secured in the inner aspect of the mesh with 3-0 chromic gut suture (Fig. 8.21). Any slowly resorbing membrane (i.e., type I cross-linked collagen membrane, pericardium, or allogeneic dura) can be used. The

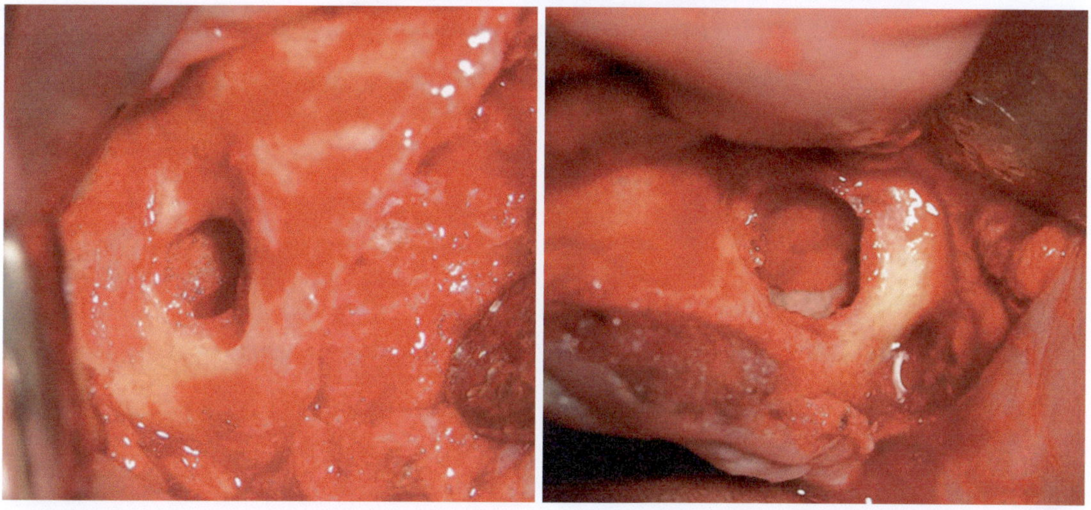

Fig. 8.19 Exposure of bilateral maxillary sinuses

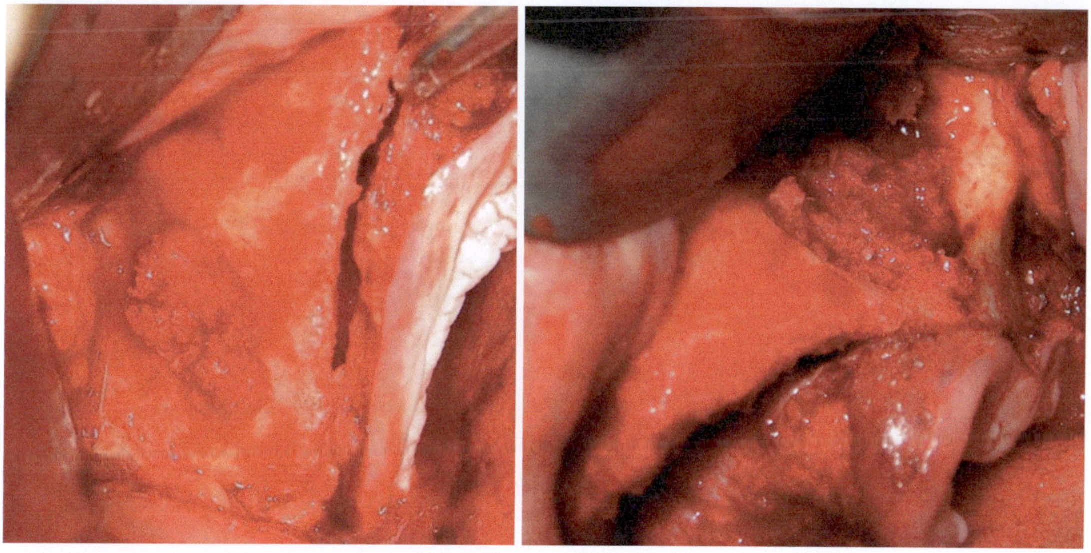

Fig. 8.20 Tissue-engineered construct packed into sinus lift cavities

authors now prefer to drape the facial surface of the contoured mesh using the chosen membrane with or without suture, whether it is sutured to the mesh or just draped.

8. The tissue-engineered construct was then packed into the titanium-dura mater crib. The composite graft with the contoured crib was then placed into the alveolar ridge of the maxilla (Fig. 8.22). The crib is then stabilized using 1.5 mm titanium screws of 5 mm of length to engage the native maxillary bone (Fig. 8.23). Additional composite graft is then packed from the anterior crib access point to further condense the material posteriorly. When adequate packing is achieved and if there is additional composite graft material, the remaining rhBMP sponges are isolated and placed directly into the graft site.
9. A submucosal dissection is then completed on the buccal flap (Fig. 8.24), and the flap is advanced. The buccal flap should effortlessly

Fig. 8.21 Titanium mesh trimmed and molded to the desired geometrical shape of the alveolar ridge. Synthetic dura mater membrane is secured in the inner aspect of the mesh with 3-0 chromic gut suture

advance into the palatal region (Fig. 8.25). Any remaining BMAC is then combined with calcium chloride and thrombin to coagulate the BMAC, and this is then delivered over the graft prior to closure (Fig. 8.26).

10. Closure is then facilitated using 3-0 nylon sutures in a horizontal mattress fashion for primary mucosal coverage. The wound is over-sutured using 3-0 chromic gut on a tapered needle. A watertight closure is absolutely critical. The use of tissue sealants is not usually needed in maxillary reconstruction yet can be used per surgeon's choice. In the noted case, a V-Y closure was utilized to minimize shortening of the upper lip (Fig. 8.27).
10. Discharge medications should be given to patient including analgesics, antibiotics, and Peridex 0.12% oral rinse. Sinus precaution instructions must also be given to the patient.

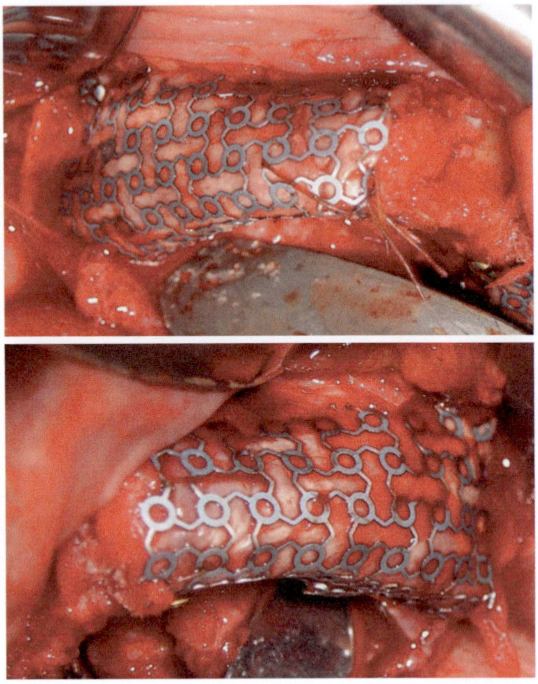

Fig. 8.22 The tissue-engineered graft is packed into the titanium mesh-dura membrane containment system and placed over the alveolar ridge of the maxilla

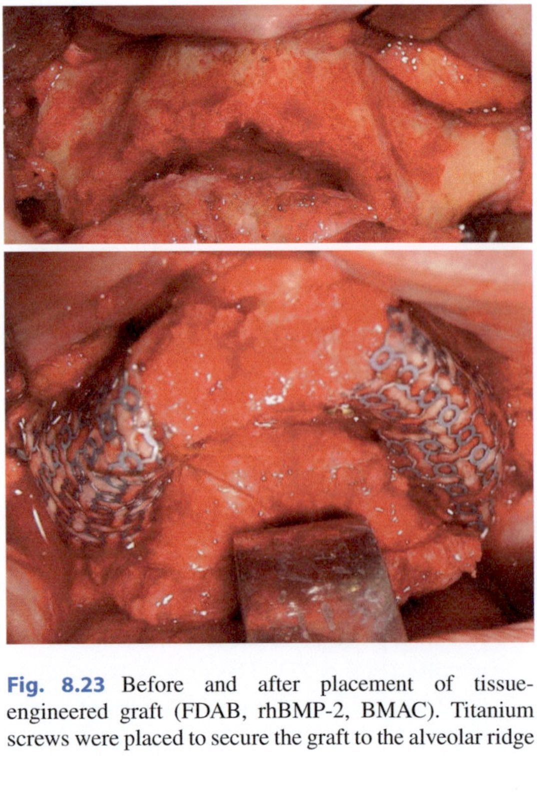

Fig. 8.23 Before and after placement of tissue-engineered graft (FDAB, rhBMP-2, BMAC). Titanium screws were placed to secure the graft to the alveolar ridge

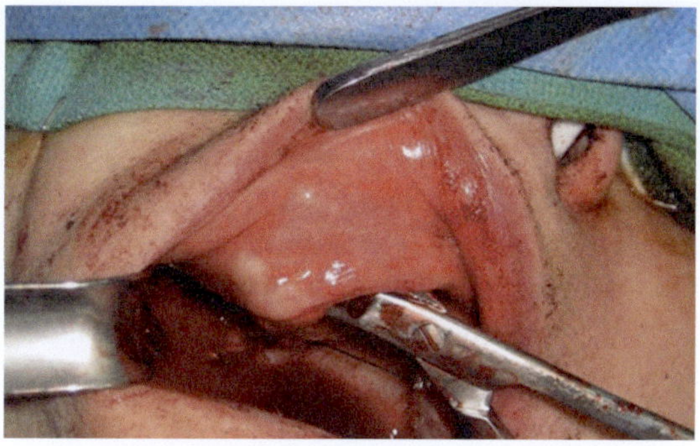

Fig. 8.24 Submucosal dissection was used in the buccal/labial flap in order to achieve primary, tension-free closure of the tissue-engineered graft. Forceps were used to hold traction of the flap, while Metzenbaum scissors were used to dissect the mucosal layer away from the underlying periosteum. Care must be taken to avoid mucosal perforation

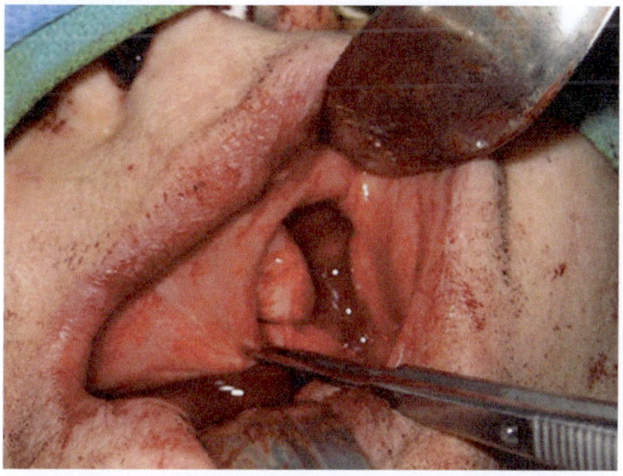

Fig. 8.25 Buccal mucosal flap is easily advanced into the palatal region using the submucosal dissection technique

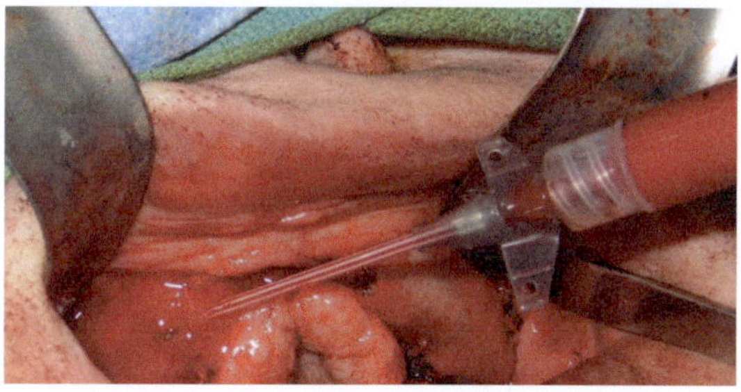

Fig. 8.26 Remaining BMAC is combined with calcium chloride and thrombin for coagulation and is then delivered over the graft with a syringe

The surgeon should also obtain an immediate postoperative radiographic image (Fig. 8.28).

11. After a period of osteointegration (authors prefer 9 months for this type of reconstruction), cone-beam imaging shows more than sufficient vertical and horizontal augmentation to facilitate dental implant placement (Fig. 8.29). The reconstruction site is exposed, and the mesh is removed (Fig. 8.30). When removing the mesh, the surgeon should be aware that the reconstructed bone is normally type IV quality bone at this point of time. As such, removal of the mesh must be performed delicately so as not to lose any of the grafted bone. After removal of fixation screws, the authors find using a #9 periosteal elevator to gently pry the mesh from the bone and then using a wire driver to peel the

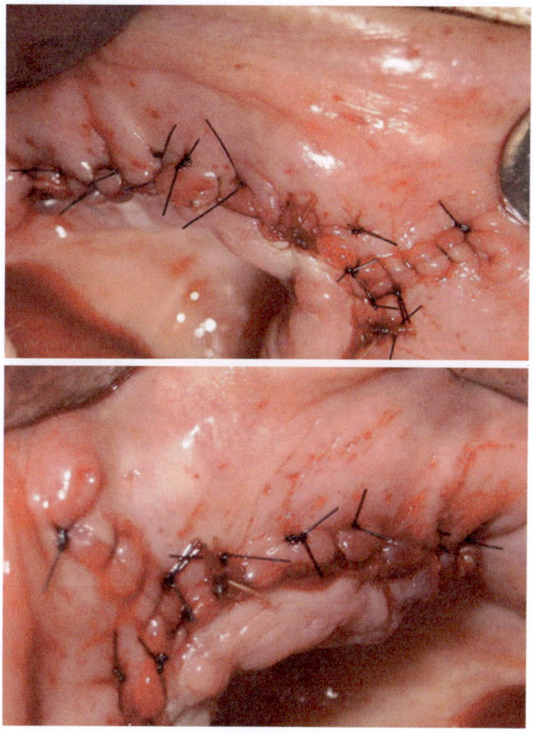

Fig. 8.27 Watertight closure of the mucosa using chromic gut and nylon suture. A V-Y closure was done to prevent shortening of the lip

mesh in a "tuna can" fashion to be most effective (see Fig. 8.31; this is a separate example and is not of the documented case). At 9 months, the surgeon will also note fibrinous ingrowth into the mesh, which also makes its removal slightly more challenging. Once the reconstructed maxilla is exposed, planned dental implants are completed (Fig. 8.32).

8.2.2.3 Case 3

Tissue-engineered reconstruction of the localized maxillary horizontal and vertical defect utilizing allogenic cancellous bone in particulate form, bone morphogenetic protein (rhBMP-2), platelet-rich plasma (PRP), a titanium crib, and type I collagen membrane.

A 34-year-old male presents with maxillary and mandibular defects as a result of a motor vehicle accident 3 years ago (Fig. 8.33). Facial fracture was repaired, and unsalvageable teeth were extracted at the time of accident. Patient returned to clinic for reconstruction of dentoalveolar defects. After discussion with the restorative dentist and the patient, the reconstructive plan consists of using tissue-engineered construct of allogeneic cancellous bone in particulate form, rhBMP-2, and PRP.

Reconstruction of the isolated maxillary defect follows similar principles as discussed previously. The reconstruction of a localized maxillary defect differs from the reconstruction of a generalized

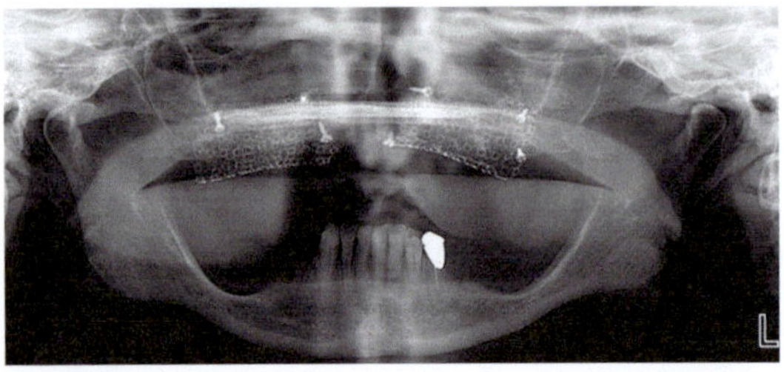

Fig. 8.28 Postoperative panoramic radiograph showed tissue-engineered graft secured by titanium mesh and screws

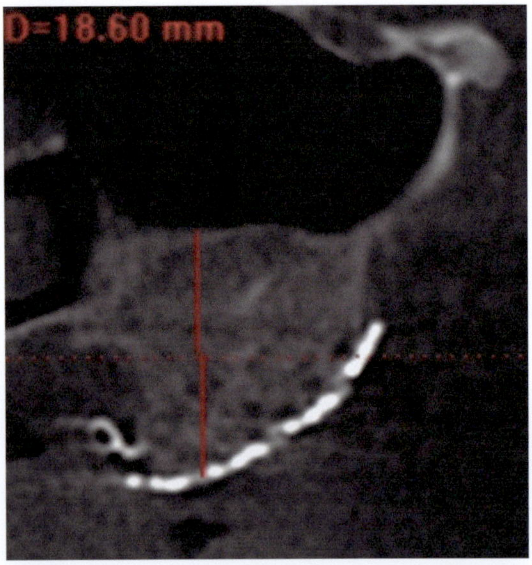

Fig. 8.29 Cone-beam CT after 6 months of healing demonstrated increased horizontal and vertical bone height

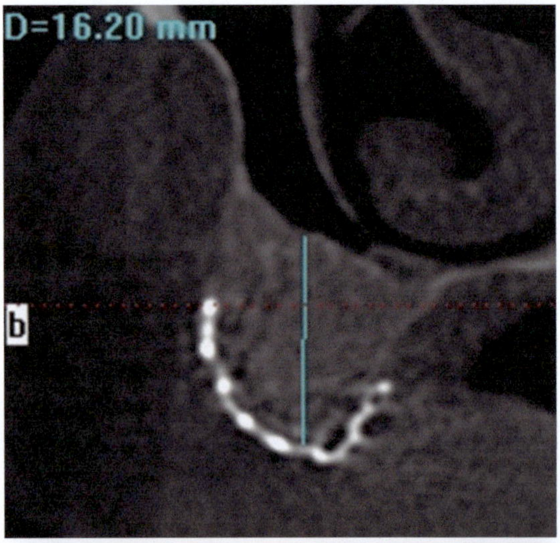

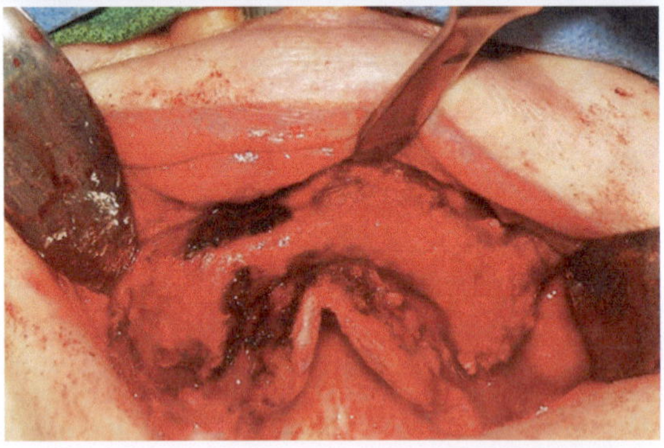

Fig. 8.30 Exposure of maxillary alveolar ridge 9 months postoperatively demonstrated excellent bone volume (titanium mesh was also removed at this appointment)

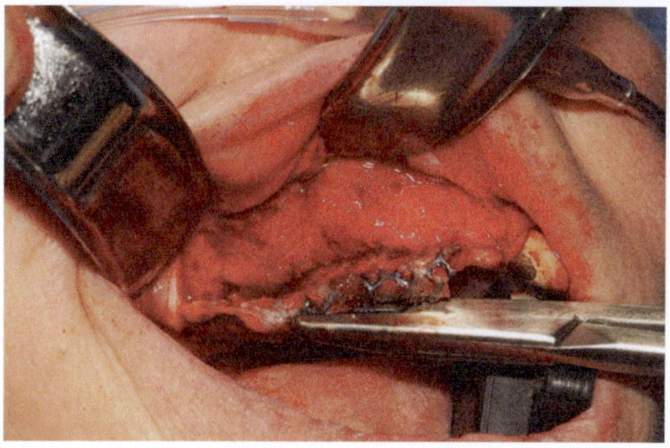

Fig. 8.31 After removal of fixation screws, a #9 periosteal elevator was used to gently pry the mesh from the bone, and then a wire driver was used to peel the mesh in a "tuna can" fashion. Note: this image is derived from a separate case and not representative of the case documented here

maxillary defect primarily due to the presence of adjacent teeth and the need to preserve their associated periodontium. In addition, due to the smaller nature of the area of reconstruction, the authors find that the use of PRP can be utilized instead of BMAC.

Armamentarium

1. Small rhBMP-2 (INFUSE) kit (total graft volume 2.8 cc, total rhBMP 4.2 mg).
2. 15 cc unit of cancellous bone cubes from a major tissue bank (the authors have access to this bone from either the Musculoskeletal Tissue Foundation or the University of Miami Tissue Bank). The cancellous cubes are milled down to 1.0–2.0 mm particulate graft.

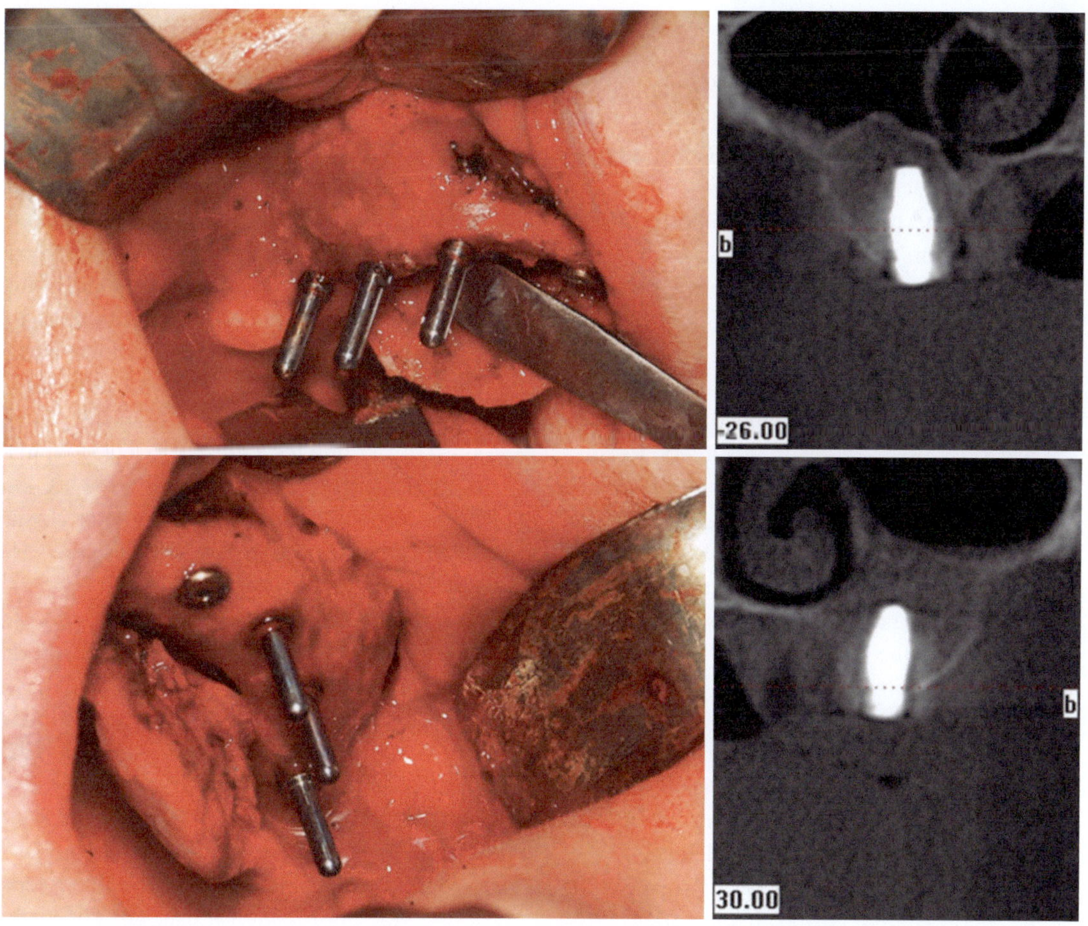

Fig. 8.32 Dental implants were placed in accordance with the restorative plan. Postoperative CBCT demonstrates good placement of implants in well-integrated, horizontal- and vertical-augmented bone

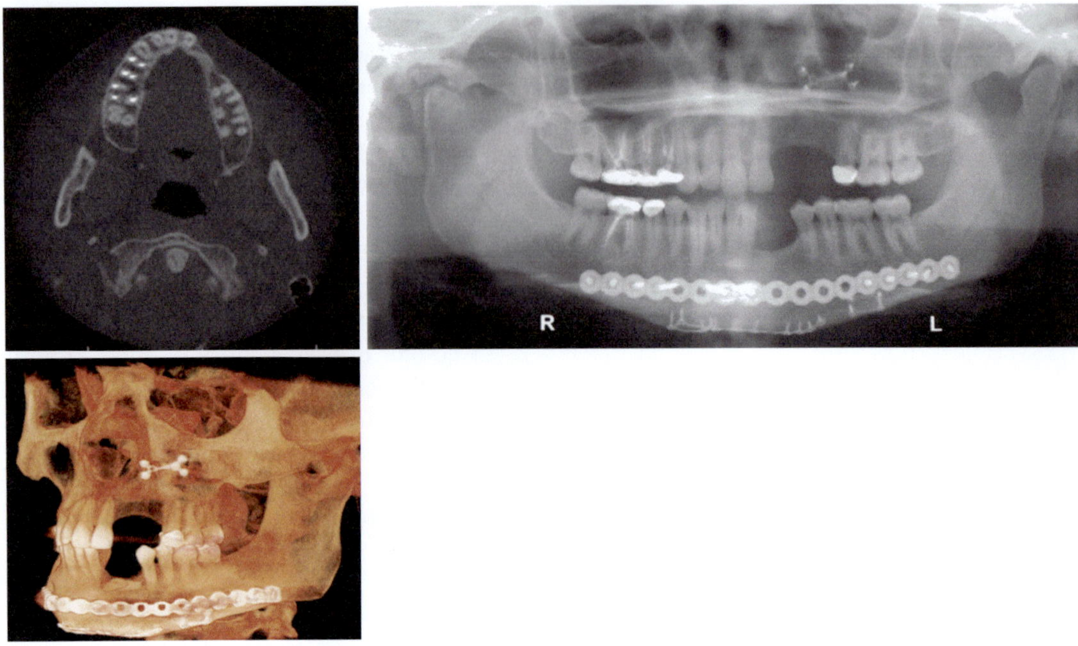

Fig. 8.33 Preoperative radiographic images demonstrate insufficient bone in both vertical and horizontal dimensions

3. 60 cc Harvest® Platelet-Rich Plasma Procedure Pack.
4. Titanium mesh 0.4 mm malleable (preferred by the authors).
5. Type I collagen cross-linked membrane.

Protocol

1. The patient is screened and medically optimized for tissue engineering surgery. Relative and absolute contraindications to PRP and rhBMP-2 use are assessed and ruled out. Additionally, off-label use to BMP outside of FDA-approved procedures, i.e., sinus augmentation and extraction socket preservation, must also be informed and consented by the patient. The surgeon must also advise the patient of postoperative sequelae such as moderate swelling and need for meticulous oral hygiene.
2. Peridex 0.12% is poured into the mouth and left for a minute prior to suctioning in order to decrease intraoral bacterial load. 2 g of ampicillin or Ancef can be administered in the perioperative period. If the patient has a penicillin allergy, 900 mg of IV clindamycin can be used.
3. Figure 8.34 shows a localized vertical and horizontal maxillary defect of both soft and hard tissues. The planned incision is marked using a surgical marking pen. The surgeon must account for the adjacent teeth and associated periodontium, as well as the need to keep the vertical releasing incisions away from the planned mesh edges. A crestal incision is made with oblique releasing incisions, providing for a greater vascular supply to the mobilized flap as well as allowing for tissue mobilization to facilitate local advancement and primary closure of the surgical site (Fig. 8.34).
4. Freeze-dried particulated allogenic cancellous graft is first hydrated with a portion of the PRP (Fig. 8.35). Any unabsorbed PRP is recollected for future use. The reconstituted rhBMP-2 is cut into small segments using a tissue pickup and tissue scissor. The reconstituted allogeneic bone, PRP, and the segmented pieces of rhBMP are then combined into a tissue engineering construct and kept to the side until ready for use. The construct can be kept non-implanted, in room temperature for up to 4 h and still maintain viability.

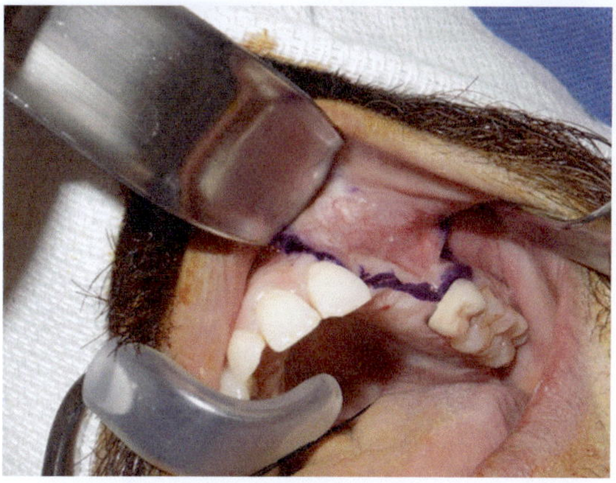

Fig. 8.34 Incision was outlined using a marking pen. Mid-crestal-type incision with distobuccal releasing incisions is performed in this case. The surgeon must be mindful of adjacent teeth and their associated periodontium to avoid iatrogenic injury

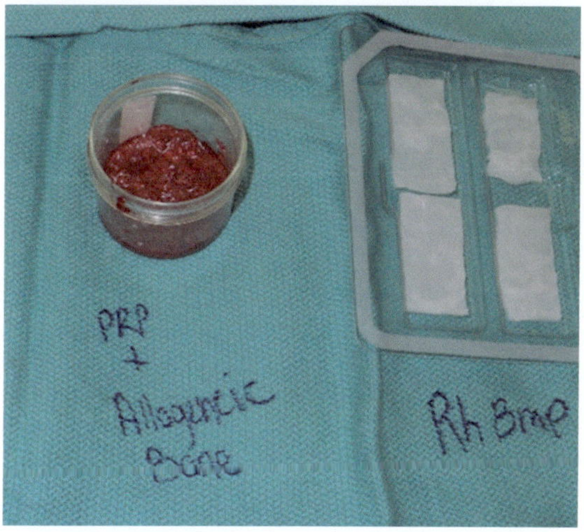

Fig. 8.35 Freeze-dried particulated allogenic cancellous graft is first hydrated with a portion of the PRP. rhBMP is also shown here and is cut into smaller segments

5. Titanium mesh was trimmed and contoured to the desired geometrical shape of the alveolar ridge in both horizontal and vertical bone dimensions. To protect the marginal gingiva of the adjacent teeth, the proximal and distal extents of the mesh should lie short of the adjacent teeth. The mesh must have stable points of contact on the palate as well as superiorly on the facial bone to facilitate mesh stabilization. The surgeon will locate

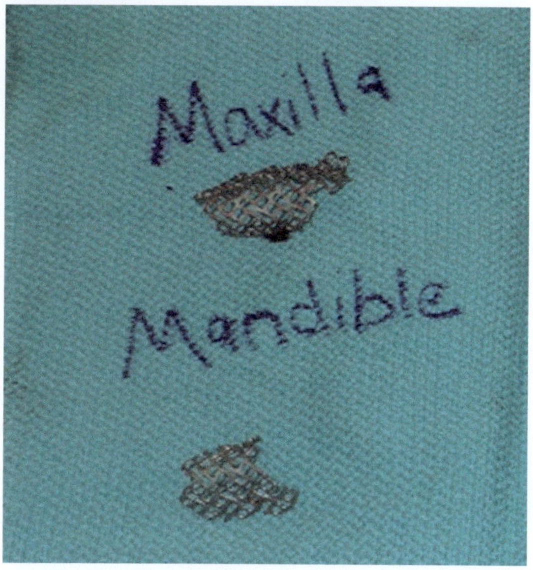

Fig. 8.36 Titanium mesh is trimmed and shaped to the correct geometrical shape of the maxillary and mandibular defects. Sharp edges must be removed using a sharp wire cutter, an oval bur, or a bovie scratch pad

areas that are acceptable for screw fixation. It is important to remove any sharp edges from the mesh to prevent tissue trauma and graft exposure in the post-op period. This can be facilitated using a sharp wire cutter, an oval bur, and bovie scratch pad (Fig. 8.36).

6. The tissue-engineered construct was then packed into the titanium crib. The composite graft with the contoured crib was then placed into the alveolar ridge of the maxilla. The crib is then secured using 1.5 mm titanium screws of 5 mm of length to engage the native maxillary bone (Fig. 8.37). Additional composite graft material is then packed from the anterior and posterior crib access points condensing the material as it is placed. If there is excess composite graft material, the remaining rhBMP is localized and is placed into the graft.
7. A submucosal dissection is then completed on the buccal flap and the flap is advanced. The buccal flap should effortlessly advance into the palatal region (Fig. 8.38).
8. Cross-linked collagen membrane was then placed to cover the mesh/graft composite making sure to cover both facial and palatal

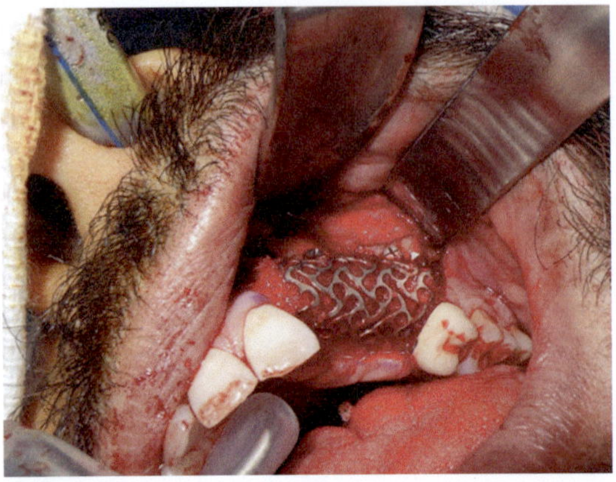

Fig. 8.37 The tissue-engineered construct is packed into the titanium mesh and placed in the target defect. Titanium screws were used to secure the mesh and must adequately engage the native bone

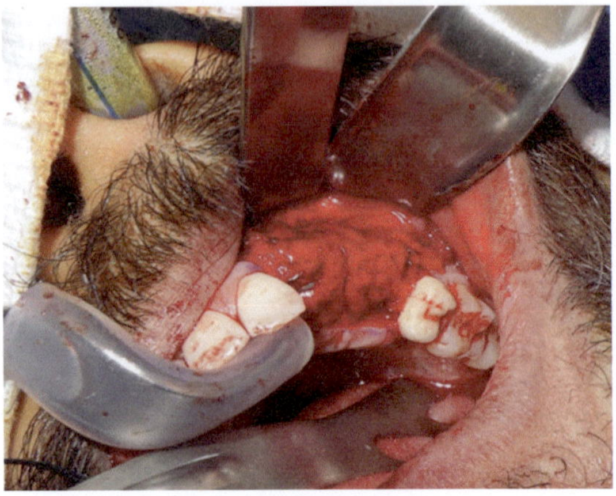

Fig. 8.39 Remaining PRP is mixed with calcium chloride and thrombin and then delivered onto the graft construct prior to closure

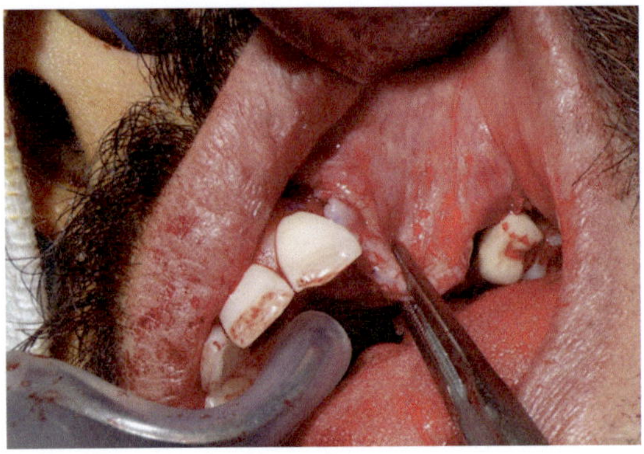

Fig. 8.38 Submucosal dissection allows for extension of the buccal mucosal flap toward the palatal bone. The tissue-engineered construct is adequately covered with the flap in a tension-free manner

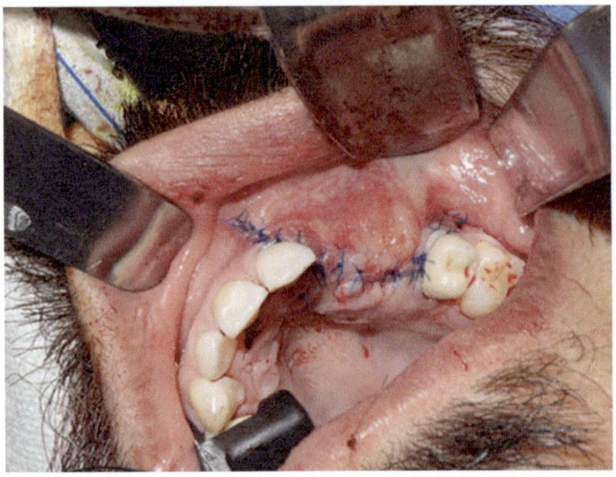

Fig. 8.40 Primary closure using 3-0 nylon and 3-0 chromic gut was done in a watertight manner. There must not be excessive tension on the mucosa due to the expected postoperative edema. If more release is needed, submucosal dissection should be extended for recruitment of additional tissue

surfaces. Any remaining PRP is then combined with calcium chloride and thrombin to coagulate and activate the PRP, and this is then delivered over the graft prior to closure (Fig. 8.39).

11. Closure is then facilitated using 3-0 nylon sutures in a horizontal mattress fashion for primary mucosal coverage. The wound is over-sutured using 3-0 chromic gut on a tapered needle. A watertight closure is critical (Fig. 8.40). The use of tissue sealants is not usually needed yet can be used per surgeon's choice.
12. Discharge medications should be given to patient including analgesics, antibiotics, and Peridex 0.12% oral rinse. Sinus precaution instructions must also be given to the patient. The surgeon should also obtain an immediate postoperative radiographic image (Fig. 8.41).
13. After 4 months of osteointegration, follow-up cone-beam imaging shows more than sufficient vertical and horizontal augmentation

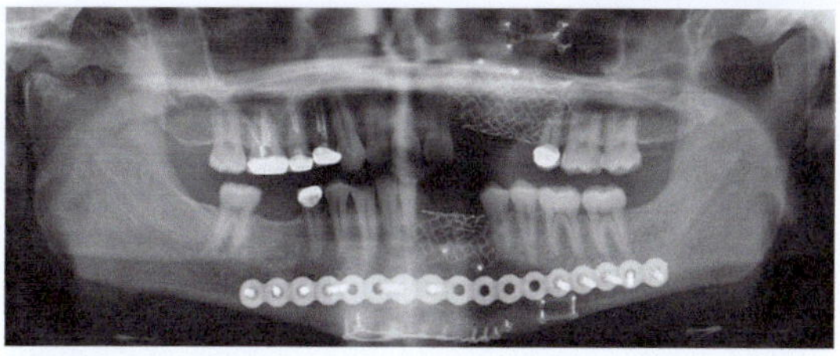

Fig. 8.41 Postoperative panoramic shows tissue-engineered graft secured by titanium mesh and screws

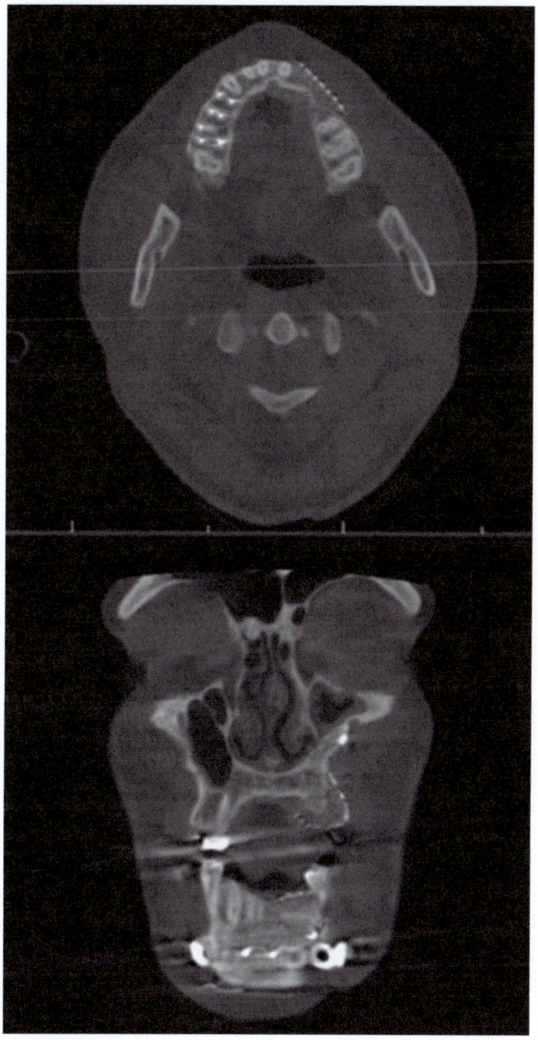

Fig. 8.42 Cone-beam CT image taken after 4 months of healing demonstrates integrating bone graft with adequate vertical and horizontal volume. Patient will return to the clinic in 3–5 months for removal of mesh and placement of dental implants

to facilitate dental implant placement (Fig. 8.42). Patient will return to clinic in 3–5 months for placement of dental implants and removal of titanium mesh.

8.2.3 Discussion

Historically there have been many approaches to the reconstruction of the fully edentulous and atrophic maxilla as well as the extensive localized maxillary defect. These included autogenous block grafting, Le Fort osteotomy with autogenous interpositional grafts, and mesh reconstruction with particulate autogenous bone with or without sinus lifts. Although outcomes can be satisfactory using these techniques, one consistency with each of the aforementioned techniques is the need for harvesting usually large bone volumes from patient donor sites. Traditional harvest sites include anterior iliac crest and posterior iliac crest, as well as calvarium. Autogenous bone grafting has traditionally been the gold standard and the only way to maximize the probability of a successful outcome. With the advent of tissue engineering principles and sound surgical techniques, the authors have found and demonstrated here that maxillary reconstruction can be predictably achieved utilizing this innovative approach.

References

1. Emes Y, Aybar B, Yalcin S. On the evolution of human jaws and teeth: a review. Bull Int Assoc Paleodont. 2011;5(1):37–47.
2. Pape HC, Evans A, Kobbe P. Autologous bone graft: properties and techniques. J Orthop Trauma. 2010;24:S36–40.
3. Boyne PJ, Mikels TE. Restoration of alveolar ridges by intramandibular transposition osseous grafting. J Oral Surg. 1968;26(9):569.
4. Boyne PJ. Regeneration of alveolar bone beneath cellulose acetate filter implants. J Dent Res. 1964;43:827.
5. Sheikh Z, et al. Natural graft tissues and synthetic biomaterials for periodontal and alveolar bone reconstructive applications: a review. Biomater Res. 2017;21(1):9.

6. Rispoli L, et al. Surgery guidelines for barrier membranes in guided bone regeneration (GBR). J Otolaryngol Rhinol. 2015;1:1–8.
7. Anitua E, Andia I, Ardanza B, Nurden P, Nurden AT. Autologous platelets as a source of proteins for healing and tissue regeneration. Thromb Haemost. 2004;91(1):4–15.
8. Marx RE, Carlson ER, Eichstaedt RM, Schimmele SR, Strauss JE, Georgeff KR. Platelet-rich plasma: growth factor enhancement for bone grafts. Oral Surg Oral Med Oral Pathol Oral Radiol Endod. 1998;85(6):638–46.
9. Marx RE. Platelet-rich plasma (PRP): what is PRP and what is not PRP? Implant Dent. 2001;4(4):225–8.
10. Horowitz R, Holtzclaw D, Rosen PS. A review on alveolar ridge preservation following tooth extraction. J Evid Based Dent Pract. 2012;12(3):149–60.
11. Hämmerle CHF, Araújo MG, Simion M. Evidence-based knowledge on the biology and treatment of extraction sockets. Clin Oral Implants Res. 2012;23(s5):80–2.
12. Cawood JI, Howell RA. A classification of the edentulous jaws. Int J Oral Maxillofac Surg. 1988;17(4):232–6.
13. Sharan A, Madjar D. Maxillary sinus pneumatization following extractions: a radiographic study. Int J Oral Maxillofac Implants. 2008;23(1):48.
14. Varela-Centelles P, et al. Detection of the posterior superior alveolar artery in the lateral sinus wall using computed tomography/cone beam computed tomography: a prevalence meta-analysis study and systematic review. Int J Oral Maxillofac Surg. 2015;44(11):1405–10.
15. Kang S-J, et al. Anatomical structures in the maxillary sinus related to lateral sinus elevation: a cone beam computed tomographic analysis. Clin Oral Implants Res. 2013;24(A100):75–81.
16. Mardinger O, et al. Morphologic changes of the nasopalatine canal related to dental implantation: a radiologic study in different degrees of absorbed maxillae. J Periodontol. 2008;79(9):1659–62.
17. Rossetti PHO, Bonachela WC, Rossetti LMN. Relevant anatomic and biomechanical studies for implant possibilities on the atrophic maxilla: critical appraisal and literature review. J Prosthodont. 2010;19(6):449–57.
18. Cavallaro J, et al. Management of the nasopalatine canal and foramen associated with dental implant therapy. Compend Contin Educ Dent. 2016;38(6):367–72.
19. Peñarrocha M, et al. The nasopalatine canal as an anatomic buttress for implant placement in the severely atrophic maxilla: a pilot study. Int J Oral Maxillofac Implants. 2009;24(5):936.
20. Peñarrocha D, et al. Implants placed in the nasopalatine canal to rehabilitate severely atrophic maxillae: a retrospective study with long follow-up. J Oral Implantol. 2014;40(6):699–706.

9 Tissue Engineering Solutions for Dental Implantology, Dentoalveolar Surgery, and Maxillary Sinus Floor Augmentation

R. Gilbert Triplett, Marianela Gonzalez, and Francisco Curiel

Tissue engineering represents a new era for oral and maxillofacial surgery, and discovery is moving at a rapid pace with the goal of regeneration whereby tissues are re-established into their pre-injury state with minimal to absent native tissue scar formation while trying to minimize risk and morbidity [1]. The use of cells, scaffolds, and growth factors can be employed in innovative ways to reconstruct and regenerate anatomical deficits resulting from developmental deformities, trauma, disease, and atrophy resulting from tooth loss. Current tissue engineering goals encourage the re-establishment of three-dimensional form and function while using cells, scaffolds, biological factors, and both biomechanical and biophysical stimulation to assist in regeneration of tissues through manipulation of the host environment [2].

Bone has a natural capacity for self-repair. However, a number of variables including operative location as well as patient age, sex, and health can limit the potential for auto-regeneration of human bone. In such cases, a bone graft can be used to stimulate bone formation at the operative site. The ideal bone graft combines osteogenic cells (either endogenous or introduced) and associated bioactive osteoinductive factors with an osteoconductive structural matrix [3]. Although the properties and interactions of these three components determine the characteristics and bone-forming capacity of a graft, osteogenesis is ultimately a cellular process that begins with a unique type of multipotential adult stem cell (mesenchymal stem cell), which gives rise to osteoblasts [4]. More than 900,000 bone graft procedures are performed in the United States each year [5, 6].

Although a wide variety of products and techniques are used in bone graft surgery, all are intended to stimulate bone formation and promote bony unions at the operative site. The seminal event for all bone formation is the mesenchymal stem cell's differentiation into an osteoblast that is essential for bone formation [2].

R. Gilbert Triplett (✉) · M. Gonzalez
Department of Oral and Maxillofacial Surgery, Texas A&M University College of Dentistry, Dallas, TX, USA

Department of Surgery, Baylor University Medical Center, Dallas, TX, USA
e-mail: gtriplett@tamhsc.edu; GTriplett@bcd.tamhsc.edu

F. Curiel
Department of Periodontics, Texas A&M College of Dentistry, Dallas, TX, USA
Private Practice, Seattle, WA, USA

J. C. Melville et al. (eds.), *Regenerative Strategies for Maxillary and Mandibular Reconstruction*,
https://doi.org/10.1007/978-3-319-93668-0_9

9.1 Bone Growth Factors

Growth factors have been explored for over 50 years for specific applications to manipulate the host environment for promotion of repair and regeneration. Specific growth factors for the maxillofacial region include rhBMP-2 and rhBMP-7, platelet-derived growth factor (PDGF), vascular endothelial growth factor (VEGF), and TGF-beta [3].

Osteoinductive demineralized bone matrix (DBM), for example, stimulates bone fusion through the properties of naturally occurring bone morphogenetic proteins (BMPs), which broadcast chemical signals that induce osteogenic activity among endogenous cells [1]. However, the osteoinductive activity of various DBM preparations is inconsistent due to variation in the amount and types of BMPs contained in the preparation. Alternatively, recombinant human BMPs provide a more consistent source of osteoinductive signals, but at a much higher cost [6].

Bone morphogenetic proteins, particularly BMP-2 and BMP-7, are the most reliable of the BMPs for osseous regeneration. Rh BMP-2/ACS (Infuse)® Medtronics is FDA approved for use in the maxillofacial region for buccal wall defects in alveolar dental socket and in maxillary sinus floor. Rh BMP-7 OP1® (Stryker Corp, 2815 Airview Blvd, Kalamazoo, MI, 49002) is FDA approved for use in orthopedic indications [3]. These signaling proteins work by the recruitment and differentiation of bone-forming cells inducing new bone formation or healing existing bone. They must be paired with a suitable carrier, and numerous potential materials have been studied; however, some materials released the rhBMP-2 molecules too fast and were removed from the site before adequate chemotaxis could be realized, and some bound the molecules too strongly and released the molecules too slowly. The absorbable collagen sponge that was eventually chosen has ideal release kinetics and is biphasic—with an initial burst release of some of the protein and then a slower release as the sponge is resorbed over a period of about 20 days. The initial release in the first few hours after implantation is sufficient for the initial induction of chemotaxis. This chemotactic recruitment brings mesenchymal stem cells from adjacent bone, muscle, and periosteum to infiltrate the graft site. As the sponge degrades, additional stem cells are recruited and proliferate and differentiate into osteoblasts and begin to form trabecular bone. The MSCs increase in number, and it has been shown in vitro that the BMP-2 can increase the proliferation of several multipotent stem cells [1].

Preclinical studies support that the bone formation started by rhBMP-2/adsorbed onto an absorbable collagen sponge (ACS) is self-limiting, forming a predictable amount of bone at the site of implantation. The ability of rhBMP-2 to induce new bone formation depends on its concentration. The rate of bone formation, the amount of bone formed, and the density of the resulting bone are positively correlated with both the concentration of rhBMP-2 and the length of time that rhBMP-2 is present at the implant site [3].

The only rhBMP currently FDA approved for maxillofacial grafting is recombinant human BMP-2 (rhBMP-2). RhBMP-2 is marketed by Medtronic Biologics as Infuse® (Medtronics Inc., 710 Medtronic Pkwy, Minneapolis, MN, 55432), a device available in various kit sizes that include lyophilized rhBMP2, a diluent, and an acellular collagen sponge (ACS) carrier/matrix. The rhBMP2 is reconstituted with the diluent and then applied to the ACS at a 1.5 mg/ml concentration with a collagen binding time (minimum) of 15 min before use. Significant preclinical testing and human clinical controlled trials led to its approval [7].

Clinical experience has shown that significantly less material can be used than was used in the controlled randomized studies and by pairing it with other bone products, it becomes even more economically feasible for routine use. Dr. Robert Marx has suggested a guide similar to the one below for choosing Infuse® grafting material:

Approximate size of space	Kit size	Quantity of liquid (ml)	Dose (mg)
1–2 tooth segment	XX-small kit	0.7	1.0
2–4 tooth segments	X-small kit	1.4	2.1
4–6 tooth segment	Small kit	2.8	4.2
Segmental osseous defect 4–5 cm	Medium kit	5.6	8.4
Segmental osseous defect ≥6 cm	Large kit	8.0	12
Maxillary sinus floor grafts	X-small to small (depending on volume of augmentation needed)	1.4 – 2.8	2.1 – 4.2
Bilateral sinus grafts with significant pneumatization	Medium kit	5.6	8.4

9.2 Contraindications for Infuse® Bone Graft Surgery

Infuse® Bone Graft should not be used:

- In patients with a known hypersensitivity to rhBMP-2, to bovine type I collagen, or to any other components of the formulation
- In the vicinity of a resected or extant tumor
- In patients with any active malignancy or in patients undergoing treatment for a malignancy
- In patients with an active infection at the operative site
- In pregnant women [8]

9.3 GEM 21s®

Platelet-derived growth factor (PDGF) is stored in the alpha granules of platelets and also produced by osteoblasts, macrophages, and monocytes. It was discovered in 1974 [9, 10]. Three different isomers of PDGF exist: PDGF-αα, PDGF-ββ, and PDGF-αβ (1). PDGF-ββ has chemotactic and mitogenic effects on mesenchymal stem cells (2), cementoblasts (3), periodontal ligament (PDL) fibroblasts (4), and osteoblasts (5). Because of these properties, PDGF-ββ has been investigated as a potential biologic material to aid in periodontal soft and hard tissue regeneration [9, 11] (Luitpold Pharmaceuticals, Inc. Osteohealth Dev. One Luitpold Dr, PO Box 9001, Shirley, NY, 11967). It has been reported as a graft in maxillary sinus floor grafts and appears to be a more robust graft than other non-biologic bone substitutes. However, while some case reports have demonstrated efficacy in maxillary sinus floor grafting, controlled studies on its use have been limited to its use in periodontal defects [12].

9.3.1 Mesenchymal Stem Cells (MSCs)

Mesenchymal stem cell therapy is at the forefront of regenerative investigation given the ability to direct differentiation and enhance tissue regeneration through the provision of specific cell lines. Stem cells located throughout the body can be harvested and cultured to stimulate cell-specific differentiation and expansion. Hard tissue regeneration focuses on mesenchymal stem cells, hematopoietic stem cells, and adipose stem cells to differentiate into chondrocytes and osteoblast [3, 13].

There are three essential bone growth properties that are necessary for remodeling and repair: osteogenic cells, osteoinductive signals, and an osteoconductive scaffold. MSCs are especially amenable to therapeutic applications because they have been shown to be hypoimmunogenic and anti-inflammatory. The number and concentration of local MSCs determine the capacity for bone formation at the operative site. Therefore, direct delivery of MSCs may result in more rapid, reliable, and uniform bone formation. Through decades of intensive research, scientists and clinicians have generated an increasingly comprehensive understanding of the cellular and molecular events underlying bone formation. Drawing from this expansive knowledge base, scientists have developed path-breaking technology related to the isolation, manufacture, and therapeutic use of mesenchymal stem cells [13].

Trinity Evolution® (Orthofix International NV, Boston, MA, USA) is a proprietary allograft formulation specifically created to contain three separate elements to induce bone healing. These include (1) living osteogenic cells (both mature osteoblasts and osteoprogenitor cells), (2) an osteoconductive matrix, and (3) osteoinductive cytokines. The material is essentially a mixture of a consistent and very high concentration of mesenchymal stem cells, a fine cancellous bone matrix, and an admixture of demineralized cortical bone. In concept, osteogenesis, osteoconductivity, and osteoinduction are supplied by the three components, respectively [6, 14].

The graft itself is processed sterilely from the donor, and the mesenchymal stem cell component is concentrated and can be stored on site after shipping at −70 °C to −80 °C for up to 3 months. Studies have not been conducted in the maxillofacial region; however, individual reports of off-label use seem to be positive. It is FDA approved for use in orthopedic indications. At this time there is no intention by the manufacturer to conduct studies in the maxillofacial region.

9.4 Selection of Graft Material and Planning for Implantation

The type of graft that the surgical team select will also influence the methods of containment or stabilization. The use of study models or computer-generated models can be of significant assistance in selecting the quantity of graft needed and can assist in preparing a material for containing and stabilizing the graft. Titanium mesh is a common choice and the defect on the model can be augmented in wax and used to develop a mesh tray to confine the graft material (Fig. 9.1). Titanium bone plates and screws can be used for stabilization and can be configured in such a manner that it creates and maintains space for the graft material (Fig. 9.2). Sonic Weld Membrane® KLS Martin (Jacksonville, FL) is another method of creating a tray or barrier for confinement and stability of the graft. It is a resorbable material (poly-D-L-lactic acid) that serves as an adaptable mesh membrane secured by fixation pins of the same resorbable material in order to assist the surgeon to confine and protect the bone graft material during the maturation and regeneration phases. The ultrasonic process allows the polymer material of the pins to flow into the trabecular structures of the bone; a second ultrasonic vibration is used to weld the membrane onto the pins. The membrane can be bent to go over the ridge from lingual to buccal, making a tunnel where a bone graft material can be placed to augment the alveolar crest to the desired dimensions (Fig. 9.3).

In some instances, a site has a natural space-making configuration (intrabony defects) (Fig. 9.4), but in most cases the surgical team will need to choose a method to create space and maintain stability for the graft. The use of a barrier membrane is also an acceptable method of confining the graft, but the membrane or the graft will need enough structure or support to resist compressive forces of the repositioned flap onto the underlying graft.

9.5 Management of Soft Tissue Flaps

Augmentation of the alveolar bone requires soft tissue release in order to achieve a tension-free closure. This can be accomplished by releasing incisions and selective incision through periosteum (periosteal release). This should be performed as judicially and atraumatically as possible. The need to undermine the soft tissue flaps should be anticipated and ideally should be released prior to placing the graft. Tension on the flap may compromise the blood flow to the flap margins and contribute to ischemia and wound dehiscence.

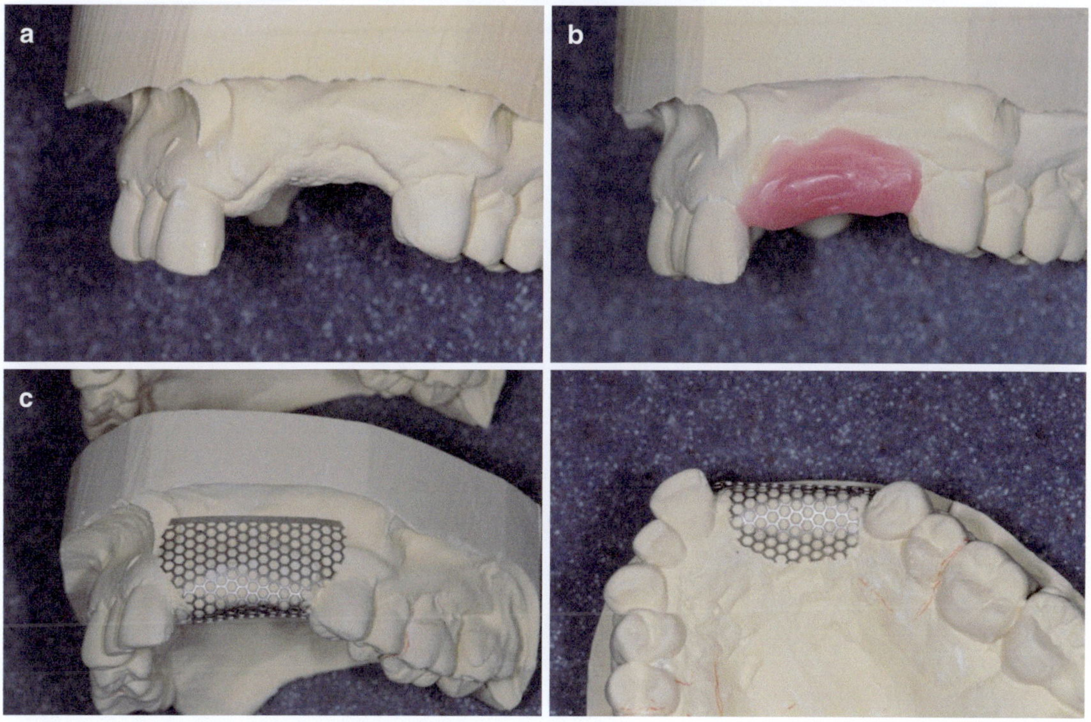

Fig. 9.1 (**a**) Stone model of maxilla demonstrating an osseous defect in sites 9–11. (**b**) Model waxed to the desired reconstruction dimensions. (**c**) Adaptation of titanium mesh to form a tray for bone graft

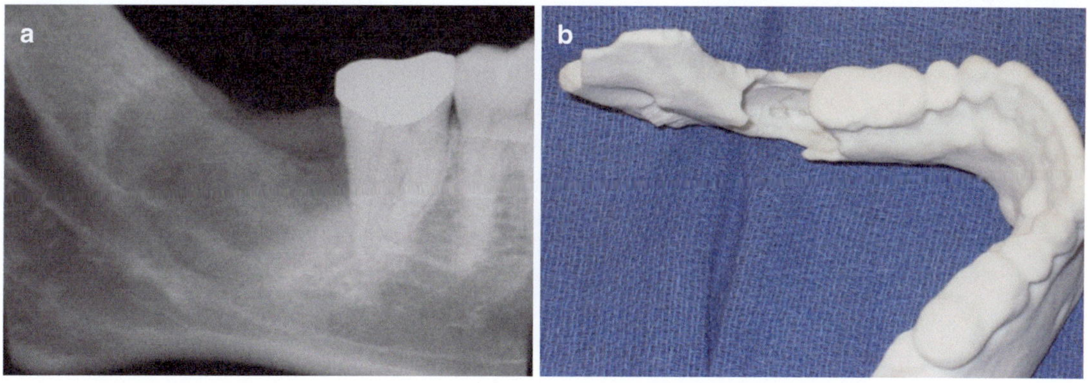

Fig. 9.2 (**a**) Radiolucent lesion from a unicystic ameloblastoma. (**b**) Computer-generated model of site which was used to help plan surgery and pre-bend reconstruction bone plate

9.6 General Guidelines for Selection of a Graft Material

The choice of graft material should be made based on the need of the site in question. Sites that have good healthy tissue—with little to no scarring—and are naturally space making and have protecting architecture will not require as robust a graft material as one without these features. For intraosseous and naturally space-making deficits, bone substitutes (allograft, xenografts, or alloplast) may perform satisfactorily, depending on osteoconduction and

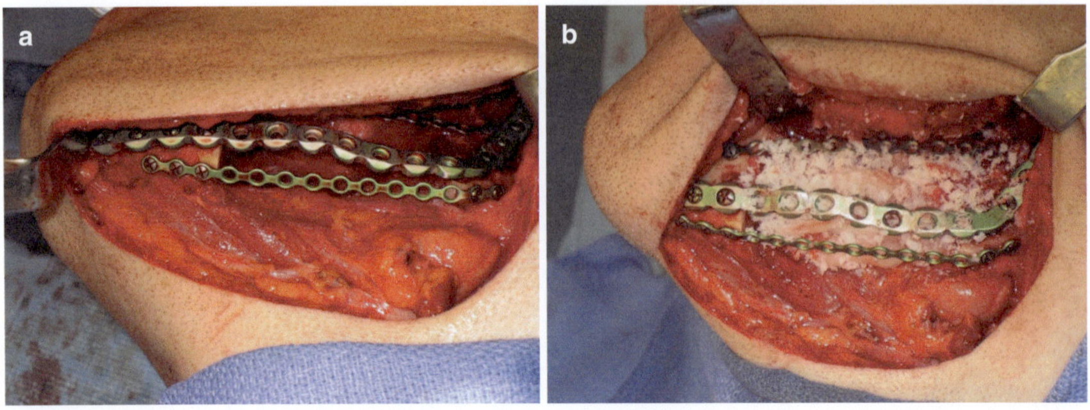

Fig. 9.3 (**a**) Mandibular resection with bone plates to stabilize segments and hold space for infuse/allograft composite. (**b**) Infuse Bone Graft is interspersed with the cancellous particulate allograft

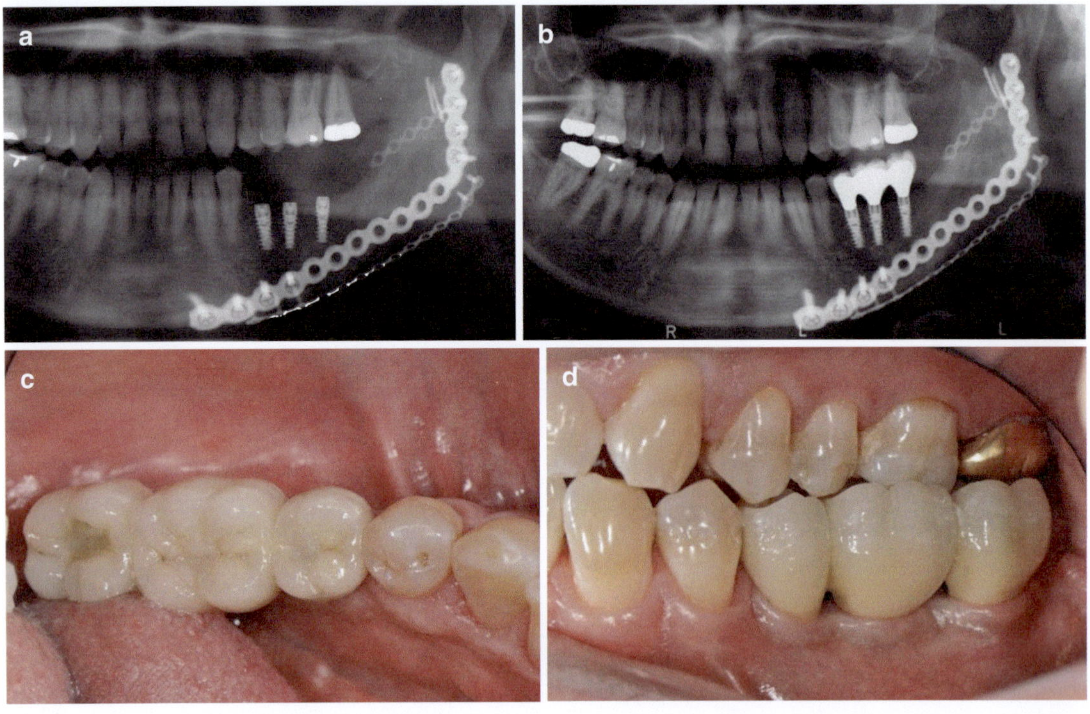

Fig.9.4 (**a**) Radiograph 6 months after grafting. Implants were placed in regenerated mandible. (**b**) Radiograph after final prosthesis. (**c**) Final prosthesis (occlusal view). (**d**) Final prosthesis (lateral view)

creeping substitution for repair and regeneration. More complex and larger deficiencies require a more active repair process. In these cases, autologous grafts or genetically engineered growth factors are indicated. It is also important to note that not all autologous bone grafts are equal. Their efficacy depends on the viable cellular elements as well as the quality and quantity of viable cells. Due to its scarcity of viable cells, a graft of dense cortical bone may not perform any better than an allograft, so intelligent informed decisions are necessary to predictably achieve the goal of viable osseous repair or regeneration.

9.7 Treatment Planning

Augmentation and regeneration of alveolar bone defects have become much more in demand and predictable because of the reconstructive options that have been made possible because of the wonder of osseointegration and implant dentistry. The success of bone regeneration is dependent on basic concepts such as site preparation (debridement, minimizing contamination, adequate vascularity to support and nourish the regenerate material) and then adaptation of the graft to the site, confinement, space maintenance, stability of the graft material, tension-free primary closure, and appropriate postoperative care.

The choice of a grafting material should be based on the need at the recipient site. Where a robust material is needed, autologous material or a biologic such as a bone growth factor, i.e., rhBMP-2 (Infuse), is preferred. In intrabony defects, an allograft or xenograft may be appropriate, whereas deficient vertical and horizontal dimensions of the alveolar or base bone may require autologous bone, a biologic material (i.e., rhBMP-2), or a combination of materials such as bone allograft + a biologic.

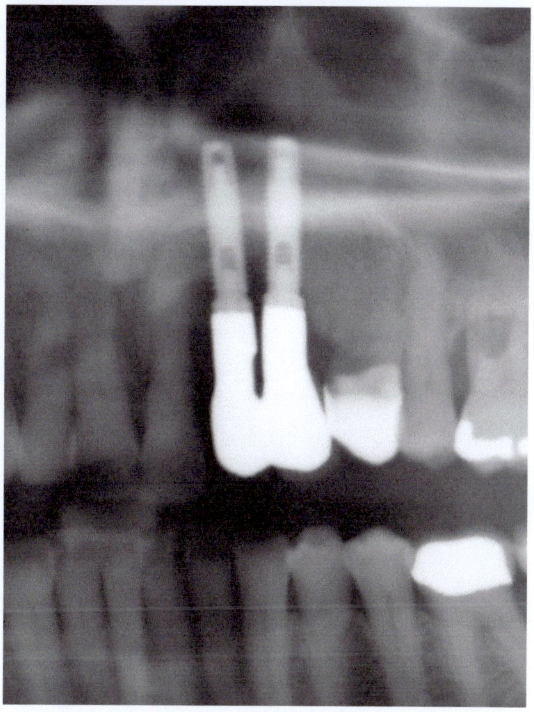

Fig. 9.5 Radiograph showing bone loss on implant in sites 9 (tooth) and 10 and 11 (implants)

Case 1 A 45-year-old male in good general health was referred for evaluation of a radiolucent lesion in the left mandibular third molar region (Fig. 9.4). The lesion was previously biopsied and diagnosed as unicystic ameloblastoma. The recommendation for treatment was resection of the lesion with immediate osseous reconstruction including removal of teeth 18–19 (Fig. 9.5). The patient was an executive who had a high-pressure job and had to be on his feet much of the day. He was given the option of autologous bone graft from iliac crest or an Infuse Bone Graft in combination with cancellous allogenic bone from the bone bank. The decision was for Infuse allograft combination. Resection was carried out, and the segments were stabilized with a pre-contoured reconstruction bone plate. Additional small (2.0 mm) bone plates were applied to the superior lateral border of mandible and at the superior crestal border to confine the graft and maintain space. A combination of cancellous particulate allograft and Infuse bone graft was used to augment this osseous defect (8.4 mg rhBMP-2) (Fig. 9.6). Patient was discharged on the first postoperative day. Postoperative course was unremarkable except for prolonged swelling. He was planned for implant placement at 6 months post resection/reconstruction. Guide splints were constructed, and implants were placed. It was noted that subjective bone density was type 3 (Branemark scale). Prosthesis was delivered at 8 months postoperatively, and patient has been seen on yearly follow-up with continued function (Fig. 9.7).

Case 2 A 48-year-old male in satisfactory general health was referred for evaluation of failing implants in sites 10 and 11 with significant crestal bone loss and loss of bone on distal root of tooth #9 (Fig. 9.8). Work-up was performed to include CBCT imaging, models, and photographs (Fig. 9.9). Plan was approved by the patient and referring restorative dentist for staged approach with removal of implants and soft tissue healing

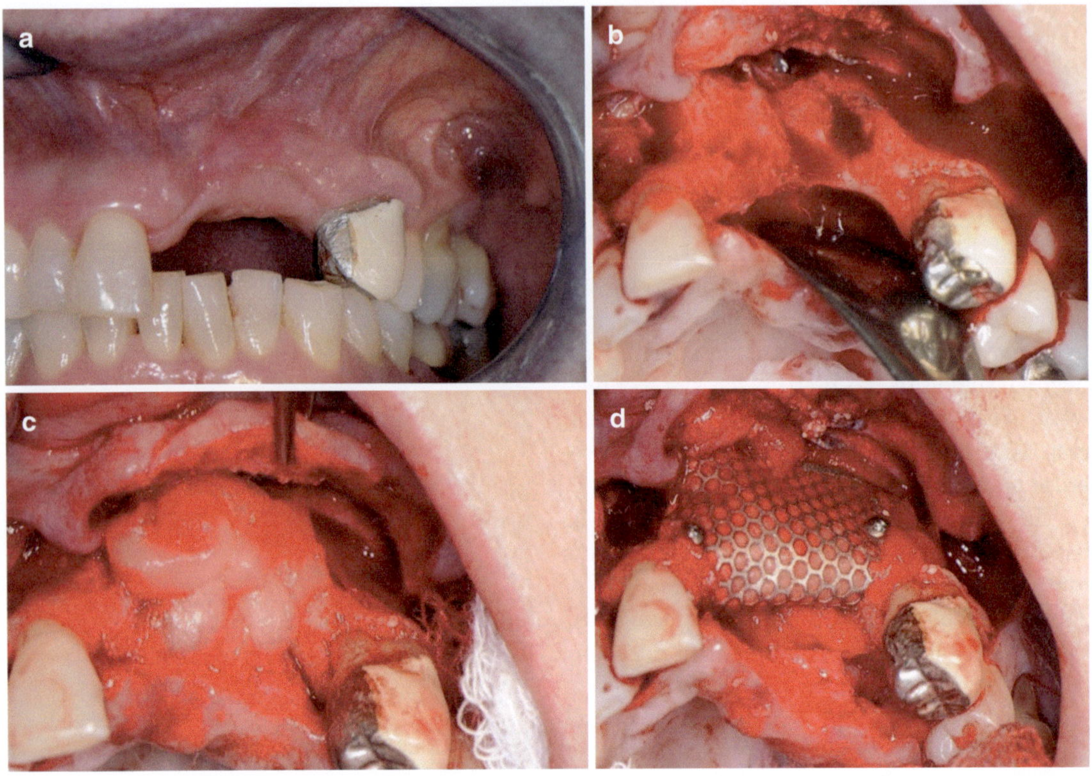

Fig. 9.6 (**a**) Residual ridge in site 2 months after implant and tooth removal. (**b**) Osseous defect from failing implants. (**c**) Infuse Bone Graft material placed in osseous defect. (**d**) Titanium tray is secured to contain and protect the graft

followed by grafting and stabilization/confinement of the Infuse graft and bone allograft (particulate cancellous) with pre-adapted titanium mesh tray for augmentation of the ridge width and height (Fig. 9.10). Postoperative CBCT made to document position of mesh and augmented ridge (Fig. 9.11). Reentry surgery for mesh removal and implant placement occurred at 7 months post grafting. After mesh removal there was a soft tissue "scar" overlying the augmented bony alveolus typical of this method, and it was removed. Using a surgical guide, implants were placed in sites 9 and 11. Subsequently patient was restored with a fixed bridge replacing teeth 9, 10, and11 (Fig. 9.12).

Case 3 The patient was a 62-year-old female in good general health with no known drug allergies who had been edentulous for 42 years. She had a mandibular overdenture supported by a 3 implant bar and clips which had been in place for 4 years. She was frustrated with the maxillary prosthesis and was seeking a stable solution to her ill-fitting maxillary denture. She was advised of a randomized clinical trial we were conducting comparing an autologous bone graft (the control) to an investigational genetically engineered bone growth factor to the maxillary sinus floor in order to create bone to support endosseous dental implants similar to the one she previously received in the mandible. After a lengthy explanation and determination that she meet the inclusion criteria, she consented and was accepted to enter the trial. By randomization she was selected for the study group with recombinant bone morphogenetic protein-2/absorbable collagen sponge (rhBMP-2/ACS) dose of 0.75 mg/ml. The total dose was 12 mg, 6 mg in each maxillary sinus floor graft. Her postoperative course was uneventful, and implants were placed 6 months after the sinus floor grafts (Figs. 9.13 and 9.14). She was followed for 20 years with an excellent outcome.

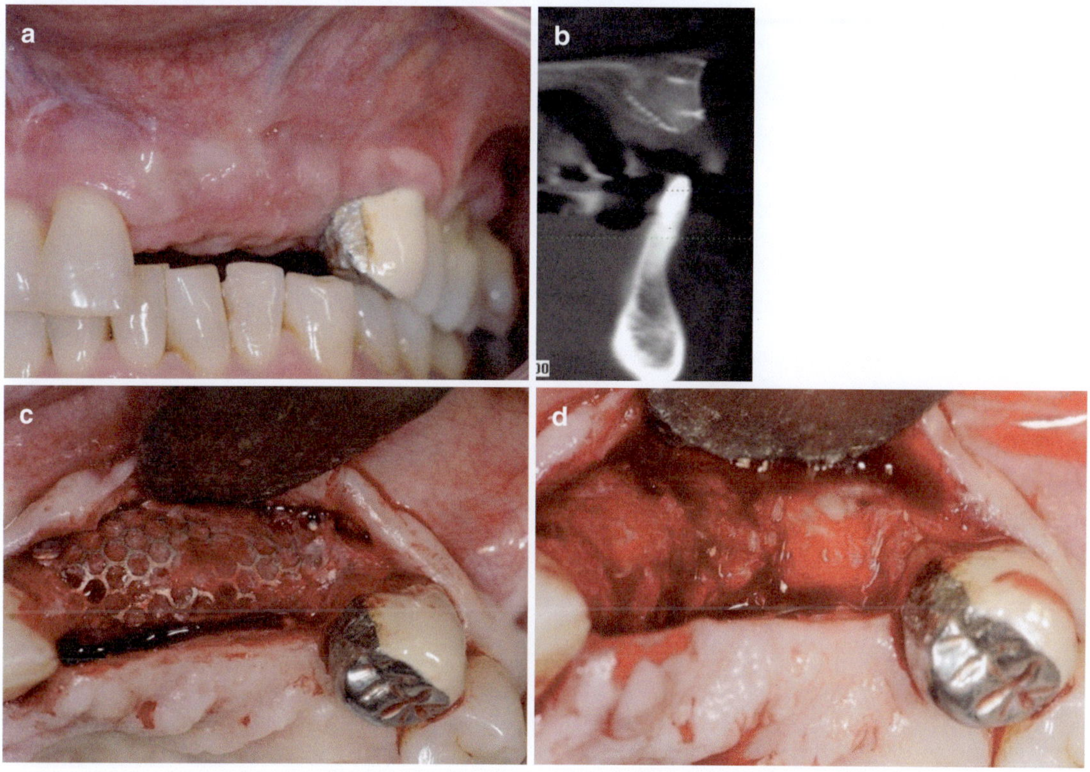

Fig. 9.7 (**a**) Clinical ridge demonstrating graft material and titanium mesh. (**b**) CT image demonstrating graft maturation. (**c**) Reentry to grafted site exposing titanium mesh. (**d**) Augmented bone in site of previous osseous defect

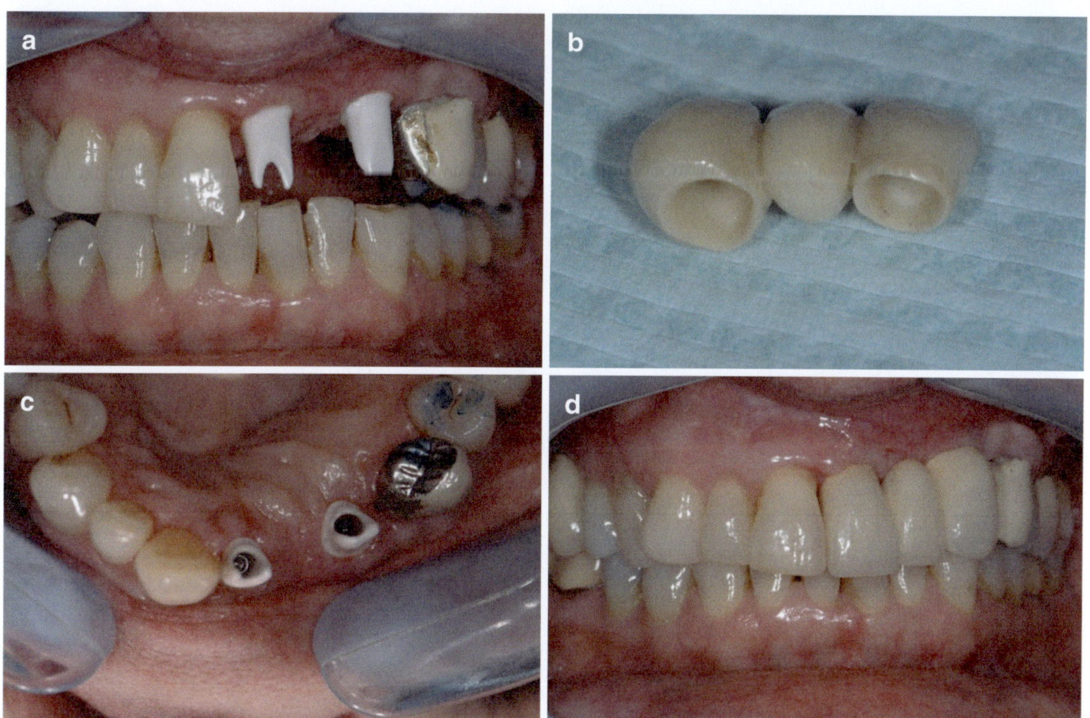

Fig. 9.8 (**a–d**) Abutments, prosthesis, and final restoration

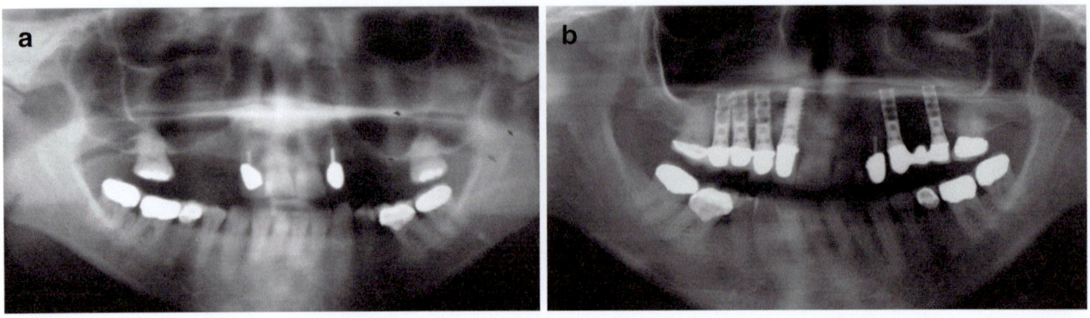

Fig. 9.9 (**a**) Pre-op radiograph of bilateral sinus floor graft with rhBMP-2 ACS. (**b**) 20-year post grafting

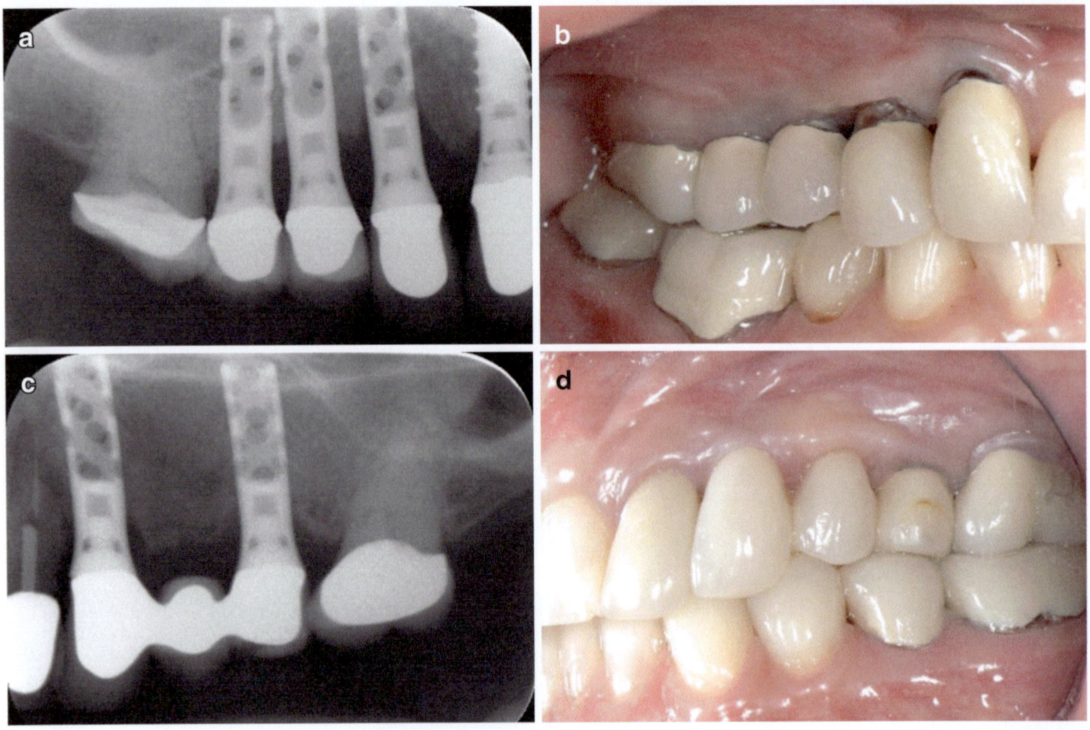

Fig. 9.10 (**a**) Radiograph at 20 years post-op right posterior quadrant. (**b**) Clinical photo at 20 years post-op right posterior quadrant. (**c**) Radiograph at 20 years post-op left quadrant. (**d**) Clinical photo at 20 years post-op left quadrant

Case 4 A 65-year-old man, heavy smoker for four decades, presented with chief complaint: "my lower teeth are loose." Clinical examination revealed that the lower anterior incisor #24 was missing and #23, #25, and #26 were indeed mobile due to loss of periodontal attachment. Periapical radiographs demonstrated vertical bone loss of 7 mm.

After evaluation and consultation, it was decided to extract teeth #23, #25, and #26 and augment the deficient alveolar ridge in this area with a bone graft consisting of Infuse® Bone Graft combined with demineralized freeze-dried bone allograft and utilization of Sonic Weld Membrane® to confine the graft. The membrane was 0.3 mm in thickness, and the pins were 1.5 × 5 mm. The resorption rate for the mesh membrane is approximately 2–6 months dependent on metabolism in individual patients. Teeth

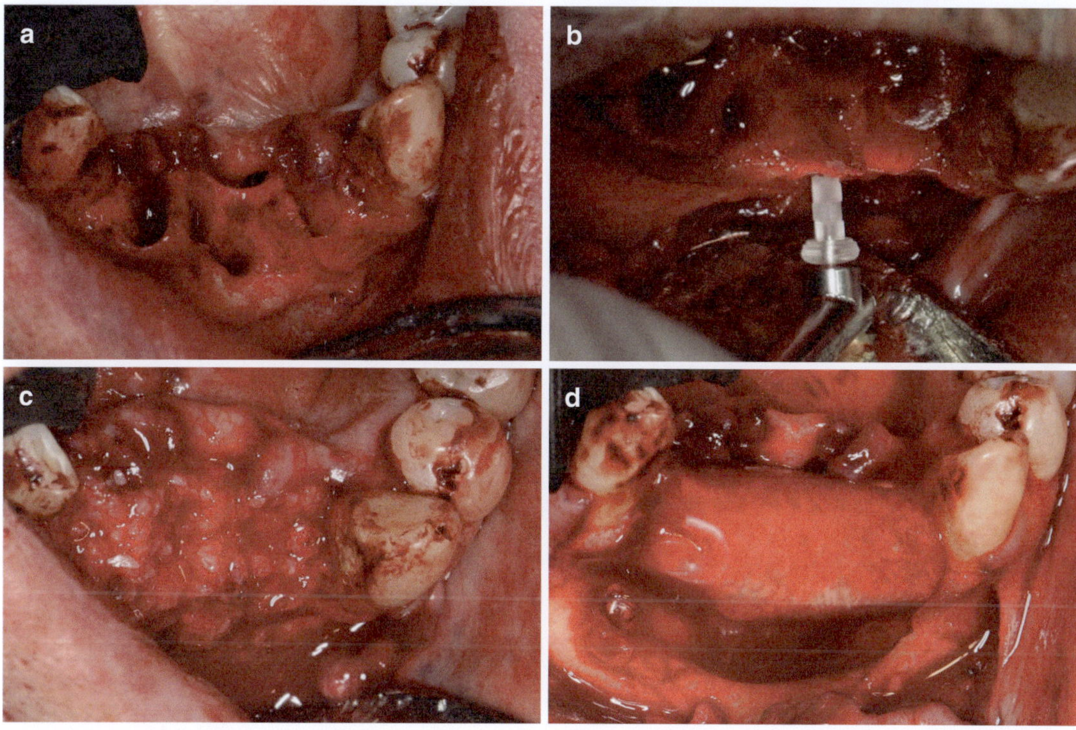

Fig. 9.11 (**a**) Osseous defect in anterior mandible 23–26 sites. (**b**) Defect with Sonic Weld pin® placed. (**c**) Infuse Bone Graft/allograft in defect. (**d**). Resorbable (Sonic Weld®) membrane in position

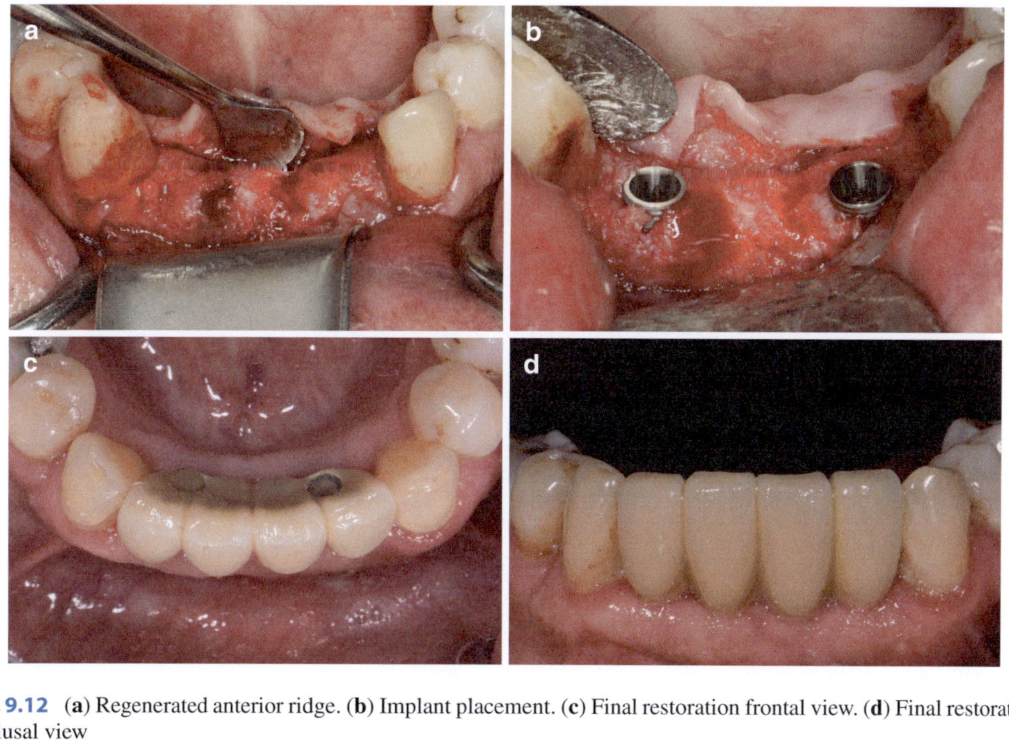

Fig. 9.12 (**a**) Regenerated anterior ridge. (**b**) Implant placement. (**c**) Final restoration frontal view. (**d**) Final restoration occlusal view

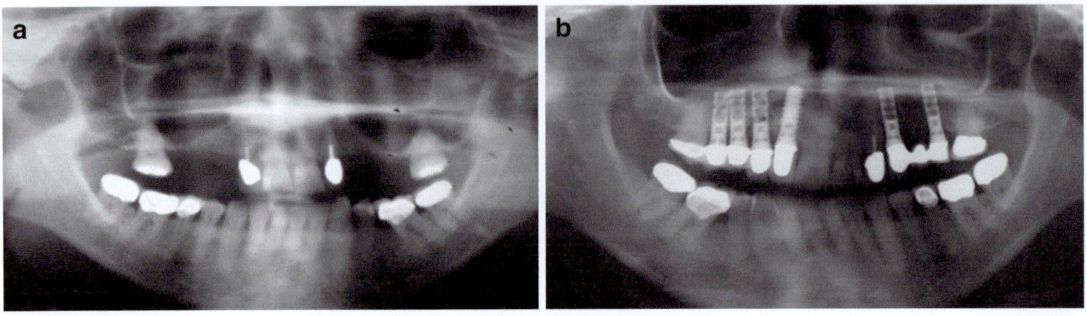

Fig. 9.13 (**a**) Pre-operative radiograph. (**b**) Radiograph post implant placement

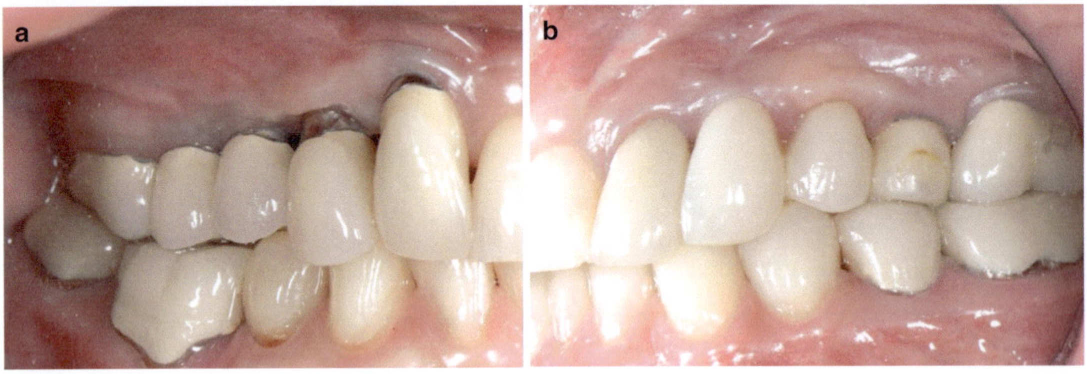

Fig. 9.14 (**a**) Final restoration on the right side. (**b**) Final restoration on the left side

were extracted; a mucoperiosteal flap was elevated; the Sonic Weld Membrane® was cut, adapted, and secured with sonic weld pins and Infuse® Bone Graft; and DFDBA (particle size 400–700 mμ) was placed. The mucoperiosteal flap was re-approximated, released where indicated, and closed. After 6 months of healing, a full-thickness mucoperiosteal flap was reflected. There was no evidence of Sonic Weld Membrane®, and two dental implants were placed in sites #23 and #26 (3.5 × 18 and 3.5 × 15, respectively). After a healing period of 4 months, the site was opened, the implants were exposed, and healing abutments were placed. An implant-level impression was made by open tray technique. A screw-retained final restoration was fabricated as a four-unit implant bridge from #23 to 26. Based on clinical examination and periapical radiographs, it was possible to determine that approximately 6 mm of alveolar bone height was regenerated. Some alveolar width was also gained but was not quantified. The increase of bone height provided sufficient quantity of bone for ideal implant placement, with sufficient soft tissue coverage over the grafted sites. No clinical complications were recorded over the 4-year follow-up period.

References

1. Smith MH, Izumi K, Feinberg SR. Tissue engineering. In: Bagheri SC, Bell RB, Khan HA, editors. Current therapy in oral and maxillofacial surgery. St. Louis: Saunders; 2012. p. 79–91.
2. Aghaloo TL, Felsenfeld AL. Principles of repair and grafting of bone and cartilage. In: Bagheri SC, Bell RB, Khan HA, editors. Current therapy in oral and maxillofacial surgery. St. Louis: Saunders; 2012. p. 19–26.
3. Stevens MR, Emam HA. Dental implant prosthesis rehabilitation: allogenic grafting/bone graft substitutes

in implant dentistry. In: Bagheri SC, Bell RB, Khan HA, editors. Current therapy in oral and maxillofacial surgery. St. Louis: Saunders; 2012. p. 157–62.

4. Wikesjo UM, Huang YH, Polimeni G, Qahash M. Bone morphogenetic proteins: a realistic alternative to bone grafting for alveolar reconstruction. Oral Maxillofac Surg Clin North Am. 2007;19(4):535–51, vi-vii.
5. Hollinger JO, Hart CE, Hirsch SN, Lynch S, Friedlaender GE. Recombinant human platelet-derived growth factor: biology and clinical applications. J Bone Joint Surg Am. 2008;90(Suppl 1):48–54.
6. Nevins M, Giannobile WV, McGuire MK, et al. Platelet-derived growth factor stimulates bone fill and rate of attachment level gain: results of a large multicenter randomized controlled trial. J Periodontol. 2005;76(12):2205–15.
7. Young CS, Ladd PA, Browning CF, et al. Release, biological potency, and biochemical integrity of recombinant human platelet-derived growth factor-BB (rhPDGF-BB) combined with Augment(TM) Bone Graft or GEM 21S beta-tricalcium phosphate (beta-TCP). J Control Release. 2009;140(3):250–5.
8. Digiovanni CW, Baumhauer J, Lin SS, et al. Prospective, randomized, multi-center feasibility trial of rhPDGF-BB versus autologous bone graft in a foot and ankle fusion model. Foot Ankle Int. 2011;32(4):344–54.
9. Hollinger JO, Onikepe AO, MacKrell J, et al. Accelerated fracture healing in the geriatric, osteoporotic rat with recombinant human platelet-derived growth factor-BB and an injectable beta-tricalcium phosphate/collagen matrix. J Orthop Res. 2008;26(1):83–90.
10. Caplan AI. Why are MSCs therapeutic? New data: new insight. J Pathol. 2009;217(2):318–24.
11. Vaccaro AR. The role of the osteoconductive scaffold in synthetic bone graft. Orthopedics. 2002;25(5 Suppl):s571–8.
12. Rush SM. Trinity evolution: mesenchymal stem cell allografting in foot and ankle surgery. Foot Ankle Spec. 2010;3(3):140–3.
13. Vanichkachorn J, Peppers T, Bullard D, et al. A prospective clinical and radiographic 12-month outcome study of patients undergoing single-level anterior cervical discectomy and fusion for symptomatic cervical degenerative disc disease utilizing a novel viable allogeneic, cancellous, bone matrix (trinity evolution) with a comparison to historical controls. Eur Spine J. 2016;25(7):2233–8.
14. Baboolal TG, Boxall SA, El-Sherbiny YM, et al. Multipotential stromal cell abundance in cellular bone allograft: comparison with fresh age-matched iliac crest bone and bone marrow aspirate. Regen Med. 2014;9(5):593–607.

Alveolar Ridge Augmentation (Vertical and Horizontal) with Bioengineering

10

Tibebu Tsegga and Brian Rethman

10.1 Horizontal Ridge Augmentation with Bioengineering

Although autogenous bone grafting is arguably still considered the gold standard in dental alveolar reconstruction, there are multiple well-known disadvantages to autologous block grafts, with donor site morbidity being the most remarkable. Comparatively, allogenic corticocancellous block graft (ACCBG) can become a predictable tool for width regeneration in preparation for dental implant placement. With the clinical application of ACCBG, operative risk becomes minimized, and with careful patient selection and site placement, the success rates can approach the gold standard. There is an expanding body of literature supporting the use of allogenic block grafts as a predictable alternative to autologous grafting which also reinforces anecdotal experience.

As these case illustrations demonstrated, the replacement resorption of allogenic corticocancellous block grafts (ACCBG) varies based on initial corticocancellous ratio in response to host physiology. The ratio of the cortex to ratio in these grafts varies even from the same procurement tissue bank. The recipient sites are routinely prepared with intermarrow penetration and grouted with particulate bone graft to eliminate perimeter void defects. Chronologic progression of healing in the postoperative period usually demonstrates a gradual loss of distinction between the cortex and marrow of the allogenic block (Fig. 10.1d, e). In particular, in the anterior maxilla, the allogenic CCB graft can provide initial primary stability with a simultaneous or staged implant placement protocol and allow for maintenance of labial gingival soft tissue architecture.

10.1.1 Horizontal-Vertical Augmentation (Fig. 10.2)

10.1.2 Horizontal Augmentations with Simultaneous Implant Placement (Fig. 10.3)

10.1.3 Section Summary

As demonstrated in these clinical scenarios, the cortical portion of the ACCBG allows for a rigid point of anchorage with titanium fixation screws with a comparatively longer resorption kinetics. In addition, the cancellous portion allows for compressive contact with the basal bone which facilitates vascular infiltration [1, 2]. The long-

T. Tsegga (✉) · B. Rethman
San Antonio Military Oral and Maxillofacial Surgery, Lackland Air Force Base, United States Air Force, Texas, USA

J. C. Melville et al. (eds.), *Regenerative Strategies for Maxillary and Mandibular Reconstruction*,
https://doi.org/10.1007/978-3-319-93668-0_10

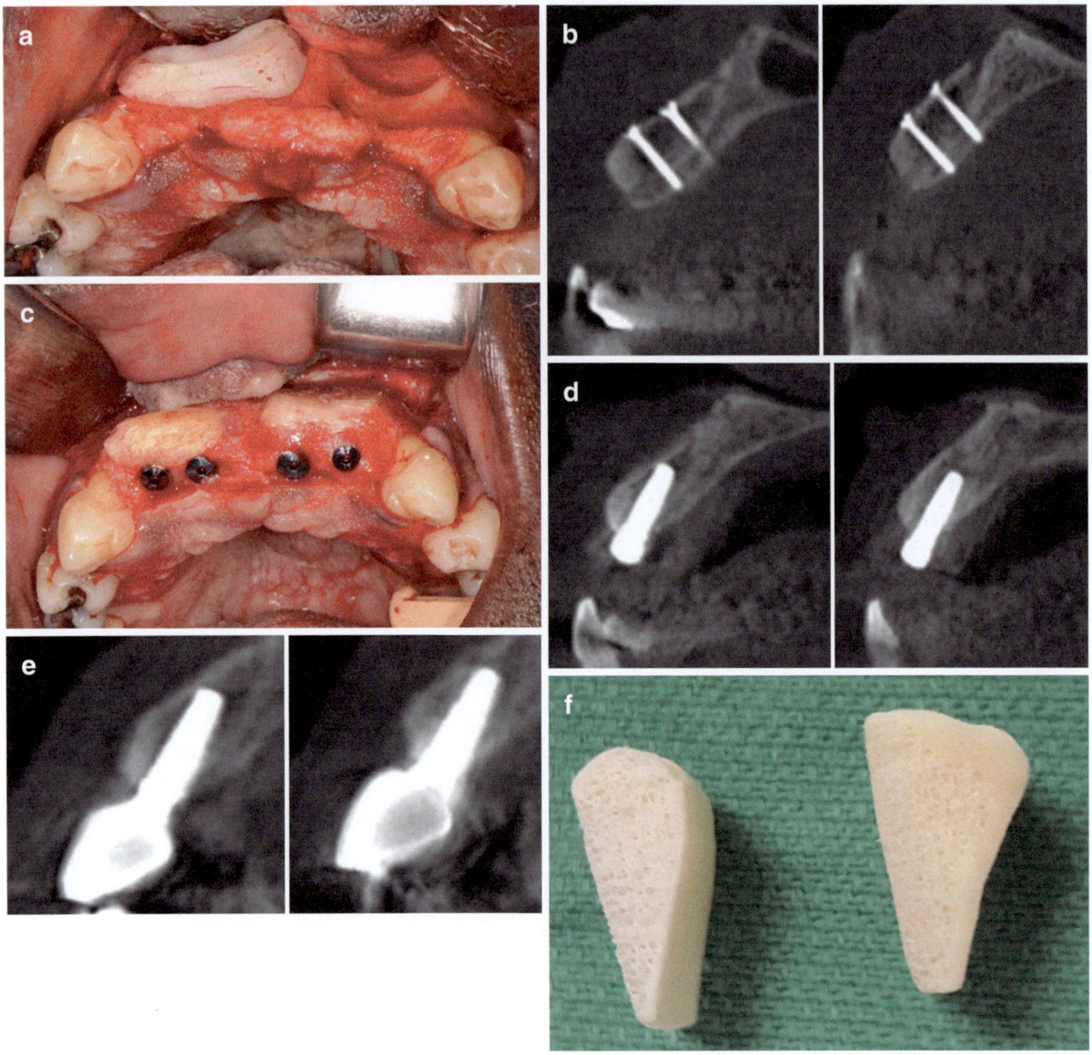

Fig. 10.1 (**a**) Edentulous anterior maxillae with placement of bone wax as anatomical template. (**b**) Radiographic cross section illustrating compression of allogenic marrow onto basal bone using lag screw fixation. (**c**) Eight months after allogenic block graft placement, note the amount of residual allograft present. (**d**) Fourteen months after block graft placement and placement of healing abutments. (**e**) Two years after placement of allogenic block graft. Note the beginning replacement resorption of the outer cortex. (**f**) Cross section of the allograft from the same manufacturer demonstrating cortical thickness variation

term sustainability of allogenic block bone has been investigated clinically and histologically. Al-Abedalla et al. conducted a cohort study to determine whether the quality of bone augmented with allograft onlay was sufficient to place dental implants and achieve success rates comparable to those in ungrafted bone. There are two cohort studies conducted: one looking at the histological quality and quantity of bone augmented with allograft onlay and the second one comparing the success rates of implants placed into allograft-augmented bone compared with autograft-augmented bone and native alveolar bone. The final conclusion demonstrates no clinical difference in the quantity and quality of allograft-augmented bone in comparison with host native alveolar bone with similar success rate of implants placed in both cohort groups [3]. In

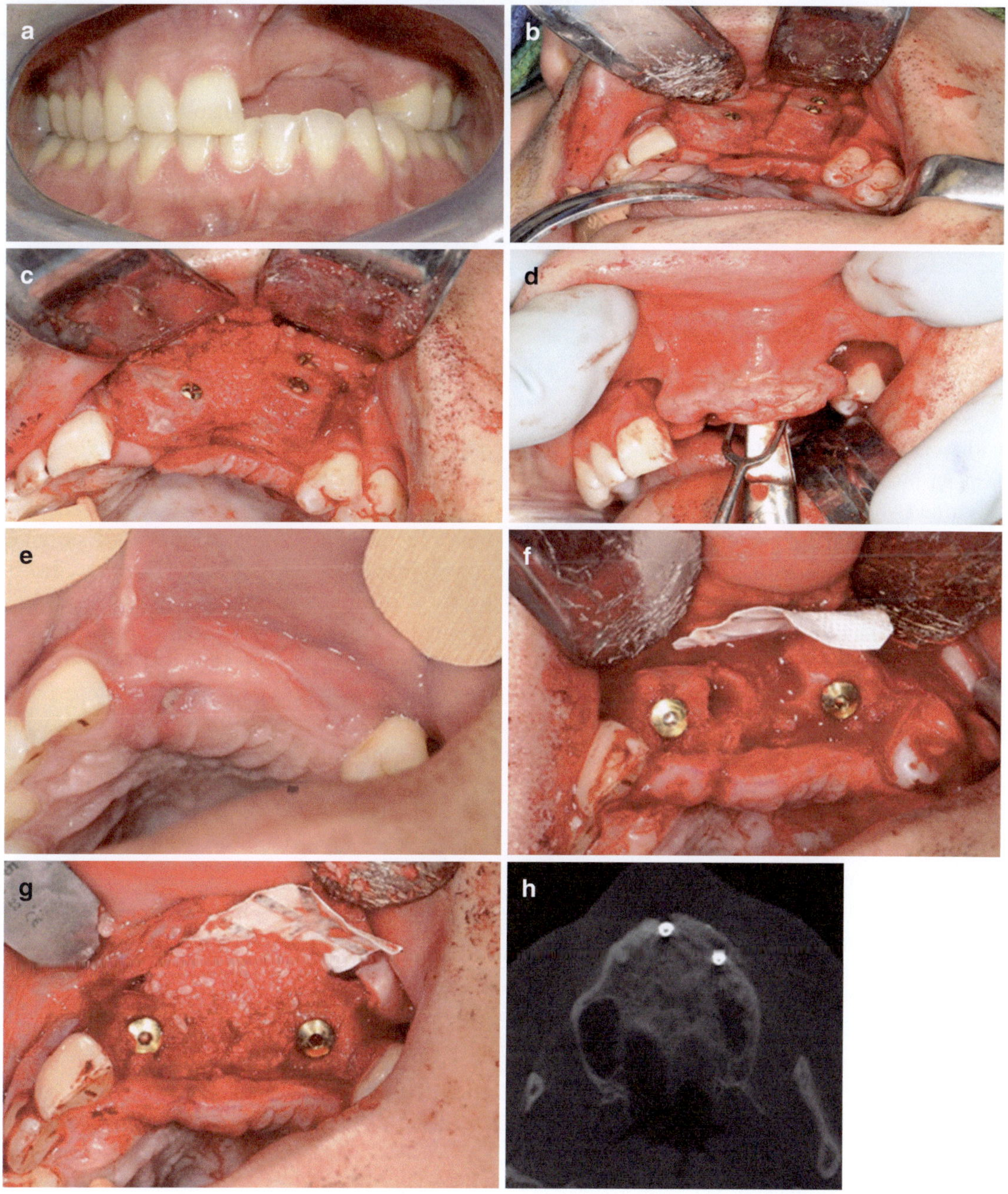

Fig. 10.2 (**a**) Clinical picture showing the combine bone loss spanning from # 9 to 12. (**b**) Two allogenic cortico-cancellous block grafts fixated using lag screw technique. Note the J-configuration of the anterior block allowing a saddlelike adaptation on the existing vertically deficient portion of the basal bone. (**c**) Interblock graft zone compacted with demineralized particulate allograft mixed with PRP. (**d**) Submucosal dissection is accomplished to allow for passive closure of expanded envelope. (**e**) Site 7 months after allogenic block graft. Note exposure in the area of the graft that was positioned for vertical augmentation. (**f**) Two root form implants placed and attempt to position one more implant were unsuccessful due to the poorly consolidated nature of the particulate bone graft that was placed in between the block grafts. (**g**) Site was regrafted using mineralized allogenic particulate bone and covered with a non-resorbable titanium-supported cytoplast membrane (Osteogenics Biomedical, Lubbock, TX). (**h**) Axial CBCT showing reestablishment of hard tissue arch form. (**i**) Six months after placement of implant at site of particulate bone graft

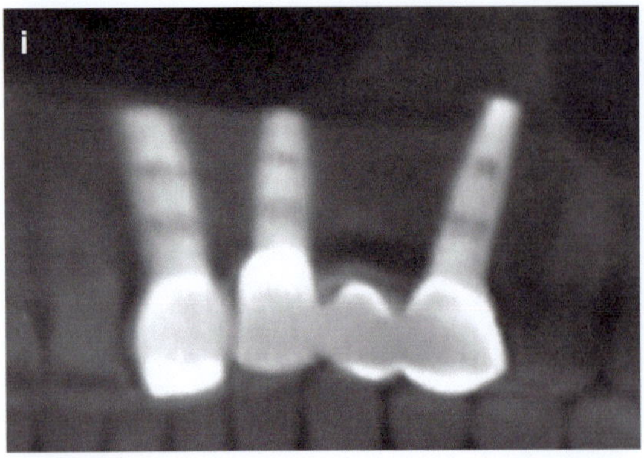

Fig. 10.2 (continued)

addition, it is important to keep in mind that implant placement using allogenic block is usually after 6–8 months of healing compared to 3–4 with an autograft. In our experience, we have noticed a slight increase in failure rates in the mandible compared to the maxillae. It is presumed that this difference is not primarily due to the variation in blood supply but more likely attributable to the oftentimes lack of keratinized marginal gingiva and increased functional muscular motion that is a characteristic of the peri-mandibular alveolus which oftentimes leads to soft tissue dehiscence.

As noted in the literature, even 6 months after placement of an allogenic block graft, there is still remnant allogenic matrix that has not undergone replacement resorption. The degree of replacement resorption at the recipient site is of course multifactorial and dependent on the volume of cortical composition of the block graft, patient-specific resorption kinetics, genetics, vascular perfusion, addition of biological adjuncts, etc. The predictable manufactured size of allogenic block grafts allows for reproducible volume expansion that can sometimes be very difficult to achieve with an autograft. Waasdorp and Reynolds conducted a systematic review to ascertain the efficacy of allogenic bone onlay and highlighted several poignant details. The basic surgical technique employed in their review was comparable with respect to defect preparation, including intramarrow penetration, graft adaptation, rigid fixation, and primary closure [4]. Graft failure most often involved mandibular posterior defects (71%), typically manifested with soft tissue dehiscence and/or infection, and occurred within 12 months after placement (86%) [5]. In the reviewed studies, the majority of the ACCBG were covered with barrier membranes, although unclear to what extent the success was attributed to guided tissue regeneration.

10.2 Vertical Ridge Augmentation

Vertical augmentation of the atrophic dental alveolus can be challenging and fraught with multiple complications. Idealized occlusal clearance to allow for a well-formed and aesthetic prosthesis is the intended goal and a prerequisite for root form implant placement. Long-standing edentulous sites are often fraught with disuse atrophy and progressive resorption, which provides insufficient vertical bone height. As with any hard tissue augmentation, the surgical pitfalls that lead to suboptimal outcomes include soft tissue dehiscence, occlusal forces (soft tissue envelope pressure), and hardware failure. More precisely, with vertical augmentation, there is an emphasis on maintaining an expanded soft tissue envelope to allow for bone regeneration either by using an autogenous or alloplastic containment device.

10.3 Alloplastic Containment Devices-Cytoplast Membrane (Osteogenics Biomedical, Lubbock, TX)

Non-resorbable containment devices have demonstrated a broad application as adjunctive tools for vertical ridge augmentation. The second-generation Gore-Tex and titanium-supported non-resorbable membranes have several features that make them very versatile and simple to utilize. Their advantage as illustrated in this case involves a 32-year-old male with vertical ridge atrophy in areas # 29 and 30 (Fig. 10.4).

The following case illustrates a 71-year-old male with multiple attempts at autogenous grafting with subsequent graft failure. In this case an attempt to recruit and concentrate the relevant

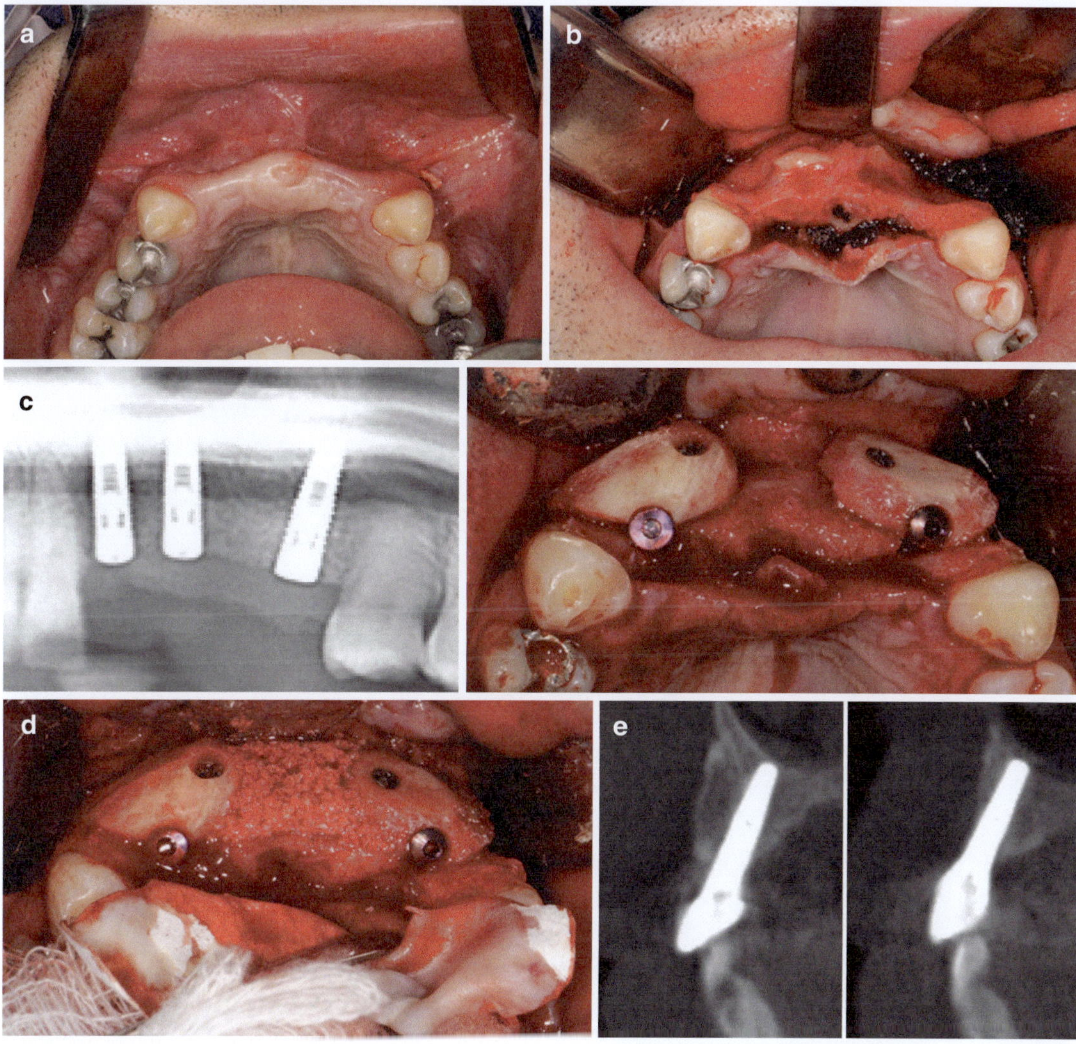

Fig. 10.3 (**a**, **b**) Anterior maxillae with horizontal ridge deficiency and failed autogenous block graft. (**c**) Recipient site prepared for placement of two allogenic CC blocks. Lag screws first placed mesial to anticipated position of endosseous implant. Implant osteotomy gradually made to full dimension and narrow platform implants placed with a minimum of 25 Ncm of torque. (**d**) Interblock region packed with mineralized allogenic particulate bone combined with Bio-Oss and PRP. Resorbable membrane placed over entire graft and closed primarily. (**e**) Cone-beam CT cross section 2 years after placement of ACCBG (sites # 7 and # 10)

osteoprogenitor mesenchymal stem cells was done by utilizing bone marrow aspirate concentrate (BMAC) [6]. The technique which is described elsewhere in this book allows for the concentration or supra-concentration of the necessary bone progenitor cells to allow for alveolar regeneration. In this case, the bone marrow aspirate volume was 60 ml; the concentrate was a total of 6 ml and was admixed with allogenic mineralized corticocancellous bone. A small amount of the activating thrombin and calcium chloride is admixed into the composite bone graft to facilitate handling and minimize spill-off BMAC. The containment device was the previously noted dPTFE cytoplast non-resorbable membrane. Nine-month consolidation period was observed in this case, and a demonstrable alveolar volume enhancement is appreciated to allow for placement of four root form implants (Fig. 10.5).

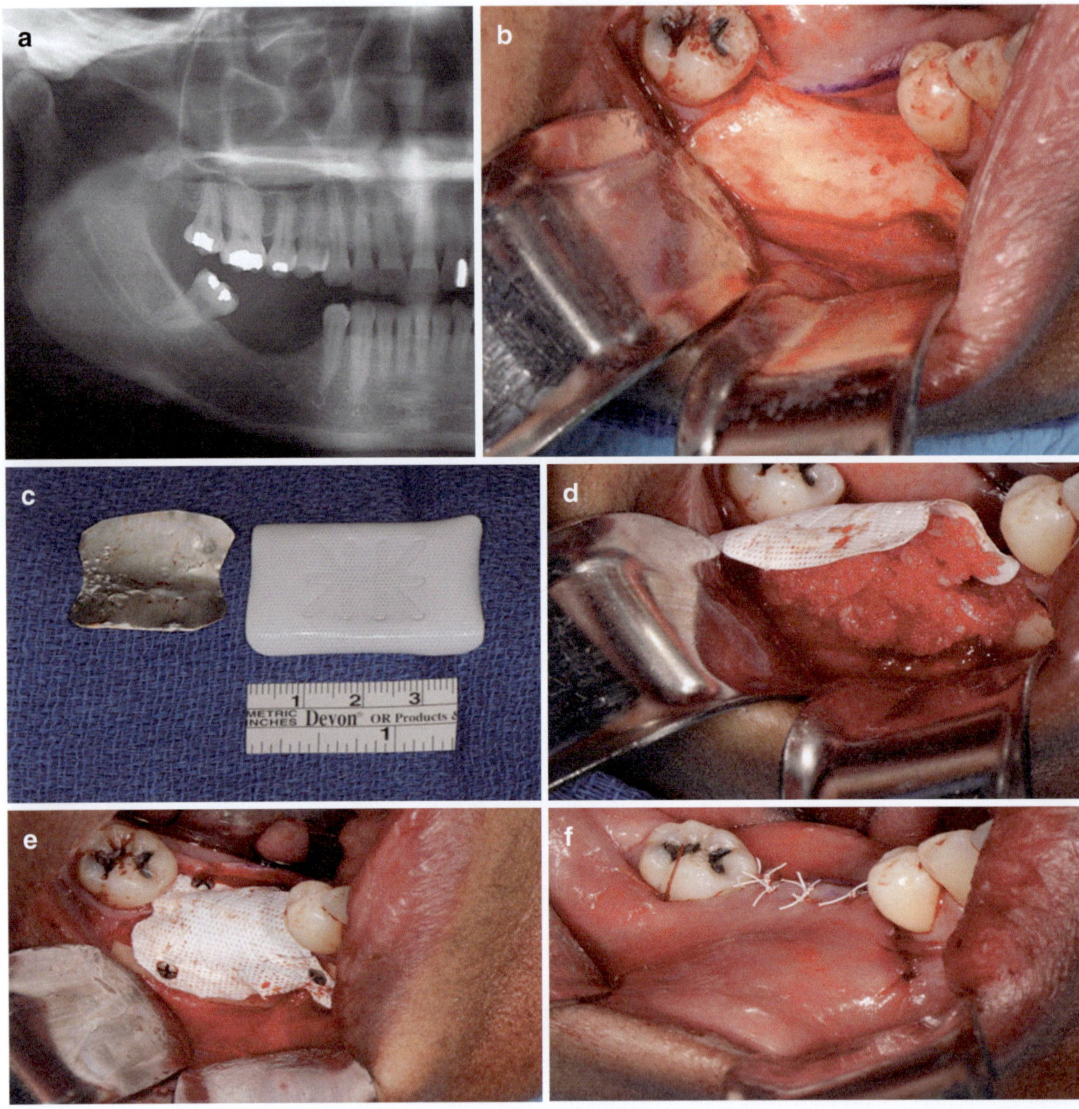

Fig. 10.4 (**a**) Panoramic X-ray with an emphasis on relevant anatomical structure. (**b**) Crestal incision with the two vertical diverging release incisions to expose the recipient site. (**c**) Suture package is used as a template prior to final structuring of the cytoplast membrane (Osteogenics Biomedical, Lubbock, TX). (**d**) Bone graft composite consists of demineralized particulate allograft and PRP. Not viewed in the picture is the presence of one tacking pin on the posterior lingual aspect of the membrane. (**e**, **f**) Membrane secured with three low-profile 3 mm fixation screws and closed with 3.0 Gore-Tex suture after significant submucosal dissection to allow passive release and minimal tension of the coronal-repositioned flap. (**g**) Six-month postoperative picture showing augmented dentoalveolar morphology and marginal amount of transposed keratinized attached tissue. (**h**) Exposure and removal of membrane are easily accomplished leaving behind a vascular pseudoperiosteal layer which is left adherent over the bone graft. (**i**, **j**) Idealized position of bone level for endosseous implants

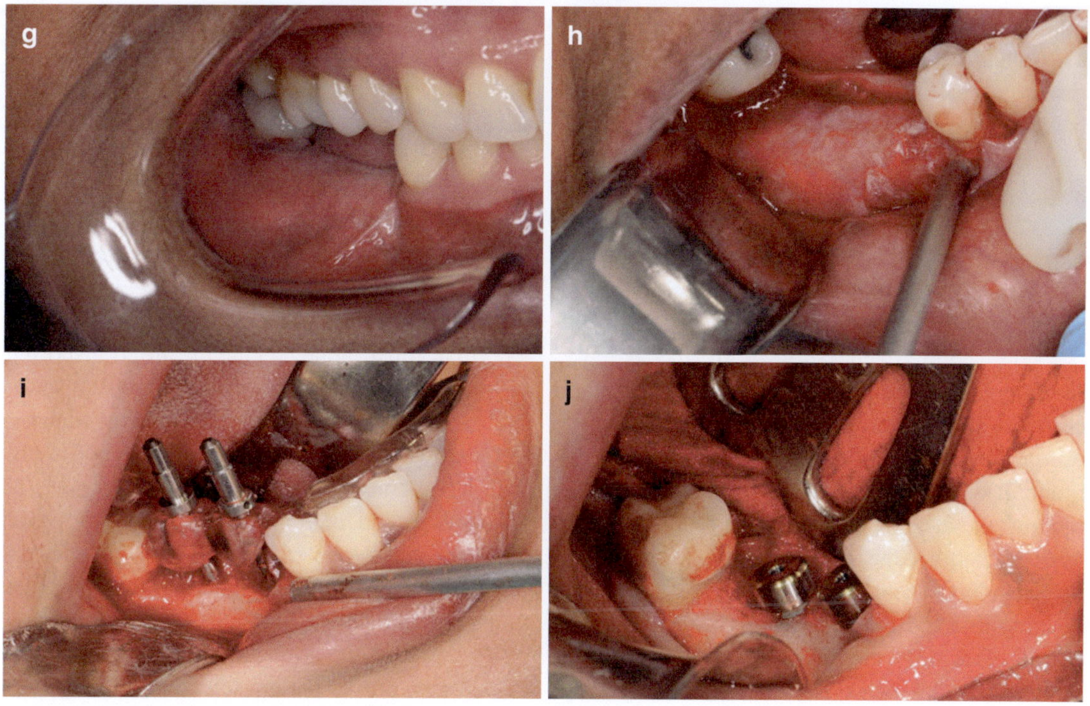

Fig. 10.4 (continued)

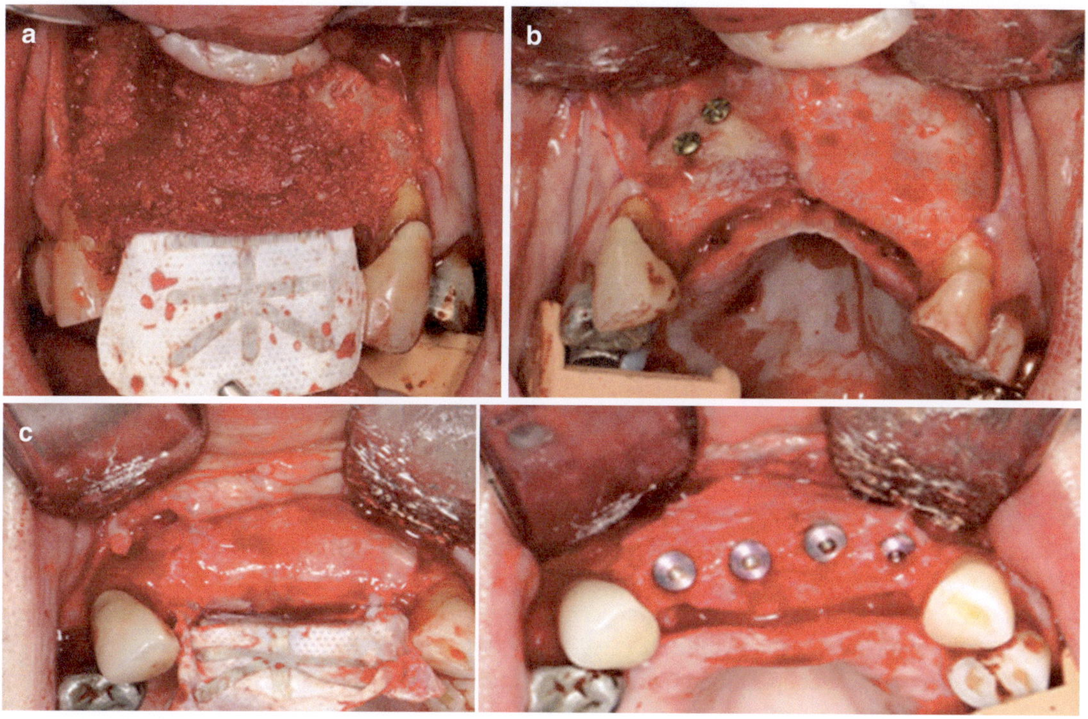

Fig. 10.5 (**a**, **b**) Atrophic anterior maxillae. Handling and molding of graft according to reconstructive demands. (**c**, **d**) Uncovering of membrane with placement of root form implants through vascular pseudoperiosteum

Pertinent details of this technique involve the use of a dense polytetrafluoroethylene non-resorbable membrane dPTFE cytoplast membrane (Osteogenics Biomedical, Lubbock, TX). Small alveolar vertical augmentation can predictably be achieved with the use of mineralized allograft and some form of autologous blood derivative, (PRP, fibrin glue, etc.). The utility of the secondary additive is twofold, the additive cellular elements to assist in wound healing and potential bone regeneration with the equally important role in creating a congealed, intact composite bone graft that facilitates handling during the placement of the containment membrane. It is recommended that there is a 1–2 mm margin of clearance between membrane and adjacent tooth edge, preferably a minimum of three monocortical screws/tacks for stability keeping in mind that the density of mandibular cortex often does not facilitate the ease of use with most tacking systems. Upon secondary entry for implant placement, the soft tissue interface between the membrane and bone should be left undisturbed as this vascular layer is functionally very different from what is usually observed around titanium mesh.

10.4 Vertical Augmentation-Osteoperiosteal Flap

Obtaining proper occlusal clearance to allow for a well-formed and aesthetic prosthesis is a fundamental prerequisite for dental implant restoration. There are several predictable techniques described to address these pre-prosthetic alveolar deficiency dilemmas [7–9]. The osteoperiosteal flap technique has recently made a strong contribution toward management of these defects. Essentially, mobilizing a segment of alveolus attached to the overlying soft tissue obtains a uni- or bidirectional augmentation.

The osteoperiosteal flap or "bone flap," commonly used in segmental orthognathic surgery, is a bone fragment moved in space without detachment of the investing periosteum [9]. Animal studies which have investigated the biology of small segment wound healing have noted that after 2 weeks, revascularization of the small dento-osseous segment was noted [10]. The cross application of such animal studies are helpful, but do not completely capture the additional challenges in the broad technique illustrated in the following case report. The technical variation from representative animal studies include the fact that most bone segments are transported away from their basal bone and secured in space using some form of stabilization device. Depending on the level of displacement, there will be a conversely proportionate amount of pressure placed on the surrounding soft tissue envelope and expected luminal stress of the terminal arterioles which support viability of the transport segment.

The procedure is fairly straightforward and begins with proper patient selection. When determining this technique for the anterior maxillae, residual bone height from the residual crest to the nasal floor should be determined. Soft tissue architecture and integrity overlying edentulous zone should be inspected for evidence of previous surgical site, remaining keratinized attached epithelium and vertical nature of bone mesial to adjacent intact dentition (Fig. 10.6).

Once the vertical dimension has been established, the more predictable horizontal augmentation can be accomplished in a staged manner. Ideally it would be useful to create a labial trajectory to the transport segment by incorporating a bend in the vertical limb of the stabilization plate to allow for more labial bone projection. This modification has been described in the literature, but it can sometimes be technically challenging. In addition, it is not always mandatory to place a fixation device. If the bone graft of choice within the interpositional zone is rigid enough to withstand the compressive pressure on the surgical site during the gradual phase of replacement resorption, then perhaps additional hardware might not be necessary.

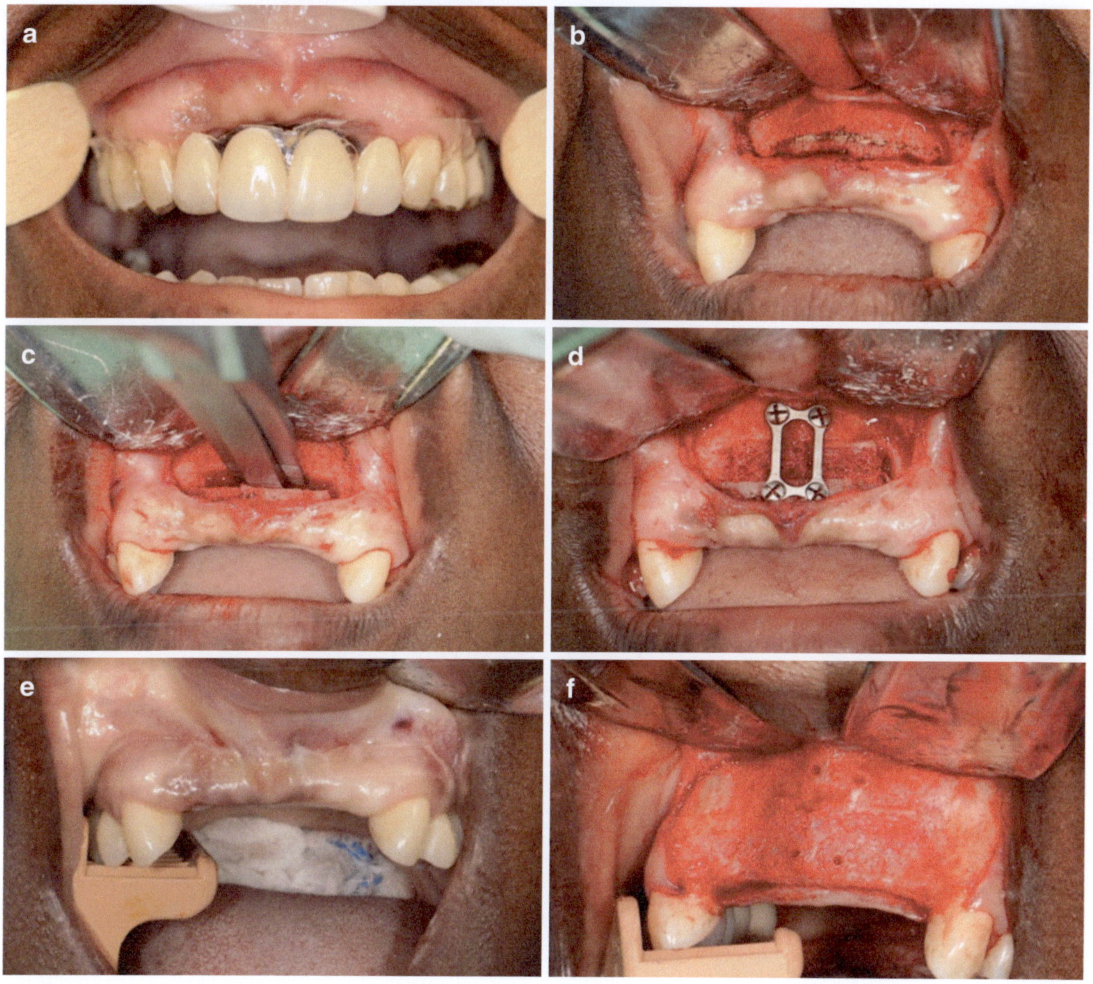

Fig. 10.6 (**a**) Clinical picture demonstrating vertical bone deficiency in the region from #7 to 9, adequate keratinized tissue and established low smile line. (**b**) Horizontal vestibular incision at least 3 mm cephalad to mucogingival junction. (**c**) Full-thickness horizontal and diverging vertical osteotomy to allow full release of transport segment which is gradually mobilized with a Smith spreader. (**d**) Corticocancellous allogenic bone shaped and wedged between the transport segment and basal bone. Titanium plate (0.5 mm) profile, four holes with a 10 mm vertical limb positioned with four monocortical fixation screws. Slightly excess vertical dimension established. (**e**, **f**) Six months after grafting surgery, clinical picture demonstrating stable vertical dimension. Fixation device removed with well-consolidated block graft

10.5 Simultaneous Osteoperiosteoal Flap and Implant Placement

An amplification of the osteoperiosteal flap would include simultaneous implant placement. The following case illustrates mobilizing of a bone segment without complete divestment of investing periosteum preceded by placement of two root form implants and direct sinus lift. In this case the liability would include obtaining immobilization of the mobilized portion of the maxillae during the critical phase of bone healing. The inevitable amount of labial soft tissue manipulation that is

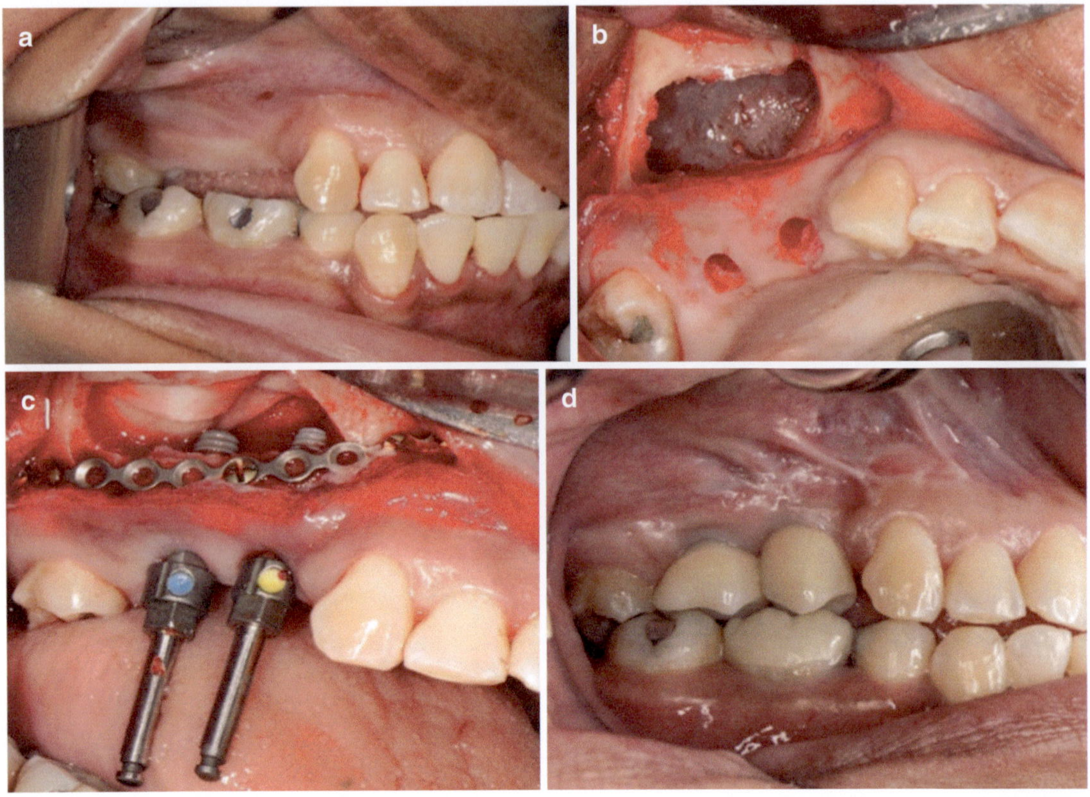

Fig. 10.7 (a) Edentulous site in the area of maxillary antrum without sufficient occlusal clearance for prosthetic crowns. (b) Direct sinus lift with transmucosal osteotomy for respective dental implants using a surgical guide at existing vertical bone level. (c) Bone flap with positioned dental implants was separated from the posterior maxilla and repositioned apically using a predicted surgical guide. (d) Eighteen months after implant placement, note the labial bone recession within the bone flap

required for simultaneous implant placement in cases like these with the long-term labial soft tissue recession would preclude such a strategy in the aesthetic zone [11] (Fig. 10.7).

10.6 Alloplast Containment Devices (Titanium Mesh)

This case illustrates the advantage of using titanium mesh in areas in which two-dimensional dentoalveolar reconstruction is required with the inclusion of a direct sinus lift. Oftentimes the direct sinus lift precludes the use of long fixation screws or any type of onlay lag screw technique. The use of titanium mesh allows both the vertical dimension (direct sinus lift) and the horizontal graft to become contained within the titanium mesh. Needless to say, one of the most common complications of using titanium mesh is hardware exposure and subsequent graft infection. It has been postulated that placement of PRP overlying the titanium mesh and meticulous tension-free closure is one way to minimize that devastating complication [12]. Figure 10.8f demonstrates three containment device options for alveolar reconstruction. All the options are nonresorbable, with a composition of e-PTFE or titanium mesh. The use of prefabricated ridge form titanium mesh allows for a more predictable application, less requirement for modification, and burnished edges that might decrease soft tis-

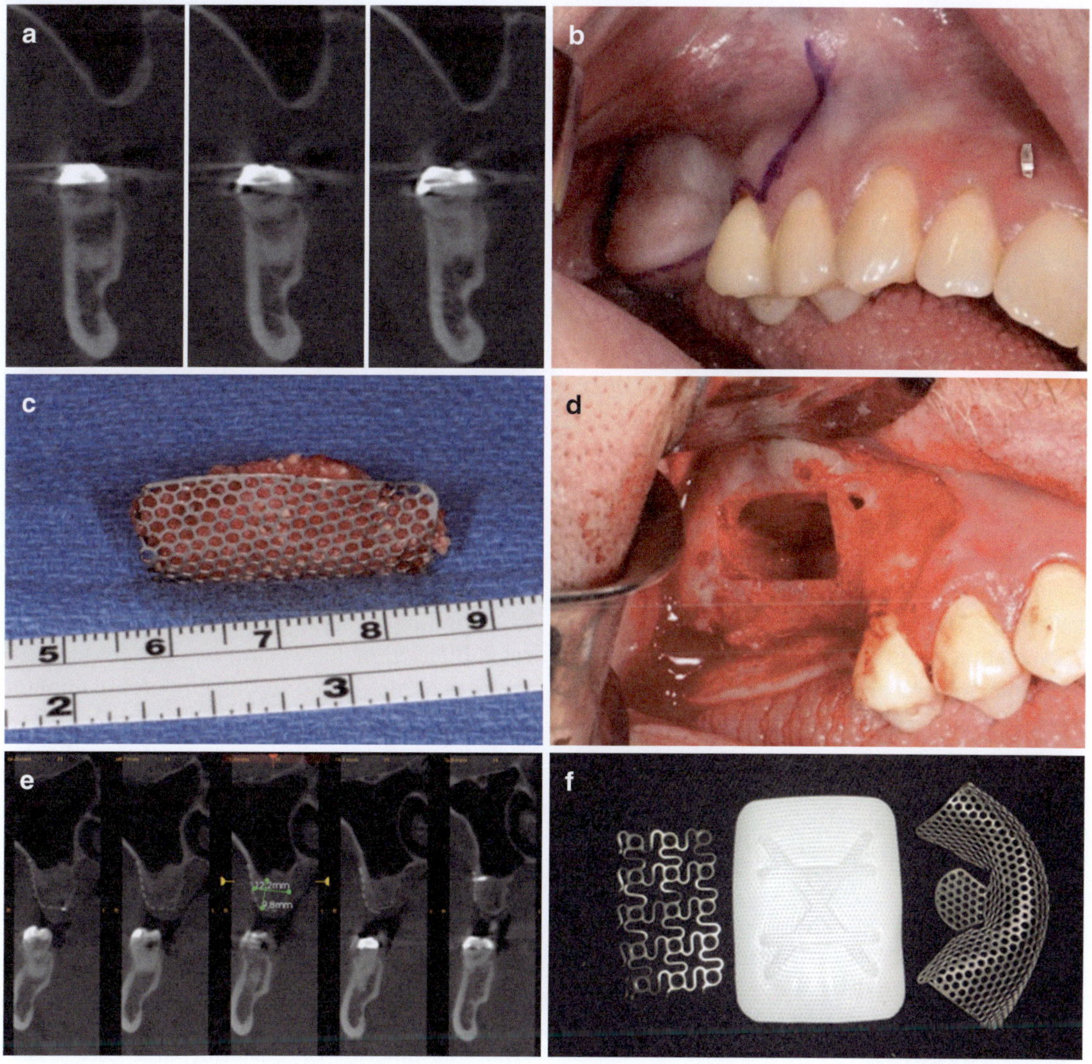

Fig. 10.8 (**a**, **b**) Clinical-radiographic imaging which highlights the combined vertical/horizontal alveolar deficiency. (**c**) Surgical site exposed, resorbable suture being passed through two lateral sinus wall holes to allow resuspension of iatrogenic rent of the sinus membrane. (**d**) Composite bone graft placed into Osteomed, Addison Texas, 0.3 mm profile malleable titanium mesh containment device. (**e**) Six months' postoperative radiograph demonstrating vertical and horizontal augmentation. (**f**) Containment device options

sue dehiscence. The perforation allows for vascular ingrowth while maintaining dimensional expansion. Louis et al. retrospectively analyzed 44 sites with that use of titanium mesh and highlighted several instructional directives. For cross-arch maxillary reconstruction, a vestibular incision was used for access, and fixation screw was placed at the nasomaxillary and zygomatic-maxillary buttress. For segmental reconstruction, a crestal incision was utilized. There was a higher incidence of dehiscence in the segmental reconstruction cohort versus the complete arch. The investing soft tissue that usually encompasses the titanium mesh upon removal was composed of dense fibrous connective tissue with numerous fibroblasts and a mean thickness of 0.7 mm [13].

10.7 Tissue Engineering in the Management of Congenital Alveolar Clefts

Tissue engineering has played a significant role in the management of alveolar cleft defects. Considering the numerous craniofacial operations that these unique group of young patients encounter, the opportunity to eliminate or minimize the necessity for autogenous iliac bone graft harvest with varied forms of tissue engineering composite bone grafts has become relatively common [14–16]. The following illustrations demonstrate a bilateral alveolar cleft with an extremely hypoplastic premaxilla housing a single adult incisor. Considering the size of the premaxillae, this patient was treated in a staged manner as if the patient had two unilateral alveolar clefts. The right side was managed with a combination of BMAC 5 ml (30 cc aspirate) and allogenic mineralized bone. The left side was treated with a combination of PRP, rhBMP-2 (x-small), and allogenic mineralized corticocancellous bone. The long-term axial cone-beam CT demonstrates adequate fusion of the maxillae at the alveolus level (Fig. 10.9).

10.8 Horizontal and Vertical Augmentation of the Complete Maxillary Alveolus

Pre-prosthetic surgery in preparation of the severely atrophic maxillae to receive endosseous dental implants can be a challenge. Oftentimes these patients present with not only a functional impairment but also lack of proper paranasal soft tissue architecture secondary to poor skeletal support. This pattern of progressive resorption gradually leads to a Class III skeletal profile and accentuates aging of the face. Demographically, most patients with severely atrophic maxillae are beyond the fifth decade of life and concurrently have other systemic comorbidities that can preclude them from being a candidate for an extended surgical encounter. Even in spite of being adequate surgical candidates, the quality and quantity of autologous osteocompetent cells gradually decline with aging.

There have been several different surgical options prescribed for the management of the severely atrophic maxillae for the prosthetic goal of implant-supported or implant-retained prosthesis. A non-exhaustive list would include LeFort I osteotomy advancement and downgraft with interpositional autogenous cancellous marrow or corticocancellous block graft, autogenous corticocancellous block onlay graft combined with bilateral posterior direct sinus lift, microvascular free fibula transfer, and selective placement of tilted implants including recruitment of the zygomatic bone as a point of anchor [17–19]. The first three options require a second extraoral harvest site with varying degrees of surgical morbidity. The fourth option is beginning to hold an expanding role in the rehabilitation of the edentulous patient but still requires some degree of basal bone for implant placement.

In this particular discussion, the LeFort I osteotomy, more appropriately named the full maxillary osteoperiosteal flap, has the advantage of repositioning skeletal form, enhancing vertical dimension of occlusion, and allowing for adequate bone height throughout the entire maxillary arch [9]. Historically, this has demanded a significant autogenous bone volume graft that involved a second extraoral harvest site either from the anterior or posterior iliac crest [20]. From a practical standpoint, the most important factor in autogenous bone transfer is the amount of cellular marrow transplanted with the bone graft, the vascularity of the tissue bed, and the maintenance of graft stability [21]. When considering these three basic parameters of autogenous bone grafting, it becomes evident that the quantity of transplanted cellular marrow and the vascularity of the recipient tissue bed are primarily dependent on the patient's physiology, with surgical technique being the primary determinant of graft stability. In light of these basic foundational principles, any advancement in surgical therapy would have to provide the same quantity and quality of osteoprogenitor cells, minimally impair tissue vascularity, decrease surgical mor-

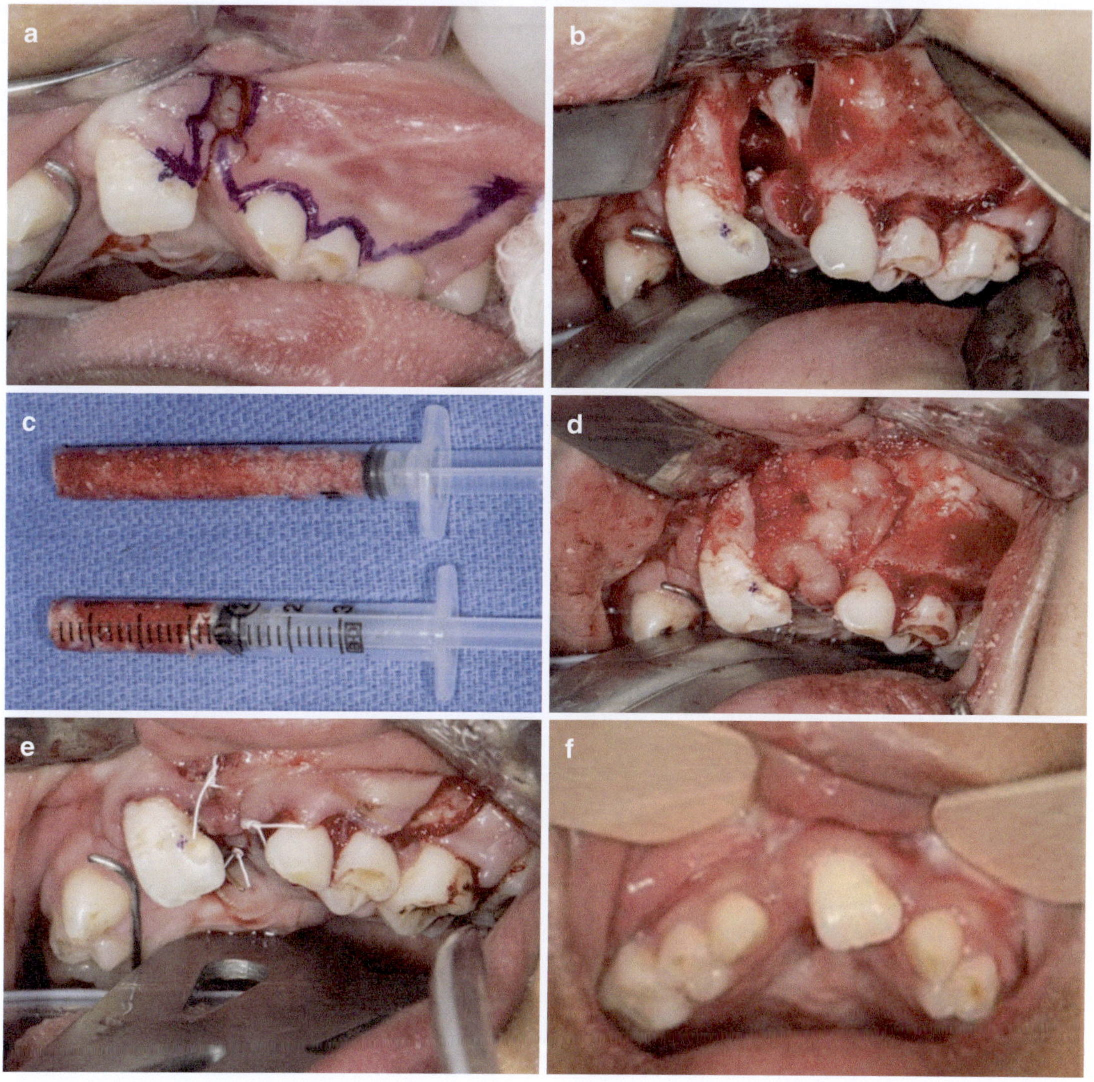

Fig. 10.9 (**a**) Left alveolar cleft with planned incision design. (**b**) Surgical access with complete separation of nasal and oral cavity with demonstration of residual bone architecture. (**c**) Preparation of bone graft composite within 3 cc delivery syringe (rhBMP-2, mineralized allogenic bone, PRP). (**d**) Deposition of bone graft with separate small cut sheet ACS (acellular collagen sponge/rhBMP-2) placed at incision junction to aid in soft tissue healing. (**e**) Labial advancement flap and primary closure. (**f**) Eight-month postoperative clinical picture demonstrating well-consolidated bone and closure of bilateral alveolar cleft

bidity, and be compatible with current stabilization techniques. With the progression of regenerative surgical medicine, this following case illustrates a very well-understood and well-utilized reconstructive option for the severely atrophic maxilla (LeFort I osteotomy), in combination with a modern composite bone graft technique that utilizes tissue engineering principles (Fig. 10.10).

10.8.1 Section Summary

The 4.2 mg rhBMP-2 (Medtronic Sofamor Danek, Warsaw, Indiana) used in this case report was prepared in the prescribed fashion [22]. One of the two absorbable collagen sponge (ACS) was then cut and placed adjacent to the bone of the intaglio aspect of the downfractured maxillae. This application was theorized to support signal transduction

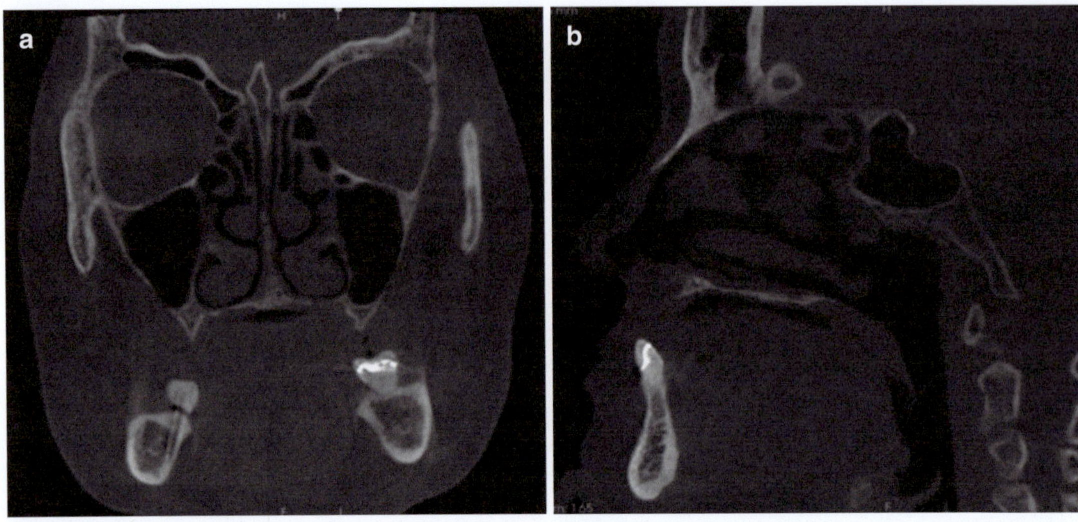

Fig. 10.10 (**a**, **b**) Preoperative coronal and sagittal view. Note the maxillomandibular anteroposterior discrepancy. (**c**) Incision and elevation of sinus membrane and nasal floor. (**d**) Fixation plates with repositioned maxillae. (**e**) Interpositional bone graft consisting of 10 ml BMAC, 4.2 mg rhBMP-2, and allogenic freeze-dried particulate corticocancellous bone. (**f**) Placement of tensor fascia lata membrane on facial aspect of graft. (**g**, **h**) Immediate postoperative coronal and sagittal view. (**i**) Ten months after bone graft surgery, nine root form 4.3-mm-diameter implants with 11.5 mm length were placed with one having to be switched to 5.0 mm diameter in order to obtain primary stability. (**j**, **k**) Twenty-four months after bone graft surgery, magnified coronal and sagittal view demonstrated well-consolidated bone graft and osseointegrated endosseous implant with improved A–P projection of the maxillae

of the osteoprogenitor cells that would be found emanating from the bleeding bone of the maxillae. The remaining ACS was then admixed with the bone marrow aspirate concentrate and allogenic bone. The allogenic bone, satisfying the osteoconductive matrix with the BMAC, provides the necessary mesenchymal stem cells (MSC) that would induce osteogenesis. Both bone marrow aspirate and trabecular bone have been identified as sources of mesenchymal stem cells [23]. The bone marrow aspirate technique followed by a density centrifuge concentration of the aspirate allows for a much higher concentration of total nucleated cells (TNC), but, more importantly, the MSC necessary for bone regeneration [6]. Although this is a very rudimentary explanation, BMAC have been shown to decrease autologous bone graft requirement for critical size osseous defect [24].

10.9 Traumatic Alveolar Defects

Traumatic alveolar defects of the jaws are often fraught with a loss of both hard and soft tissue. Additionally, in most circumstances there is also a corresponding amount of scar formation that leads to a hypovascular, minimally pliable soft tissue envelope. This undoubtedly leads to the requirement of viable, pliable, analogous soft tissue that can serve to expand the existing soft tissue envelope and enhance inflow of vascularity to assist in hard tissue regeneration. For large defects related to ablative surgeries or significant trauma, microvascular tissue transfer may be the best choice given the versatility, quantity, growing popularity, and trained surgeons in the field. However, most intraoral defects are amenable to regional and local tissue transfers. Previously described regional flaps including pectoralis major, supraclavicular, and submental flaps have fallen out of favor due to donor site, excess bulk, donor site weakness, partial flap necrosis, and transfer of hair-bearing tissue into the oral cavity. Intraoral tunneling of nasolabial flaps into the oral cavity has been described in literature but has limited use in the anterior maxilla due to a short pedicle length. The facial artery myomucosal (FAMM) flap is one predictable regional axial pattern flap that can provide the necessary environment to achieve predictable rehabilitation of small- to moderate-size alveolar defect.

10.9.1 **Case Illustration** (Figs. 10.11 and 10.12)

10.9.2 Section Summary

The FAMM was first described in 1992 by Pribaz et al. for closure of oronasal defects [25]. The FAMM has multiple applications for reconstruction of oromucosal defects after tumor resection, trauma, oroantral fistula repair and defects of the alveolar ridge, floor of mouth, and vermilion border. The flap pedicle can be based either inferiorly via anterograde flow through the facial artery or superiorly via retrograde flow through the lateral nasal artery making it a versatile flap with application over a wide variety of defects.

The face is supplied by three main arteries, the facial artery, transverse facial, and infraorbital artery. Although variation exists among the population, the facial artery is typically the dominant artery [26]. Classic descriptions of the branches of the facial artery include the following from proximal to distal: inferior and superior labial, inferior alar, lateral nasal, and angular. Significant variation occurs in the terminal artery as indicated by results from Koh et al., Mitz et al., and Niranjan [27–29]. A more recent cadaveric study by Bulter et al. consisting of 201 facial arteries showed 49% terminated as the lateral nasal artery and 20% as the angular artery. Of the specimens, the terminal branch was symmetric in 53%. Of note, seven of specimens had absence of the facial artery with a large, dominant transverse facial artery that branched to supply all the territories of the facial artery [30]. This is an important point as the surgeon evaluating a patient for application of the FAMM. Preoperative Doppler studies must be completed before the patient enters the operating room to verify the presence of the artery.

The author recommends nasoendotracheal intubation for intraoral reconstruction with a FAMM. The path of the artery is marked with a surgical marker from the retromolar pad to the superior buccal gingival sulcus with the assistance of an arterial Doppler. The artery should also be mapped and outlined on the facial skin from the antigonial notch to the lateral ala of the nose (Fig. 10.11c). In order to facilitate dissection and keep tension on the buccal mucosa, two temporary silk sutures may be placed in full thickness through the upper and lower lips on the ipsilateral side of the flap.

The superiorly based flap is based on retrograde blood flow through the lateral nasal or angular artery. Dissection begins at the distal most port of the flap; for the superiorly based flap, this would be in the retromolar pad. Sharp dissection is made through the mucosa, the submucosa, and the buccinator muscle to identify the facial artery. A small cuff of muscle should be left around the vessel during dissection to prevent damage to the vessel wall. The distal end of the facial artery is ligated, and dissection is carried bluntly superiorly and anteriorly to the maxilla gingivobuccal sulcus. The dissection is stopped at least 1 cm lateral to the oral commissure in order to leave a sufficient amount of orbicularis oris for oral competency and facial memetics. The pedicle length may be up to 10 cm in length and 3 cm wide (Fig. 10.11d). Care must be taken to avoid Stenson's duct during dissection. Once the flap is fully mobilized, the distal end can be evaluated for capillary perfusion and inset into the recipient site. Once the flap is inset, the distal portion should be evaluated with an arterial Doppler to ensure the arterial supply is not kinked or twisted. The donor site is then closed primarily with a two-layered closure.

In this case illustration, a rigid titanium mesh was contoured and pre-stabilized onto the recipient site. This was then followed by creating a cutout on the alveolar portion of the mesh to serve as an entry port for the bone graft composite (Fig. 10.11e). Intramarrow fenestrations are created, and the composite bone graft to include multipotent hematopoietic stem cells form the anterior ilium (BMAC) [6], and allogenic mineralized particulate bone is compacted within the confines of the titanium mesh. The pliability and vascularity of the FAMM flap allow for primary closure with very little concern of dehiscence over the underlying hardware (Fig. 10.11f). Subsequent to implant placement, a vestibuloplasty was done utilizing a fully resorbable collagen matrix (Geistlich Mucograft).

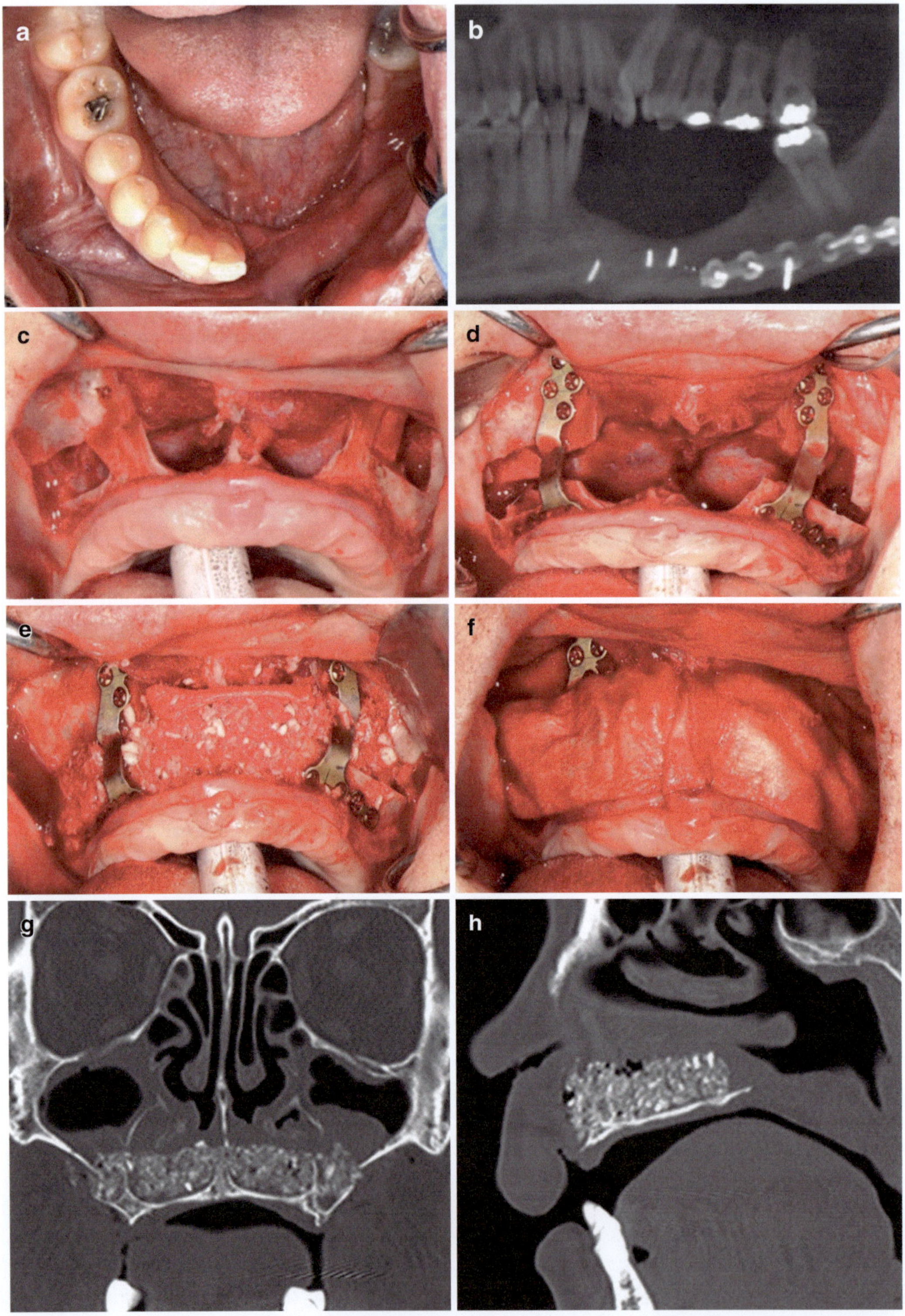

Fig. 10.11 (**a**) Clinical photo of left mandibular body with loss of vestibule due to soft tissue avulsion and scar contracture secondary to blast injury. (**b**) X-ray of the left mandibular body with vertical bone loss and retained hardware. (**c**) Identification of left facial artery and mapping on the skin. (**d**) Isolation and elevation of the left superiorly based FAMM facial artery myomucosal flap. (**e**) Inset of the flap into the recipient zone. (**f**) Exposure of the recipient bed and preparation of recipient basal bone

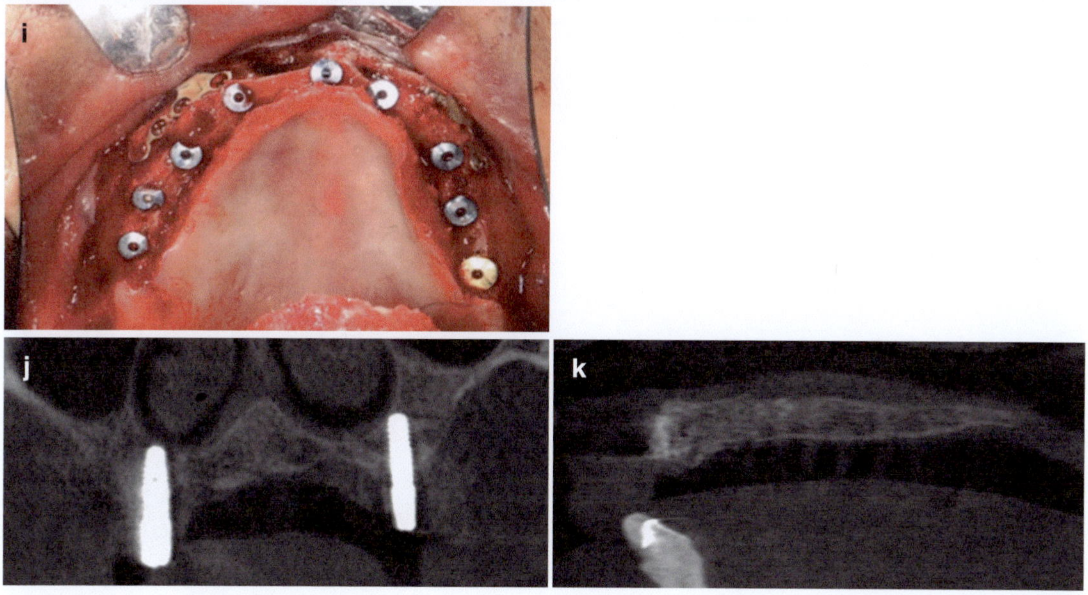

Fig. 10.11 (continued)

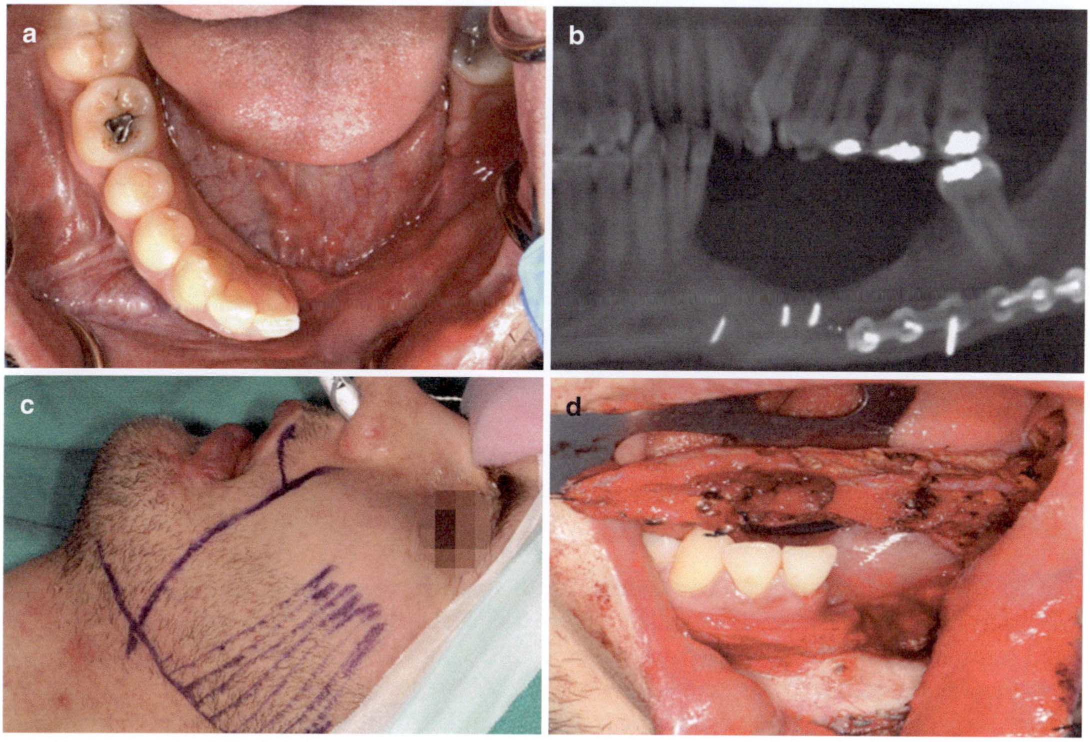

Fig. 10.12 (**a**) Titanium mesh used as containment device for composite bone graft (BMAC, allogenic bone). (**b**) Previously placed FAMM flap allowing for tension-free closure of expanded soft tissue envelope. (**c**) Placement of four endosseous implants and vestibuloplasty with mucograft. (**d**) Clinical picture with final prosthetic and expanding vestibule

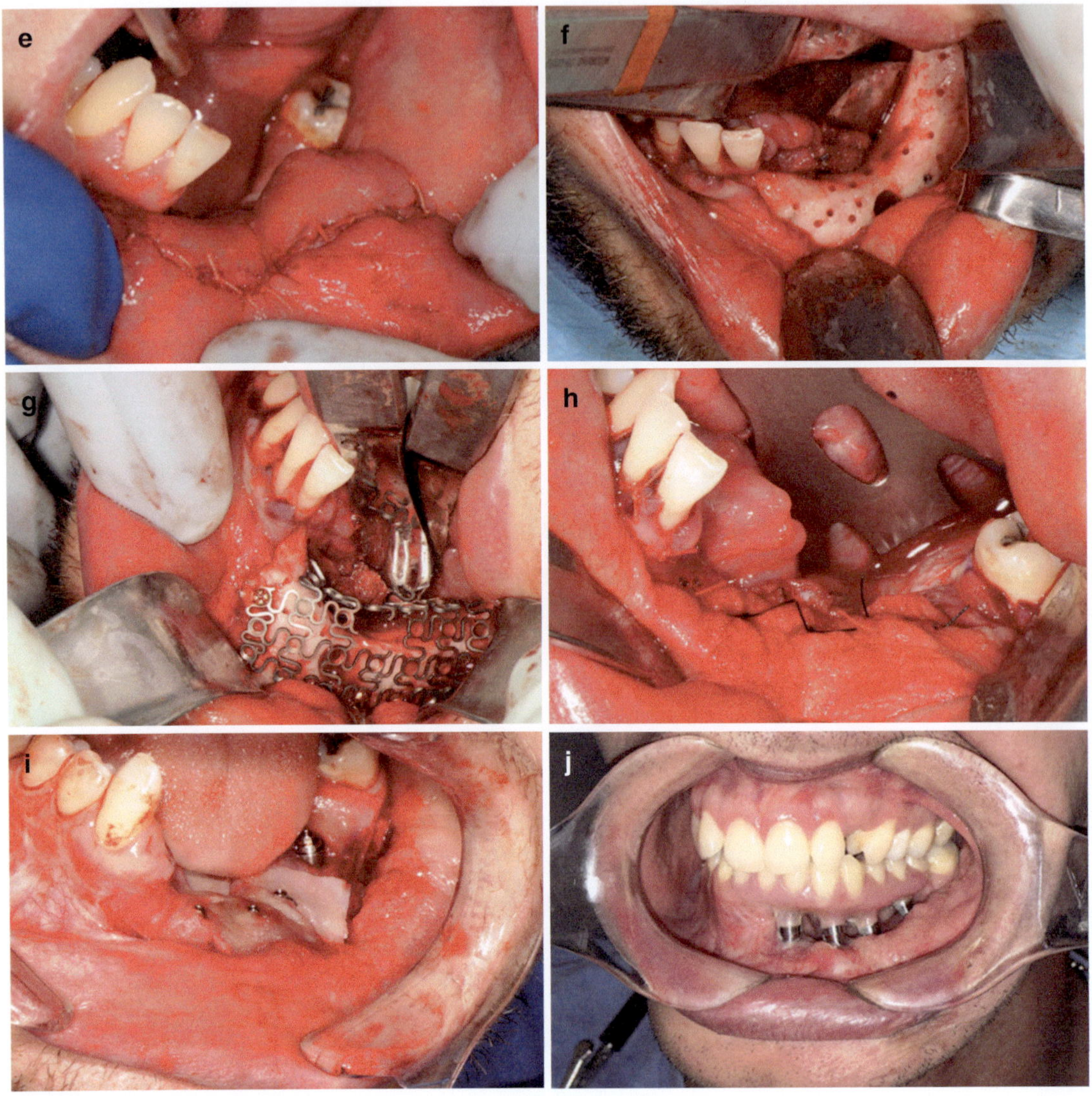

Fig. 10.12 (continued)

10.10 Chapter Summary

Understandably, the autologous corticocancellous grafts have been historically considered the "gold standard" in pre-prosthetic reconstruction of the deficient alveolus. The proposed benefits have to do with autologous bone being able to predictably satisfy the requirement for a successful bone graft transplant. Its osteogenic ability is linked to its ability in transporting viable osteoblast. The matrix scaffold that is afforded by the cortical mineral component provides the osteoconductive arm. The osteoinductive component is still being elucidated, but some pertinent biochemical constituents include the release of BMPs, the PDGF, the TGF, and the IGF [31].

Although the burden of proof and empiric data lies in favor for autologous bone harvest, the emerging field of regenerative medicine continues to challenge some of our presuppositions about bone regeneration. Decision-making is based upon multiple factors including patient's physiology, expectation, and available resources. In the current societal marketplace desire for minimally invasive surgery and expeditious end results, novice innovation will continue to address our patient's demands. The compilation of cases presented in this chapter revives some very relevant questions that are left unanswered. Considering the cost and suggested liability involved with using rhBMP-2, what is the minimum amount required for various alveolar reconstructions? Furthermore, with the expanding knowledge of the cellular constituents of BMAC to induce BMP liberation, is there a need for any exogenous BMP to be incorporated in these types of reconstruction? In the vast majority of illustrated cases, the authors utilized a crushed freeze-dried allogenic bone as a scaffold and carrier for the BMAC; is there a better alternative? The orthopedic literature has shown various hyaluronic acid-base scaffold carriers that have shown improved migration, proliferation, and differentiation of osteogenic cells compared to allogenic bone [32, 33]. In summary this chapter demonstrates the movement of the pendulum in which we can bridge the gap between the "gold standard" attributed to autogenous bone and the reconstructive models utilizing tissue engineering principles.

References

1. Leonetti JA, Koup R. Localized maxillary ridge augmentation with a block allograft for dental implant placement: case report. Implant Dent. 2003;12:217.
2. Petrungaro PS, Amar S. Localized ridge augmentation with allogenic block grafts prior to implant placement: case reports and histologic evaluations. Implant Dent. 2005;14:139.
3. Al-Abedalla K, Torres J, Cortes AR, Wu X, Nader SA, Daniel N, Tamimi F. Bone augmented with allograft onlays for implant placement could be comparable with native bone. J Oral Maxillofac Surg. 2015;73(11):2108–22.
4. Waasdorp J, Reynolds MA. Allogeneic bone onlay grafts for alveolar ridge augmentation: a systematic review. Int J Oral Maxillofac Implants. 2010;25:525–31.
5. Keith JD Jr, Petrungaro P, Leonetti JA, et al. Clinical and histologic evaluation of mineralized block allograft: results from the developmental period (2001–2004). Int J Periodontics Restorative Dent. 2006;26:321–7.
6. Marx RE, Turson R. A qualitative and quantitative analysis of autologous human multipotent adult stem cells derived from three anatomic areas by marrow aspiration: tibia, anterior ilium, and posterior ilium. Int J Oral Maxillofac Implants. 2013;28(5):e290–4.
7. Irinakis T. Efficacy of injectable demineralized bone matrix as graft material during sinus elevation surgery with simultaneous implant placement in the posterior maxilla: clinical evaluation of 49 sinuses. J Oral Maxillofac Surg. 2011;69:134–41.
8. Chiapasoo M, Casentini P, Zaniboni M. Bone augmentation procedures in implant dentistry. Int J Oral Maxillofac Implants. 2009;24:237–69.
9. Jenson OT. The osteoperiosteal flap: a simplified approach to alveolar bone reconstruction. Hanover Park, IL: Quintessence; 2010.
10. Jenson OT, Bell W, Cottam J. Osteoperiosteal flaps and local osteotomies for alveolar reconstruction. Oral Maxillofac Surg Clin North Am. 2010;22:331–46.
11. Tsegga T, Wright T. Maxillary segmental osteoperiosteal flap with simultaneous placement of dental implants: case report of a novel technique. Int J Implant Dent. 2017;3:2.
12. Torres JT, Tamimi F, Alkhraisat MH, et al. Platelet-rich plasma may prevent titanium-mesh exposure in alveolar ridge augmentation with inorganic bovine bone. J Clin Periodontol. 2010;37:943–51.
13. Louis JP, Gutta R, Said-Al-Naief N, Bartolucci AA. Reconstruction of the maxilla and mandible with particulate bone graft and titanium mesh for implant placement. J Oral Maxillofac Surg. 2008;66:2356–245.
14. Fengzhou D, Huanhuan W, Haidong L, et al. Bone marrow mononuclear cells combined with beta-tricalcium phosphate granules for alveolar cleft repair: a 12-month clinical study. Sci Rep. 2017;7:13773.
15. Van Hout WM, Mink van der Molen AB, Breugem CC, et al. Reconstruction of the alveolar cleft: can growth factor-aided tissue engineering replace autologous bone grafting? A literature review and systematic review of results obtained with bone morphogenetic protein-2. Clin Oral Investing. 2011;15(3):297–303.
16. Francis CS, Mobin SS, Lypka MA, et al. rhBMP-2 with a demineralized bone matrix scaffold verses autologous iliac crest bone graft for alveolar cleft reconstruction. Last Reconstr Surg. 2013;131(5):1107–15.
17. Schmidt BL. Maxillary reconstruction using zygomaticus implants. Atlas Oral Maxillofac Surg Clin North Am. 2007;15:43–9.
18. Sevetz EB. Treatment of the severely atrophic fully edentulous maxilla: the zygoma implant option. Atlas Oral Max Surg clin N Am. 2006;14:121–36.
19. Att W, Bernhart J, Strub JR. Fixed rehabilitation of the edentulous maxilla: possibilities and clinical outcome. J Oral Maxillofac Surg. 2009;67(11 Suppl):60–73.
20. Gil JN, claus JDP, Campos FEB, Lima SM Jr. Management of the severely resorbed maxilla using Le Fort I osteotomy. Int J Oral Maxillofac Surg. 2008;37:1153–5.

21. Marx RE. Bone and bone grafting. Oral Maxillofac Surg Clin North Am. 2007;19:455–66.
22. Block SM, Achong R. Bone morphogenetic protein for sinus augmentation. Atlas Oral Maxillofac Surg Clin North Am. 2006;14:99–105.
23. Weissman IL. Stem cells: units of development, units of regeneration, and units in evolution. Cell. 2000;100:157–68.
24. Jager M, Herten M, Fochtmann U, et al. Bridging the gap: bone marrow aspiration concentrate reduces autologous bone grafting in osseous defects. J Orthop Res. 2011;29(2):173–80.
25. Pribaz J, Stephens W, Crespo L, et al. A new intraoral flap: facial artery musculomucosal (FAMM) flap. Plast Reconstr Surg. 1992;90(3):421–9.
26. Lasjaunias P, Berenstein A, Doyon D. Normal functional anatomy of the facial artery. Radiology. 1979;133:631–8.
27. Mitz V, Ricbourg B, Lassau JP. Facial branches of the facial artery in adults. Typology, variations and respective cutaneous areas. Ann Chir Plast. 1973;18:339–50.
28. Koh KS, Kim HJ, Oh CS, et al. Branching patterns and symmetry of the course of the facial artery in Koreans. Int J Oral Maxillofac Surg. 2003;32:414–8.
29. Niranjan NS. An anatomical study of the facial artery. Ann Plast Surg. 1988;21:14–22.
30. Bulter P, Lohn J, Penn J, Norton J. The course and variation of the facial artery and vein: implications for facial transplantation and facial surgery. Ann Plast Surg. 2011;67(2):184–8.
31. Clementini M, Boniello R, Gasparini G, Moro A, Pelo S. Surgical treatment of severe atrophic maxilla by means of multiple extraoral harvesting. Oral Implantol. 2009;2(3):4–10.
32. Sasaki T, Watanabe C. Stimulation of osteoinduction in bone wound healing by high-molecular weight hyaluronic acid. Bone. 1995;16:9–15.
33. Huang L, Cheng YY, Koo PL, et al. The effect of hyaluronan on osteoblast proliferation and differentiation in rat calvarial-derived cell cultures. J Biomed Mater Res. 2003;66:880–4.

Tissue Engineering for Transoral Reconstruction of Large Mandibular Continuity Defects

11

James C. Melville, Huy Q. Tran, Dhafer Alsuwied, Srinivasa R. Chandra, and Jonathan W. Shum

11.1 Introduction

Tissue engineering has garnered attention over the years as an important technique in reconstructive surgery. In the early 1970s, W. T. Green, a pediatric orthopedic surgeon at Boston Children's Hospital, first introduced the concept by undertaking several experiments aimed to generate cartilage from chondrocytes seeded in bone spicules [1]. Since then, a number of major research centers around the world including MIT (Massachusetts, USA), Rice University (Texas, USA), and Nagoya University (Japan) have continued to explore and advance the field of tissue engineering. In 1993, Drs. Robert Langer and Joseph Vacanti described tissue engineering as "an interdisciplinary field that applies the principles of engineering and the life sciences toward the development of biological substitutes that restore, maintain, or improve tissue function" [2]. The formation of the Tissue Engineering Regenerative Medicine International Society (TERMIS) in 2003 highlighted a major milestone and has united worldwide scientists to share and collaborate on their research. These joint efforts have translated to significant improvements in the clinical setting by increasing efficiency and postoperative outcome while decreasing patient morbidity and surgical complications.

Prior to the advances of tissue engineering, the reconstruction of maxillomandibular defects has met many challenges due to the infection risk associated with the polymicrobial environment of the oral cavity, insufficient soft tissue volume, and lack of native bony regenerative potential [3]. Traditionally, these defects are reconstructed using autogenous bone or osteocutaneous free flaps, requiring a secondary surgical site that poses additional risk for complication and greatly increases morbidity for the patient. Furthermore, these patients often require longer length of hospital stay due to the need for physical therapy and monitored wound care. In autogenous bone grafting, the potential for oral bacterial contamination leads to most surgeons opting for a staged second surgery to allow for complete intraoral healing. Additionally, they often elect an extraoral surgical approach that comes with the risk of nerve injury and unaesthetic scarring. This chapter will

J. C. Melville (✉) · D. Alsuwied · J. W. Shum
Department of Oral and Maxillofacial Surgery
University of Texas Health Science Center,
Department of Oral and Maxillofacial Surgery, Oral, Head & Neck Oncology and Micro-vascular Reconstructive Surgery,
Houston, TX, USA
e-mail: James.C.Melville@uth.tmc.edu; Jonathan.Shum@uth.tmc.edu

H. Q. Tran
Oral and Maxillofacial Surgery, University of Texas Health Science Center at Houston,
Houston, TX, USA

S. R. Chandra
Head and Neck Surgery, Oral and Maxillofacial Surgery, University of Nebraska Medical Center,
Omaha, NE, USA

J. C. Melville et al. (eds.), *Regenerative Strategies for Maxillary and Mandibular Reconstruction*,
https://doi.org/10.1007/978-3-319-93668-0_11

explore the role of tissue engineering in maxillomandibular reconstruction through immediate, one-staged surgery using an intraoral approach. This technique serves as an alternative reconstructive modality that can decrease surgical morbidity and decrease inpatient hospitalization while maintaining excellent postoperative outcome and effective bone regeneration.

In this technique, grafting of defects consists of the use of recombinant human bone morphogenic protein (rhBMP-2), bone marrow aspirate concentrate (BMAC), and particulate bone allograft mixed in a homogenous manner. The basic principles of osteogenesis are fulfilled by all three materials. The rhBMP-2/acellular collagen sponge (ACS) mixture is osteoinductive, which provides recruiting signals for migrating osteogenic cells and activation signals for osteoblasts. The BMAC has osteogenic potential by providing a significant amount of growth factors and mesenchymal stem cells that can differentiate into osteoblasts and osteocytes. Finally, the allogenic bone chips provide a framework that is conducive for cells to lay down new bone. All three components are mixed together, placed into the recipient site, and secured by a biocompatible membrane. Suturing technique and material play a major role to achieve adequate primary soft tissue closure.

11.2 Indications for Surgery

Patient selection is important in this technique, and there are critical requirements for both graft recipient and BMAC donor sites to ensure tissue engineering success. First, the maxillofacial defect must not be the result of an oncologic resection. Second, the patient must not have a history of chemotherapy or radiation to the grafting site. Third and most importantly, there must be adequate soft tissue volume for primary, watertight closure. If the native soft tissue is lacking, the surgeon has the option of using mucocutaneous free flap to fulfill the third criteria [4]. The ideal patient should also have no worse than mild systemic disease in order to maximize healing and grafting potentials. If the patient uses tobacco, they must also be counseled to avoid tobacco use during the perioperative period. Uncontrolled diabetes mellitus is also a contraindication to the single-stage approach.

11.3 Bone Marrow Aspirate Concentrate (BMAC)

Initially described in 1952, bone marrow aspirate concentrate (BMAC) was first studied and utilized in various orthopedic procedures [5]. It has since become widely popular among maxillofacial reconstructive surgeons because the aspiration procedure is relatively fast and simple, with minimal morbidity to the patient. Approved by the US Food and Drug Administration, BMAC can be obtained from the tibia, or anterior or posterior iliac crests, to deliver the patient's own mesenchymal stem cells (MSCs) to the site of defect. However, it is important to note that only 0.001% to 0.01% of all nucleated cells in the BMAC are MSCs, thus requiring a dual centrifugation process to further increase its concentration [6]. It can be argued that the growth factors contained in the aspirated solution, including platelet-derived growth factor (PDGF), transforming growth factor-β (TGF-β), and vascular-endothelial growth factor are all equally important in bone formation.

Contraindications to BMAC harvest include (1) bone diseases including congenital disorders (osteogenesis imperfecta), metabolic (osteopetrosis), or malignancy (multiple myeloma) and (2) a history of radiation to the harvest site. Caution must be used when there is known anatomical deformity or trauma history [7].

In general, a BMAC kit consists of a trocar, aspiration needle, and syringe with anticoagulant solution. For each donor site, up to 120 cc of bone marrow aspirate can be obtained. Patient age must be accounted for as the total number of viable mesenchymal cells has been found to be inversely proportional to age [8]. If more bone marrow volume is needed, it is recommended to make a separate stab incision for additional harvest sites. Additionally, this process can be repeated on the opposite ilium or tibia. The aspirate is then immediately filtered and processed with a dual centrifugation process accordingly by the manufacturing company. This process serves to further concentrate mesenchymal stem cells and growth factors for application in the defect.

11.4 Recombinant Human Bone Morphogenic Protein-2/ Acellular Collagen Sponge (rhBMP-2/ACS)

In March 2007, the US Food and Drug Administration (FDA) approved the use of rhBMP-2 in two maxillofacial surgical procedures: (1) maxillary sinus floor augmentation and (2) localized alveolar ridge augmentation for defects associated with dental extraction [9]. It is important to let the patient know that the utilization of BMP in any other way is off-label, and the patient should consent to such procedure. The current consensus among multiple randomized, controlled clinical trials is that a concentration of 1.5 mg/mL of rhBMP-2/ACS is most efficacious for bone grafting in oral surgery. Both histological quality and volume of bone induced by rhBMP-2/ACS at 6- and 12-month period are comparable to those of the gold standard autogenous graft [10, 11].

Contraindications to the use of rhBMP-2 include (1) patients with known hypersensitivity to any formulation component of rhBMP, including bovine collagen, (2) application in resected or extant tumor, (3) active malignancy or infection at recipient site, and (4) patients who are pregnant. Cautions must be made in patients less than 18 years old, as well as women who are nursing [12]. A negative pregnancy test must always be made before proceeding with the procedure.

The rhBMP-2 comes in a lyophilized powder that can be mixed with prepackaged sterile water. The rhBMP-2 solution is then applied to the acellular collagen sponge (ACS), which serves as both carrier for the protein and scaffold for bone formation. The mixture is left alone for 15 min to maximize molecular binding. It is recommended that the rhBMP/ACS mixture is used within 2 h to avoid desiccation and loss of proteins. Significant swelling is expected with the use of rhBMP, and the patient must be forewarned. In addition, tension-free closure needs to be achieved over the graft to avoid wound dehiscence due to increasing edema postoperatively.

11.5 Allogenic Bone Graft

An allograft is collected from a member of the same species, and over the years, it has become widely accepted among clinicians as an excellent bone graft substitute. Although inferior to autografts due to the lack of osteogenetic property, allogenic bone graft is readily available and poses no donor site morbidity to the patient. Previous concerns of disease transmission have led to tighter processing regulation by the American Association of Tissue Banks, such that the current risk of contracting diseases is nil to none. The risk of contracting HIV in allogenic graft is 1 in 1.6 million compared to 1 in 450,000 in blood transfusion [13]. Recipient immunogenic response was an issue with allograft application, especially with fresh or frozen specimen. This led to increased tissue processing including lyophilization and antibiotic washing. Freeze-dried allograft retains good osteoconductive property and significantly decreases immunogenic reaction but at the cost of inferior osteoinductive and mechanical properties compared to fresh tissue [14]. Due to the excellent osteogenesis and osteoinductive properties already provided by BMAC and rhBMP-2/ACS, the author prefers freeze-dried allograft in this combination.

11.6 Membrane Carrier

Non-resorbable titanium mesh or resorbable poly(L-lactide) and poly(D, L-lactide) membrane can be used as a crib for containment of the mixture. Current membranes in the market allow for excellent penetration of host vascular tissue and migrating osteocompetent cells into the site of defect. The author prefers the use of the titanium mesh in most reconstructive cases due to its strength and rigidity, which prevent deformity and soft tissue collapse during the healing period [15].

Titanium also has high plasticity, allowing easy shape bending and contouring intraoperatively. Furthermore, its superior biocompatibility prevents additional inflammatory reaction compared to that of resorbable membranes. In the pediatric population, the author recommends using resorbable membrane due to the patient's growing facial bones, better inflammatory resiliency, and healing potential. In the author's experience, Resorb x (KLS Martin North America, Jacksonville, FL) had the least amount of bone reabsorption in radiographic studies 1 year postoperatively.

Above all the most important components from in situ tissue engineering are derived from growth factors upregulated and increased by BMP + BMAC (VEGA, EGF, etc.) which allows for an increase in mucosa/soft tissue healing for optimal infection-free host bed for bone regeneration. With adequate soft tissue, successful bone regeneration is likely.

11.6.1 Armamentarium

1. Large to medium rh-BMP2 (INFUSE) kit (large vs. medium kit) depending on the size of the defect. The author prefers a large kit if the defect is >6 cm.
2. Cortical cancellous bone (MTF (musculoskeletal tissue foundation) or the University of Miami Tissue Bank. Milled down to 1.0–2.0 mm particulate graft.
3. 120 cc harvest® bone marrow aspirate concentrate (BMAC®) system.
4. Titanium mesh 0.4 mm used as an undercarriage to inferiorly contain the graft material. The reabsorable mesh is an alternative in the pediatric patient.
5. DERMABOND (2-octyl cyanoacrylate) Ethicon USA, LLC.
6. TISSEEL (fibrin sealant) Baxter Healthcare Corporation.

11.6.2 Protocol

1. The patient is fully screened to have adequate mucosa for watertight primary tension free closure after ablation. The patient should be aware that bone graft will be aborted if tumor resection encompasses unexpected amounts of soft tissue not allowing for watertight primary tension free closure. The patient is also informed of maxillomandibular fixation for 3 weeks and NPO status/Dobhoff feeding tube for 2 weeks.
2. 3.0 g of Unasyn and 500 mg of Flagyl IV should be requested to be given >20 mins prior to the placement of the bone graft. If penicillin allergic, Levaquin is a good alternative.
3. After resection of the lesion and performing the grafting, soft tissue closure should not be under tension. Glycopyrrolate 0.2 mg was given 15 min before placement of the bone graft to minimize saliva pooling. If the anterior portion of the mandible is excised, care should be taken to anchor/suture the anterior digastric muscle to the reconstruction plate to resuspend the tongue into proper position.
4. Peridex 0.12% was poured into the defect site for a minute and then irrigated with saline thoroughly to decrease bacterial load in the graft site. A crib for the bone mixture was using either a titanium mesh or a poly(L-lactide) (PLLA) or poly(D, L-lactide) (PDLLA) reabsorable mesh (the latter having the advantage of being biodegradable).
5. A small strip of BMP and a gel-like mixture composed of platelet-poor plasma, calcium chloride, and thrombin is placed underneath the mucosa prior to closure in order to promote faster soft tissue healing.
6. Suturing technique and suture material used for closure play a major part in achieving watertight seal in the soft tissue. Our preferred technique for closure is a running vertical mattress suture with 3–0 prolene suture, oversewn with 3–0 chromic gut suture in an interrupted or continuous fashion.
7. A thin layer of DERMABOND was placed to act as a temporary sealant around questionable margins such as around teeth. Surface area must be dry.
8. TISSEEL fibrin sealant is placed over suture margins as another barrier against saliva.

9. The patient is placed in maxillomandibular fixation for 3 weeks unless the patient has a contraindication to MMF such as obstructive sleep apnea, seizures, etc.

11.6.3 Postoperative Management

1. Patient be NPO status with Dobhoff feeding tube for 2 weeks.
2. MMF for 3 weeks unless contraindicated or edentulous.
3. A course of antibiotics should be given for 7 days after surgery.
4. If perforation or tear is noticed within the first 2 weeks, the graft has a higher rate of failure. Local irrigation with Peridex and NPO status is the protocol for noticeable perforation and tears.
5. If noticeable signs of infection are noted (purulence with swelling, abscess formation, etc.), the graft should be debrided and soft tissue allowed to heal. Tissue should be allowed to heal infection-free for at least 3 months before reattempting to graft via extraoral approach or immediate composite microvascular flap.
6. Bone graft should allow to mature for 6 months before placement of dental implants (Figs. 11.1, 11.2, and 11.3).

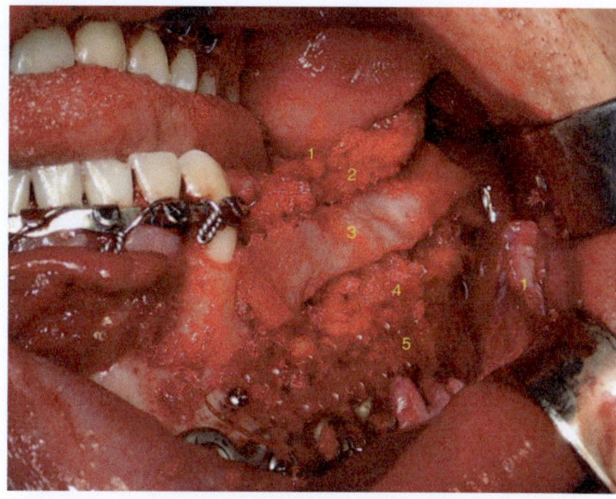

Fig. 11.1 Layers of tissue-engineered graft from bottom to top. (1) Preserved mucosa undermined and ready for tension-free closure. (2) Platelet-poor plasma coagulated with calcium thrombin mixture to form a biological membrane. (3) Strip of rhBMP-2 soaked with BMAC to act as a secondary membrane and stimulate growth factors to stimulate mucosal healing. (4) A mixture of crushed cortical cancellous bone, rhBMP-2, and (5) resorbable crib

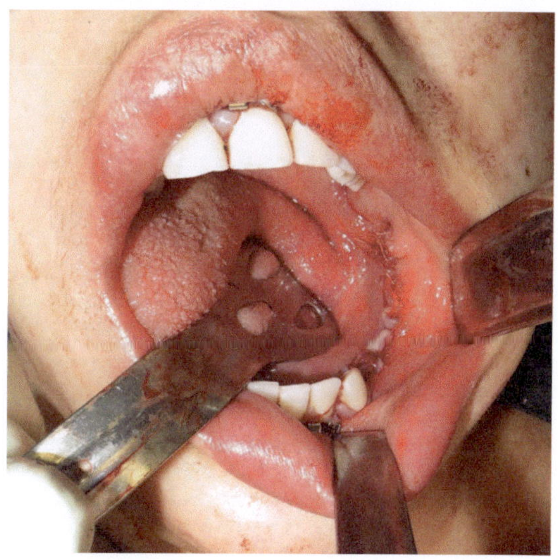

Fig. 11.2 "Watertight" tension-free closure

11.6.4 Management of Complications

1. Early dehiscence and exposure of bone within 3 weeks will need to be debrided and closed primarily. A delayed transcervical neck approach autogenous bone graft or immediate microvascular composite free flap is recommended.
2. Dehiscence after 3 weeks should be locally irrigated with chlorhexidine (Peridex). The patient should be treated with a 10-day course of amoxicillin 500 mg tid or Levaquin daily. The primary closure should not be attempted unless robust health margins are present, and the surgeon can achieve a tension-free closure. Usually, the tissue will granulate over with healthy mucosa over time. Debride tissue if the graft has signs of purulence or fistula development. A delayed transcervical autogenous bone graft or immediate microvascular composite free flap is recommended.
3. If bone regeneration is not significant for implant placement or stable prosthesis

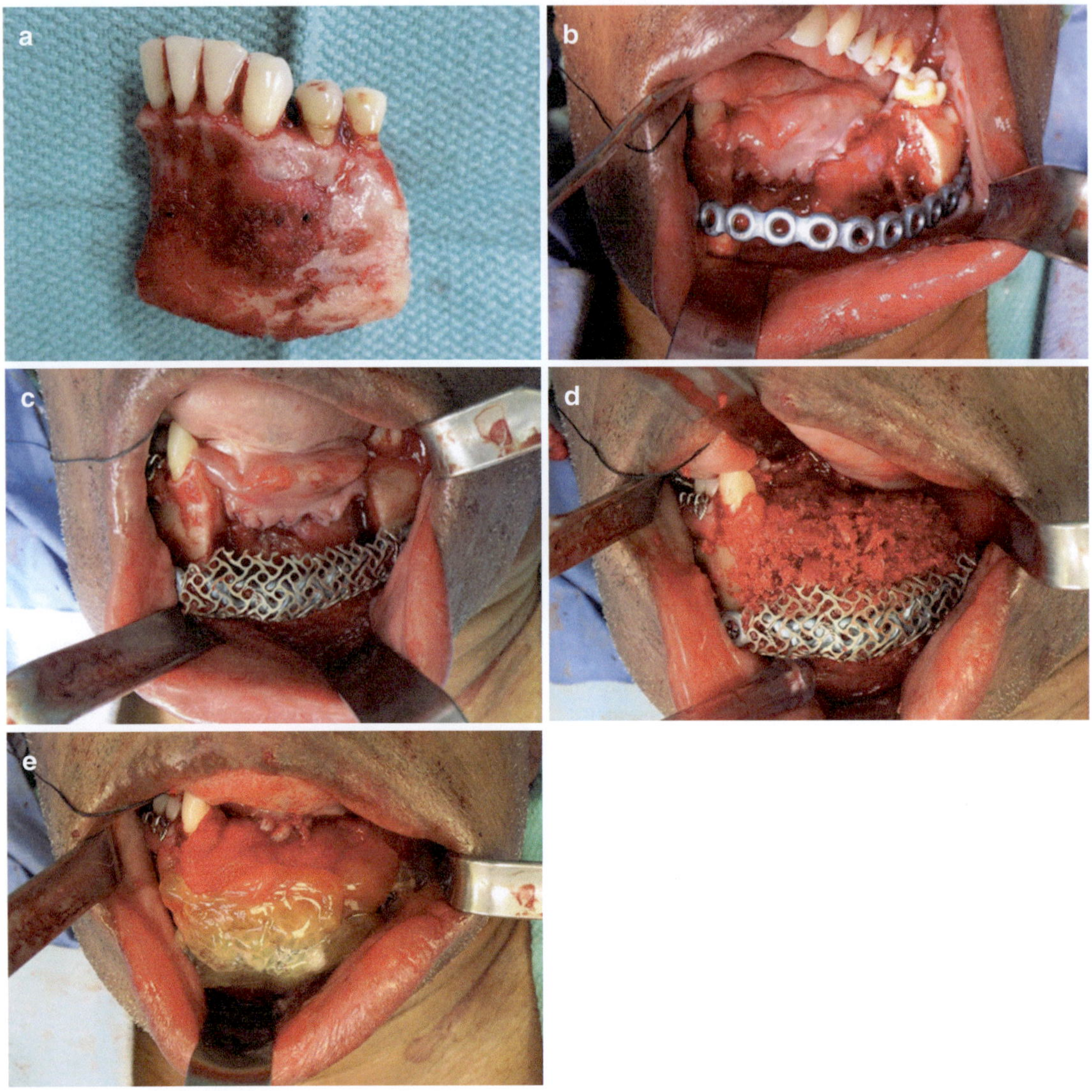

Fig. 11.3 (**a**) Transoral removal of ameloblastoma involving the anterior portion of the mandible. (**b**, **c**) Placement and reconstruction plate and mesh container. (**d**) Placement of tissue-engineered bone graft (BMAC + BMP + Allogeneic bone). (**e**) Placement of platelet-poor plasma coagulated with calcium thrombin mixture to form a biological membrane, on the most superior layer strip of rhBMP-2 soaked with BMAC to act as a secondary membrane and stimulate growth factors to stimulate mucosal healing

placement, a transcervical autogenous bone graft (vascular or nonvascular) is recommended (Figs. 11.4, 11.5, 11.6, 11.7, 11.8, 11.9, 11.10, 11.11, 11.12, 11.13, 11.14, and 11.15).

The advantage of a combined transcervical and transoral resection and immediate reconstruction is the ability to graft to the height of the adjacent resected mandible. Due to the need for tension-free mucosal closure, the bone is not

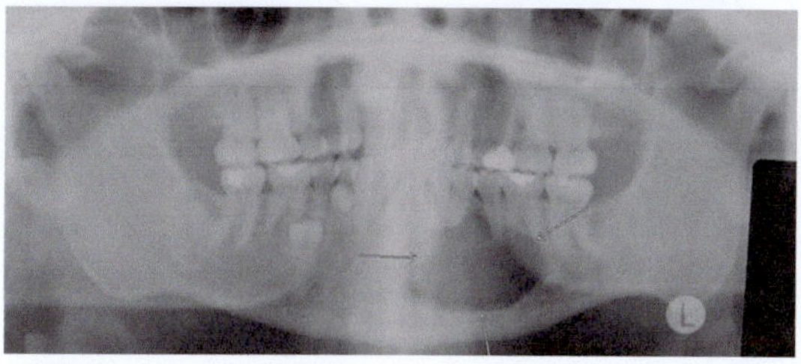

Fig. 11.4 Ameloblastoma of the anterior mandible on a healthy 32-year-old

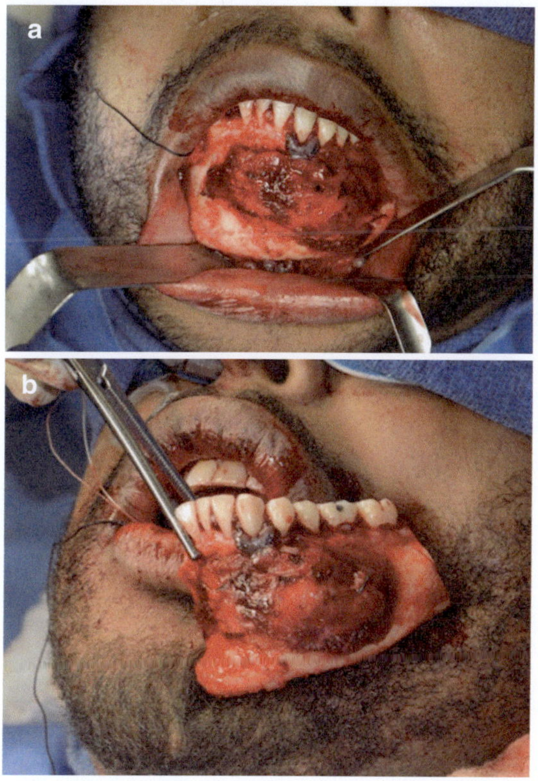

Fig. 11.5 (**a**, **b**) Resection of tumor transorally with 1 cm margins. Adequate soft tissue for tension-free watertight closure

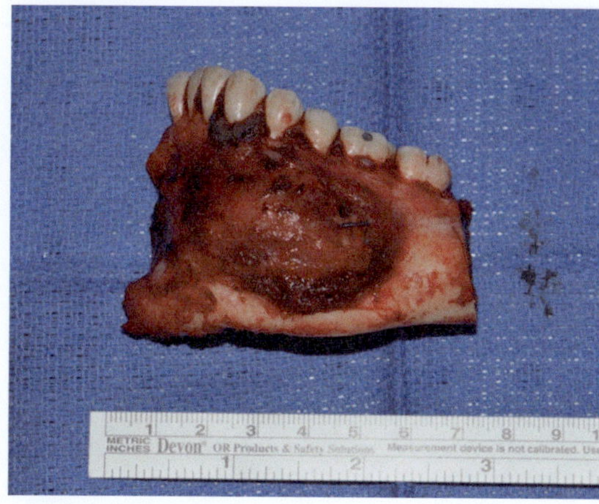

Fig. 11.6 Specimen

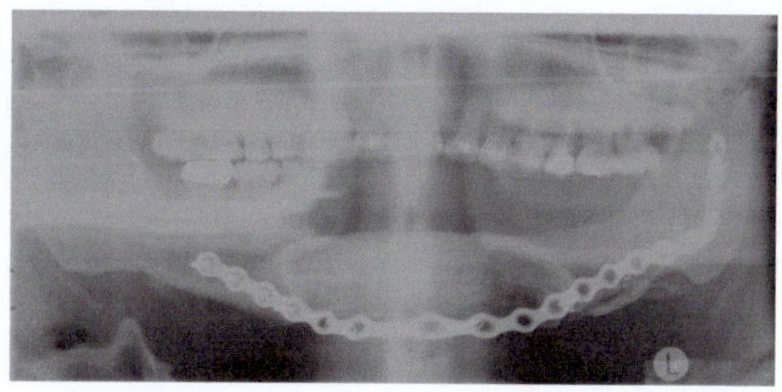

Fig. 11.7 6-month panorex demonstrating excellent bone consolidation

packed all the way to the top of the defect. In a transcervical approach, the wound is closed in the oral cavity first, and the bone graft is placed through the neck which tents up the graft to match the height of the adjacent mandibular bone. The following panorex display the bone height in immediate reconstruction in using a combination transcervical and transoral approaches (Figs. 11.16, 11.17, and 11.18).

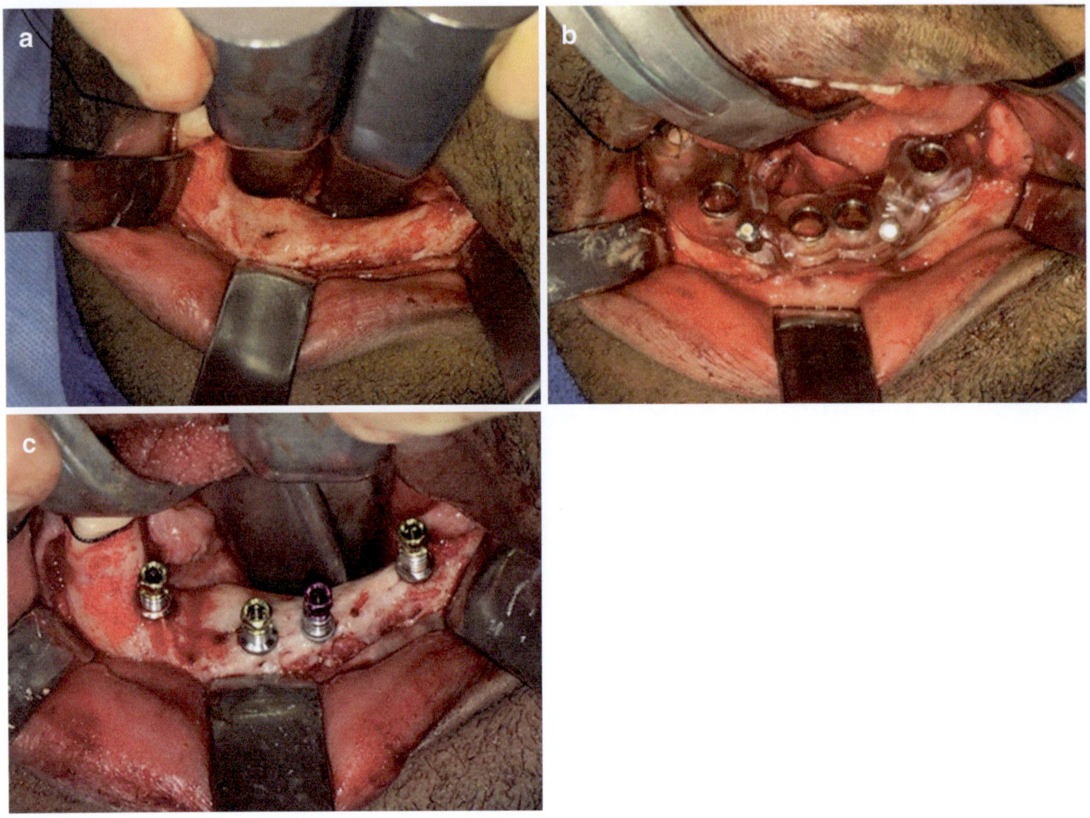

Fig. 11.8 (a–c) Placement of dental implants. Alveolar ridge demonstrating excellent height and width for placement of 4 × 5 × 13 mm implants

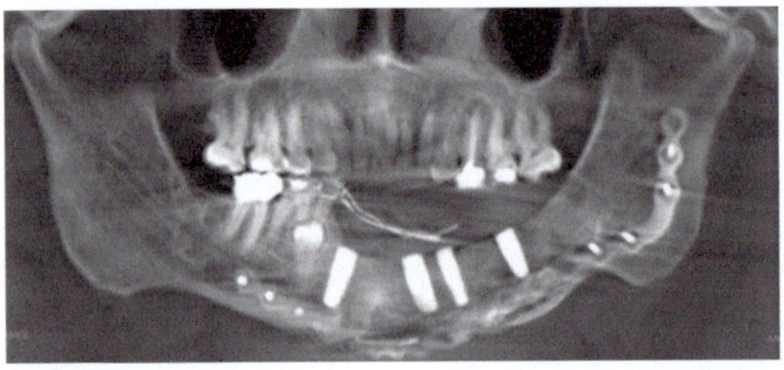

Fig. 11.9 Dental panoramic radiograph 1 year after the implant placement

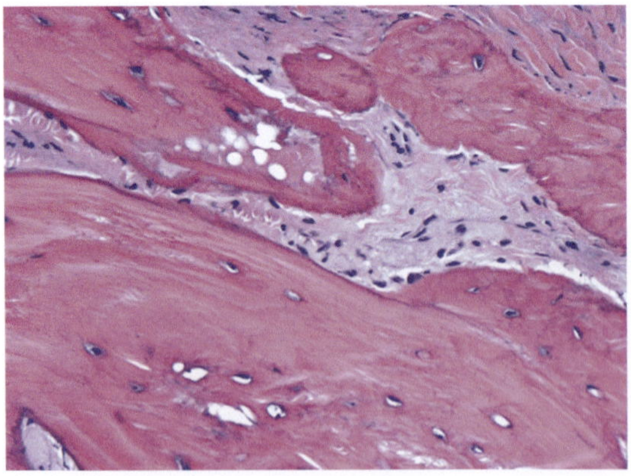

Fig. 11.10 Bone specimen was taken by a trocar during implant placement. H&E stain. 20× magnification: normal reactive bone with regular trabecular pattern with fibrosis. No remnants of cadaver bone after 8 months

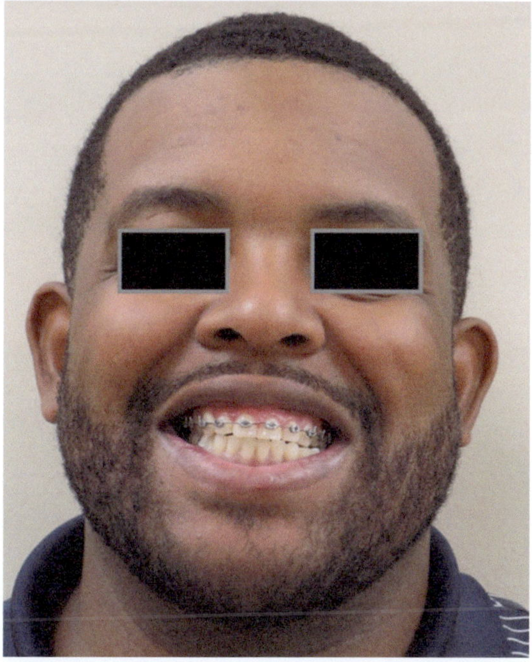

Fig. 11.11 Patient with final implant supported prosthesis

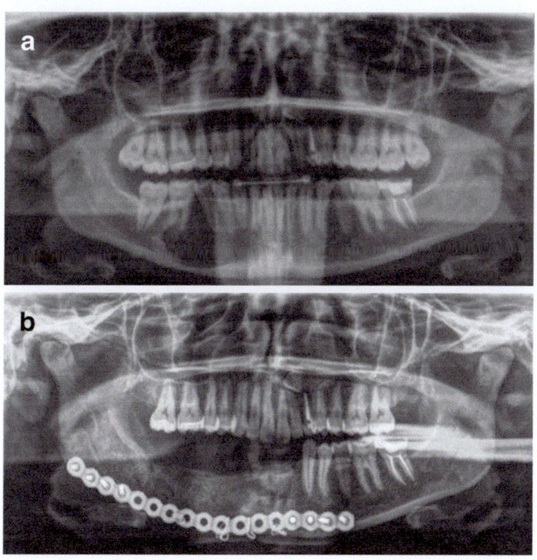

Fig. 11.12 (**a**) A 32-year-old female with ameloblastoma. Cone-beam computed tomography. (**b**) Eight months postop cone-beam computed tomography

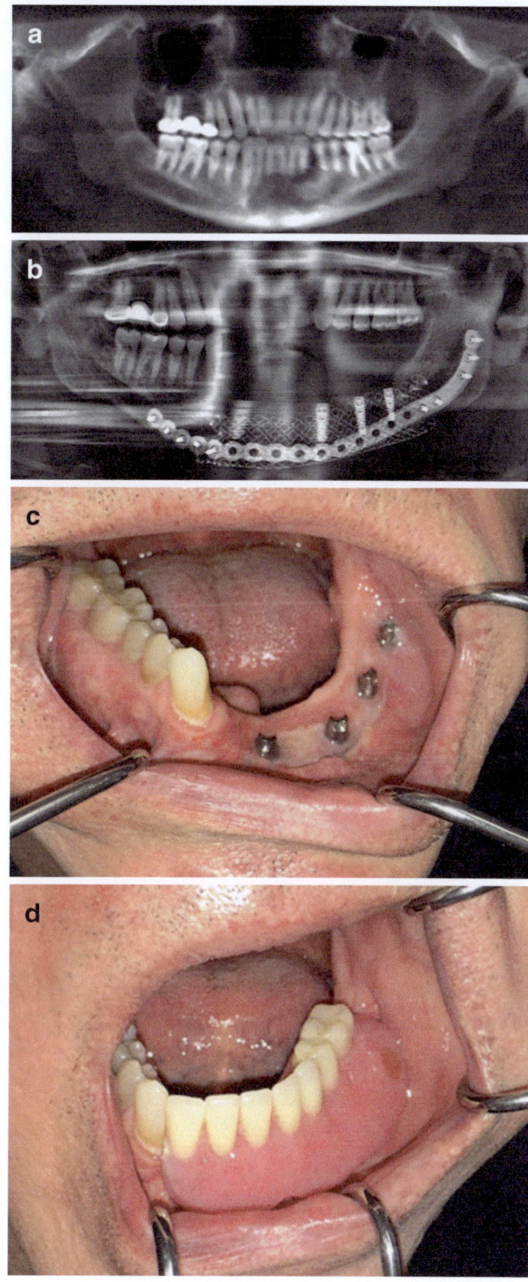

Fig. 11.13 (**a**) A 65-year-old male with ameloblastoma of the anterior mandible. (**b**) Cone-beam computed tomography of 14-month follow-up. (**c**) Fourteen-month follow-up after placement of dental and split-thickness skin graft. (**d**) Patient with final removal denture

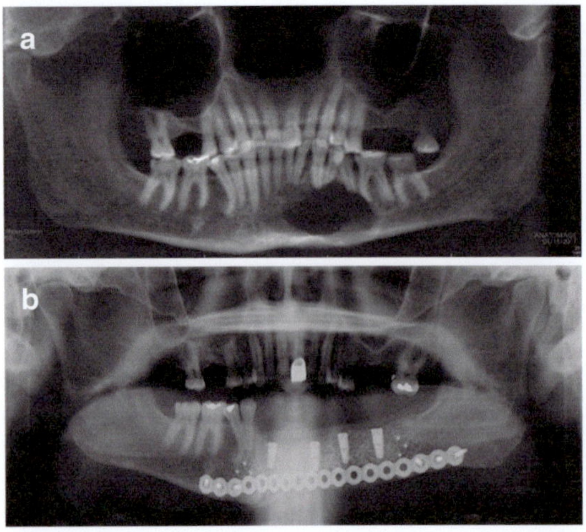

Fig. 11.14 (**a**) A 63-year-old male with ameloblastoma of the anterior mandible. (**b**) Nine- month follow-up after placement of four dental implants

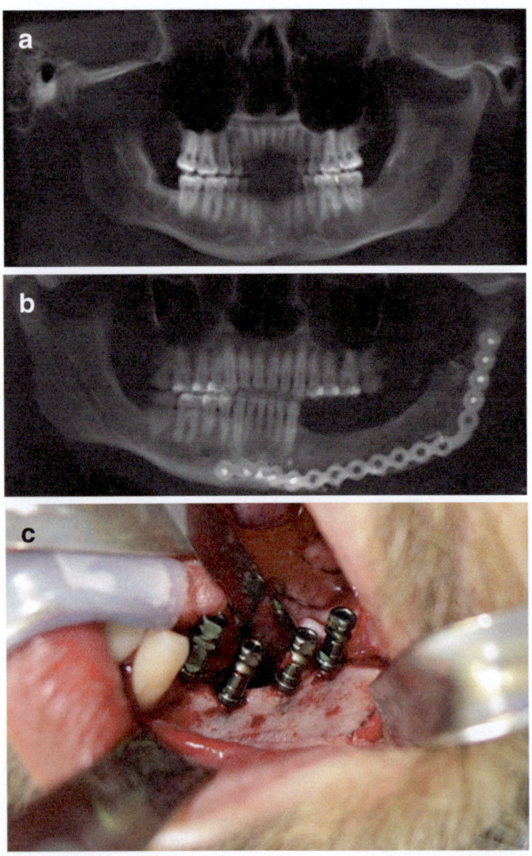

Fig. 11.15 (**a**) A 16-year-old male with ameloblastoma of the left body and ramus, extending into the condyle. An autogenous rib was harvested and used along with the tissue-engineered graft. Initial patient panorex. (**b**) One-year follow-up showing excellent bone formation and integration of rib into bone graft. (**c**) Placement of 4× dental implants. (**d**) Immediate panorex after implant placement

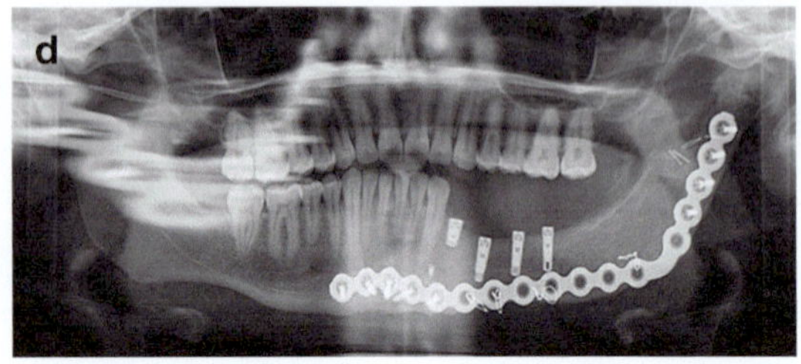

Fig. 11.15 (continued)

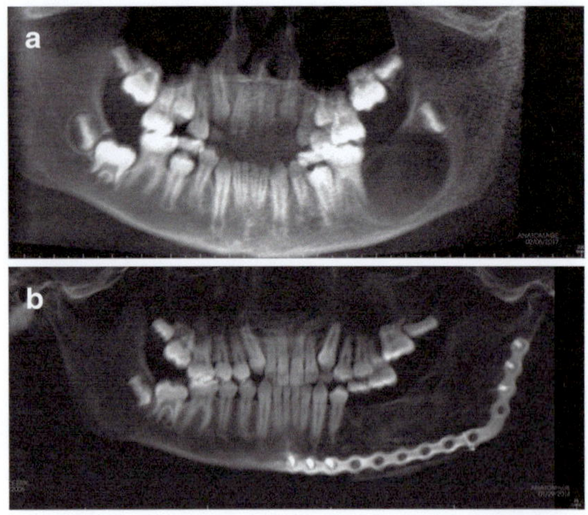

Fig. 11.16 (**a**) An 11-year-old male with ameloblastoma encompassing the right body and ramus of the mandible. Resection and reconstruction via combine transoral and transcervical neck approach. (**b**) CBCT 1-year follow-up.

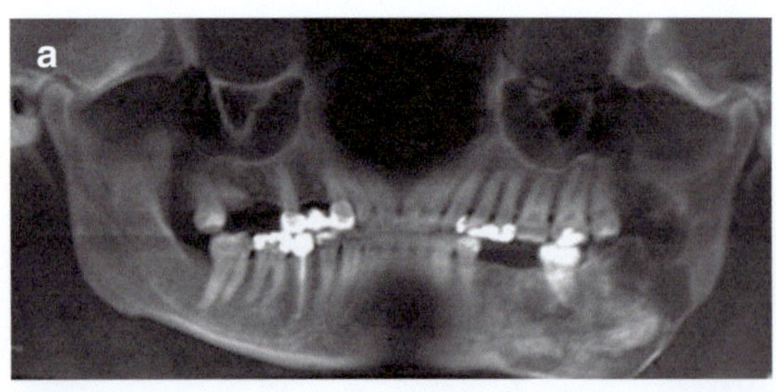

Fig. 11.17 (**a**) A 38-year-old female with aggressive ossifying fibroma involving left body and ramus. Combination approach was used to resect and immediately reconstruct. (**b, c**) Resection and placement of tissue-engineered bone graft. (**d**) Placement of implants 14 months from original surgery

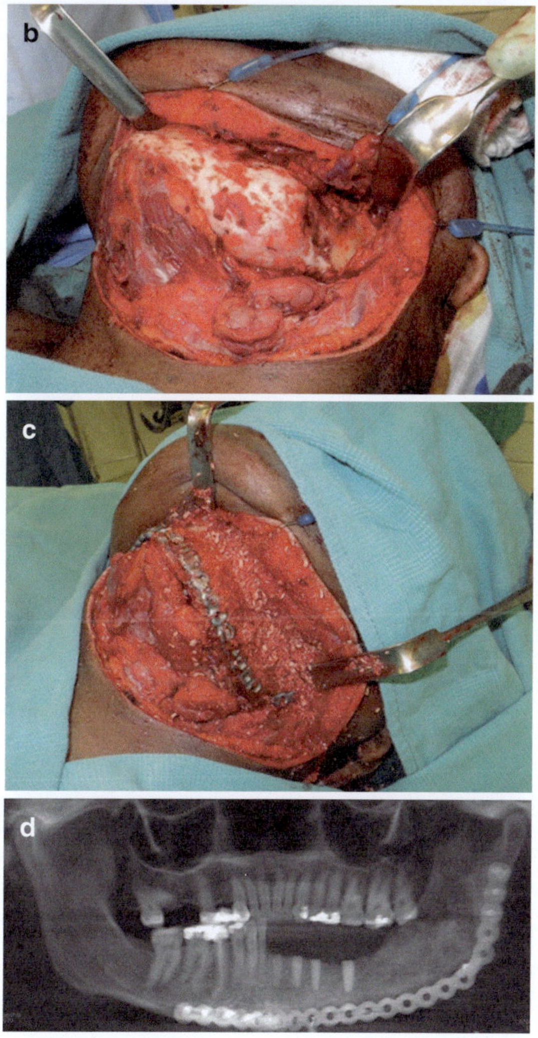

Fig. 11.17 (continued)

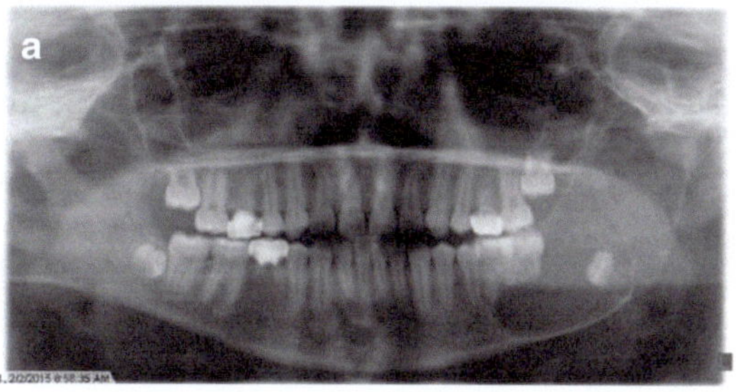

Fig. 11.18 (**a**) An 11-year-old female with ameloblastoma of the left body and ramus extending into the condyle. Combined approach with immediate resection and reconstruction. (**b**) Panorex 2-year follow-up

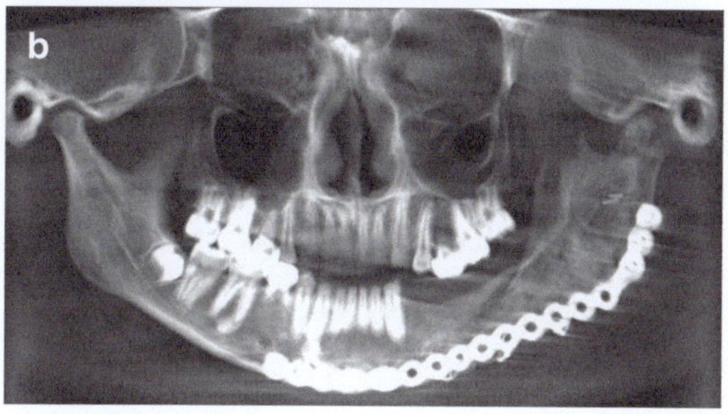

Fig. 11.18 (continued)

References

1. Vacanti CA. The history of tissue engineering. J Cell Mol Med. 2006;10(3):569–76.
2. Langer R, Vacanti J. Tissue engineering. Science. 1993;260:920.
3. Lindeboom JA, van den Akker HP. A prospective placebo-controlled double-blind trial of antibiotic prophylaxis in intraoral bone grafting procedures: a pilot study. Oral Surg Oral Med Oral Pathol Oral Radiol Endod. 2003;96(6):669–72.
4. Melville JC, Tursun R, Green JM, Marx RE. Reconstruction of a post-traumatic maxillary ridge using a radial forearm free flap and immediate tissue engineering (bone morphogenetic protein, bone marrow aspirate concentrate, and cortical-cancellous bone): case report. J Oral Maxillofac Surg. 2017;75(2):438.e1–6.
5. Smith B, Goldstein T, Ekstein C. Biologic adjuvants and bone: current use in orthopedic surgery. Curr Rev Muscoskelet Med. 2015;8(2):193–9.
6. Pittenger MF, et al. Multilineage potential of adult human mesenchymal stem cells. Science. 1999;284(5411):143–7.
7. Marx R. Atlas of oral and extraoral bone harvesting. Hanover Park: Quintessence Pub Co; 2010.
8. Payne KA, Didiano DM, Chu CR. Donor sex and age influence the chondrogenic potential of human femoral bone marrow stem cells. Osteoarthr Cartil. 2010;18(5):705–13.
9. McKay WF, Peckham SM, Badura JM. A comprehensive clinical review of recombinant human bone morphogenetic protein-2 (INFUSE® Bone Graft). Int Orthop. 2007;31(6):729–34.
10. Boyne PJ, Lilly LC, Marx RE, Moy PK, Nevins M, Spagnoli DB, Triplett RG. De novo bone induction by recombinant human bone morphogenetic protein-2 (rhBMP-2) in maxillary sinus floor augmentation. J Oral Maxillofac Surg. 2005;63:1693–707.

11. Fiorellini JP, Howell TH, Cochran D, Malmquist J, Lilly LC, Spagnoli D, Toljanic J, Jones A, Nevins M. Randomized study evaluating recombinant human bone morphogenetic protein-2 for extraction socket augmentation. J Periodontol. 2005;76(4):605–13.
12. Medtronic package insert no. M704819B001. INFUSE® Bone Graft for Certain Oral Maxillofacial and Dental Regenerative Uses. 2007.
13. Khan SN, Cammisa FP Jr, Sandhu HS, et al. The biology of bone grafting. J Am Acad Orthop Surg. 2005;13(1):77–86.
14. Kao ST, Scott DD. A review of bone substitutes. Oral Maxillofac Surg Clin North Am. 2007;19(4):513–21.
15. Rakhmatia YD, Ayukawa Y, Furuhashi A, Koyano K. Current barrier membranes: titanium mesh and other membranes for guided bone regeneration in dental applications. J Prosthodont Res. 2013;57(1):3–14.

Contemporary Management of Mandibular Resection-Related Nerve Injury

12

Shadi Alzahrani, Arshad Kaleem, Hisham Marwan, and Ramzey Tursun

12.1 Introduction

Trigeminal nerve sensory disturbance can result from trauma or pathology. The discomfort or pain due to sensory disturbance can be debilitating such that they cause great distress and affect the quality of life of the patients. Outcomes of nerve injury and recovery of normal function depend upon the age of the patient, the severity and type of injury, the injury time to surgical repair, the anatomy and alignment of the nerve fascicles, the method of surgical repair, and the surgical skills of the surgeon. End-to-end nerve coaptation remains the gold standard. However, in pathological conditions that require mandibular resection, the nerve is transected with a gap, and direct primary nerve repair may not be possible without putting undue tension at the site of repair. Such instances necessitate surgical repair with graft to establish continuity and function. In this chapter we will focus on surgical techniques and principles of nerve grafts for the inferior alveolar nerve.

Recovery following nerve injury is a complex phenomenon. It is influenced by the changes at the proximal and distal end of the nerve injury and the distance from the cell body. Once injured, inflammatory changes ensue, with the release of series of prostaglandins at the site of injury. This leads to sensitization of the nerve that causes allodynia and hyperalgesia [1]. If these inflammatory conditions continue for a longer time, the neurons undergo degeneration making recovery very difficult.

Microneurosurgery with direct repair, nerve graft, nerve transfer, and neural conduit is the mainstay of nerve injury treatment. Microneurosurgery of the trigeminal nerve is still in its infancy stages, but over the past few years, the technique has gained momentum by providing reliable and significant neurosensory functional results after nerve injury [2, 3]. Injury to the terminal branches of the trigeminal nerve is a well-known risk associated with a wide array of dental and surgical procedures. These injuries often heal spontaneously without medical or surgical intervention [3]. However, they can sometimes result in a variety of symptoms including pain and altered or loss of sensation which may have an impact on speech, mastication, and social interactions. These injuries can also cause significant morbidity when accompanied by neuropathic pain.

S. Alzahrani
Division of Oral and Maxillofacial Surgery, Miller School of Medicine, University of Miami, Miami, FL, USA

A. Kaleem · H. Marwan · R. Tursun (✉)
Oral/Head and Neck Oncologic and Microvascular Reconstructive Surgery, Division of Oral and Maxillofacial Surgery, Miller School of Medicine, University of Miami, Miami, FL, USA
e-mail: r.tursun@med.miami.edu

J. C. Melville et al. (eds.), *Regenerative Strategies for Maxillary and Mandibular Reconstruction*,
https://doi.org/10.1007/978-3-319-93668-0_12

A variety of conditions, such as mandibular tumors, osteomyelitis, and osteonecrosis caused by medication or radiation, require resection of the mandible; the inferior alveolar nerve may be sacrificed as a result [4]. Options for management of the associated nerve injury include (a) resection without nerve repair; (b) delayed nerve repair that carries the risk of extensive scarring around the nerve ends, making reconstruction more challenging; and (c) immediate nerve repair when possible with direct end-to-end neural tubes or use of autografts and allografts depending upon the nerve defect.

In general, a nerve defect greater than 1 cm cannot be repaired with direct sutures, without putting undue tension at the site of the repair. In such case nerve graft is recommended. Outcomes of neurosensory functions are variable and dependent upon many factors. The younger the patient, the higher the chances of successful nerve repair. The severity and mechanism of the nerve injury also influence the recovery, injuries resulting from GSW, blast injuries, and tractions have poor prognosis because they affect longer segment. The time between the initial injury and surgical repair immensely affects the return of normal sensation. With time nerve stumps degenerate along with fibrosis and scarring in the area, and surgical repair becomes less likely to succeed. This requires resection and freshening of the nerve stumps, making final nerve defect even larger.

The two most common autografts to repair trigeminal nerve injuries are sural nerve and greater auricular nerve. Autologous nerve grafts do have certain disadvantages, like donor site neurological deficit, scarring, wound infection, and second surgical site. Over the years, the use of allogenic (cadaveric) nerve graft has become popular.

Allogenic decellularized nerve grafts are devoid of cellular components and provide a scaffold for axonal regeneration. Nerve tissue regenerates at a rate of 1 mm/day; therefore, functional recovery of the nerve is related to the length of the gap between proximal and distal ends of the nerve and to the length of the nerve graft [5–7].

12.2 General Surgical Technique for Trigeminal Nerve Repair

1. The patient is placed in a supine position on the operating table. General anesthesia via nasal endotracheal intubation is initiated. The surgical table is positioned such that it allows the surgical microscope to be shared by two operators at either side of the table. The patient is prepped in the usual sterile manner.
2. Inferior alveolar nerve (IAN) can be approached surgically from an extraoral or an intraoral approach depending upon the surgeon's preference and the site of the nerve injury. In cases that require mandibular resection, access is mostly gained through a transcervical approach.
3. The proximal nerve stump can be prepared by either lateral cortical window or by "pull through" of the nerve at the lingula. Decortication of the medial bone distal to lingula may help to ease the pull without injuring the nerve (Figs. 12.1, 12.2, 12.3, 12.4, 12.5, 12.6, and 12.7).
4. In the pre-planned (virtual surgical planning) cases, often an accurate measurement is provided which helps in preparing the nerve graft, thus minimizing the time in the operating room and increasing the success of the reconstruction [8].
5. An allogenic nerve graft of appropriate diameter and length is prepared. For better coapta-

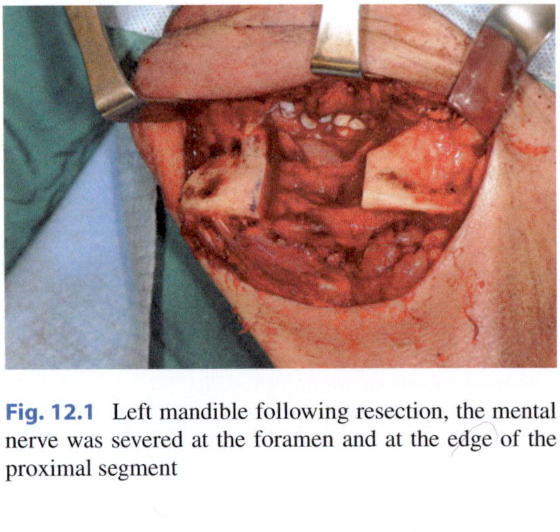

Fig. 12.1 Left mandible following resection, the mental nerve was severed at the foramen and at the edge of the proximal segment

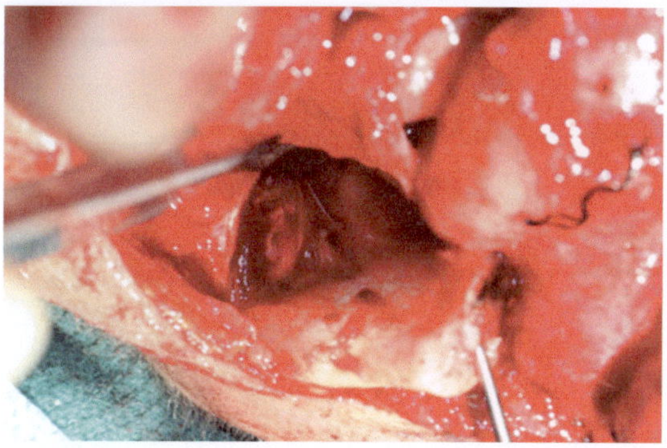

Fig. 12.2 Proximal segment of mandible, showing the IAN entering the foramen

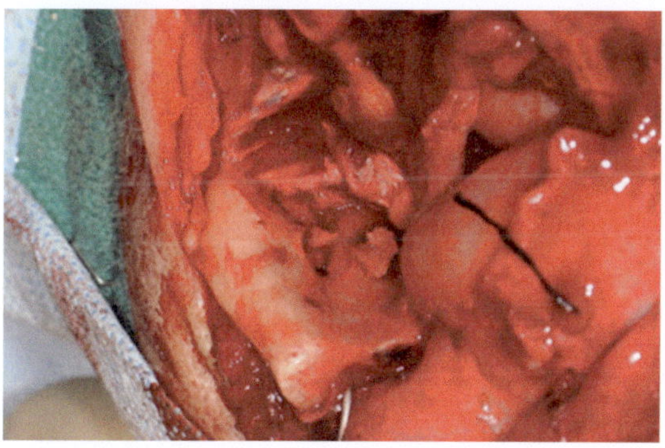

Fig. 12.3 The lingual cortex is decorticated at the lingula, which provides more length of the proximal nerve stump for easier coaptation

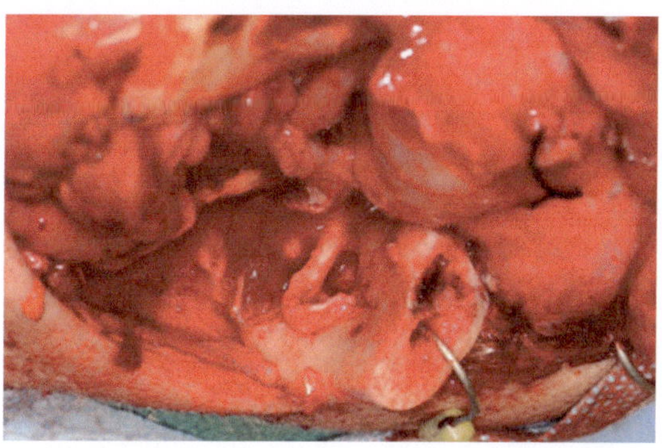

Fig. 12.4 The "pull-through" technique at the lingula has been accomplished

tion, a nerve conduit can be used to approximate the nerve ends with two epineural sutures at 3 o'clock and 9 o'clock positions.

6. Next, the whole construct will be brought to the defect site; the proximal nerve stump is

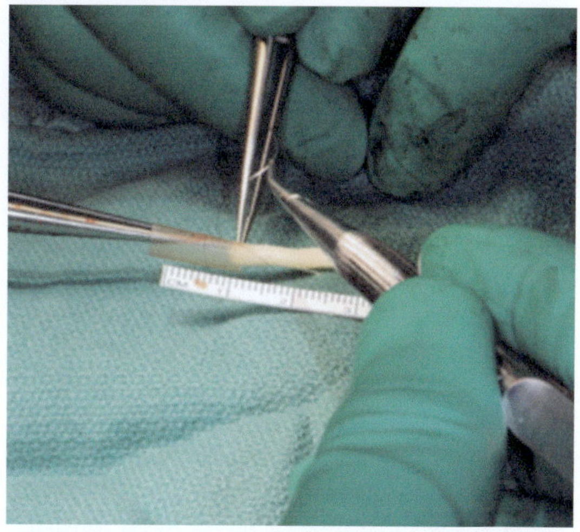

Fig. 12.5 The nerve graft is prepared by securing a conduit at either end of the nerve graft

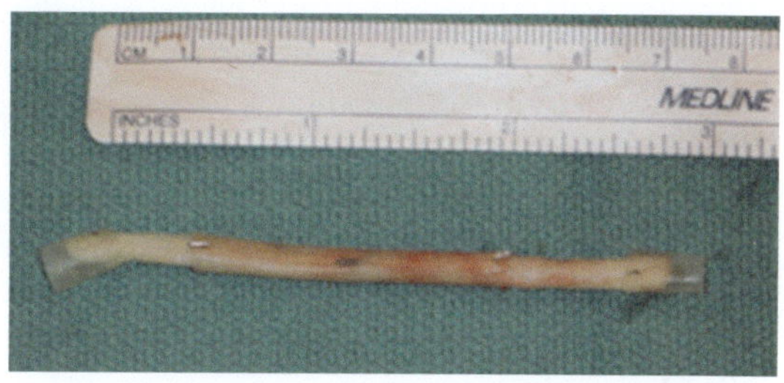

Fig. 12.6 The nerve "construct" is completed with conduits secured at both ends of the nerve graft. The nerve protector can be placed prior to bringing the construct to the surgical site or once the nerve repair has been completed

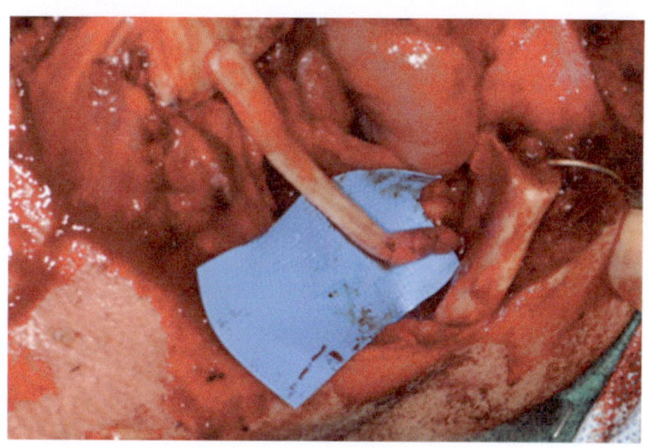

Fig. 12.7 The coaptation at the proximal end has been completed utilizing the nerve conduit. Two epineural sutures on each side of the conduit are used to secure the conduit to the nerve and aid in the alignment of the fascicles to minimize scar formation

intubated into the nerve conduit and secured with two epineural sutures. The same process is repeated at the distal nerve stump.

7. It is recommended to place a nerve protector around the allogenic nerve graft and secure the edges with micro clips. This may aid in identifying the position of the nerve on post-operative imaging when placing implants. In addition, placement of the nerve protector reduces postoperative scarring and improves recovery of sensory function (Figs. 12.8, 12.9, and 12.10).
8. If mandibular reconstruction with vascularized or non-vascularized bone graft is carried out at the same time of the nerve reconstruction, the choices are to:
   (a) Complete the nerve repair at both proximal and distal ends. In addition, secure the nerve protector to the lateral soft tissue bed. This places the nerve construct away from the mandibular defect, minimizing disturbing the coaptation of the nerve graft.
   (b) Delay the distal coaptation till after mandibular reconstruction is performed; both are useful to prevent injury or detachment of the newly coaptated nerve ends.

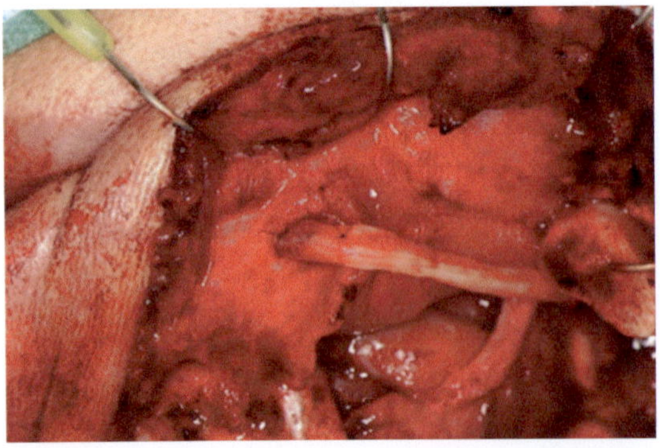

Fig. 12.8 Finally, the distal nerve coaptation is completed

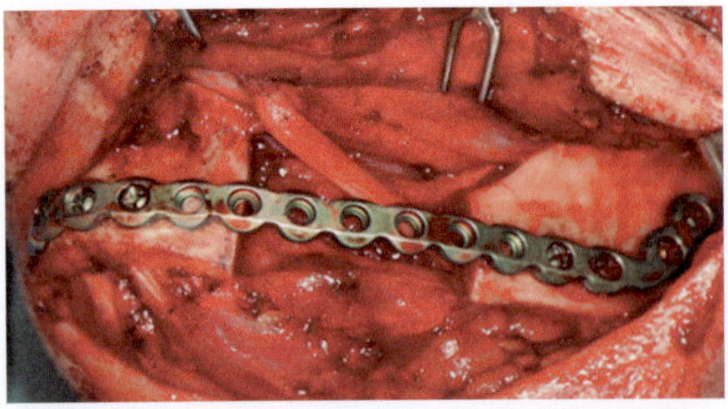

Fig. 12.9 The plate is secured at this stage

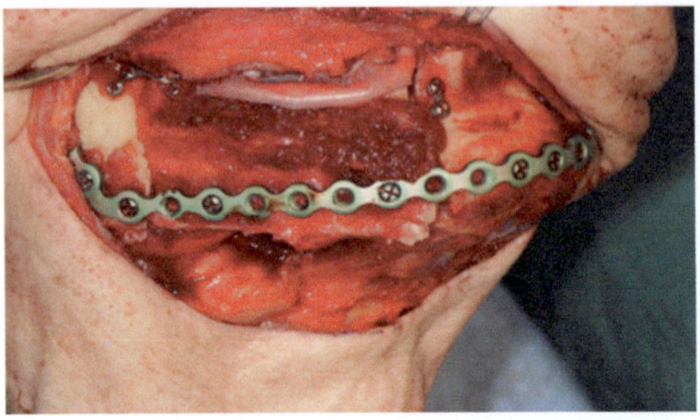

Fig. 12.10 Fibula free flap is secured in the defect. Notice the relationship of the nerve to the fibula. The nerve protector is secured to the buccal flap, moving the nerve graft buccally away from the reconstructed mandible, to prevent nerve injury if future implant placement is desired

12.3 Conclusion

Losing sensory function of the trigeminal nerve, after ablative head and neck procedures, greatly affects one's ability to function properly. Socially it can be stressful and embarrassing. The function recovery of the nerve after injury is dependent upon many factors.

- Age is an important indicator of nerve recovery; children and teenagers do better than adults.
- Timing of surgical repair is directly related to the return of neurosensory functional.
- Direct neural repair should be attempted when possible, without placing the sutures under tension. Defects larger than 1 cm should be repaired with a nerve graft.
- If possible, larger nerve defects should be reconstructed with grafts immediately without delay.
- Shorter nerve grafts do better than longer ones. Nerve tissue grows at a rate of 1 mm/day.

- If performed as a two-team approach, communication between the ablative and the reconstructive team is of paramount importance; this include tagging the nerve, appropriate lateral mandibular cortical window, cutting of the nerve ends leaving enough nerve stumps for coaptation, and finally appropriate positioning of the nerve graft in relation to the reconstructed mandible, if performed at the same time.
- Meticulous microsurgical technique and repair are the most important factor to determine the surgical success and return of neurosensory function of the inferior alveolar nerve after tumor ablative resection of the mandible (Figs. 12.11, 12.12, 12.13, 12.14, and 12.15).

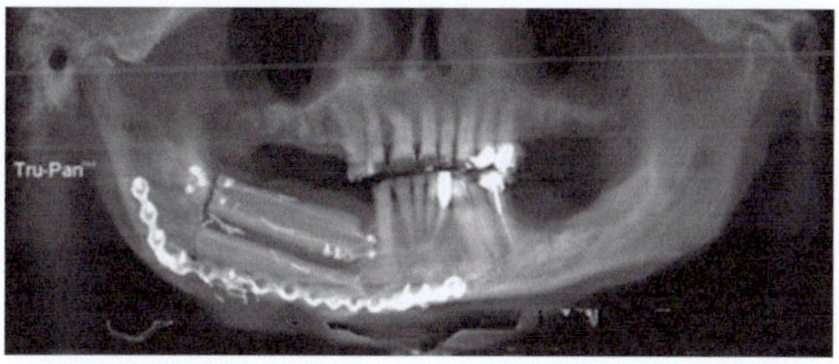

Fig. 12.11 Surgical clips aiding in future implant placement

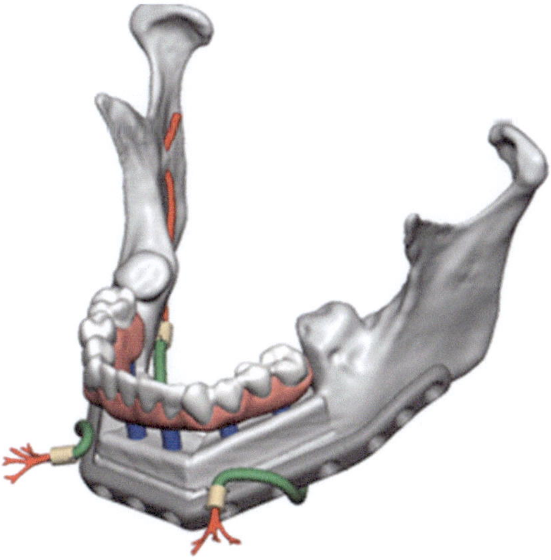

Fig. 12.12 Presurgical planning with virtual surgical planning of the whole case from resection to reconstruction of bilateral IANs

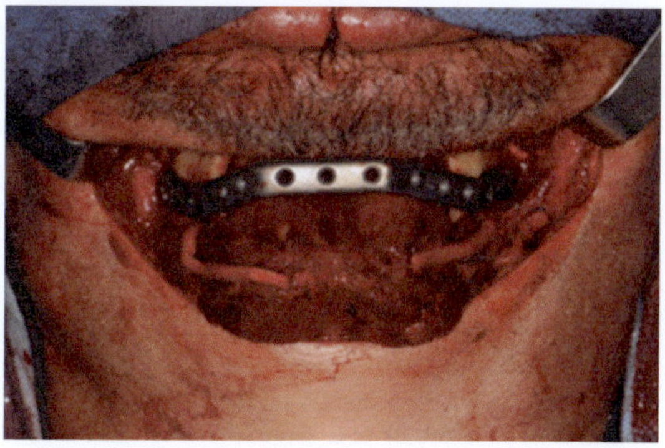

Fig. 12.13 Proximal coaptation nerve graft is completed bilaterally. Custom plate is secured to the mandible. Distal nerve coaptation is delayed until the end of the bone reconstruction

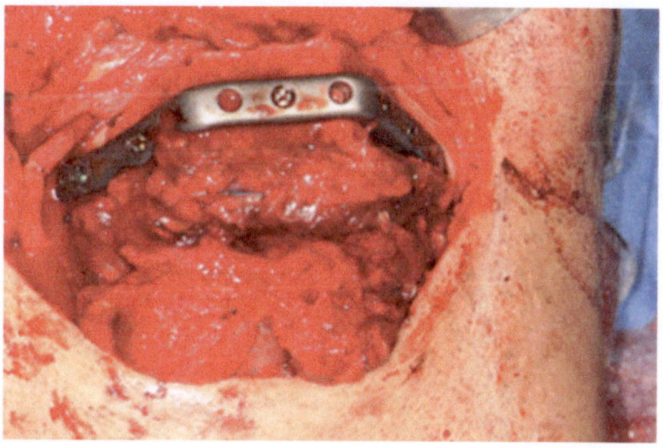

Fig. 12.14 Once fibula free flap is secured, the distal nerve coaptation is completed

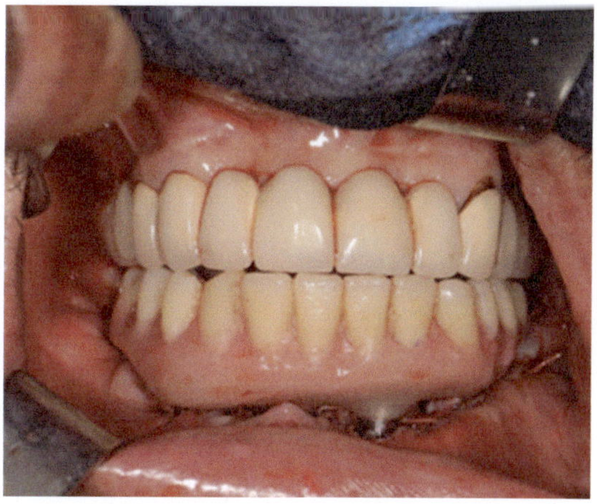

Fig. 12.15 Reconstructed mandible with immediate dental rehabilitation at the end of the surgery

Reference

1. Rosén A, Tardast A, Shi T-J. How far have we come in the field of nerve regeneration after trigeminal nerve injury? Curr Oral Heal Rep. 2016;3:309–13.
2. Ziccardi VB. S121: microsurgical repair of trigeminal nerve injuries. J Oral Maxillofac Surg. 2009;67:114.
3. Erakat MS, Chuang SK, Shanti RM, et al. Interval between injury and lingual nerve repair as a prognostic factor for success using type I collagen conduit. J Oral Maxillofac Surg. 2013;71(5):833–8.
4. Ziccardi VB. Microsurgical techniques for repair of the inferior alveolar and lingual nerves. Atlas Oral Maxillofac Surg Clin North Am. 2011;19:79.
5. Isaacs J. Treatment of acute peripheral nerve injuries: current concepts. J Hand Surg Am. 2010;35(3):491–7. quiz 498.
6. Johnson PJ, Newton P, Hunter DA, Mackinnon SE. Nerve endoneurial microstructure facilitates uniform distribution of regenerative fibers: a post hoc comparison of midgraft nerve fiber densities. J Reconstr Microsurg. 2011;27(2):83–90.
7. Karabekmez FE, Duymaz A, Moran SL. Early clinical outcomes with the use of decellularized nerve allograft for repair of sensory defects within the hand. Hand (New York, NY). 2009;4(3):245–9.
8. Miloro M, Markiewicz MR. Virtual surgical planning for inferior alveolar nerve reconstruction. J Oral Maxillofac Surg. 2017;75(11):2442–8. https://doi.org/10.1016/j.joms.2017.04.001.

Use of Skin and Soft Tissue Substitutes in Maxillofacial Reconstruction

13

Jonathon Jundt and Nagi Demian

13.1 Introduction

Soft tissue reconstruction of defects involving the scalp, face, intraoral cavity, or neck may be achieved through a combination of local advancement and rotation flaps, split thickness skin grafting, full-thickness skin grafting, microvascular free tissue transfer, negative pressure therapy, and dermal substitutes. The complexity of the wound defect often dictates the complexity of the reconstruction. This chapter focuses on the use of dermal substitutes in maxillofacial reconstruction and includes a review of the indications, selected current products, methods for application, and future developments.

Maxillofacial reconstruction for the purpose of this chapter includes the scalp, face, neck, and oral cavity. Each region requires a different approach due to anatomical and environmental differences. This chapter will discuss the most commonly used skin and soft tissue substitutes in use for maxillofacial reconstruction drawing similarities and differences in application and procedural steps. An exhaustive examination of all available products is not the aim. As of the publication of this book, a list of skin and soft tissue substitutes is provided.

Autologous skin grafting reliably reconstructs maxillofacial defects. Disadvantages include donor site morbidity, postoperative pain management, and inpatient requirements. Advanced bioengineered skin substitutes present an alternative option. Cultured epithelial autografts (CEA), micrografts, use of stem cells, or the combination of techniques has the theoretical advantage of decreased rejection, rapid healing, and treatment of larger defects. Disadvantages of these techniques include increased cost, expanded procedural complexity, and prolonged duration of treatment [1, 2]. Additionally, engineered epithelial constructs are too fragile on their own and often require additionally dermal or other scaffolding for mechanical strength.

Dermal substitutes play an important role in reconstructing full-thickness epithelial defects. In addition to improving pliability and decreasing scar formation, they act as scaffold for cell growth and differentiation [3].

J. Jundt (✉) · N. Demian
Department of Oral and Maxillofacial Surgery, University of Texas Health Science Center at Houston, Houston, TX, USA
e-mail: Jonathon.Jundt@uth.tmc.edu; Nagi.Demian@uth.tmc.edu

J. C. Melville et al. (eds.), *Regenerative Strategies for Maxillary and Mandibular Reconstruction*, https://doi.org/10.1007/978-3-319-93668-0_13

13.2 Characteristics of Soft Tissue Substitutes

The ideal substitute would have the following properties:

1. Biocompatible.
2. Lack of rejection and antigenicity.
3. Nontoxic.
4. Protection of exposed tissue.
5. Conforms to the host architecture.
6. Infection resistant.
7. Degrades or incorporated in growing tissue.
8. Easy to store and transport.
9. Cost-effective.

13.3 Types of Soft Tissue Substitutes

Skin and soft tissue substitutes may be classified as biologic or bioartificial. Biologic substitutes may be either xenograft or allograft based. Bioartificial substitutes are partially derived from living organisms, produced using living organisms, or entirely synthetic.

13.3.1 Biologic

Biologic substitutes are derived from previously animate objects and undergo processing to render the product safe for implantation. Processing involves lyophilization, gamma irradiation, and de-cellularization to reduce the potential for transmissible disease.

13.3.2 Bioartificial

A bioartificial skin or soft tissue substitute is a highly complex biological structure that is partially or completely synthetic in nature and performs the tasks of the structure it is intended to replace.

13.3.3 Indications

Indications for utilizing dermal substitutes in maxillofacial reconstruction depend on the chronicity of the wound. Dermal substitutes may be used for cicatrix prevention in acute wounds, granulation bed development in avulsive wounds, moisture loss and basement membrane replication in burn wounds, and closure in chronic wounds associated with diabetes, radiation, high velocity, or other multiply operated non-healing site [1, 3, 4, 5, 6, 7, 8, 9].

13.3.4 Common Skin Substitutes and Protocol for Usage

A complete description of all currently available skin substitutes is outside the scope of this chapter. The most commonly used skin substitutes in maxillofacial reconstruction will be covered in this chapter. Most skin substitutes contain extracellular matrix proteins that facilitate fibroblast migration, inosculation, imbibition, and proliferation of tissue. An exaggerated healing response is desired, and regeneration of the dermis or a neodermis is the desired result. This should simulate the process, whereby split-thickness skin grafts heal beginning with a fibrin network that insures graft adherence, plasmatic imbibition, inosculation, revascularization, lymphatic circulation, and hyperplasia of dermal fibroblasts to increase the thickness and/or replace the new epidermis. Initial immobility and lack of shearing forces are paramount in both skin grafts and skin substitutes. No skin substitute should be placed in an infected site as this could result in loss of the skin matrix system [10].

13.4 Integra Wound Matrix System

Integra was the first commercially available skin substitute developed in the 1980s by Yannas, primarily as an alternative to skin grafting for use in burn patients. It is a bilaminate, bioartificial dermal substitute with an inner layer of bovine collagen matrix that facilitates inosculation and subsequent development of a neodermis. The outer layer, comprised of silicone, functions as an environmental barrier that reduces moisture loss from the wound during the first 2–3 weeks after placement. Once the silicone layer is removed, the site is often grafted with a thin split thickness skin graft. Integra has been noted for a lack of significant scar tissue and a smooth neodermis that readily incorporates secondary grafting.

Applications in maxillofacial reconstructive surgery include revision of neck scar contractures, buccal defects from mucosal graft harvests for urethral reconstruction, wound coverage in partial glossectomy, osteoradionecrosis, and medication related osteonecrosis defects.

13.4.1 Protocol for Use

(Without Negative Pressure Therapy)

1. Using aseptic technique, peel open the outer pouch, and gently drop the inner foil pouch onto a sterile field or surface.
2. Using sterile technique, place foil pouch flat and peel open the inner pouch.
3. Remove the product, including the protective polyethylene sheets.
4. While holding the product with the tab, remove one polyethylene cover sheet. Turn the product and remove the second polyethylene cover sheet.
5. Using the tab, the product can now be placed into a basin containing a sterile saline solution. Carefully remove the tab from the product while rinsing for 1–2 min.
6. Keep the product in the basin until application.
7. Prepare wound bed using standard methods to ensure the wound is free of debris and necrotic tissue. If necessary, surgically debride the wound to ensure the wound edges contain viable tissue.
8. Cut the device to size, and apply immediately following wound bed preparation. Note: It is critical that the collagen layer be in direct contact with the prepared wound. The silicone layer, identified by the black threads, must be placed out (away from the wound bed). Do not apply upside down as the black threads must be clearly visible.
9. Integra Wound Matrix should be firmly secured using surgical staples, sutures, or other mechanical means. Any air bubbles should be carefully removed to maintain contact with the wound.
10. After application, use appropriate secondary dressings to maintain device adherence and protect the wound area. The optimum secondary dressing is determined by wound location, size, depth, and user preference.

13.4.2 Postoperative Management

1. Change the secondary dressings as needed. Frequency of secondary dressing change will be dependent upon the volume of exudate produced, type of dressing used, and clinician's need to inspect the wound bed for signs of infection or healing.
2. Removal of silicone layer from Integra Wound Matrix.
   (a) If edges of Integra Wound Matrix are loose before full healing has occurred, the silicone can be trimmed away from the loose areas until the entire wound has healed.
   (b) Remove the silicone layer of the Integra Meshed Bilayer Wound Matrix when the tissue underneath is healed, typically 14–28 days. The Matrix may be loose in spots.

(c) Remove by starting at one corner and pull gently. The silicone layer will peel off healed tissue relatively easily.
(d) If bleeding occurs, or if patient complains of excessive pain, stop and wait 1–2 additional days. Forced removal may result in wound reinjury (Figs. 13.1, 13.2, 13.3, 13.4, 13.5, 13.6, and 13.7).

(With Negative Pressure Therapy)

1. Using aseptic technique, peel open the outer pouch, and gently drop the inner foil pouch onto a sterile field or surface.
2. Using sterile technique, place foil pouch flat and peel open the inner pouch.
3. Remove the product, including the protective polyethylene sheets.
4. While holding the product with the tab, remove one polyethylene cover sheet. Turn the product and remove the second polyethylene cover sheet.
5. Using the tab, the product can now be placed into a basin containing a sterile saline solution. Carefully remove the tab from the product while rinsing for 1–2 min.

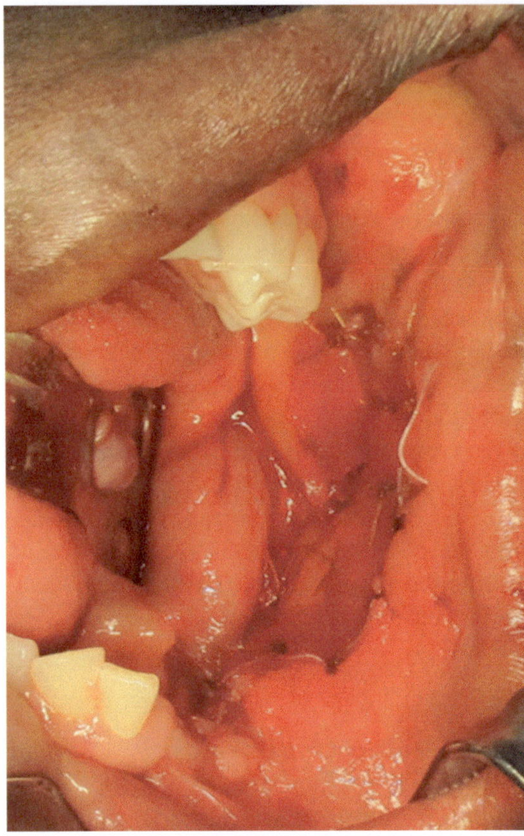

Fig. 13.2 Integra sutured in situ

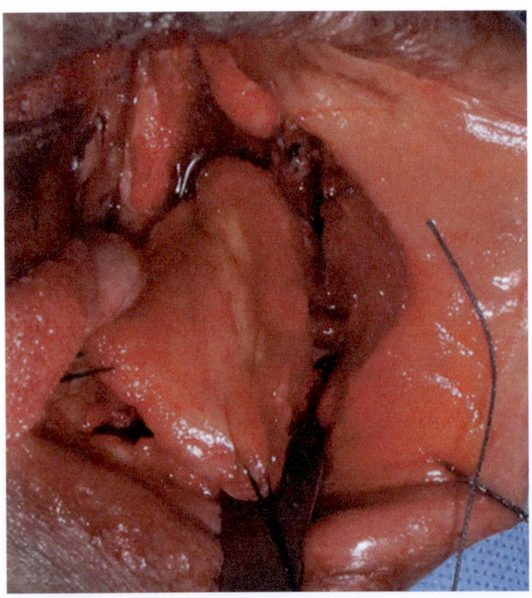

Fig. 13.1 Buccal mucosal defect

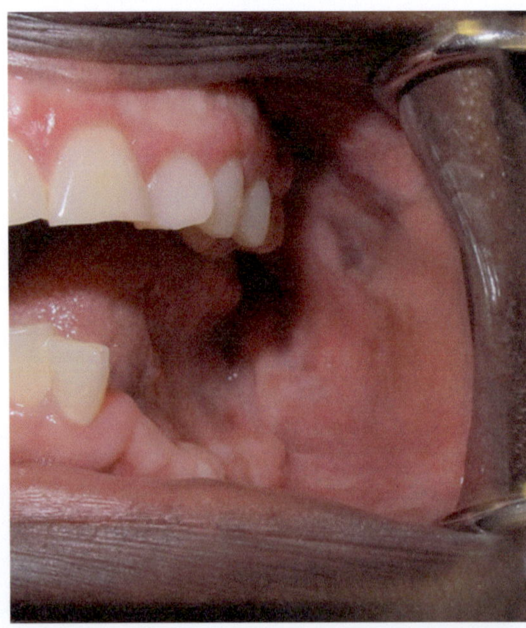

Fig. 13.3 Healing of buccal mucosal wound with Integra

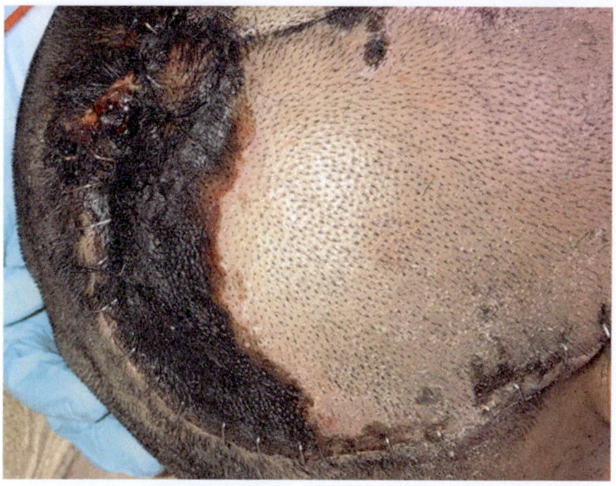

Fig. 13.4 Scalp wound

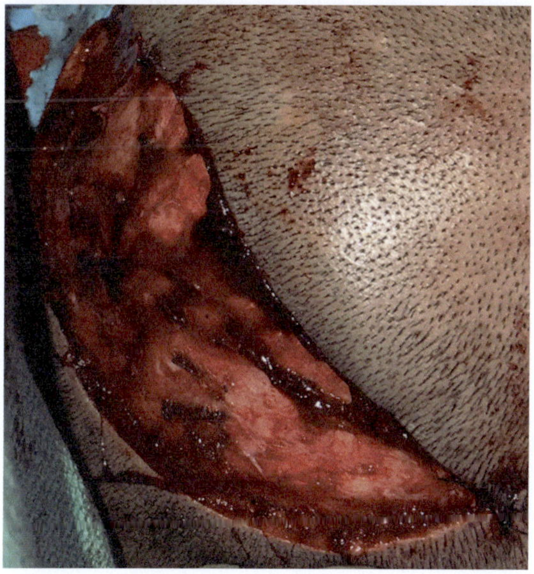

Fig. 13.5 Scalp debrided

6. Keep the product in the basin until application.
7. Prepare wound bed using standard methods to ensure wound is free of debris and necrotic tissue. If necessary, surgically debride the wound to ensure the wound edges contain viable tissue.
8. Cut the device to size, and apply immediately following wound bed preparation. Note: It is critical that the collagen layer be in direct contact with the prepared wound. The silicone layer, identified by the black threads, must be placed out (away from the wound bed). Do not apply upside down; the black threads must be clearly visible.
9. Integra Wound Matrix should be firmly secured using surgical staples, sutures, or other mechanical means. Any air bubbles

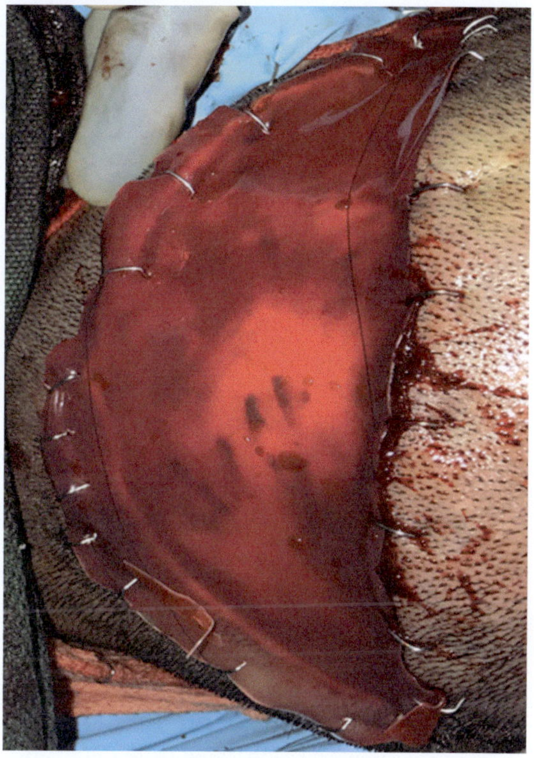

Fig. 13.6 Integra applied

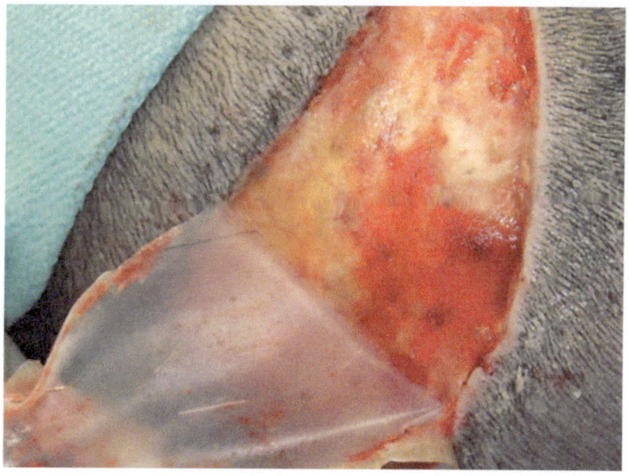

Fig. 13.7 Integra removed with healthy underlying tissue bed

should be carefully removed to maintain contact with the wound.

10. Apply negative pressure wound therapy device according to manufacturer's instructions for use, paying specific attention to contraindications, warnings, and precautions.

13.4.3 Postoperative Management

1. Utilize negative pressure wound therapy device according to manufacturer's instructions paying special attention to wound monitoring.
2. During dressing changes, carefully remove the negative pressure wound therapy device and secondary dressing, foam or film, using caution to not disrupt the Integra Wound Matrix.
3. Discontinue negative pressure wound therapy when the collagen matrix has been integrated and the silicone has separated from the Integra Meshed Bilayer Wound Matrix (typically 14–28 days).
   (a) When using negative pressure wound therapy, this time frame may be reduced. Removal of silicone layer from Integra Wound Matrix.
   (b) If edges of Integra Meshed Bilayer Wound Matrix are loose before full healing has occurred, the silicone can be trimmed away from the loose areas until the entire wound has healed.
   (c) Remove the silicone layer of the Integra Wound Matrix when the tissue underneath is healed, typically 14–28 days. The Matrix may be loose in spots.
   (d) Remove by starting at one corner and pull gently. The silicone layer will peel off healed tissue relatively easily. Caution: If bleeding occurs, or if patient complains of excessive pain, stop and wait 1 to 2 additional days. Forced removal may result in wound reinjury.

13.5 AlloDerm

AlloDerm is indicated for repair or replacement of damaged or inadequate integument tissue or for other homologous uses of human integument. First introduced in 1994, it has been a key tool for tissue reconstruction with a wide range of applications.

13.5.1 Protocol for Use

AlloDerm has two distinct sides, the "dermal" side and the "basement membrane" side. The dermal side absorbs blood. The basement membrane side repels blood. When applied to the wound bed in a grafting procedure, the dermal side should be placed against the wound bed, with the basement membrane side facing up. When applied as an implant, the dermal side should be placed against the most vascular tissue. To determine proper orientation once the graft has been rehydrated, add a drop of blood to both sides of the graft, and rinse with rehydration solution. The dermal side will have a bloody appearance, whereas the basement membrane side will appear pink.

13.5.1.1 Rehydration Step 1

1. Tear open the foil bag at the notch and remove the inner pouch. (Keep both the foil bag and inner pouch out of the sterile field.) Peel open the inner pouch and aseptically remove the tissue. Do not peel printed paper backing at this point in the process.
2. Submerge the tissue completely, and soak for a minimum of 5 min or until the backing separates from the AlloDerm.
   (a) Warming saline up to 37 °C and using gentle movement of AlloDerm in the solution speed the rehydration process. However, do not heat saline above 37 °C.
   (b) When rehydrating multiple pieces, ensure the pieces are not overlapping or clumping together as this may slow down the process. Use multiple bowls if necessary.

(c) Keep AlloDerm fully submerged by weighing it down, e.g., with sterile forceps.
(d) If you are having a problem with rehydration, gently wipe/rub both sides of AlloDerm, with a sterile gloved hand, to remove any excess cryoprotectant that may be creating a barrier between the AlloDerm and the rehydration fluid.

13.5.1.2 Rehydration Step 2

1. Using forceps, remove and discard the backing once it separates from the tissue. Then, aseptically transfer the tissue to a second bath sufficiently filled with rehydration fluid. Submerge completely, and soak until the tissue is fully rehydrated (thicker grafts may take up to 40 min).
2. Keep AlloDerm fully submerged by weighing it down, e.g., with sterile forceps. When AlloDerm is fully rehydrated, it is soft and pliable throughout. At this stage, it is ready for application to the surgical site.
3. AlloDerm may be aseptically trimmed to required dimensions.
4. Use AlloDerm within 4 h of rehydration.

13.5.2 Postoperative Management

1. The inner dressing of petrolatum-impregnated gauze must not be changed for at least 7 days. Optimum take of the AlloDerm and overlying autograft requires that both remain undisturbed during this period to allow revascularization and reepithelialization.
2. The outer layers of the dressing may require frequent changing during the first few days to prevent accumulation of fluid and bacteria. When changing, take extreme care not to disturb the ointment-impregnated gauze or the composite graft.
3. On or about day 7, the inner layer may be removed. A generous application of petrolatum-based ointment prior to takedown will prevent adherence and stress to the grafted area. Saline soaks may be effective during dressing removal.
4. It is normal for some areas of the AlloDerm graft to appear white/yellow during this time period.
5. Until the grafts are fully revascularized and reepithelialized, they should be re-dressed with antibiotic-impregnated fine mesh gauze or other nonadherent dressing. The damp saline layer may be eliminated at this time.
6. On day 7, the autograft may appear whiter than the surrounding epidermis and may only weakly adhere. This is normal.
7. Occasionally, all or part of the autograft may come off with the dressing. This does not necessarily mean regrafting is required. Enough epidermal cells may have migrated from the autograft to the basement membrane to make regrafting unnecessary.
8. With a meshed graft, the growth of epithelial cells under the thin dermal layer of the autograft may cause it to detach. However, the autograft may have already seeded the surface of the AlloDerm with enough keratinocytes to provide partial covering. Careful examination of the surface will help determine if regrafting is needed.
9. As the epidermis establishes over the entire AlloDerm surface and keratinocytes differentiate to form a cornified layer, protective dressings may be eliminated. Once a cornified layer has been established (10–14 days), bathing with mild soaps and limited activity may commence (Figs. 13.8, 13.9, and 13.10).

13.6 ACell (Cytal, MatriStem)

ACell is derived from the lamina propria and basement membrane of a porcine urinary bladder matrix. This combination acts as a scaffold for cell migration and enhancement of tissue growth.

Cytal Wound Matrix 1-Layer and Cytal Wound Matrix 2-Layer are intended for the management

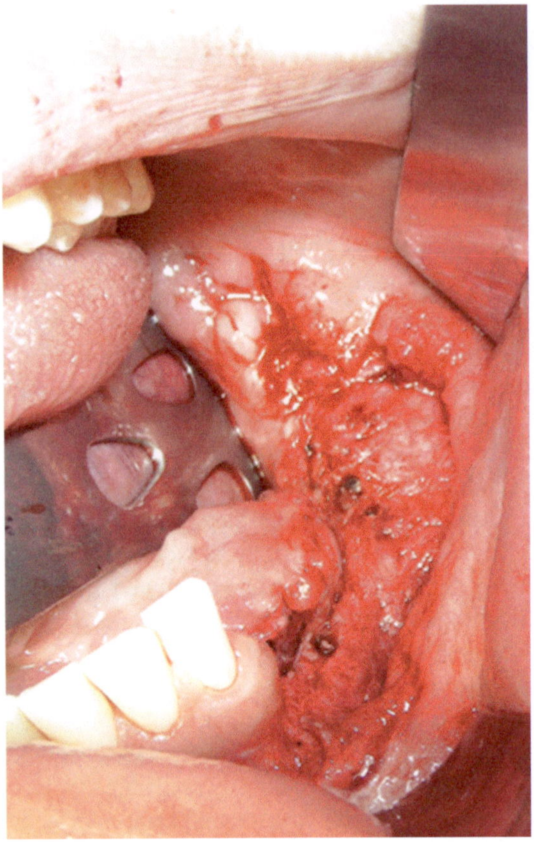

Fig. 13.8 Vestibuloplasty defect

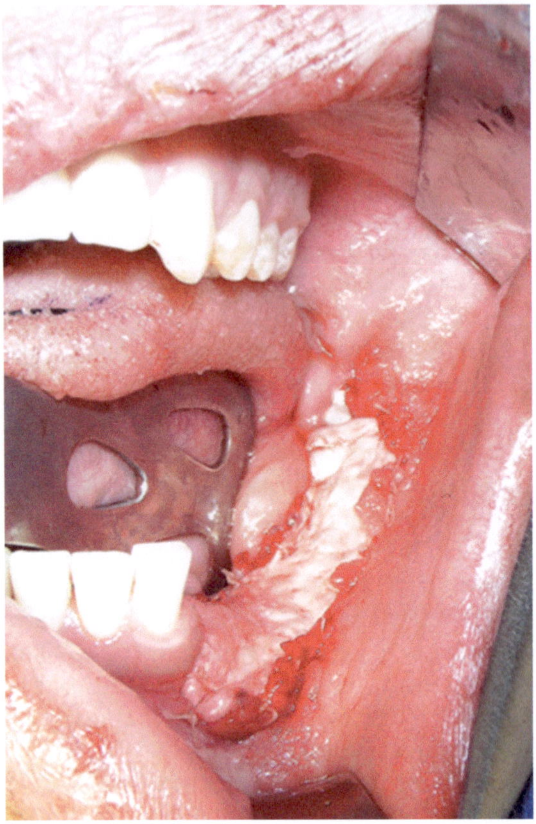

Fig. 13.9 AlloDerm insertion

of wounds including partial- and full-thickness wounds, pressure ulcers, venous ulcers, diabetic ulcers, chronic vascular ulcers, tunneled/undermined wounds, surgical wounds (donor sites/grafts, post-Mohs surgery, post-laser surgery, podiatric, wound dehiscence), trauma wounds (abrasions, lacerations, second-degree burns, skin tears), and non-infected draining wounds.

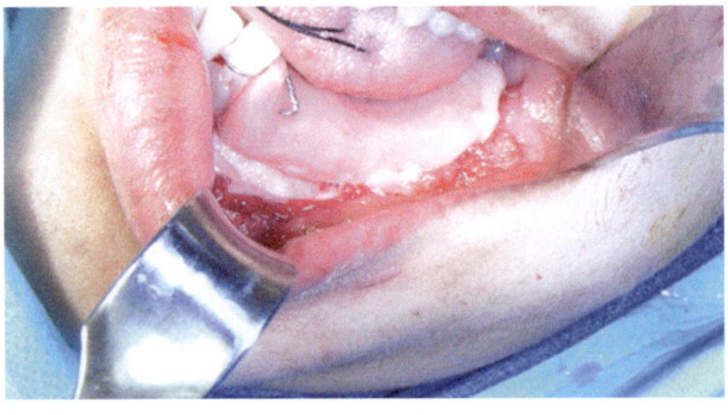

Fig. 13.10 Stent in situ

13.6.1 Protocol for Use

1. Clean wound bed and irrigate to remove exudate and debris.
2. Perform standard debridement procedures in line with internal institutional protocols.
3. Using standard aseptic/sterile technique, open device pouches, and remove device from inner pouch.
4. Hydrate device in a sterile dish with room temperature sterile saline (0.9%) or sterile lactated Ringer's solution prior to use.

Minimum hydration time: 2 min.
Maximum hydration time: 45 min

5. Cut device to the desired size, ensuring complete wound coverage.
6. Apply device directly to the wound bed.
7. Cover device with nonadherent dressing. Secure dressings to wound site.

Wet Wound: Apply an absorptive dressing. Secure dressings to wound site.

Dry Wound: Apply a hydrogel dressing to keep wound moist. Secure dressings to wound site.

13.6.2 Postoperative Management

1. Inspect primary dressing at least every 7 days.
2. Remove exudate, and apply new device to any non-covered wound areas as necessary.
3. During new device application, cover device with a new and appropriate secondary dressing, as described above.
4. Change all secondary dressings as appropriate for the wound type.
5. During secondary dressing changes, do not remove any remaining device that is present on wound surface.
6. As the device is resorbed and incorporated, it may form a caramel-colored gel and emit a pungent odor.
7. Rinse wound surface gently leaving any existing caramel-colored gel on wound surface.
8. Repeat wound care management process weekly until wound has epithelialized or until desired wound state is achieved (Figs. 13.11, 13.12, 13.13, 13.14, and 13.15).

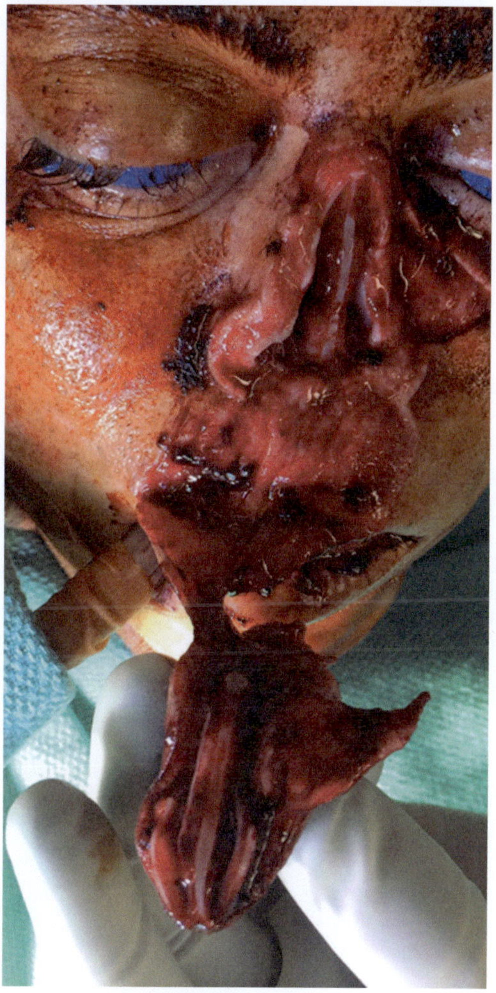

Fig. 13.11 Nasal avulsion

13.7 AmnioFix

Human amniotic membrane is a unique, thin, collagenous membrane derived from the submucosa of the placenta, the area in which the human fetus grows and develops within the mother's uterus. Human amniotic membrane consists of collagen layers including (1) basement membrane and (2) stromal matrix.

AmnioFix allografts are processed, dehydrated, and sterilized human amniotic membrane tissue grafts. Each allograft is packed in a double peel-pouch packaging configuration.

13.7.1 Protocol for Use

1. Carefully peel open the corner of the outer pouch, and present the inner pouch onto the sterile field.
2. In the sterile field, carefully peel a corner of the inner peel pouch, and allow the surgeon to

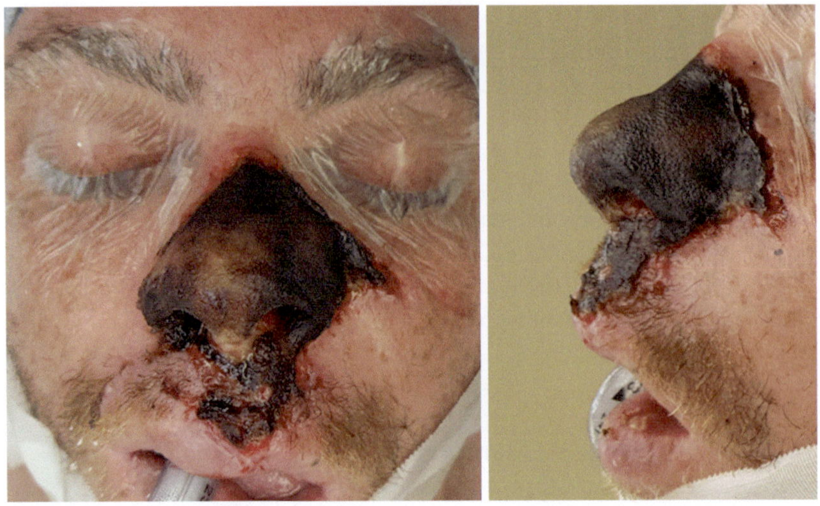

Fig. 13.12 Necrotic nasal tissue after surgical reattachment

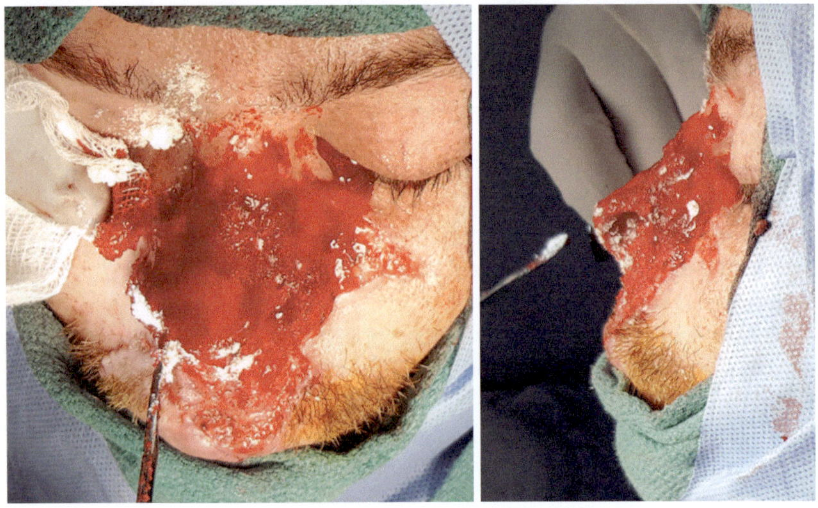

Fig. 13.13 Debridement and application of ACell MicroMatrix

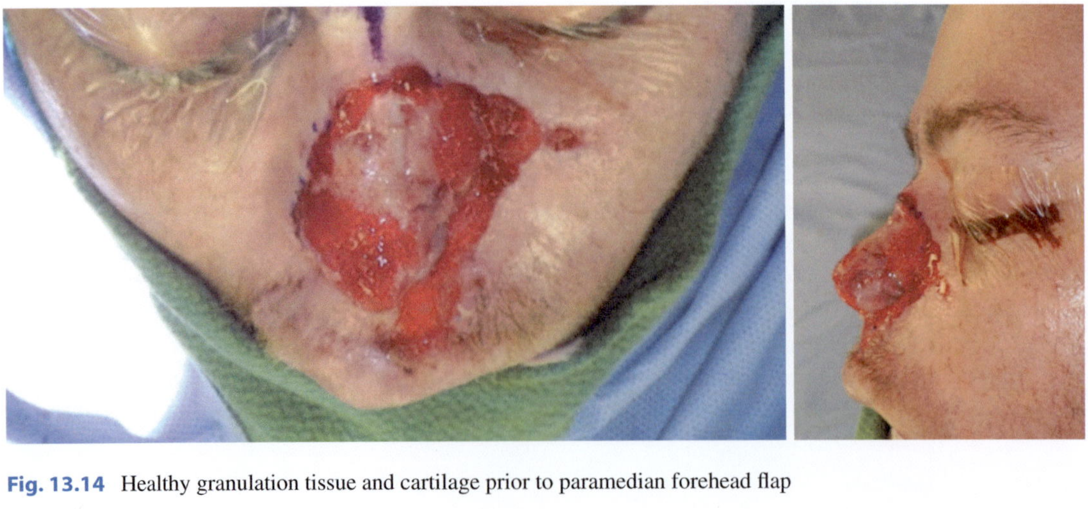

Fig. 13.14 Healthy granulation tissue and cartilage prior to paramedian forehead flap

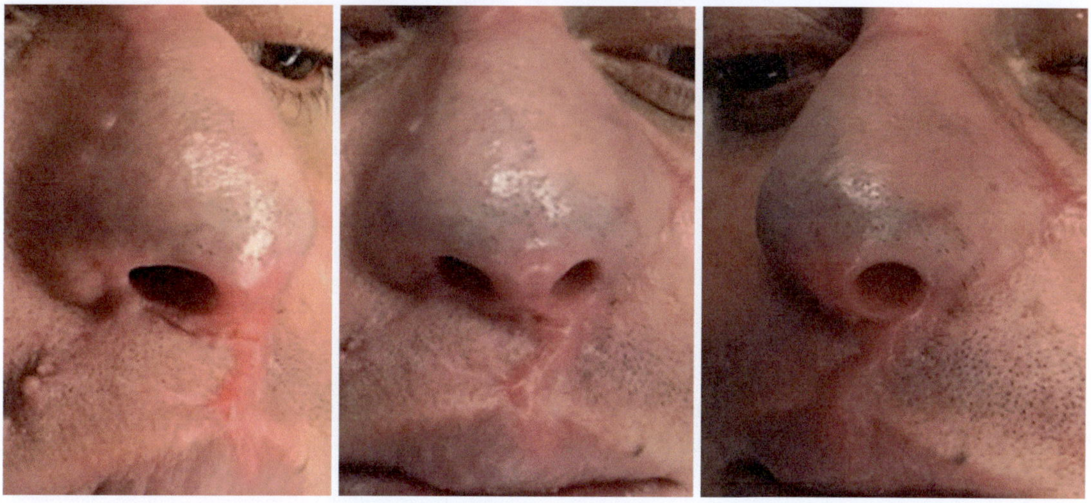

Fig. 13.15 Final result prior to laser hair removal and scar revision

grasp the allograft with fingers or non-toothed, sterile forceps.

3. In its dry state and prior to hydration, the surgeon may cut the allograft with sharp scissors to the appropriate and approximate size required.
4. The allograft should then be placed on the surgical site, using the orientation of the embossment lettering as a guide. Proper orientation of the allograft can be noted when the embossment nomenclature reads correctly from left to right. The orientation of the allograft at the surgical site will vary based on the surgical indication.
5. The allograft can then be hydrated while on the surgical site with sterile saline solution. Apply several drops of sterile solution to the allograft at 1- to 2-min intervals for a period of 5–10 min.
6. During and following hydration, the embossment on the allograft will begin to fade. It may take several minutes for the embossment to completely fade.
7. Following the recommended hydration period, the surgeon should monitor the effect of hydration by visual inspection and while manipulating the allograft. Some allografts may take slightly longer to hydrate.
8. Absorbable and nonabsorbable suture material and/or tissue adhesives can be used to fixate AmnioFix allografts to the surgical site.

13.8 Future Development

Development in tissue-engineered products will likely result from changes in delivery and application methods, site preparation, and adjunctive methods in improving wound healing.

As three-dimensional printing of biological scaffolds continues to improve, a change from packaged bioartificial products to on-demand site-specific applications may develop. This would be subject to quality control and cost limitations but has the potential to improve patient-specific outcomes.

A skin printing and autografting patent was issued on May 21, 2015. Claim 14 describes a wound treatment product comprising a resorbable material shaped as a honeycomb structure, the honeycomb structure being relatively flat, a quantity of living cells, and optionally one or more selected from the group consisting of (a) collagen, (b) collagen matrix, (c) collagen matrix proteins, and (d) extracellular matrix proteins, atop the honeycomb structure.

It is likely that dermal substitutes will continue as a mainstay of treatment for chronic and acute wounds of the head and neck region.

References

1. Feinberg SE, Aghaloo TL, Cunningham LL Jr. Role of tissue engineering in oral and maxillofacial reconstruction: findings of the 2005 AAOMS research summit. J Oral Maxillofac Surg. 2005;63(10):1418–25.
2. Chua A, Khoo Y, Tan B, Tan K, Foo C, Chong S. Skin tissue engineering advances in severe burns: review and therapeutic applications. Burns Trauma. 2016;4:3.
3. Shahrokhi S, Anna A, Jeschke M. The use of dermal substitutes in burn surgery: acute phase. Wound Repair Regen. 2014;22(1):14–22.
4. Halim AS, Khoo TL, Yussof SJM. Biologic and synthetic skin substitutes: an overview. Indian J Plast Surg. 2010;43:23–8.
5. Debels H, Hamdi M, Abberton K, Morrison W. Dermal matrices and bioengineered skin substitutes: a critical review. Plast Reconstr Surg Glob Open. 2015;3(1):284.
6. Girod DA, Sykes K, Jorgensen J, Tawfik O, Tsue T. Acellular dermis compared to skin grafts in oral cavity reconstruction. Laryngoscope. 2009;119:2141–9.
7. Al-Rubaiy L, Al-Rubaiy K. Skin substitutes: a brief review of types and clinical applications. Oman Med J. 2009;24(1):4–6.
8. Ferreira MC, Paggiaro AO, Isaac C, Neto NT, Dos Santos G. Skin substitutes: current concepts and a new classification system. Rev Bras Cir Plást. 2011;26(4):696–702.
9. Ehrenreich M, Ruszczak Z. Update on dermal substitutes. Acta Dermatovenerol Croat. 2006;14(3):172–87.
10. Fahrenbach EN, Qi C, Ibrahim O, Kim JY, Alam M. Resistance of acellular dermal matrix materials to microbial penetration. JAMA Dermatol. 2013;149(5):571–5.

Part III

State of the Art Free Vascular Transfer for Maxillomandibular Reconstruction

14 Reconstruction of Maxillary Defects

Baber Khatib, Ashish Patel, Eric J. Dierks, R. Bryan Bell, Melissa Amundson, and Allen Cheng

14.1 Introduction

The maxilla is a pyramidal-shaped bone that is at the crossroads of several compartments of the viscerocranium. It provides the floor of the orbit, supporting its position and serving as the attachment for the inferior oblique muscle. It houses the nasal cavity and maxillary sinuses. It is the superior boundary of the oral cavity and forms a portion of the anterior skull base, separating the neurocranium from the viscerocranium. It provides the bony foundation for the maxillary dentition. It provides the key vertical and horizontal buttresses of the midface. It serves as the insertion points of mimetic muscles and pharyngeal muscles. Therefore, the maxilla's integral role in both facial form and function cannot be overstated.

B. Khatib
Fellow in Microvascular Reconstructive Surgery, Providence Cancer Center, Portland, OR, USA

M. Amundson
Oral and Maxillofacial Surgery, Legacy Emanuel Medical Center, Portland, OR, USA

A. Patel · E. J. Dierks · R. Bryan Bell · A. Cheng (✉)
Oral and Maxillofacial Surgery, Legacy Emanuel Medical Center, Portland, OR, USA

Head and Neck Surgery, Providence Cancer Center, Portland, OR, USA

Oral/Head and Neck Oncology, Legacy Good Samaritan Cancer Center, Portland, OR, USA
e-mail: Chenga@hnsa1.com

Total and partial defects of the maxilla can be due to a variety of disease processes. These include ablative defects for neoplasms, both benign and malignant, trauma, osteonecrosis (radiation or medication related), congenital defects, and necrotizing infections.

As with all reconstructive surgeries, the fundamental goals are to establish a return-to-normal or near-normal form and function. For maxillary defects, this includes the following specific goals. Creating adequate support for the globe is important to maintain normal appearance and range of motion. Establishing barriers between the oral cavity, nasal cavity, and paranasal sinuses is essential for preserving normal speech, swallowing, and breathing. Recreating the horizontal and vertical buttresses of the midface is critical for facial projection and facial height, which in turn is necessary for soft tissue support and normal facial appearance. Recreating the maxillary dentition is an important step to establishing the ability to efficiently chew, speak, smile, and return to normal social interactions.

This chapter will review an approach to evaluation of an anticipated maxillary defect, treatment planning for deciding how to best manage the defect, and key principles in executing the treatment plan.

J. C. Melville et al. (eds.), *Regenerative Strategies for Maxillary and Mandibular Reconstruction*,
https://doi.org/10.1007/978-3-319-93668-0_14

14.2 Surgical Anatomy of the Maxilla

The maxillary bones contribute to the face's horizontal and vertical buttresses by articulating with nine surrounding bones, providing an integral support system of the face. These bones include the lacrimal, palatine, frontal, nasal, zygomatic, ethmoid, vomer, sphenoid, and the adjacent maxillary bones. With its central location and intricate articulations, the maxilla also supports the orbit, houses the maxillary antrum, and forms the floor and lateral wall of the nose. Maxillectomy has the potential of disrupting all three cavities; thus a thorough understanding of maxillary anatomy is key to adequately reconstructing form and function after maxillectomy.

14.2.1 Bony Anatomy

The maxilla can be simply described as a cube: a roof that represents the orbital floor which supports the globe; the floor which represents the palate, separating the oronasal cavities; the lateral wall composed of the zygomatic buttress; and the medial wall representing the lateral nasal wall between the nasal cavity and the maxillary antrum. Anteriorly it incases the maxillary sinus and posteriorly separates the infratemporal fossa. Depending on the Brown class of maxillectomy, each of these theoretical cube walls may be involved.

There are four processes that extend from this cube: the zygomatic process, a lateral pyramidal projection that articulates with the zygomatic bone; the frontal processes, which articulate with the frontal, nasal, and lacrimal bones; the palatine process, a thick horizontal process that articulates with the palatine bone to form the hard palate and floor of the nasal cavity; and, lastly, the alveolar process, which houses the maxillary arch's dentition and is in occlusion with the mandibular dentition.

Medially, the paired maxillae articulate with one another, the ethmoid, vomer and palatine bones, and the inferior concha.

Classically described, the maxilla has three vertical buttresses; these consist of the nasomaxillary, zygomaticomaxillary, and pterygomaxillary buttresses. The nasomaxillary buttress is located between the anterior maxillary alveolus, including the lateral piriform rim, and terminates at the nasofrontal suture. The zygomaticomaxillary buttress, the strongest and thickest of the three, includes the posterior maxillary alveolus, the lateral portion of the zygoma up to the zygomaticofrontal suture. The pterygomaxillary buttress is the junction between the posterior maxilla and the sphenoid bone. In addition, the maxilla is one of the three horizontal buttresses. Accurate reconstruction of these is critical.

14.2.2 Muscles of Facial Expression/Fat Pads

Key muscles of facial expression including the levators, zygomaticus, and orbicularis oris/oculi muscles are attached to the maxilla by the superficial musculoaponeurotic system and the retaining ligaments of the face (e.g., McGregor's patch). The midface houses deep and superficial fat pads that can be used to help reconstruct defects (e.g., buccal fat pad) or mask bony asymmetry (malar fat pad). Preservation of these tissues when possible and adequate resuspension can help facilitate an esthetic outcome.

14.2.3 Soft Palate

The soft palate, composed of the tensor veli palatine, levator veli palatine, palatoglossus, palatopharyngeus and muscle of the uvula, insert onto the maxillary process of the palatine bone. The palatine bone fuses with the palatine process of the maxillary bones to form the hard palate. Disruption of the muscle insertion effects deglutition and phonation. Continuity defects of the maxilla result in hypernasality and misarticulations secondary to distortion of the oronasal resonance balance and tongue to palate contacts [1].

14.2.4 Blood Supply

The major vascular supply of the maxilla comes from terminal branches of the maxillary artery and ascending pharyngeal artery both of which branch off the external carotid artery. Due to the close proximity of the maxillary artery branches to the pterygoid plates and posterior maxilla, they are the most commonly injured vessels in LeFort osteotomies [2, 3]. Care must be taken when performing the posterior osteotomy of the pterygomaxillary fissure, a cut ideally left to last, so the maxilla can be quickly mobilized should the need for hemorrhage control be required. In addition, the descending palatine arteries can be a source of delayed bleeding due to tears or pseudoaneurysms at the time of surgery. The descending palatine artery branches from the third portion of the maxillary artery and enters the greater palatine canal and divides into the greater and lesser palatine arteries. Adequate perfusion to the maxilla remains mainly from the ascending pharyngeal artery when the descending palatine sutures are divided. Preservation of this vessel is preferable during lymphadenectomy.

14.3 Classification of Maxillary Defects

An ideal classification scheme for maxillary defects should be simple, account for both effects on form and function, and guide reconstruction. This ideal remains elusive, evidenced by how many different classification schemes that have been published by radiation oncologists, surgeons, and prosthodontists, none of which having been universally adopted.

The earliest classification scheme was by Dr. G. Ohngren in the 1930s. Dr. Ohngren described a plane that bisects the maxillary sinus that starts at the medial canthus and extends to the mandibular angle. Lesions anterior and inferior to "Ohngren's line" were thought to represent a more indolent or benign disease. Lesions posterior and superior to this plane were thought to be more aggressive or malignant [4]. This scheme focused primarily on anticipating disease behavior and biology as, at that time, it was felt that surgical resection was ineffective. Consequently, Ohngren, a proponent of radiation and "endothermy," made no effort to classify the resulting defects [5].

In 1978 Aramany, a maxillofacial prosthodontist, published two articles, the first classifying post-extirpative maxillary defects and the second on how to obturate them. This is likely the first classification based on surgical defects with the intent on guiding prosthetic rehabilitation [6, 7].

Spiro et al. popularized a classification based on the surgical ablative procedure: limited, subtotal, and complete maxillectomies [8]. The greatest strength of this schema was that it provided a descriptive framework for planning surgical resection that was simple and easy to put into practice. However, because the schema did not include stratification of the oral defect, orbital defect, or malar defect, it is limited in its usefulness for reconstructive treatment planning.

In 2000, Brown et al. published a classification system that was the first to describe the defect in a system that would lend naturally to decisions on reconstruction using contemporary techniques, including both prosthetic and surgical options [9]. In the initial iteration of the Brown classification, they included criteria for stratifying the vertical defect, the laterality, and alveolar defect, allowing for anticipation of the reconstructive challenges both surgically and prosthetically. This was widely felt to be the most useful for reconstructive surgeons because it laid down a mental framework for thinking about the problems that would have to be addressed with the reconstruction.

Dr. Okay, a maxillofacial prosthodontist, along with his surgical colleagues Drs. Buchbinder and Urken, believed that both the palate and remaining dentition should be used to describe horizontal defects, as the dentition provides support, retention, and stability to any prosthesis. In their classification, they also included descriptions of the orbit and zygoma, reflecting their role in facial appearance [10].

Cordeiro and Santamaria built on Spiro's classification and subdivided maxillectomies into partial/limited (type I), subtotal (type II), total with preservation of orbital contents (type IIIa), total with orbital exenteration (type IIIb), and orbitomaxillectomy (type IV) and provided various reconstructive options for each [11]. However, their update did not include descriptions of the maxillary alveolus.

In 2010, Brown et al. published a revised classification accounting for both the vertical and horizontal (functional) defects of the maxilla in a stepwise gradient that coincided free flap selection while also adding a section for isolated nasomaxillary defects [12]. Brown et al. further elaborated upon the relationship between the sizes of the dentoalveolar/oral defect and anticipated prosthetic challenges. These dentoalveolar defects are divided into four categories: a central palatal defect (Class A), ½ or less of the unilateral palate and alveolus (Class B), anterior maxillary defect (Class C), and greater than ½ palatal alveolar defect (Class D) (Fig. 14.1). In contrast to the algorithm published by Cordeiro et al. that primarily recommended soft tissue free flaps with occasional non-vascularized bone grafts (n = 100 74% rectus myocutaneous flap, 6% radial forearm osteocutaneous), Brown et al. advocated for the use of composite osteocutaneous free flaps (n = 147 36% soft tissue free flap, 63% osteocutaneous free flap), which allows for endosseous implant placement and a bony foundation for a fixed prosthetic solution for dental rehabilitation while also adequately separating facial compartments and maintaining support of the soft tissue envelope.

For these reasons, the authors favor the revised Brown classification of maxillectomy defects. The remainder of this chapter will discuss evaluation and treatment planning using this classification as its framework.

14.4 Evaluation

The reconstructive surgical consultation starts with taking a good history. The disease being treated is determined, paying attention to the anticipated extent of the defect given that disease's natural history. For malignant or benign neoplasms, this would entail determining the size of the tumor and the planned ablative defect. Particular attention should be paid to the magnitude of the oral defect, the areas of lost facial support, the extent of involvement of the nasal cavity and paranasal sinuses, whether the globe would be included, and whether intracranial involvement would be expected. For trauma with composite avulsive defects, in addition to the obvious extent of the injury, the mechanism of injury should be considered. Ballistic injuries, particularly those that involve explosions or high-velocity ballistics, can be complicated by a much larger zone of injury than initially observed. These may manifest with progressive dieback of hard and soft tissues. These may also affect the viability of potential recipient vessels for microvascular anastomoses. In short, the disease process needs to be

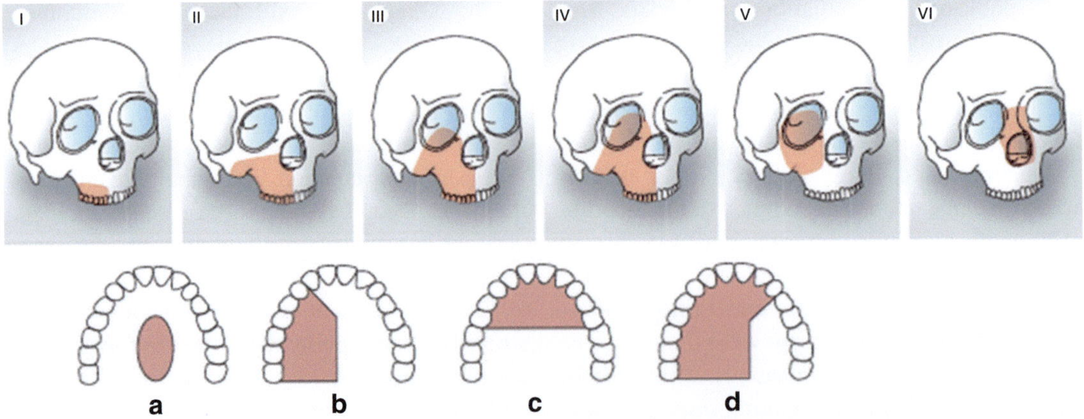

Fig. 14.1 Classification of vertical and horizontal maxillectomy and midface defect. Vertical classification: I—maxillectomy not causing an oronasal fistula; II—not involving the orbit; III—involving the orbital adnexae with orbital retention; IV—with orbitol enucleation or exenteration; V—orbitomaxillary defect; VI—nasomaxillary defect. Horizantal classification: a—palatal defect only, not involving the dental alveolus; b—less than or equal to 1/2 unilateral; c—less than or equal to 1/2 bilateral or transverse anterior; d—greater than 1/2 maxillectomy. Letters refer to the increasing complexity of the dentoalveolar and palatal defect, and qualify the vertical dimension

investigated in order to adequately anticipate and plan for the resultant defect.

A thorough review of the past medical history is important. Several comorbid conditions affect the available treatment options. Free flap surgery is lengthy, and any systemic health issues that would complicate tolerating long surgery and anesthetic times need to be evaluated and optimized. Peripheral vascular disease can affect the flap options available. Malnutrition increases the risk of poor wound healing and flap failure [13]. Early identification of severe nutritional deficiencies may allow for preoperative optimization. Cigarette smoking and alcoholism also contribute to these problems.

The physical examination includes a detailed facial examination, head and neck examination, oral examination, and examination of potential flap donor sites. For the facial examination, special attention is paid to facial proportions, facial dimensions, and shape. For the oral examination, special attention is paid to assessing the occlusion, the number of and health of remaining teeth, and the amount of gingival show on full animation. In addition to the size and shape, the components of the defect must be assessed, specifically whether the external skin, oral mucosa, the globe, and bone will be absent.

Most of these patients already have had extensive anatomic imaging for evaluation of the disease process. In addition to these, if not already ordered, a maxillofacial CT (non-contrast) with fine cuts (less than 1.25 mm) is obtained. A cone beam CT can also be used, although the quality of the information for computer surgical planning is not quite as good. At this point, the authors already have a sense of what reconstructive option will most likely be used, so the relevant imaging study for the flap donor site will be ordered as well. For example, the fibula osteocutaneous free flap is very often used. A CT angiogram of the bilateral lower extremities is ordered to assess the distal arterial runoff as well as provide patient-specific skeletal data for computer planning.

If primary placement of implants is being considered, dental impressions of both arches are also obtained. Previously, the patient would also have to have a scan while biting on occlusal bite registration with an imbedded fiducial marker to aid in registration of the dental models to the CT scan, but improvements in software and engineering have made this unnecessary.

14.5 Reconstructive Options

14.5.1 Dental Obturators

Obturation is the use of a custom dental appliance to separate the maxillectomy defect and oral cavity from the nasal and paranasal cavities. It is the most commonly used approach for maxillectomy defects worldwide [14]. In order for an obturator to be successful, it requires that it be crafted to engage enough facial and dental undercuts to maintain adequate retention during function. It must be well adapted to the mobile soft tissues to create a seal for it to adequately separate the nasal cavity and paranasal sinuses from the oral cavity. This is essential for preventing nasal regurgitation and hypernasal speech. It must have adequate support from the remaining tissues or implants such that it will not move excessively during mastication. Many advocate removal of the inferior turbinate to prevent an impediment to this support.

The definitive obturator has two main components, a metallic framework and an acrylic resin obturator bulb. The metal framework helps resist the torsion and cantilever forces between the defect and remaining retentive tissue, not unlike a removable partial upper denture [15, 16].

Traditional obturators are very similar to a tooth- and tissue-borne removable partial dentures with the addition of the obturator bulb. Support of the obturator is primarily from the remaining hard palate and partially by the remaining teeth. Retention depends upon stress-relieving clasps around the remaining teeth.

In many patients, the remaining hard palate and teeth are insufficient for adequate retention, stability, and/or support. The incorporation of traditional and zygomatic endosseous dental implants can mitigate these shortcomings by creating a fixed foundation for retention and support that can be distributed across the arch. The use of implants can allow for obturation of even Brown Class II d defects (Fig. 14.2a–f (Courtesy of Dr. Brian L. Schmidt MD, DDS, PhD, FACS)).

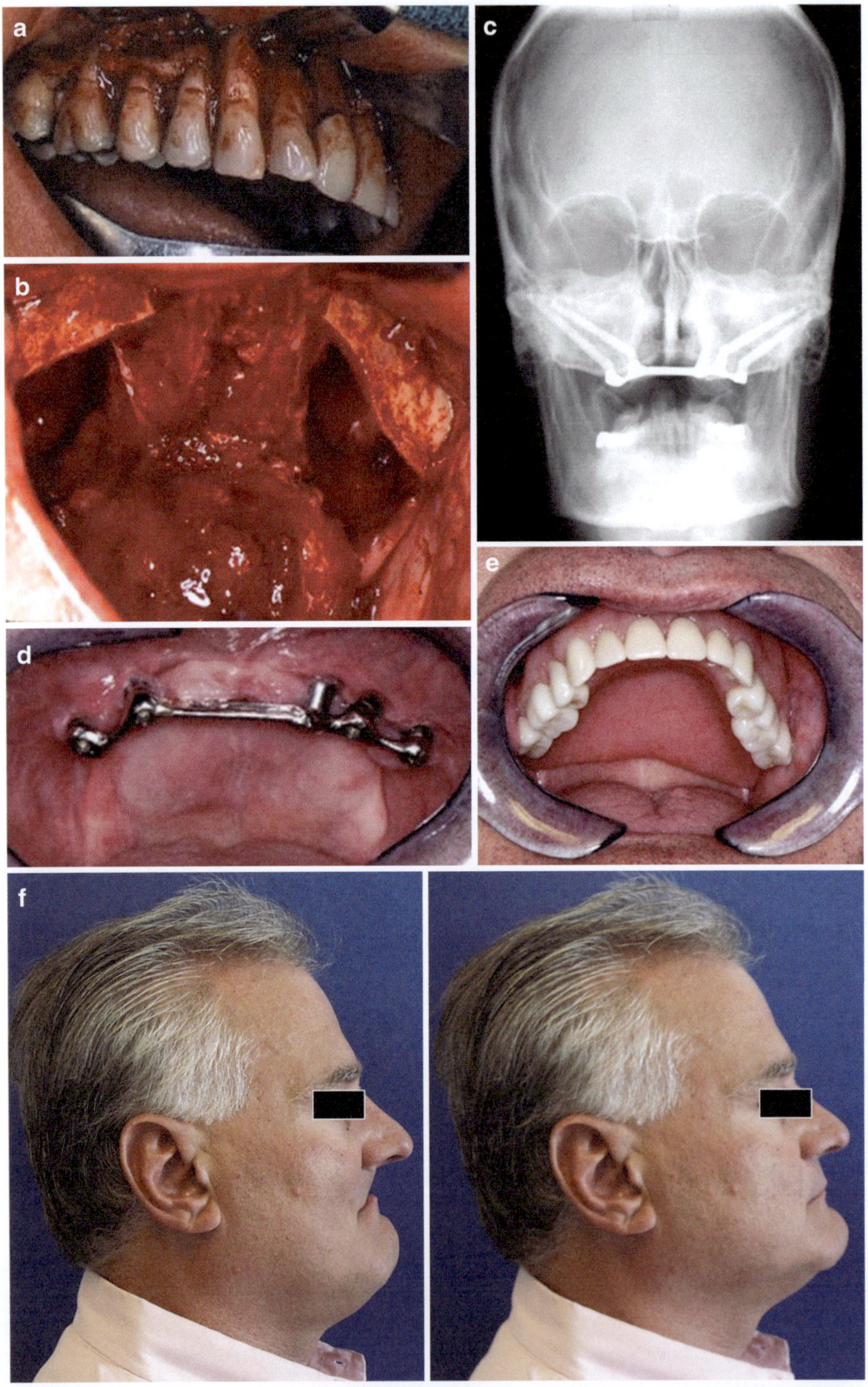

Fig. 14.2 Courtesy of Dr. Brian L. Schmidt MD, PHD, DDS. (**a**) Forty-six-year-old male with rhinocerebral mucormycosis presenting with gross mobility of maxilla. (**b**) Brown Class II d maxillectomy after debridement. (**c**) Quad-zygoma implants placed. (**d**) Custom bar fabrication for zygomatic implants. (**e**) Final prosthesis 2-year follow-up. (**f**) Excellent midface projection 2-year follow-up

There is some debate among reconstructive surgeons and maxillofacial prosthodontists around obturation versus reconstructive surgery. Proponents for obturation note the early return to function, esthetics of immediately replacing dentition, simplicity of treatment, and the ability for tumor surveillance.

Breeze et al. recently assessed quality of life and patient satisfaction between patients who received prosthetic obturation compared to reconstructive surgery. They found no difference in those metrics, regardless of the size of the vertical defect or whether or not patients received postoperative radiation [14]. However, other studies have demonstrated improved patient satisfaction, speech, and swallowing in palatal defects greater than 50% reconstructed with free flaps [17, 18]. Of note, the obturator comparison group was not stratified by implant versus tooth-tissue-borne appliances. There is also data suggesting that the age of the patient affects their acceptance, with younger patients having a harder time with obturators [19].

There are also several system factors that limit the utilization of dental obturators. One is that the expertise required to fabricate dental obturators is highly specialized. Few prosthodontists or dentists have experience with making them. Dental obturators may or may not be a covered benefit by medical insurance, depending on regional norms. If they are not a covered benefit, they can be quite costly. Even if they are a covered benefit, if the reimbursement is either too low, too cumbersome, or both, many prosthodontists who would be able to provide the care decline to do so. The same limitations apply to dental implant-based prosthetic solutions. These barriers to care must be factored into treatment planning decisions.

14.5.2 Free Flaps

Free flaps are the most common form of maxillary reconstruction. Several options are available, which allow for tailoring the reconstruction to the patients' needs. Free flap reconstruction can be performed primarily, allowing for earlier return to function for patients. Free flaps, despite involving lengthy operating room time, are generally highly successful and well tolerated.

The *fibula osteocutaneous free flap* is the flap most commonly used (by the authors) for reconstruction of composite defects of the maxilla. It offers several advantages. It provides bone that is easily osteotomized into complex shapes, especially when computer planning and 3-D printed cutting guides are used. One of the primary challenges of reconstructing Brown Class III or IV defects is restoring malar projection, the piriform buttress, and the alveolar bone, as the axes of each of these struts are in different directions (Fig. 14.3). The ability to segment the fibula into multiple axes is very helpful to addressing this. Another advantage is that the skin flap is based on septocutaneous perforators that allows for flexibility with positioning. It is also thin enough to fold such that portions can be used for oral, cutaneous, and nasal lining, when necessary. The pedicle, depending on the length of the bone required for reconstruction, is usually long enough to reach vessels in the neck without incorporating vein grafts. Furthermore, the donor site is remote from the defect, allowing for simultaneous site preparation and flap harvest.

There are a few disadvantages to the fibula flap. The arterial runoff to the lower extremity is often the first affected by peripheral vascular disease.

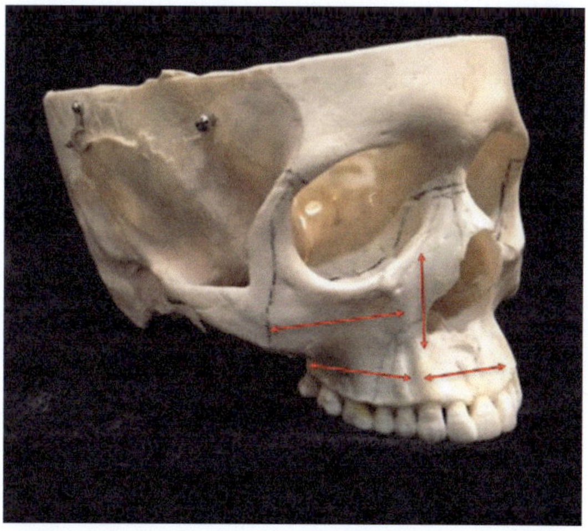

Fig. 14.3 Red arrows on the right maxilla illustrate the differing vectors of the piriform, malar prominence, and anterior and posterior alveolus

The surgical anatomy is more complicated than many of the other flap options available, involving dissection of all four compartments of the lower leg. The scar at the donor site is not easily hidden. Although the bone quality, which is primarily cortical (type I bone with respect to dental implant osseointegration), allows for excellent primary stability following implant placement, it is theoretically less ideal for osseointegration.

Flaps based off of the *subscapular system (scapular flaps, parascapular flaps, scapular osteocutaneous flaps, scapular tip flaps, latissimus dorsi flaps, and chimeric mega flaps)* have the advantage of enormous flexibility. It is ideally suited for Brown Class IV defects, particularly those that extend into the nasomaxillary region and anterior skull base. The ability to have three-dimensionally independently mobile bone, muscle, and skin of large quantities within a single flap enables the surgeon to restore large and highly complex defects. In particular the ability to obliterate the orbit and seal off the cranial fossa with large amounts of vascularized muscle is very useful. The subscapular system is usually spared from peripheral vascular disease, making it a good option when other flap donor sites are not available.

One disadvantage of the subscapular flaps is that harvest often requires placing the patient in some degree of lateral decubitus position. This can make it uncomfortable to perform simultaneous ablative surgery and flap harvest. The scapular bone can be osteotomized, but the segments do not have as many degrees of freedom compared to the fibula. The pedicle length is shorter, especially when using a chimeric flap, making reaching the neck vessels challenging. The bone volume is of variable thickness and is less ideal for placement of dental implants. Although the shape of the scapula is ideal for malar defects, it is challenging to reconstruct both the malar region and the maxillary alveolus.

The *radial forearm free flap* can be used for Brown Class I or II defects that are lateralized and where few teeth are lost and there are no plans to replace them. The radial forearm can be raised as a fasciocutaneous flap or an osteocutaneous flap. The primary advantages are that the radial forearm is a very straightforward flap with a long vascular pedicle and highly reliable skin perfusion and is usually thin and pliable. As a soft tissue flap, it can be used to seal off the nasal cavity and/or maxillary sinus from the oral cavity, restoring normal speech and preventing nasal/sinus leakage. When used as an osteocutaneous flap, some amount of malar and lip support can be provided. When used as a fasciocutaneous flap only, it can be combined with other flaps where the soft tissue of the second flap may be too bulky for the indication. The ideal defect for this is a Brown Class I/II a/b, where the goal is to only seal off the nasal/paranasal spaces, but not to restore bony support.

However, the utility of the radial forearm flap is limited to smaller defects. The bone is insufficient for restoring support of the maxillary buttresses and upper lip for any defect greater than a Brown Class II b. The bone is also insufficient for implant placement. When used as a fasciocutaneous flap only, attempts have been made to place zygomatic implants through the soft tissue into the zygoma [20]. Although this has been done successfully, anecdotally, these implants often suffer from peri-implantitis and require constant vigilance for maintenance. It also doesn't offer a clear advantage over having the implants alone without the soft tissue provided from the flap. When used without the bone, the soft tissue exists somewhat as a tarp over the bony defect. Therefore it also hinders the retention and stability of prosthetic solutions.

The *deep circumflex iliac artery flap (DCIA)* is a very commonly used flap for maxillary reconstruction, heavily favored by Brown et al. It offers several advantages. The shape of the iliac crest is well matched to that of the infraorbital rim and malar region. Creative osteotomies can allow for reconstruction of alveolus as well. Dr. Batstone has presented his modification of the DCIA flap where he uses an osteotomy on a segment of the harvested iliac crest, leaving it pedicled to the iliacus and DCIA artery, and medializes this portion so it approximates the position of the desired neo-maxillary alveolar bone. In this setup, the iliac is oriented with the crest toward the mouth and the cut edge of the iliac wing (inferior) positioned as the inferior orbital rim (Fig. 14.4a–d, courtesy of Dr. Martin Batstone MBBS, BDSc, MPhil(Surg), FRACDS(OMS), FRCS(OMFS)). The internal oblique muscle provides a sizeable muscle flap for obturating defects and/or oral lining. Simultaneous harvest and site preparation are possible.

There are a few disadvantages. The DCIA flap has a comparatively short vascular pedicle, with the upper limit being 6–8 cm depending where the bone is harvested (the further the bone is harvested from the anterior superior iliac spine, the longer the available pedicle). This can still usually reach either the facial or superficial temporal arteries but can be a bit of a stretch. Although a skin paddle can be harvested with the DCIA based on myocutaneous perforators, the skin is fixed to the iliac crest and not independently mobile, reducing its usefulness. The internal oblique muscle can be used for oral tissue lining and will eventually mucosalize. However, the mucosa that grows over the muscle is of the unattached and nonkeratinized variety, which makes it less optimal for implants. The bone quantity is very good but of type IV quality. Therefore, primary stability of implants is not

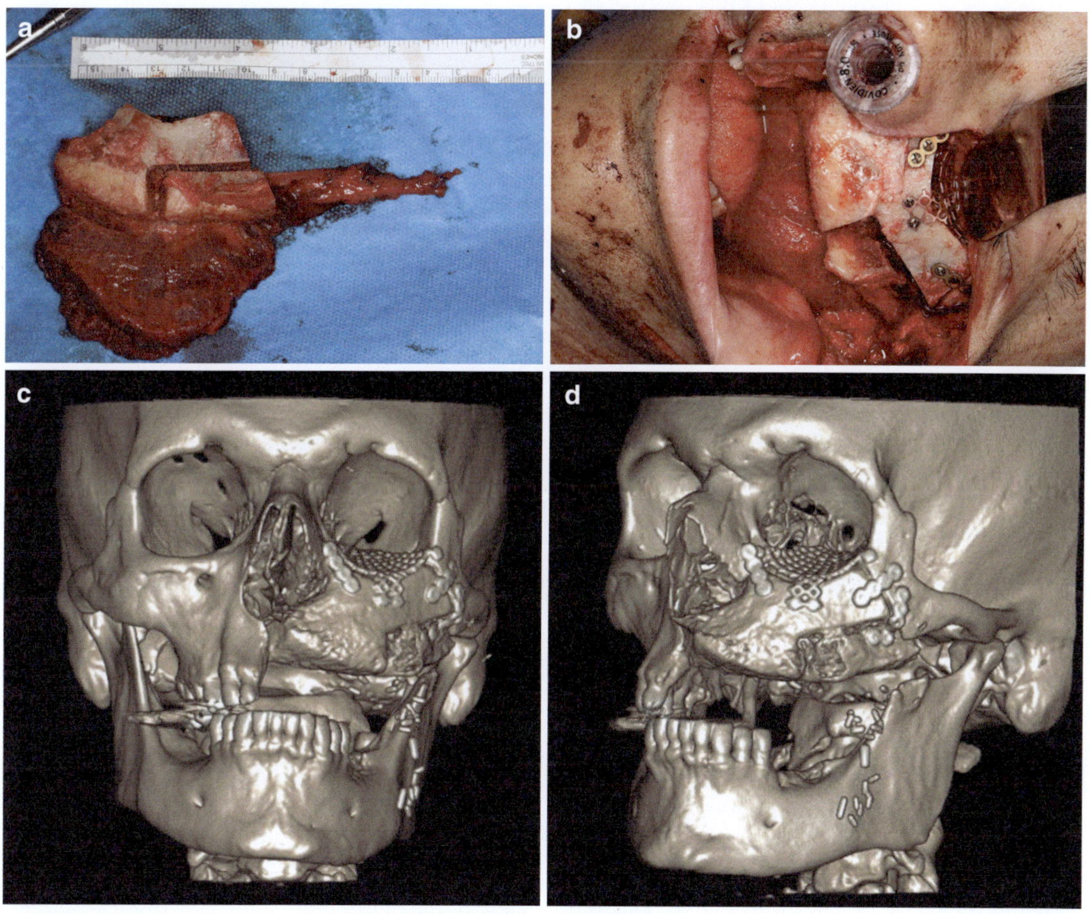

Fig. 14.4 Courtesy of Dr. Martin Batstone MBBS, BDSc, MPhil(Surg), FRACDS(OMS), FRCS(OMFS). (**a**) DCIA harvest with osteotomy of iliac crest, pedicled on the iliacus muscle. (**b**) Reconstruction of anterior and posterior alveoli with thick iliac crest and remainder of ilium harvest reconstructing Brown Class III b defect. (**c**) CBCT showing excellent projection and reconstruction. (**d**) Lateral view

as good as with a fibula. Immediate loading protocols are potentially riskier, although this concern has not been investigated. Interestingly, a recent study suggests that the scapula and DCIA free flaps had higher implant success rates (at 97.6 and 96.9%, respectively) than the fibula free flap (91.7%), although not statistically significant [21].

The rectus abdominis free flap is a commonly used free flap, popular among Cordeiro et al.'s group. It is also a fairly simple flap to elevate with minimal donor site morbidity. It can include a large amount of both the muscle and skin, making it useful for obturating the orbit, sinus, and skull base while also replacing the skin.

The main disadvantage is the lack of bone. As such, it does little to establish the midfacial buttresses. In Brown Class III, IV, or II c and d defects, the loss of bone support in the midface results in a sunken in appearance when only reconstructed with soft tissue alone (Fig. 14.5a–c).

14.6 The Role of Technology in Maxillary Reconstruction

For several reasons, maxillary reconstruction is one of the more challenging areas of the face to reconstruct. As described in the section on surgical anatomy, the maxilla has an irregular shape. The plane of the infraorbital rim, piriform rim, zygomatic maxillary buttress, and maxillary alveolar bone are along three different axes. As described above, the reconstructive surgeon will have to also manage separation of the different facial and cranial compartments from each other.

Historically, the ability to successfully reconstruct the complex anatomy of the maxilla required a level of artistic intuition that was achieved by only a small number of very experienced and talented surgeons. Although many surgeons might have been able to perform the essential elements of microvascular reconstructive surgery, few had the depth and breadth of experience that allowed them to craft a flap on the table into an appropriate shape and use it to accurately recreate normative midface appearance.

Over the last decade, the advent of both *computer surgical planning and 3-D printing* has made the ability to reproducibly and predictably achieve high-level reconstructive results accessible to even novice surgeons. Computer surgical planning allows a surgeon to freely plan and adjust both the defect and flap, with unobstructed three-dimensional visualization. 3-D printed cutting guides based on the computer-generated plans transfer the virtual into the real and have simplified the complex closing osteotomies required to recreate the planned shape. Using data sets created from the computer plan, most plating manufacturers offer the ability to fabricate patient-specific *custom reconstruction plates*. These custom plates allow for more rigid hardware to be used despite the irregular shape of the maxilla. They also adapt well to the patient's native bone with much less bending and adjustments required (Fig. 14.6). Computer surgical planning and 3-D printed cutting guides can also aid in primary placement of dental implants. During the planning session, ideal locations for dental implants, with the end restorative goal in mind, can be precisely placed virtually. This position is translated to the patient using the same cutting guides for shaping the bone flap. Taken a step further, combining advances in computer surgical planning for maxillofacial reconstruction with guided dental implant surgery and immediate loading protocols on splinted implants with cross arch stabilization, select patients can have their ablative procedure, facial reconstruction, dental implant placement, and delivery of a fixed provisional hybrid restoration all within the same day. Discussion of this protocol goes beyond the scope of this chapter, but it is described in detail in the literature [22].

Another significant challenge is surgical access. The space in which the reconstructive surgeon must work is small and very tight, even when using a wide-open access with a Weber-Ferguson approach. Once the flap is inset, the view of the entire recipient site is often obscured. This makes it very difficult to properly position any bone flap in three dimensions. In particular, for defects that include the alveolar segment, it is very easy to create yaw, roll, and pitch errors

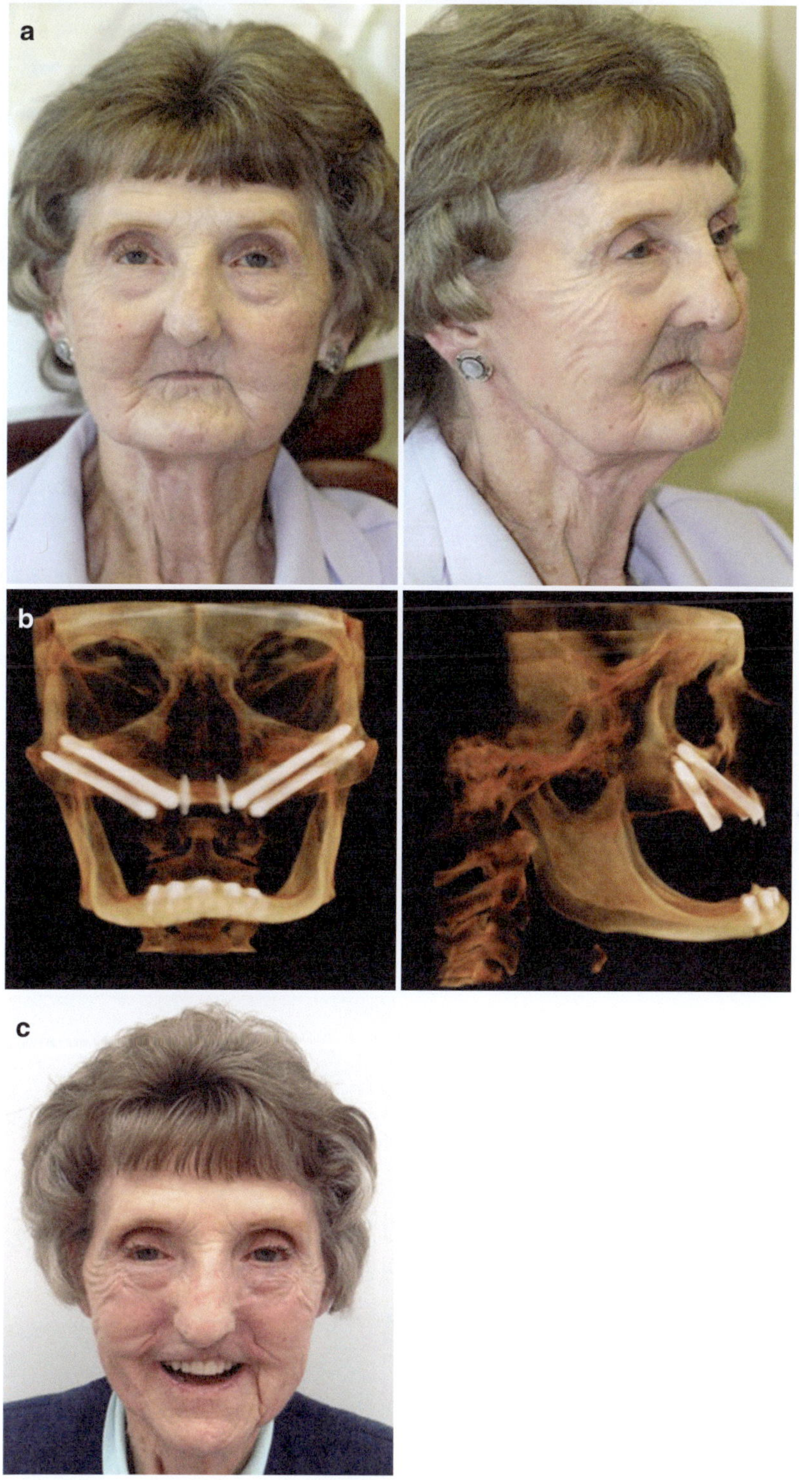

Fig. 14.5 Seventy-eight-year-old female s/p anterior maxillectomy for SCC reconstructed with radial forearm soft tissue flap. (**a**) Note sunken in appearance. (**b**) Later reconstructed with zygomaticus/endosseous implants. (**c**) Three-year postoperative follow-up with excellent function and prosthesis retention

because visual cues that a surgeon would normally use (i.e., during repair of maxillary trauma or orthognathic surgery) to position the neomaxilla are not available (Fig. 14.7).

To a small degree, custom reconstruction plates help with some of the uncertainty. However, with the complex shape of the maxilla and lack of exposure, they only partially mitigate the challenge. *Intraoperative CT* scans and *intraoperative navigation* have been both used, either alone or together. Navigation is somewhat helpful, especially with placement of the bone flaps against the pterygoid maxillary buttress. This is an area where the bone flap is easily placed too lateral (most commonly), too inferior, or, on some occasions, too superiorly. However, navigation becomes less reliable as the surgery goes along. This is because the initial registration is inevitably moved or displaced, inadvertently, during the length of the case. For the authors, intraoperative CT has been the most helpful in identifying errors intraoperatively. Although there are several inconveniences with intraoperative CT (using a special head rest, requiring a lengthy stop in surgery, radiation dosage), the value of confirmation of accurate flap placement while it still can be adjusted is tremendous.

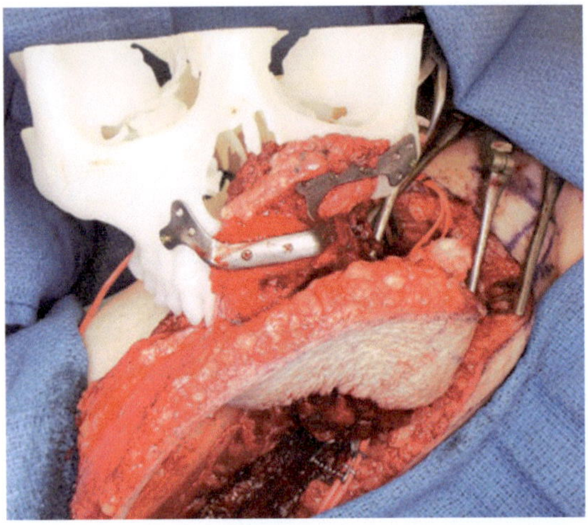

Fig. 14.6 Custom plate used for fibula free flap reconstruction of a high-left Brown II b maxillectomy

14.7 Strategies for Maxillary Reconstruction Using Brown Classification

14.7.1 Brown Class I a: Palatal Defect Only

These defects do not involve an alveolar segment or loss of teeth. As such, the support of the dentition and facial soft tissues is not compromised. The resulting defect does create an oroantral and/

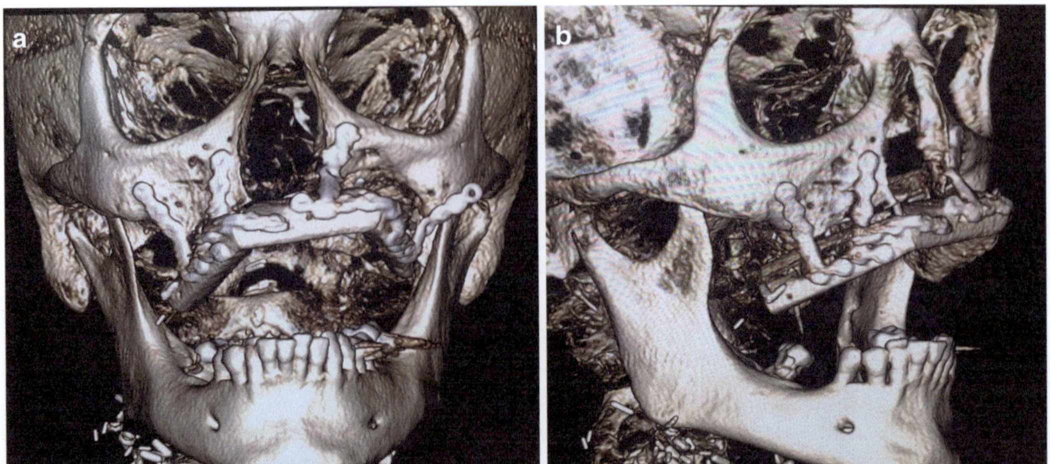

Fig. 14.7 Poor intraoperative position of maxillary fibula segments secondary to loss of anatomical references from advanced MRONJ

or oronasal communication that results in hypernasal speech and nasal regurgitation.

This can be addressed either prosthetically or surgically. Prosthetically, a tooth- and tissue-borne or tissue-borne removable obturator prosthesis can very successfully seal this off. This has the benefit of surgical simplicity but in some patients is not as well tolerated.

This can also be reconstructed using surgical techniques. Depending on the size of the defect, these can be closed by local flaps, regional flaps, or free flaps. Local flaps are better suited for patients who have not received or will not be receiving radiation therapy. For local flaps, Von Langenbeck palatal flaps combined with a septal finger flap can be used effectively for smaller defects. For lateral defects in edentulous patients, buccal fat pad can also be used. A buccal facial artery myomucosal flap is also very helpful. For radiated patients, the authors favor a free flap, with the radial forearm being the simplest.

14.7.2 Brown Class I b–d: Dentoalveolar Defect Only

Brown Class I b–d defects, that is, those that involve the maxillary alveolus without oroantral or oronasal communication, require reconstruction of the dentoalveolar restructures to provide a functional dentition and upper lip support. The tissues lost will be the bone, teeth, and attached gingiva.

For I b and c defects, there are several options. The segment can be replaced entirely by a removable partial denture (Kennedy Class II) if there is remaining native maxillary dentition. Implant-supported solutions are also possible, although they involve creative implant placement and sophisticated prosthetic design. Often there is a dearth of available bone height for implants. Also, there is usually missing keratinized tissue. This can be mitigated by use of short implants splinted together (Fig. 14.8). For defects including the posterior segment, 1–2 zygomatic implants combined with an implant in the anterior maxilla can be used. Palatal grafting on the buccal/labial aspect of the implants may be required to preserve peri-implant health.

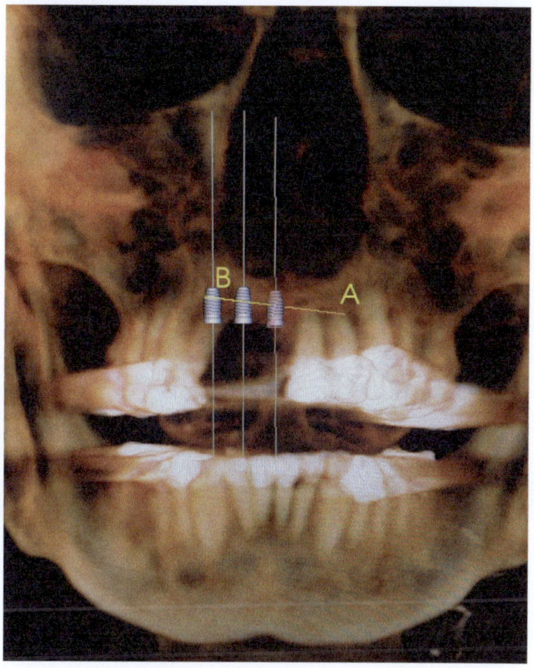

Fig. 14.8 Seventy-year-old male with pT2M0N0 maxillary SCC s/p partial maxillectomy, planned for short splinted dental implants

For Brown Class I d defects, this essentially mimics the clinical scenario of the atrophic maxilla. A removable partial denture remains an option, but retention will be an issue. More likely an implant-supported prosthesis will be required. Four or more splinted dental implants (either angled or including zygomatic implants) can provide a platform for a fixed prosthesis (fixed hybrid, fixed bridge, or overdenture on a bar). Soft tissue may be an issue, and creative use of local palatal flaps or grafts may be necessary to ensure a cuff of attached gingiva around transmucosal portions of the implants.

14.7.3 Brown Class II b: Posterolateral Infrastructure Maxillectomy with Posterior Dentoalveolar Defect

This defect results in an oroantral and oronasal defect in the posterior region and loss of posterior dentoalveolar support. Because the contralateral teeth (if present), hard palate, and alveolus remain, a prosthetic option is still possible.

An obturator can very elegantly be used to obturate the defect, replace the missing dentition, and restore facial support (Fig. 14.9). The advantages and disadvantages have been already described. Implants can aid in retention and support, when necessary.

For very posterior defects where the patient retains premolars, a soft tissue-only flap can be used. This simplifies the reconstruction. Because the piriform rim, the infraorbital rim, and anterior

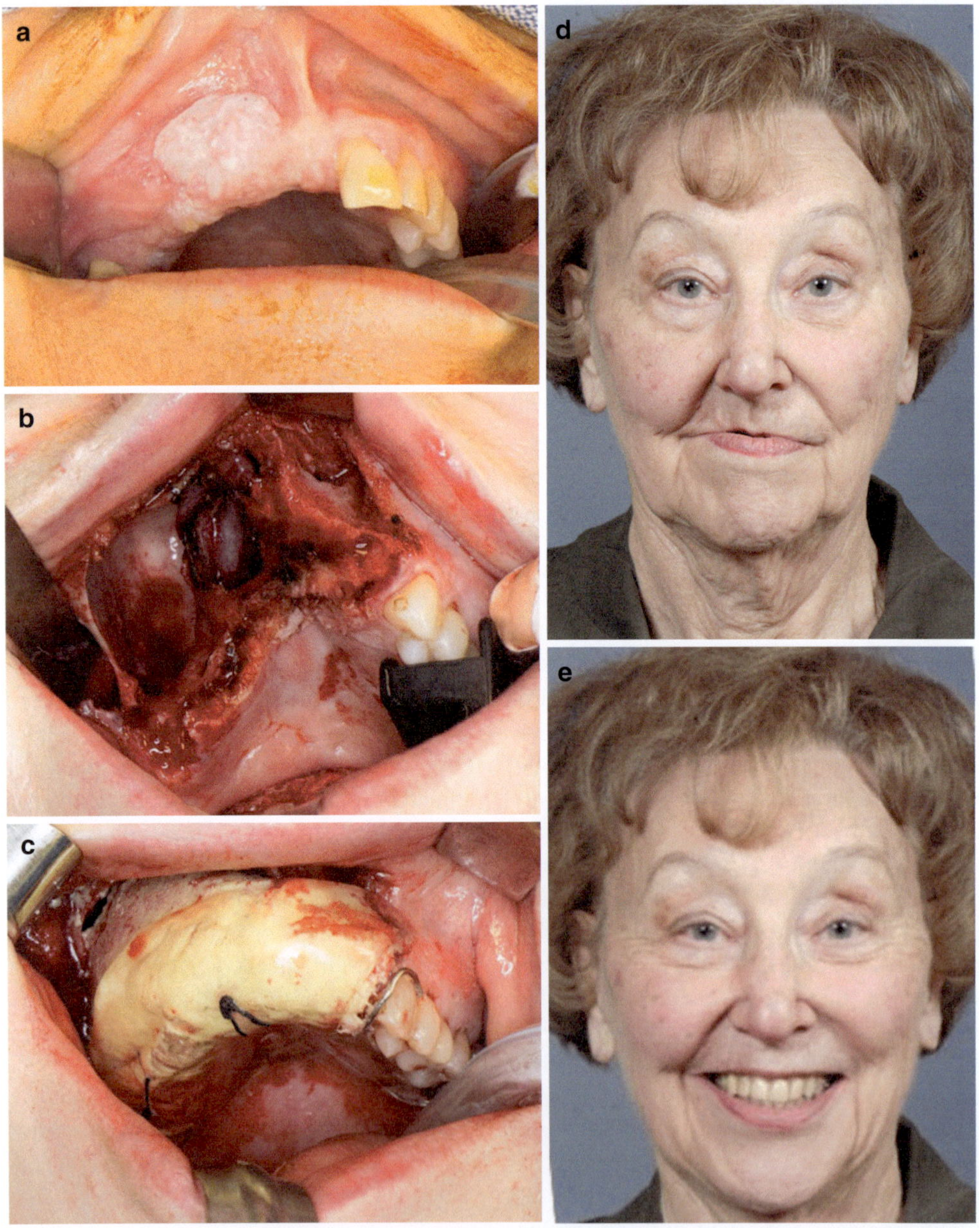

Fig. 14.9 Eighty-three-year-old female with pT4aN0M0 maxillary SCC. (**a**) Lesion. (**b**) Maxillectomy defect. (**c**) Intraoperative temporary obturator. (**d**) One-year postoperative photo without obturator. (**e**) One-year postoperative photo with obturator

teeth are not lost, there is no loss of facial support. A soft tissue flap can therefore sufficiently separate the maxillary sinus from the oral cavity without the need for bony reconstruction. This does make prosthetic replacement of the posterior teeth much more challenging. A soft tissue flap lacks support for a removable partial denture and will depend entirely on the remaining hard palate to resist chewing forces on the same side of the defect. As mentioned above, Hirsch et al. have proposed the placement of zygomatic implants through a radial forearm free flap in order to provide a platform for an implant-supported prosthesis. Although this can be done successfully, we have largely abandoned this approach, as the soft tissue does not provide much benefit to an obturator supported by zygoma implants, while a flap does complicate implant placement and maintenance significantly.

14.7.4 Brown Class II c–d: Infrastructure Maxillectomy Involving Most of Dentoalveolar Arch

Defects that extend to over half of the maxillary arch, wrapping around and/or including the incisors, require some form of reconstruction of the facial support provided by the piriform buttresses and maxillary teeth to avoid facial disfigurement. Failure to do so leaves patients with a witch's chin deformity, with loss of lower facial height, midface projection, lip support, and nasal support (Fig. 14.5a). The loss of support results in an over-closed mandible that gives the patient a very aged face, relative prognathism, and severely downturned nasal tip.

For Brown Class II defects that are fairly low (infraorbital, malar, and paranasal support intact), leaving sufficient bone in the zygoma on both sides, quad-zygoma implants can be used to support a prosthetic reconstruction, usually an overdenture-type obturator prosthesis. As discussed above, these require a high level of prosthodontic sophistication to fabricate. The availability of such expertise is region dependent (Fig. 14.5b).

The more common solution would be surgical reconstruction, and in this case, a bone flap of some sort is necessary to avoid the abovementioned deformity. The authors prefer the use of a fibula free flap for this purpose. Other groups have advocated differing flaps, all of which have their merits. The scapula tip has been suggested as the ideal shape to replace the entire maxillary hard palate [23]. However, the authors have found the available bone to be insufficient for implant placement when the scapula is used in this way (despite what others have said). Others, including Brown et al., have favored the DCIA flap for these types of defects.

After the consultation visit, a CT angiogram of the lower extremities and a fine-cut CT maxillofacial study are ordered. These files are then uploaded to a third-party computer surgical planning company. Once uploaded, the files are then used to create 3-D computer models for planning. A web-based meeting is scheduled between the surgeons and computer surgical planning engineer. The defect, in the case of oncologic surgery, is planned out. In the case of trauma, an additional ablation is usually planned out to optimize the recipient site for receiving the bone flap. In this case, it is helpful to have cut edges for butt joints. Mobile and/or marginally viable bone segments should be discarded (Fig. 14.10a–f). The fibula is then virtually osteotomized to create the desired shape. For low Brown Class II defects, the focus is on reconstructing the alveolar segments, as the upper piriform rims and zygoma are still present. The maxillary arch is conceptually divided into three segments: the premaxilla and right and left posterolateral segments (Fig. 14.11) (take picture of maxillary arch and draw in these segments) (Fig. 14.10c). The fibula is segmented using closing osteotomies to recreate a similar shape. Particular attention is given to making sure the position of the fibula construct matches the desired position of the neo-maxilla along all three axes. This is very difficult to appreciate accurately in vivo but easily appreciated in simulation. Also, accurate maxillary width is another dimension that is difficult to assess intraoperatively but straightforward on the computer.

The posterior stop of the fibula construct is placed at the pterygoids. This also prevents excessive width of the neo-maxilla. If the ptery-

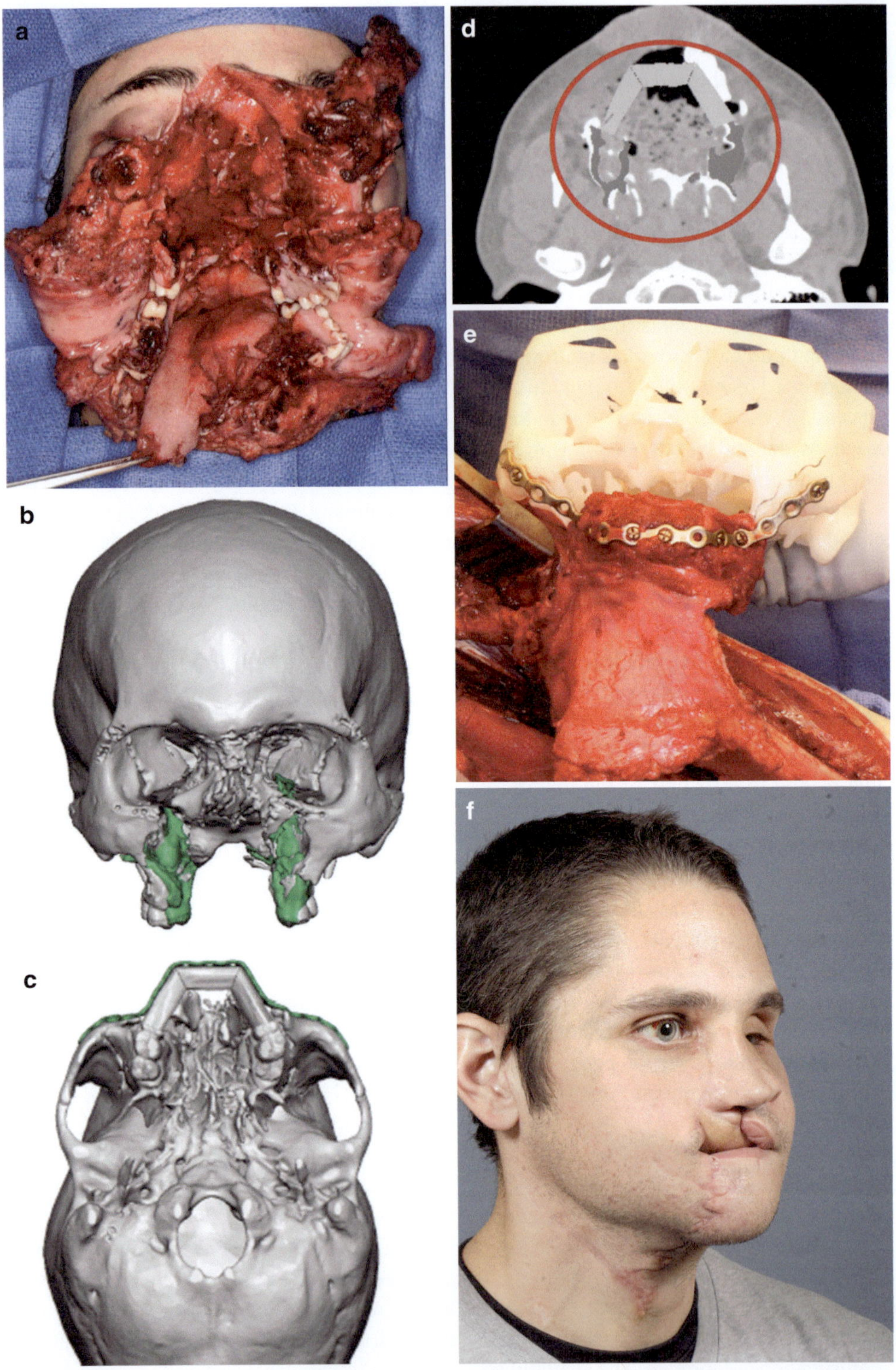

Fig. 14.10 Self-inflicted gunshot wound with shotgun. (**a**) Debridement and stabilization. (**b**) Computer surgical planning—aligning of mobile maxillary segments. (**c**) Computer surgical planning—placement of fibula against pterygoid plates to avoid widening maxilla. (**d**) Intraoperative navigation plan to verify placement. (**e**) Custom plates and stereolithographic models aid in accurate placement of fibula. (**f**) One-year post-op maxillary and mandibular reconstruction with bilateral fibulas and soft tissue revisions

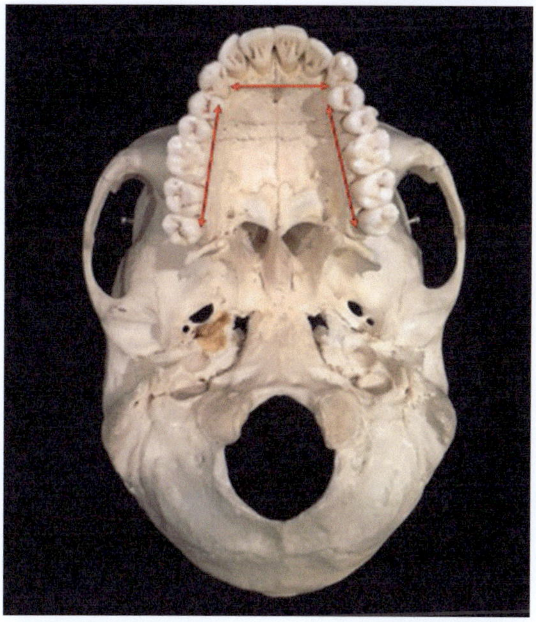

Fig. 14.11 Three segments of the maxillary arch

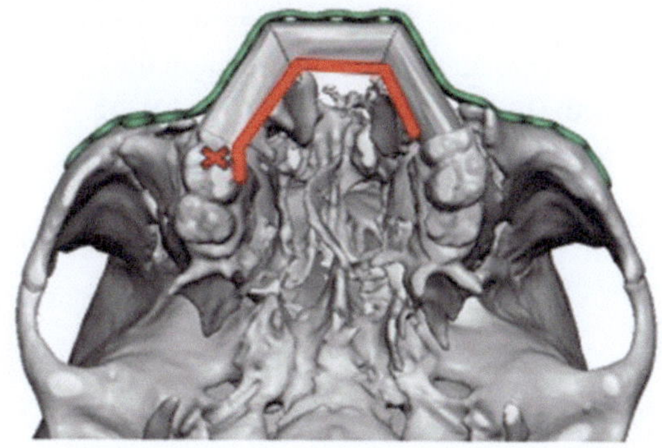

Fig. 14.12 If posterior dentition is left intact and the proximal fibula abuts it, the pedicle risks being kinked as it abuts tooth #3 in this depiction

goid plates remain, the authors often plan to remove them, as they interfere with the path of the vascular pedicle.

Similarly, having remaining posterior dentition on the side of the reconstruction complicates flap inset by creating an area of compression for the vascular pedicle. This is because the lateral aspect of the fibula is the most easily used for plating and thus is typically planned to be facing buccal/labial. The medial aspect of the fibula, where the vascular pedicle runs, is set palatal. Unfortunately, when posterior teeth are preserved they can form a barrier to the vascular pedicle's path to recipeient vessels (Fig. 14.12). The authors have addressed this technical difficulty in a number of ways. One is to simply remove all posterior teeth in that segment. The pedicle can then run posteriorly and then inferiorly along the medial side of the mandibular ramus. One can try to run the pedicle underneath the palatal mucosa along the junction of the horizontal and vertical hard palate toward the soft palate before diving medial to the mandible. However, the attached gingiva of the hard palate is thick, tight, and generally inflexible, so this creates an area where the pedicle is compressed against the bone, especially as postoperative swelling sets in. Another work around is to create a large bone window through the maxillary sinus wall. The pedicle then is run over the fibula, through the window and then through a subcutaneous tunnel on the buccal side of the mandible.

For high Brown Class II defects with loss of the piriform rim and anterior maxilla up to and including the infraorbital nerve, the loss of vertical buttressing presents a problem for fixation and support to the neo-maxillary arch. A biaxial double-barrel arrangement can be used to reconstruct the piriform buttress, infraorbital facial support, paranasal support, and maxillary arch. In designing the second axis, an adequate leash (at least 3 cm) of pedicle is allowed to make the 180-degree turn (Fig. 14.13a). The authors generally make this portion of the flap the most distal from the anastomosis. The rationale is that if the folding results in a partial flap failure, it is of the least consequence and highest potential for retention, if only as a bone graft, being that it is still enveloped in vascularized soft tissue and protected from the oral environment. Keep in mind, during flap harvest, the fibula skin paddle must be harvested off a perforator located over the more proximal segments. Otherwise, they will be facing into the maxillary sinus and will not be able to reach the oral cavity.

From this plan, 3-D printed cutting guides are made to recreate the computer plan in the operating room (Fig. 14.13b).

In the operating room, a Weber-Ferguson or similar approach is used for adequate access (Fig. 14.13c). Prior to flap inset, a tunnel is cre-

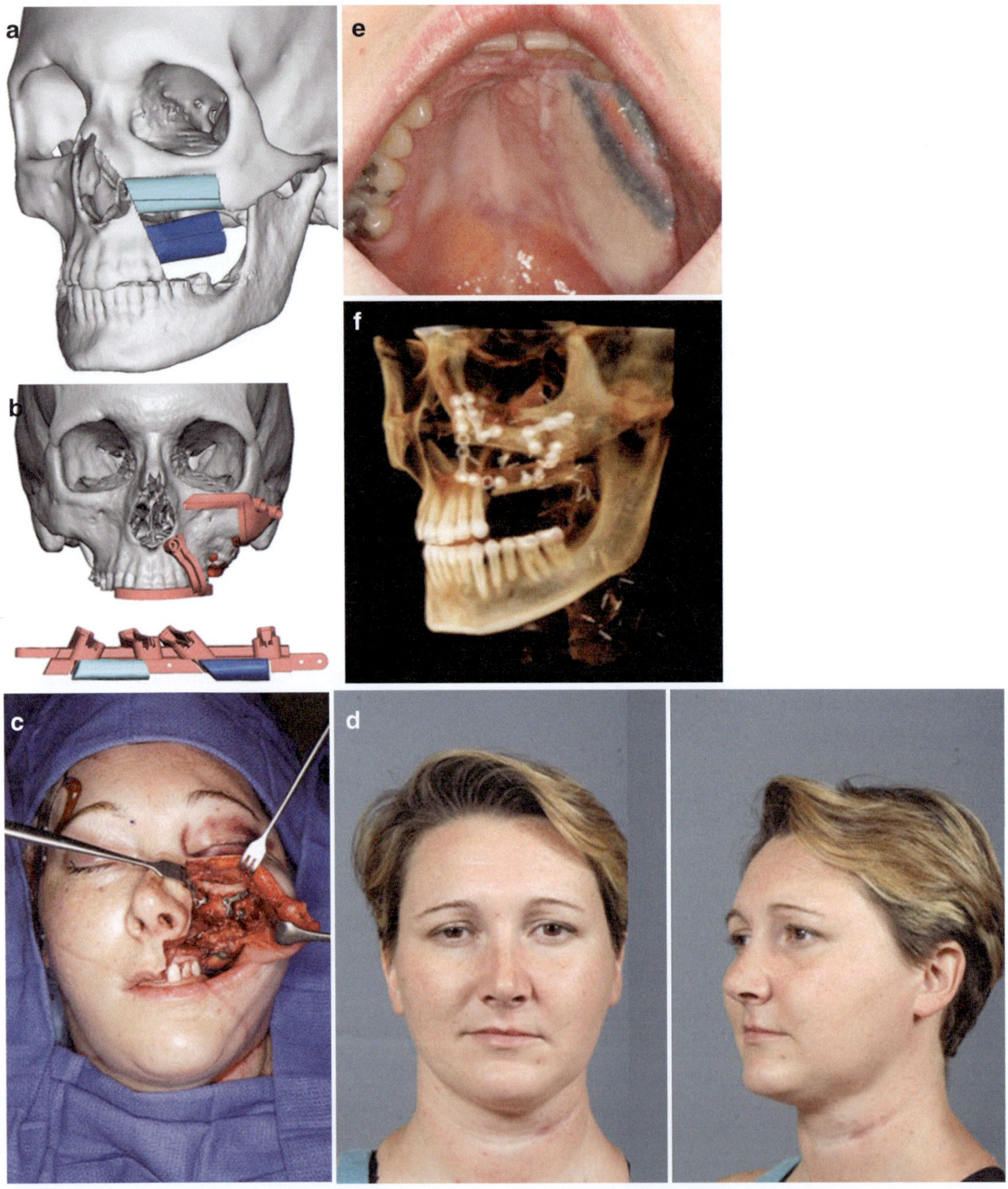

Fig. 14.13 (**a**–**f**) Thirty-three-year-old female with dentinogenic ghost cell tumor. (**a**) Biaxial double-barrel flap technique with 3 cm leash for fibula reconstruction of Class II b defect. (**b**) Maxillary (above) and fibula (below) cutting guides. (**c**) Weber-Ferguson approach to maxilla. (**d**) 1.5 years s/p Brown II b maxillectomy and reconstruction with excellent facial symmetry and projection. (**e**) Adequate bony reconstruction of alveolus for removable partial or endosseous implants. (**f**) CBCT

ated for the vascular pedicle to run from the defect into the neck. If possible, the pedicle is run along the medial aspect of the mandibular ramus, through the lateral pharyngeal space. This is the straightest and shortest route from the maxilla to the neck. It avoids the bends in the vascular pedicle when running it on the lateral aspect of the mandible and over the inferior border. During the flap inset, intraoperative navigation is used to guide the positioning of the neo-maxilla, making sure that the construct approximates the computer plan to avoid roll, yaw, and pitch deformities. After fixation of the flap and completion of the microvascular anastomosis, the final position can be confirmed using intraoperative CT or later with a postoperative CT. The end result provides both adequate facial support and a platform for future implant restoration (Fig. 14.13d–f).

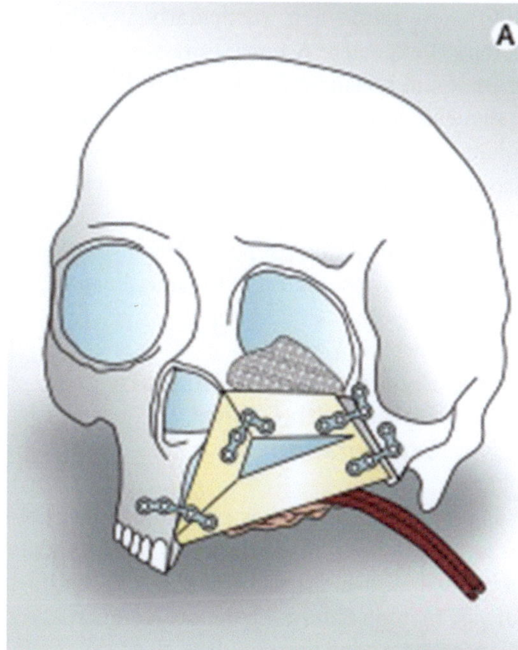

Fig. 14.14 Three-piece fibula for reconstruction of Brown III b defect—from Brown, J.S. and R.J. Shaw, *Reconstruction of the maxilla and midface: introducing a new classification*. Lancet Oncol, 2010. **11**(10): p. 1001–8

14.7.5 Brown Class III: Infrastructure Maxillectomy Involving Orbital Rim and Floor

The loss of the orbital rim and floor adds considerable challenges to the reconstruction. Removing the orbital rim adds another missing horizontal buttress for facial support. Removing the orbital floor results in a marked enophthalmos that is disfiguring.

The authors still favor the use of a fibula free flap. As mentioned above, the fibula osteotomies are planned as a biaxial double-barrel flap. Others have described creating 3–5 small segments for the inferior orbital rim, the piriform rim, the zygomatic buttress, and the maxillary alveolus (Fig. 14.14, from Brown et al.). This is enormously complicated and unnecessary. Two segments of fibula, when stacked, are typically enough to recreate the height of the piriform rim (Fig. 14.15), obviating the need to have a third small segment. This is further reinforced when using custom reconstruction plates. The zygomatic maxillary buttress, though important in reduction of fractures, is less essential when restoring defects.

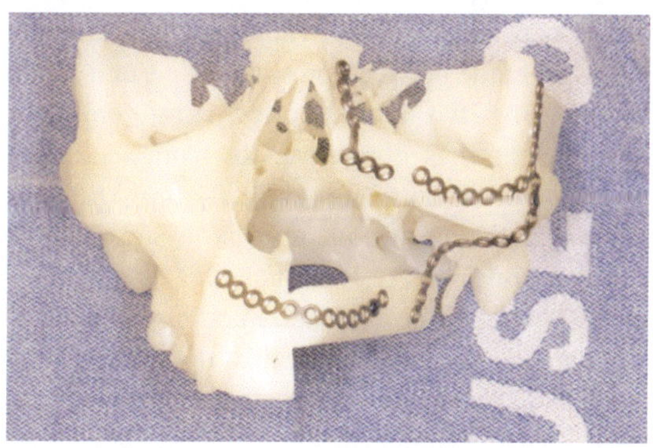

Fig. 14.15 Biaxial double-barrel flap technique with 3 cm leash for fibula reconstruction of Class III b defect

The floor can be reconstructed with a titanium plate, bone graft, or a bone flap. The simplest would be to rebuild the floor with an orbital floor fracture plate fixated to the fibula segment used for the inferior orbital rim. This is ideally covered with a soft tissue flap, particularly if radiation therapy is expected. A bone graft over a plate or by itself can be similarly successful. However, it is also vulnerable to failure due to postoperative radiation therapy. A vascularized scapula tip can also be used for floor reconstruction. If a scapula is used, the lateral border can be also used to replace the infraorbital support. However, this is inadequate for restoration of the maxillary arch.

14.7.6 Brown Class IV: Maxillary Defect with Orbital Exenteration

If an orbital exenteration is performed, the goals of reconstruction, in addition to the elements mentioned above for Brown Class II and III defects, include obturating the orbit and creating a barrier separating the cranial vault (if the resection includes the skull base). In the case of dural exposure or dural defect, a watertight closure should be obtained. A secondary goal would be to prepare the orbit for a future facial prosthesis.

Unfortunately, no one flap can achieve all of these goals satisfactorily. Therefore, the decision has to be made between prioritizing certain goals versus using two flaps. In prioritizing goals, the most essential is obturating the volume defect and separating compartments. To this end, the authors will most frequently use a subscapular system chimeric flap. The latissimus dorsi can be used to obturate the orbit and maxillary sinus and form an adequate barrier against dural repairs. If available, an anteriorly or posteriorly based pericranial flap is folded in to cover the dura as an additional layer of protection. The scapula skin can be used both externally and intraorally to reline missing skin and mucosa. The lateral border of the scapula and scapula tip can be used to reconstruct the malar region and hard palate. However, using this strategy, it is difficult to adequately reconstruct the maxillary alveolus (Fig. 14.16a–d).

If two flaps are to be used, a fibula flap can be used as described for Brown Class III defects, while a soft tissue flap (anterolateral thigh, rectus, or latissimus) is used to obturate the orbit. The challenge here is having adequate vascular access for anastomosis of two flaps that can be reached by the pedicles of both.

14.7.7 Brown Class V: Orbitomaxillary Defect Without Dentoalveolar Defect

In defects isolated to the orbit, bony reconstruction is not necessarily required. The main goal is to obturate the orbit and seal off the paranasal sinuses from the intracranial vault. If the eyelids are preserved, the reconstruction is simplified a bit.

There are two approaches to this. One is to leave a defect cavity and focus on lining the bone of the orbit. This has been referred to as an "open cavity" reconstructive approach. In these cases, a skin graft is placed over the bony walls of the orbit, leaving the orbit "open." This is a simple reconstruction and allows for easy monitoring for tumor recurrence. However, this is not suitable if there is extensive bone resection and/or if there is communication to the intracranial cavity that needs to be sealed off. In addition, if radiation therapy is anticipated, a skin graft will not be resilient enough to withstand that assault.

The more common is the "closed cavity" approach, where the orbit is obliterated. This is the strategy the authors typically use. Soft tissue flaps are ideal for this. In particular, muscle flaps, such as the latissimus dorsi or rectus, are well suited. Muscle flaps, given their vascular nature, are favorable for filling spaces and adhering to bone surfaces. Furthermore, over time these flaps will atrophy, leaving a robust, vascularized lining without the overabundance of soft tissue seen during the initial postoperative phase. This is advantageous for orbital reconstructions as it allows for a nice retentive space for a facial

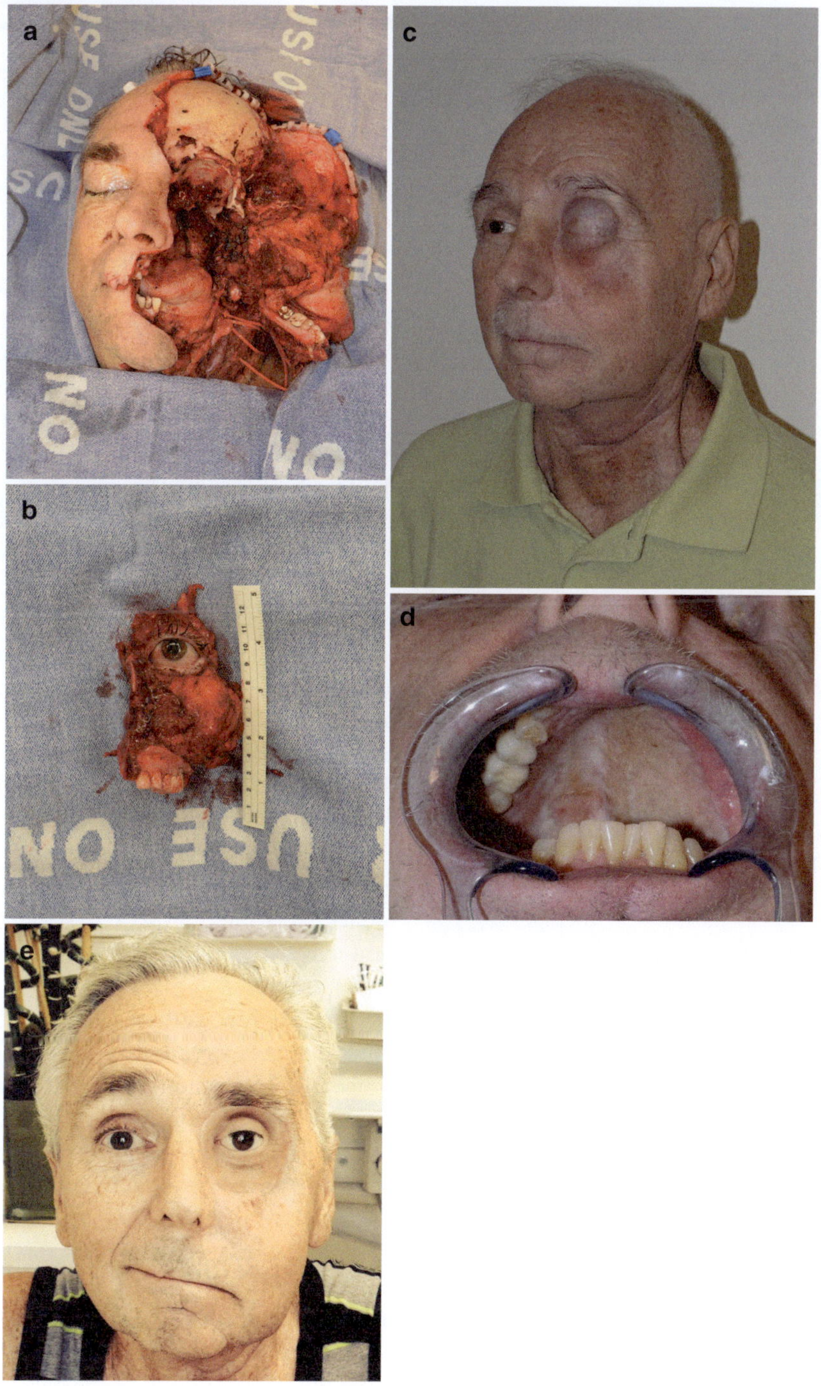

Fig. 14.16 Sixty-eight-year-old male with rpT4bN1M0 SCC of the palate, zygoma, and anterior cranium after radiation therapy s/p cranial resection and Brown Class IV b maxillectomy with orbital exenteration, reconstructed with chimeric subscapular system mega flap. (**a**) Resection. (**b**) Brown IV b maxillectomy. (**c**) Six-month postoperative follow-up. (**d**) Six-month postoperative intraoral. (**e**) Prosthetic in place

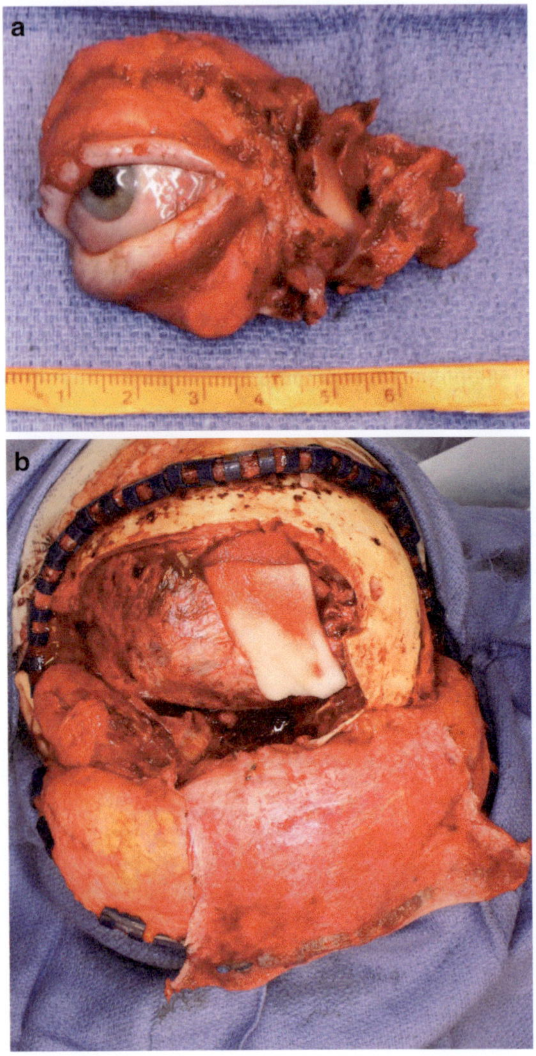

Fig. 14.17 Fifty-eight-year-old female with T4aN0M0 neuroendocrine tumor involving right eye s/p composite resection involving exenteration. (**a**) Orbital exenteration. (**b**) Exposed dura, patient underwent closed approach obliteration of orbit with latissimus free flap

prosthesis to fit into. This is also good for buttressing a dural repair (Fig. 14.17a, b). However, if atrophy of the soft tissue is not desired, an adipofascial flap, such as the anterolateral thigh flap, should be used, as fat atrophies to a much smaller degree.

14.7.8 Brown Class VI: Nasomaxillary Defect

The nasomaxillary defect, which may be limited to the nasal-orbito-ethmoid (NOE) region or extend to include a medial maxillectomy and total rhinectomy, is one of the most challenging facial defects to reconstruct.

For defects limited to the nasal-orbito-ethmoid region, separation of the anterior cranial cavity from the paranasal sinuses is essential. A thin soft tissue flap, such as the radial forearm, is very suitable. The loss of bony support at the NOE can be rebuilt with free bone grafts harvested from the cranium (Fig. 14.18a–c).

For total rhinectomy defects, as above, there are two general strategies to approaching this: prosthetic versus surgical reconstruction. Of the two, the prosthetic will more consistently lead to a result that approximates a normal appearance (Fig. 14.19a, b). A skilled maxillofacial prosthodontist can do much more with plastics than a surgeon can with a blade. The retention of this prosthesis can be aided by the strategic placement of dental implants (Fig. 14.20a–d). Because these implants are made for oral use and not intended to have transcutaneous elements, peri-implant soft tissue irritation is a problem that requires regular cleaning to keep under control.

Surgical reconstruction of the total rhinectomy defect is a topic that goes well beyond the scope of this chapter. There are several descriptions in the literature of different techniques [24–26]. Generally, there are the following components that need to be addressed: the nasal lining, the nasal framework, and the external lining. The nasal lining, for total rhinectomy defects, can be reconstructed with a radial forearm free flap. The nasal framework can be reconstructed using cartilage grafts harvested from the conchal bowl of the ear, the nasal septum, a rib, or a combination. Shaping of the grafts into a framework must include a dorsal

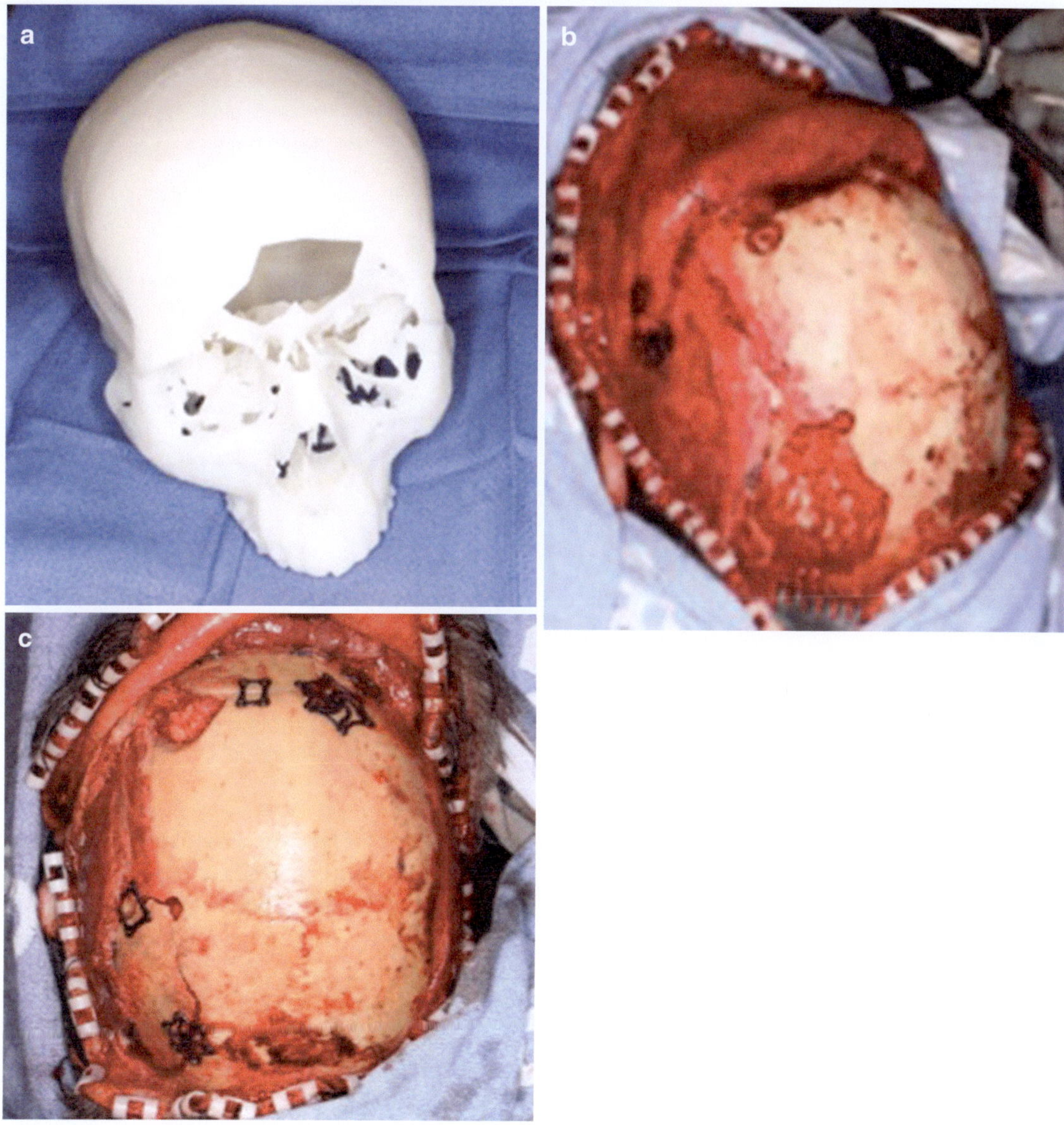

Fig. 14.18 Sixty-year-old male with sinonasal melanoma planned for composite resection of anterior frontal sinus and ethmoid sinus. (**a**) Computer surgical planning with cutting guides for anterior craniectomy. (**b**) Full-thickness cranial bone harvest using cutting guide. (**c**) Inner table at donor site, outer table with excellent fit over the anterior cranium and NOE

strut, a columellar strut, and the lower lateral nasal cartilage. The external skin is best reconstructed using a paramedian forehead flap. Depending on the size of the defect, this may benefit from tissue expansion to allow for a large enough flap to cover the new nose. Prelaminated radial forearm flaps have also been used and described.

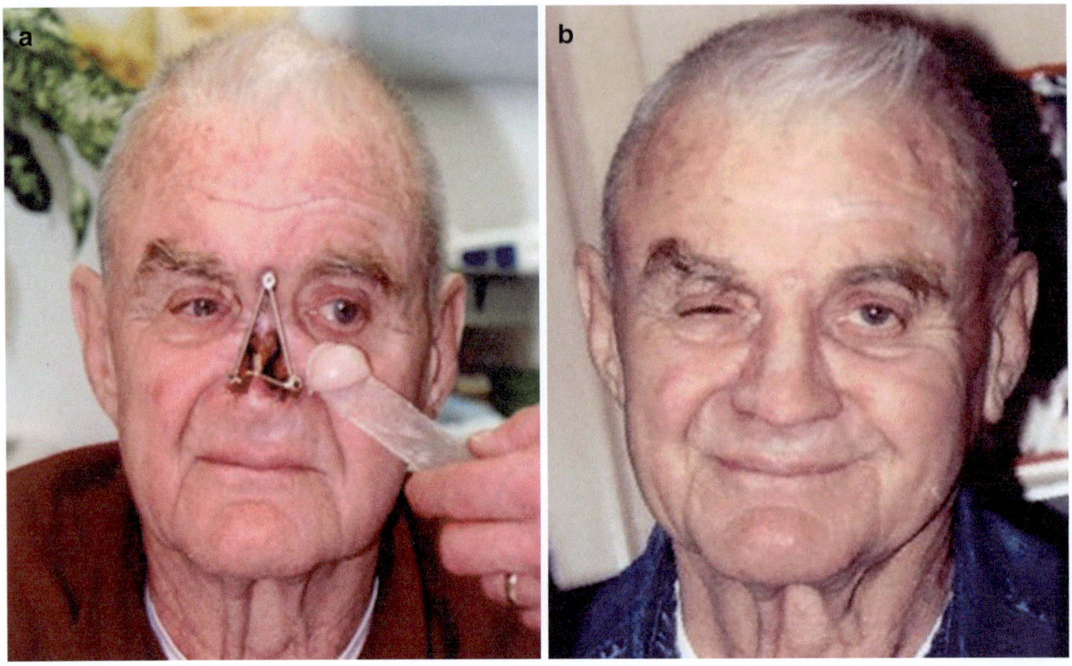

Fig. 14.19 Sixty-year-old male with melanoma of nasal dorsum s/p total rhinectomy. (**a**) Craniofacial implants with custom bar in place. (**b**) Excellent esthetics and color match of prosthesis

14.8 Conclusion

The maxilla is an area central to the face and at the crossroads of several functionally and esthetically complex areas. As such, defects of the maxilla can be quite diverse and reconstruction of these defects quite complicated. The revised Brown Classification scheme is helpful in identifying the critical elements of reconstruction for any given maxillary defect. Reconstruction of maxillary defects will employ either prosthetic-only solutions, surgical reconstructive solutions, or, in many cases, a combination of these. The reconstructive surgeon must have a diverse armamentarium of flap choices to be able to sufficiently address the wide range of possible defects. The surgeon must also be well versed in requirements for prosthetic success. Surgery is hampered by the limited access and the distance between the defect and the neck vasculature. Technological advances such as computer surgical planning, intraoperative navigation, intraoperative CT, printed cutting guides, and custom reconstruction plates all aid in improving the reproducibility of good outcomes, but do not replace thoughtful treatment planning and skillful execution.

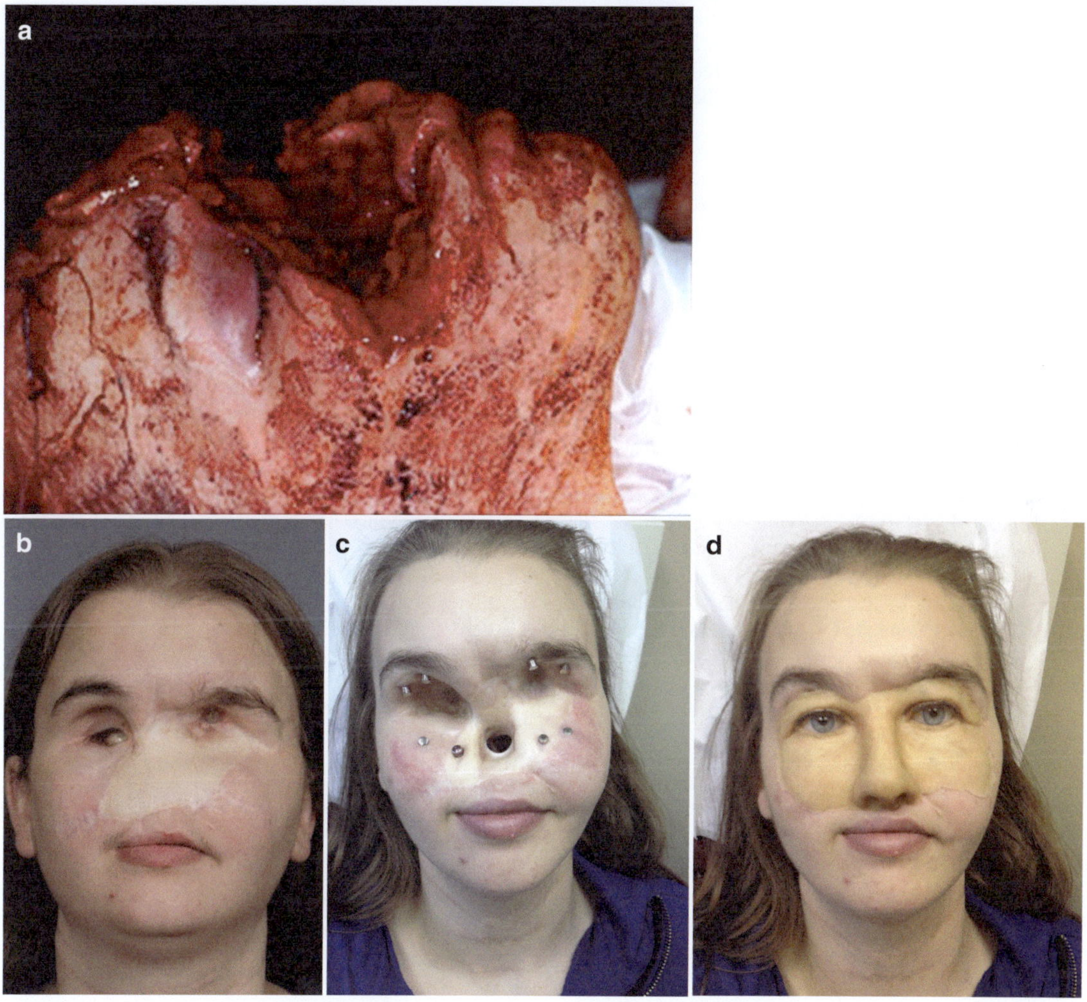

Fig. 14.20 Sixteen-year-old female s/p shotgun injury to the face. (**a**) Original injury. (**b**) Ten years after initial reconstruction. (**c**) Craniofacial implants placed after multiple soft tissue revision surgeries. (**d**) Prosthetic in place, reconstructing the Brown V and VI defects

References

1. Yoshida H, Michi K, Yamashita Y, Ohno K. A comparison of surgical and prosthetic treatment for speech disorders attributable to surgically acquired soft palate defects. J Oral Maxillofac Surg. 1993;51(4):361–5.
2. Lanigan DT, Hey JH, West RA. Aseptic necrosis following maxillary osteotomies: report of 36 cases. J Oral Maxillofac Surg. 1990;48(2):142–56.
3. Lanigan DT, Hey JH, West RA. Major vascular complications of orthognathic surgery: hemorrhage associated with Le Fort I osteotomies. J Oral Maxillofac Surg. 1990;48(6):561–73.
4. Öhngren LG. Malignant tumours of the maxillo-ethmoidal region: a clinical study with special reference to the treatment with electrosurgery and irradiation. Acta Otolaryngol (Stockolm). 1933;19:1–476.
5. Shrime MG, Gilbert RW. Reconstruction of the midface and maxilla. Facial Plast Surg Clin North Am. 2009;17(2):211–23.
6. Aramany MA. Basic principles of obturator design for partially edentulous patients. Part I: classification. J Prosthet Dent. 1978;40(5):554–7.
7. Aramany MA. Basic principles of obturator design for partially edentulous patients. Part II: design principles. J Prosthet Dent. 1978;40(6):656–62.

8. Spiro RH, Strong EW, Shah JP. Maxillectomy and its classification. Head Neck. 1997;19(4):309–14.
9. Brown JS, Rogers SN, McNally DN, Boyle M. A modified classification for the maxillectomy defect. Head Neck. 2000;22(1):17–26.
10. Okay DJ, Genden E, Buchbinder D, Urken M. Prosthodontic guidelines for surgical reconstruction of the maxilla: a classification system of defects. J Prosthet Dent. 2001;86(4):352–63.
11. Cordeiro PG, Santamaria E. A classification system and algorithm for reconstruction of maxillectomy and midfacial defects. Plast Reconstr Surg. 2000;105(7):2331–46.
12. Brown JS, Shaw RJ. Reconstruction of the maxilla and midface: introducing a new classification. Lancet Oncol. 2010;11(10):1001–8.
13. Shum J, Markiewicz MR, Park E, Bui T, Lubek J, Bell RB, Dierks EJ. Low prealbumin level is a risk factor for microvascular free flap failure. J Oral Maxillofac Surg. 2014;72(1):169–77.
14. Breeze J, Rennie A, Morrison A, Dawson D, Tipper J, Rehman K, Grew N, Snee D, Pigadas N. Health-related quality of life after maxillectomy: obturator rehabilitation compared with flap reconstruction. Br J Oral Maxillofac Surg. 2016;54(8):857–62.
15. Gay WD, King GE. Applying basic prosthodontic principles in the dentulous maxillectomy patient. J Prosthet Dent. 1980;43(4):433–5.
16. Martin JW, King GE. Framework retention for maxillary obturator prostheses. J Prosthet Dent. 1984;51(5):669–72.
17. Genden EM, Wallace DI, Okay D, Urken ML. Reconstruction of the hard palate using the radial forearm free flap: indications and outcomes. Head Neck. 2004;26(9):808–14.
18. Moreno MA, Skoracki RJ, Hanna EY, Hanasono MM. Microvascular free flap reconstruction versus palatal obturation for maxillectomy defects. Head Neck. 2010;32(7):860–8.
19. Kornblith AB, Zlotolow IM, Gooen J, Huryn JM, Lerner T, Strong EW, Shah JP, Spiro RH, Holland JC. Quality of life of maxillectomy patients using an obturator prosthesis. Head Neck. 1996;18(4):323–34.
20. Hirsch DL, Howell KL, Levine JP. A novel approach to palatomaxillary reconstruction: use of radial forearm free tissue transfer combined with zygomaticus implants. J Oral Maxillofac Surg. 2009;67(11):2466–72.
21. Burgess M, Leung M, Chellapah A, Clark JR, Batstone MD. Osseointegrated implants into a variety of composite free flaps: a comparative analysis. Head Neck. 2017;39(3):443–7.
22. Levine JP, Bae JS, Soares M, Brecht LE, Saadeh PB, Ceradini DJ, Hirsch DL. Jaw in a day: total maxillofacial reconstruction using digital technology. Plast Reconstr Surg. 2013;131(6):1386–91.
23. Clark JR, Vesely M, Gilbert R. Scapular angle osteomyogenous flap in postmaxillectomy reconstruction: defect, reconstruction, shoulder function, and harvest technique. Head Neck. 2008;30(1):10–20.
24. Rohrich RJ, Griffin JR, Ansari M, Beran SJ, Potter JK. Nasal reconstruction—beyond aesthetic subunits: a 15-year review of 1334 cases. Plast Reconstr Surg. 2004;114(6):1405–16.
25. Cannady SB, Cook TA, Wax MK. The total nasal defect and reconstruction. Facial Plast Surg Clin North Am. 2009;17(2):189–201.
26. Teichgraeber JF, Goepfert H. Rhinectomy: timing and reconstruction. Otolaryngol Head Neck Surg. 1990;102(4):362–9.

Principles of Mandibular Reconstruction with Vascularized Flaps

15

Nicholas Callahan, Joshua E. Lubek, and Donita Dyalram

15.1 Introduction

Reconstruction of the mandible continues to be one of the most common surgical challenges encountered by the maxillofacial reconstructive surgeon. The ultimate goal of mandibular reconstruction is to reestablish mandibular continuity while restoring function and appearance. Despite improvements in technologies such as virtual surgical planning, 3D models/printing, dental implants, and microvascular techniques, there are numerous factors that must be considered to provide optimal results in a timely fashion.

There are multiple techniques for mandibular reconstruction that include either the use of vascularized or non-vascularized bone grafting techniques. Considerations for choice of graft will depend on factors such as mechanism of injury/pathology, location and size of the mandibular defect, loss of overlying soft tissue, patient comorbidities/desires, surgeon experience, and need for adjuvant therapy such as radiation treatment.

Vascularized tissue transfer continues to be a reliable and predictable method for reconstruction of moderate- to large-sized defects of the mandible. These flaps have the ability to reconstruct complex defects by allowing for the restoration of both bone and soft tissues thereby achieving the desired goals of restoration: speech, esthetics, swallowing, and dental rehabilitation.

This chapter will discuss three of the most common vascularized bone flaps utilized in mandibular reconstruction: the fibula osteocutaneous flap, the deep circumflex iliac artery (DCIA) flap, and the scapula flap. Indications for type of flap selection, key aspects in harvest techniques, and avoidance of complications will be highlighted.

15.2 Approach to the Mandibular Defect

When a reconstructive surgeon is consulted to rebuild a mandible, it begins like any other surgery. A complete history and physical examination is needed with keen attention given to potential donor sites such as inguinal hernia repair, rotator cuff repair, history of fractures, DVT in extremities, and presence of peripheral vascular disease to name a few.

When considering the recipient site, the components of the mandible to be reconstructed are taken into account, such as is the condyle being replaced, is the symphysis being replaced, or is it an angle-to-angle defect or a combination? Also, the patient's occlusion plays a crucial role and needs to be reestablished and maintained.

N. Callahan · J. E. Lubek · D. Dyalram (✉)
Department of Oral Maxillofacial Surgery, University of Maryland Dental School, Baltimore, MD, USA
e-mail: jlubek@umaryland.edu; ddyalram@umaryland.edu

J. C. Melville et al. (eds.), *Regenerative Strategies for Maxillary and Mandibular Reconstruction*,
https://doi.org/10.1007/978-3-319-93668-0_15

The surgeon needs to determine whether immediate dental implants can be placed in the neo mandible and whether it is ideal for placement of an immediate interim prosthesis. The amount of soft tissue needed to be replaced also plays a role: is it a through-and-through defect or is it a mucosal defect? The condition of the neck vessels is equally important. Was there a previous neck dissection? What type of neck dissection? Is there a history of head and neck radiation? The inferior alveolar nerve restoration should not be ignored. We have the ability to restore this nerve and it needs to be discussed with the patient.

All these components need to be accessed on the first visit with the patient, for it will impact what studies are needed in the planning phase. The next step after a complete physical exam is a radiographic exam. The dental panorex, while overlooked by non-oral maxillofacial surgeon, remains an excellent starting point. It is limited since it is a 2D representation of a 3D structure, but it gives the best image of the dentition of all of the commonly used radiographic views of the mandible. A computed tomography (CT) scan is commonly undertaken. 1 mm slices with bony windows are recommended for virtual surgical planning. This allows for a detailed examination of the mandible and allows for a precise virtual surgical plan. MRI can also be a useful tool for evaluation of the extent of some lesions but cannot be used for virtual surgical planning.

15.3 Fibula Osteoseptocutaneous Free Flap

The free fibula flap dates back to 1975, when Taylor [1] first reported using this flap for repair of severe fractures of the tibia and fibula with bony defects. Gilbert [2] introduced the lateral approach in 1979, which significantly decreased the amount of time needed for harvest. In 1989 Hildago [3] reported the first 12 cases of mandibular reconstruction with a free fibula flap. The flap underwent multiple modifications, but in 1994, the osteoseptocutaneous flap as we know it today was described by Wei [4].

15.4 Advantages

There are numerous reasons why a free fibula might be ideal as a reconstructive option. First and foremost, this flap can provide ample tissue: skin, muscle, and bone. The flap can provide a large skin paddle up to 25 by 12 cm and a segment of bone up to 25 cm. This allows for reconstruction of complex mandibular defects, involving bone, mucosal, and skin. The blood supply to the fibula, the peroneal artery and its vena comitans, is endosteal and periosteal and allows for segmental osteotomies. The ability to create multiple osteotomies allows for reconstruction of complex bone shapes and even for stacking of the fibula for increased height.

The fibula is of significant distance from the head and neck region making it suitable for a two-team approach in the operating room. This can significantly decrease operative time and thus patient exposure to anesthesia.

The fibula is of adequate bone density and height to allow for the placement of dental implants and therefore improved dental rehabilitation (Figs. 15.1 and 15.2).

15.5 Disadvantages

The circulation in the lower extremities is prone to vascular disease and calcifications that can preclude this as a donor choice. The harvest leaves a large scar on the leg, and if a skin paddle is taken, it will require a skin graft. The skin from the leg is not an ideal color match for facial skin defects. There can be loss of the greater toe mobility if the flexor hallucis longus muscle is taken with the flap.

15.6 Planning for the Fibula Flap

All patients who are in need for free flap should undergo the standard work-up for their mandibular disease process. The specific work-up for a fibular free flap as always begins with a detailed history and physical exam, with attention to the lower extremities. The patient should be queried

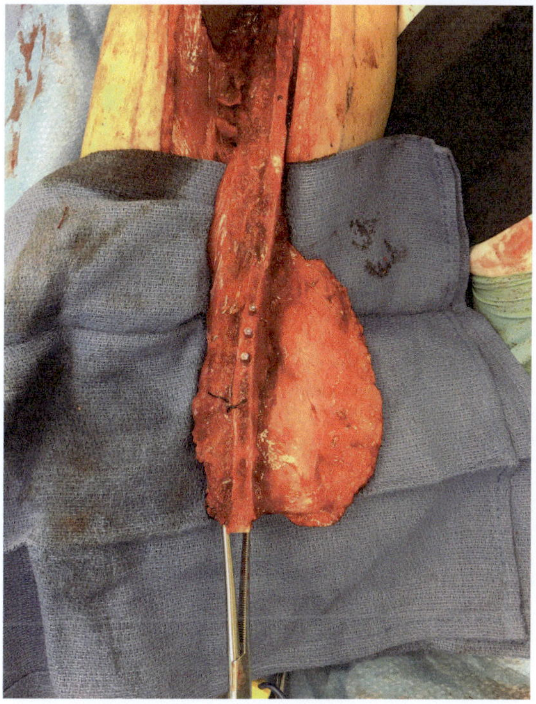

Fig. 15.1 This is a harvested fibula flap with a skip paddle. Endosseous dental implants can be planned and placed at the time of harvest. Placement of dental implants in a fibula flap before division of pedicle. It is best to place the implants prior to completion of the osteotomies

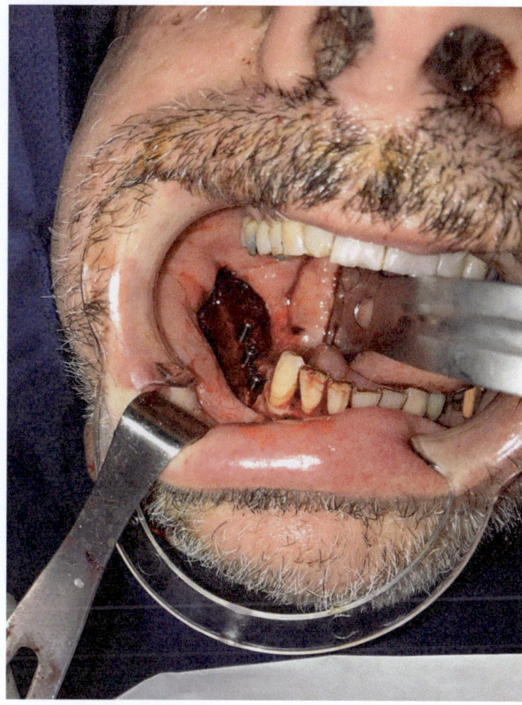

Fig. 15.2 This clinical picture shows the dental implants with healing abutments in the "neobody" of the mandible after the osteotomy of the fibula

for any injuries, surgeries, and diseases to the lower extremities.

The physical exam focuses on skin and vascular integrity. Ideal candidates should have strong dorsalis pedis and posterior tibial pulses and no pitting edema, masses, or lesions of the lower extremities. A history of melanoma, basal cell carcinoma, or squamous cell carcinoma will exclude a site. Inspection of the leg should be done to look for tattoos. If a facial skin defect is needed to be reconstructed, the potential transfer of the tattoo should be discussed with the patient.

Patients who have no physical exam findings precluding a fibular flap will undergo radiographic evaluation of the vascular anatomy of the lower extremity. Lower extremity angiogram is ideal to visualize the vessels in the leg. Patients that have atherosclerotic disease of lower extremity are not good candidates for fibula flap (Fig. 15.3). Of greater concern, 5% of the population have a vascular abnormity known as peronea arteria magna, a congenital variant where the anterior and posterior tibial arteries are hypoplastic and a large dominant peroneal artery supplies the foot [5]. This can lead to rare limb-threatening complications despite a normal vascular exam.

Once the angiogram reveals that the vascular anatomy is appropriate, the next step is to decide which leg to use. If only the bone is needed, the surgical plan becomes significantly simple, and either leg can be used. When a composite flap is needed, one must consider where the soft tissue is needed. Is this tissue needed intraorally or extraorally? Also to consider is whether the vascular pedicle needs to be positioned posteriorly or anteriorly. Lastly, which surface of the fibula will be the occlusal surface, and which surface of the fibula will the plate adhere?

Today virtual surgical planning (VSP) is common for all bony flap reconstruction. This facilitates bending of the plates prior to the case, models, and cutting guides to be used in the operating room. For a one-piece fibula segment

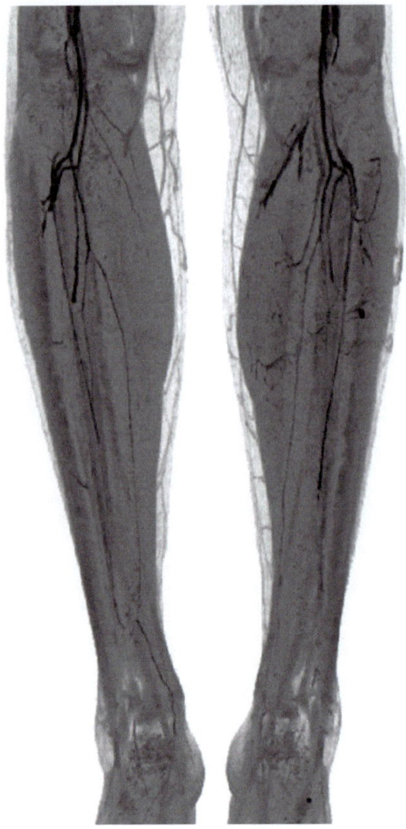

Fig. 15.3 CT angiogram of the lower extremity shows significant atherosclerotic disease making the fibula flap a poor donor choice

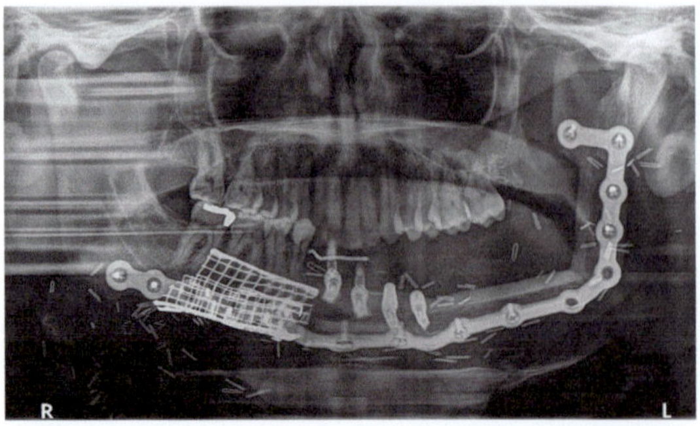

Fig. 15.4 This radiograph is of a patient who was previously reconstructed with a titanium mesh and non-vascularized bone. The left side demonstrates the use of a fibula flap with endosseous implants and the use of a custom bar

defect, that is not necessary. However for larger defect involving different aspects of the mandible, VSP is helpful and translates to time saved in the operating room (Fig. 15.4).

15.7 Key Aspects to Fibula Flap Harvest

A standard preparation for the fibula harvest is a large gel roll position such that when the knee is flexed, there is approximately the space for an outstretched hand in the underside of the knee. This gel roll is secured and will provide a space for the foot to sit and allow the knee to remain comfortably flexed. A hip bump is placed to allow inward rotation of the hip. The leg is then prepped from the upper thigh to the toes. A 3/4 sheet is placed under the leg, and this is followed with a lower extremity drape. At this time the surgical markings of the flap is outlined, and the proposed skin paddle is marked, if needed. The skin paddle typically is in the lower 1/3 of the leg as this is where the septocutaneous perforators are found. An esmarch is then used to exsanguinate the leg, and the tourniquet is placed on the upper thigh at 350 mm Hg for 120 min (Fig. 15.5).

The harvest is performed in the standard anterior/posterior approach. The osteotomies are made in the fibula using the cutting guides provided by the VSP planning, while the flap is still connected to its vascular supply. Only once the ablative surgeons have completed their portion and occlusion was established and vessels sorted is the flap moved to the recipient site.

15.8 Deep Circumflex Iliac Artery (DCIA) Flap

The iliac crest has been long used for osseous reconstruction of head and neck defects. Vascularized bone flaps were not described till 1979 when Taylor and Sander [6, 7] discovered the deep circumflex iliac artery was the dominant blood supply to the iliac crest. The first reported cases of mandibular reconstruction with this flap were reported in the following year [8]. In 1984 the addition of the internal oblique muscle with vascularized iliac crest was

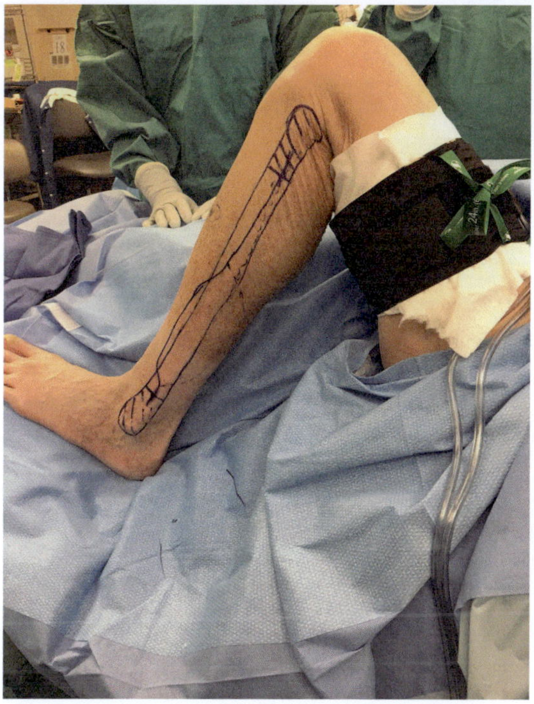

Fig. 15.5 Picture showing the setup for a free fibula flap harvest. Note the flex position of the leg as well as the tourniquet on the upper thigh. An assistant is on the opposite side to help with retraction. The lateral aspect of the fibula head and the lateral malleolus is outlined as well as the fibula

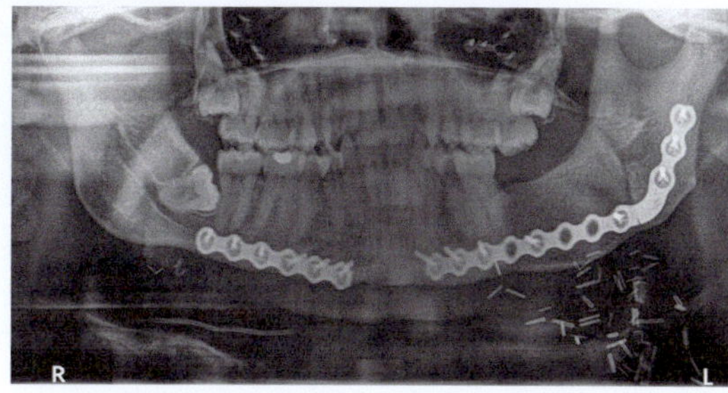

Fig. 15.6 This is a panorex of a reconstructed left body of the mandible with a DCIA flap. Note the height of the mandible that can be achieved with this flap

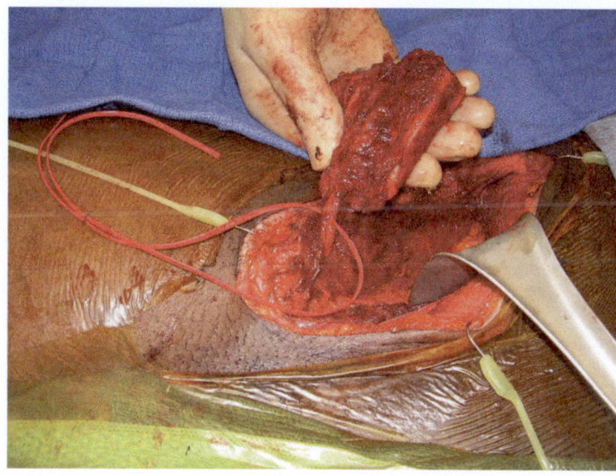

Fig. 15.7 Picture of the harvested DCIA flap. The vessel loop is around the pedicle. Note the width of the bone harvest. They are ideal for immediate dental implants

used for reconstruction of the lower extremity. Urken first used this in the head and neck reconstruction in 1989 [9].

15.9 Advantages

The DCIA provides excellent cortical and cancellous bone as well as height which makes it ideal for mandibular reconstruction. The contour of the iliac crest is similar to mandibular bone in that its natural curve can imitate the curve of the native mandible. It provides the closest height and width match to the native mandible (Figs. 15.6 and 15.7). It also provides adequate bone stock for implant placement in almost all cases. Composite flaps can be raised, allowing for harvesting of bone, muscle, and skin. A two-team approach can be used thus minimizing surgical time.

15.10 Disadvantages

The skin paddle can be bulky and difficult to inset. The bone can only provide approximately 16 cm of bone stock, which is not enough for an angle-to-angle mandibular defect. The vascular pedicle is short between 5 and 7 cm, and vein grafting may be necessary in previously treated necks. During the surgery, a hernia is created, and without adequate closure, patients can have a residual hernia and may require mesh placement.

15.11 Planning for the DCIA

A detailed history and physical exam is undertaken, as with any flap reconstruction. Although there is no imaging or preparatory work-up needed, the surgeon has to be keen regarding which patient is chosen for this flap. There are factors which can make unfavorable outcome with the DCIA flap such as abdominal/hernia repair, prior harvest of an iliac crest graft, and distal aortoiliac/femoral vascular disease. Patients who have disease process where there may be increased abdominal and intrathoracic pressure from coughing such as pulmonary problems are not ideal candidates. Patient with obesity makes the dissection challenging and has an increased mortality risk, as well as excessive thickness of the skin paddle. Suprapopliteal DVT should be considered a relative contraindication to this flap.

The lateral mandibular defect can be constructed with either DCIA flap; however the ipsilateral DCIA offers the best contour to the native mandible. If a skin paddle is planned as part of the resection, it is located over the iliac crest. A handheld Doppler is used, and they are typically found 2–3 cm cephalad to the crest and up to 7 cm from the anterior superior iliac spine (ASIS).

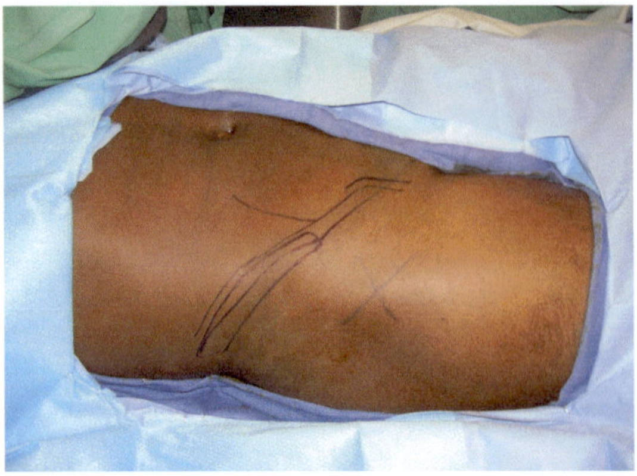

Fig. 15.8 Patient is prepped and draped for a DCIA flap. There is hip bump used to help accentuate the ASIS

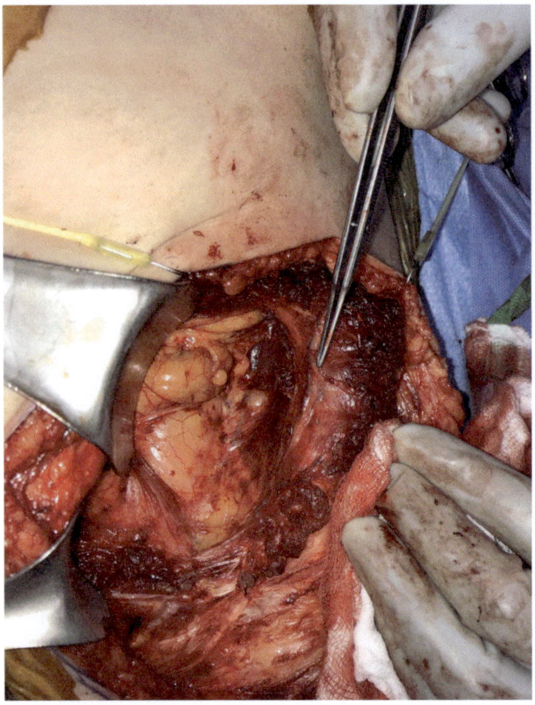

Fig. 15.9 The instrument is pointing to the ascending branch of the DCIA. This has to be preserved if a portion of the internal oblique muscle is taken with the flap harvest

15.12 Key Aspects to DCIA Flap Harvest

The patient is placed supine allowing for a two-team approach. A hip bump is placed to allow for greater access to the hip. The patient is prepped and draped from the midline of the abdomen, to the costal margin superior, the middle of the thigh inferiorly, and to the lateral surface of the bed. The anterior iliac spine is palpated, and the entire iliac crest is marked out. An incision is designed along the superior iliac crest from the femoral pulse to the midaxillary line (Fig. 15.8).

Typically the skin is not included in the flap since its arch of rotation is limited and it can be bulky in some patients. If soft tissue is needed, a portion of the internal oblique muscle is harvested. Care is taken to make sure the ascending branch of the DCIA is included to ensure blood supply to this muscle (Fig. 15.9).

The lateral cutaneous nerve to the thigh, which lies on the iliacus muscle, medial to the ASIS is vulnerable to injury. The nerve usually passes deep to the DCIA and can be dissected free from the pedicle. If the lateral femoral cutaneous nerve is superficial to the DCIA and it is necessary to transect it, it should be repaired.

Closure has to be done with care to prevent an abdominal hernia. After separation of the flap, hemostasis is obtained. The iliacus muscle and fascia are sutured to the transversalis fascia and transversus abdominis muscle. Burr holes are made in the remaining iliac crest at a 45-degree angle. A porous polypropylene mesh is used to replace the harvested internal oblique muscle. The mesh is secured to the remaining edges of the internal oblique and secured to the residual iliac crest through the burr holes. The external oblique is then closed to itself. The skin and subcutaneous tissues are closed in the traditional fashion. Two closed suction drains are used in the closure to prevent development of postoperative seroma.

15.13 Scapula Free Flap

The scapula free flap was first described by dos Santos in 1980 [10] and has since led the way to a variety of application. Dye studies conducted by Saijo [11] on cadaver showed the anastomosis of the circumflex scapular, thoracodorsal, and posterior intercostal arteries. It has been proven to be versatile giving rise to fasciocutaneous flaps, osteofaciocutaneous flaps, and musculocutaneous flaps. These multiple components of the skin, muscle, and bone can be transferred as a chimeric flap via a single pedicle for reconstruction of multiple subunits of the head and neck [12].

Gilbert and Toet [13] were the first to report using the flap for limp reconstruction. The scapula was first used in the head and neck by Swartz in 1986 who reported a case series of 26 cases [14]. There have since been numerous modifications proving that this is indeed a versatile flap [15].

15.14 Advantages

The scapula free flap had an extremely reliable vascular pedicle of good length and diameter. Multiple tissues can be harvested, including, bone, muscle, fascia, and skin. The bone is able to accept dental implants roughly 78% of the time (Fig. 15.10). A large skin paddle can be harvested of 25 × 35 cm. The skin paddle can

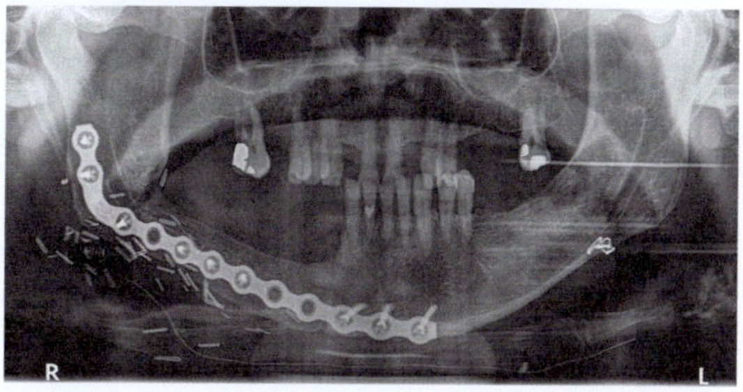

Fig. 15.10 Panorex showing reconstruction of the right lateral defect of the mandible with a scapula free flap. Note the height of the mandible achieved with this flap

also be raised in a chimeric fashion allowing coverage of multiple sites. The skin from this area is relatively hairless and is of fair color match to the head and neck. There is a relatively minor disturbance in the function of the shoulder, and complete return of function is usually achieved. The vasculature of the area is relatively unaffected by arteriosclerosis and diabetic disease, allowing for use of the flap in patients with significant disease in the extremities.

15.15 Disadvantages

The main disadvantage of the scapula free flap is that a two-team approach cannot be undertaken; this leads to increased surgical time. For instances where there is a large skin paddle needed, the donor site cannot be closed without a skin graft. This can lead to esthetic concerns.

15.16 Planning for the Scapula Free Flap

A standard history and physical exam is completed prior, and no special imaging is required. Diabetes and arteriosclerosis are not a contraindication since vessels in this area are typically spared. Obesity must be considered a relative contraindication since there is a significant amount of subcutaneous fat in this area and can make skin paddles unyielding. Any surgical procedure to the shoulder is a relative contraindication.

15.17 Key Aspects to the Harvest of the Scapula Flap

Although it has been reported that the patient in the lateral decubitus position allows for a two-team approach, it allows for a less-than-ideal access to the resection site. A SPIDER2 arm is attached to the opposite side of the bed of scapula being harvested. A beanbag is not necessary if the SPIDER arm is available. The back and shoulders are prepped from the midline to the lower border of the rib to the midaxillary line, including the entire ipsilateral arm (Fig. 15.11).

Drapes are placed and the donor arm/shoulder is laid in a sterile position during the resection. When the resection is completed, the donor site is exposed, and new drapes are laid to separate the donor site from the recipient site. The SPIDER2 arm is engaged, and care is taken during the setup to ensure that the fingers are checked for good perfusion. The arm/shoulder is laterally retracted so that the lateral border of the scapula is easily palpated.

If the scapula tip is desired for reconstruction, care has to be taken to identify and preserve the angular branch of the thoracodorsal artery. Harvest and closure of this donor site follow standard procedures.

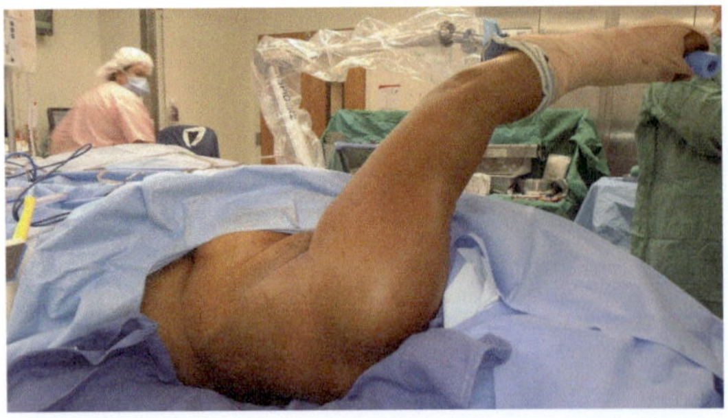

Fig. 15.11 This shows the setup for the harvest of the left scapula flap. The upper extremity is engaged with the SPIDER arm

15.18 Special Considerations

15.18.1 Approach to Condyle Reconstruction

When considering the condyle reconstruction, the fibula flap is preferred since its natural shape mimics that of the condyle. If the soft tissue defect is intraoral, then we prefer to use the same side lower extremity with the pedicle positioned anteriorly giving the option for anastomosis in the ipsilateral or contralateral neck.

This reconstruction offers the opportunity for shaping and plating the fibula to the reconstruction bar in the leg while it is still connected to the blood supply. These reconstructions are ideal for custom bars and use of VSP and the cutting guides. Once the occlusion is established, the flap can be brought to the head and neck for inset (Fig. 15.12).

The key is to make sure the condyle is in the fossa. It is very difficult to fix a laterally or medically displaced fibula at a later date. Postoperatively it is recommended that patients remain in maxillomandibular fixation for 2–4 weeks. There is always a "condylar sag" of the fibula postoperatively. Suturing the neocondyle head or using Mytex anchors has showed to make no difference. Despite this sag patients' mouth opening and function are satisfactory.

15.19 Inferior Alveolar Nerve Reconstruction

Reconstruction of the inferior alveolar nerve is often an afterthought, but given the advances in nerve repair, this should always be discussed. If it is not contraindicated and does not violate any oncological principles, the nerve can be separated from the mandible. If it is not possible to safely save the nerve, then it is recommended to tag both ends of the nerve during the resection since the nerve can be easily lost in the surgical bed.

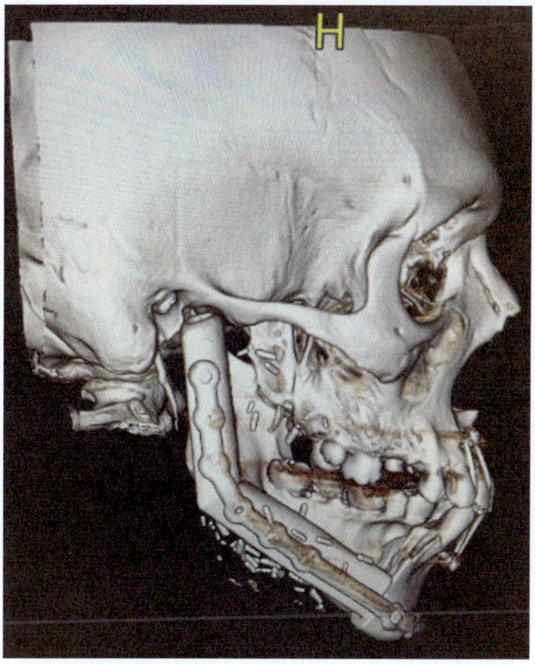

Fig. 15.12 This 3D reformatted CT scan shows the reconstruction of the condyle with a fibula flap using a custom reconstruction bar. Note that the patient is placed into maxillomandibular fixation postoperatively

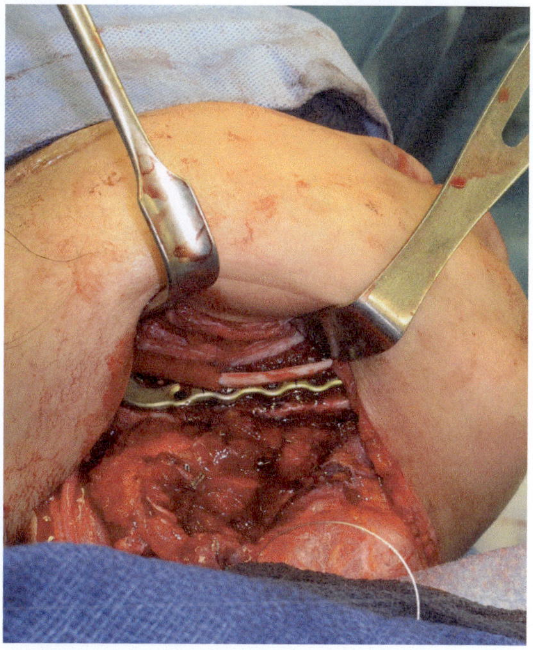

Fig. 15.13 Picture showing reconstruction of a lateral defect with a free fibula flap. There is also reconstration of the inferior alveolar nerve with an AVANCE graft. Note that the reconstructed IAN is placed lateral to the flap and the reconstruction bar

Autogenous or allogenic nerve grafts can be used. The sural nerve can be easily obtained during the harvest of the fibula flap but will result in numbness of the lateral aspect of the foot and ankle. Cadaveric nerve grafting can also be used and will not result in additional anesthesia to the patient. Nerve repair with allograph material can give approximately 86% [14] of patients some amount of return of nerve function.

Anastomosing the proximal aspect of the nerve before insertion of the flap first is best since the flap can block access to the nerve stump. The nerve only needs approximation, and this can be done with two 8–0 or 9–0 nylon sutures. A nerve connector is placed over the anastomosis to protect the repair and guide nerve regeneration. This can be secured with a 8–0 nylon suture. The distal end can be anastomosed in a similar fashion after insertion of the flap (Fig. 15.13).

15.20 Angle-to-Angle Mandibular Defect

The angle-to-angle mandibular defect poses a challenge to the reconstructive surgeon due to the length of the defect. When planning these cases, attention needs to be paid to the pedicle length. In cases where there was a previous neck dissection, the surgeon has to plan for the possibility of a vein graft and even arterial graft. Also, in designing the shape of the neo-mandible, whether the mandible is "U" shaped or "V" shaped should be considered. This will help to determine how many osteotomies are needed in the flap design (Fig. 15.14).

The fibula flap is the preferred choice since it can provide enough bone stock. This however may not always be viable if there are precluding circumstances; in such cases, the scapula flap is a

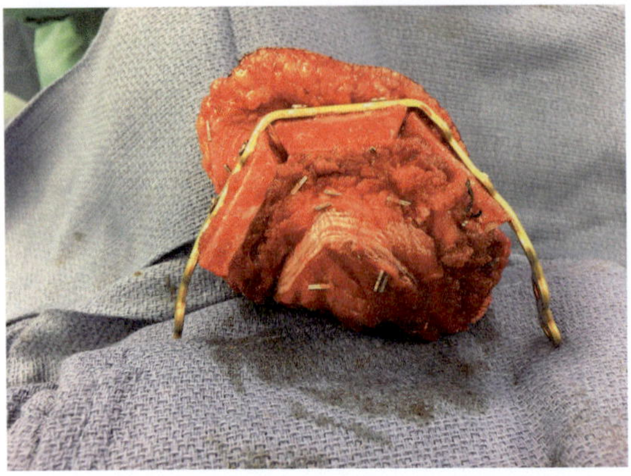

Fig. 15.14 This is a fibula flap with two osteotomies or three segments to reconstruct a body-to-body defect. Included with the fibula are the FHL muscle and a skin paddle

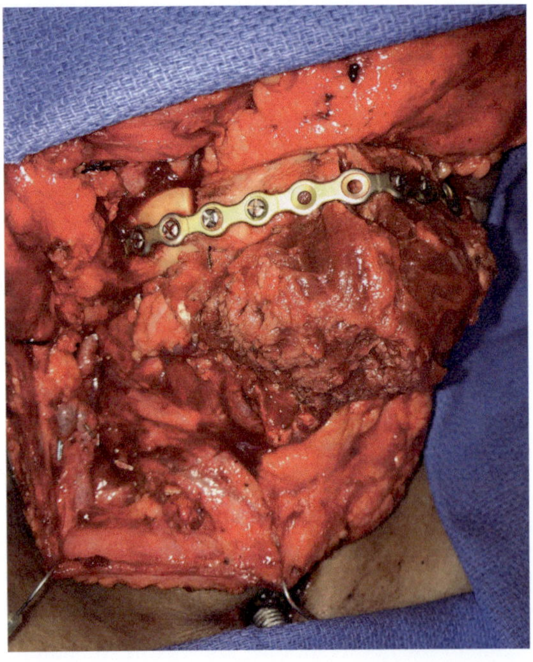

Fig. 15.15 This clinical picture shows the inset of the scapula flap with one osteotomy for reconstruction of an angle-to-angle mandibular defect

good choice (Fig. 15.15). An edentulous patient in the maxilla will make this reconstruction infinitely simpler. In these cases, the mandible can be placed in a slightly retrognathic position.

References

1. Taylor GI, Daniel RK. The free flap composite transfer by vascular anastomosis. Aust N Z J Surg. 1973;43(1):1.
2. Gilbert A. Surgical Technique: vascularized transfer of the fibulas shaft. Int J Microsurg. 1979;1:100.
3. Hildago DA. Fibula free flap: a new method of mandibular reconstruction. Plast Reconstr Surg. 1989;84(1):71.
4. Wei FC, Seah CS, Tsai YC, et al. Fibula osteoseptocutaneous flap for reconstruction of the composite mandibular defect. Plast Reconstr Surg. 1994;93(2):294.
5. Rosson GD, Singh NK. Devascularizing complications of free fibula harvest: peronea arteria magna. J Reconstr Microsurg. 2005;21(8):533–8.
6. Taylor GI, Townsend P, Corlett R. Superiority of the deep circumflex iliac vessels as the supply for the free groin grafts. Plast Reconstr Surg. 1975;64:595.
7. Sanders R, Mayou BJ. A new vascularized bone graft transferred by microvascular anastomosis as a free flap. Br J Surg. 1979;66:787.
8. Franklin JD, Shack RB, Stone JD, et al. Single stage reconstruction of the mandibular and soft tissue defects using a free osteocutaneous groin flap. Am J Surg. 1980;140:492.
9. Urken ML, Vickery C, Weinberg H, et al. The internal oblique iliac crest osteocutaeous microvascular free flap in head and neck reconstruction. J Reconstr Microsurg. 1989;5:203.
10. Dos Santos LF. Retalho escapular: um novo retalho liver microcirurgico. Rev Bras Cir. 1980;70:133.
11. Saijo M. The vascular territories of the dorsal trunk: a reappraisal for potential flap donor sites. Br J Plast Surg. 1982;35:2.
12. Fairbanks GA, Hallock GG. Facial reconstruction using a combined flap of the subscapular axis simultaneously including separate medial and lateral scapular vascularize bone grafts. Ann Plast Surg. 2002;49:104.
13. Gilbert A, Toet L. The free scapula flap. Plast Reconstr Surg. 1982;69:601.
14. Salomon D, Miloro M, Kolokythas A. Outcomes of immediate allograft reconstruction of long-span defects of the inferior alveolar nerve. J Oral Maxillofac Surg. 2016;74(12):2507–14.
15. Swartz WM, Banis JC, Newton ED, et al. The osteocutaneous scapular flap for mandibular and maxillary reconstruction. Plast Reconstr Surg. 1986;77:530.

Principles of Fibula Jaw in a Day Reconstruction: A Step-by-Step Approach

16

Mohammed Qaisi, James Murphy, and Henry Fung

16.1 Introduction

The "Fibula Jaw in a Day" is a relatively new procedure that has revolutionized the way we think about maxillomandibular reconstruction [1]. This technique involves bony reconstruction of the mandible or maxilla with immediate dental rehabilitation using a microvascular fibula free flap with immediate dental implants and a provisional prosthesis [2]. With this technique, a patient can get a jaw tumor resected and get a new jaw and new teeth all in the same setting.

The "Fibula Jaw in a Day" procedure involves occlusion-driven reconstruction, starting with the end result in mind. Everything else is positioned in space or planned based on where the anticipated occlusion will be. The technology is so sophisticated to the point that although implants are placed at the leg, one knows where every implant will end up three dimensionally in space once the fibula is transferred up to the head.

The purpose of this chapter is to provide the reader with a step-by-step approach to the "Fibula Jaw in a Day" procedure, hopefully improving the learning curve for surgeons performing this for the first time.

The first group to report on this technique was a multidisciplinary team out of New York University led by Levine and Hirsch [3]. Our group was the second group to publish their series on this technique [2]. Since then, several case reports have been published on this [4–7]. A summary of published reports as of the time of this writing is shown in the below Table.

The "Fibula Jaw in a Day" procedure offers a great advantage over conventional methods, which take 6–12 months to fully rehabilitate patients after tumor resection. Although the fibula free flap allows for immediate bony reconstruction, the process of dental rehabilitation can be very lengthy and requires numerous steps. These steps include delayed implant placement, implant uncovering, and adjunctive procedures such as flap debulking, vestibuloplasties, and palatal or skin grafts. This delayed rehabilitation can have a major psychological impact on patients who are suddenly faced with the reality of having to lose a part of their jaw and teeth due to a jaw tumor.

The evolution of the "Fibula Jaw in a Day" procedure was partly based on the success of the All-on-Four procedure with immediate provisionalization, also known as the "teeth-in-a day" procedure [8, 9]. As their name indicates, these procedures allow immediate rehabilitation of the edentulous mandible with implants

M. Qaisi (✉)
Division of Oral and Maxillofacial Surgery, Chicago, IL, USA
e-mail: mqaisi@cookcountyhhs.org

Otolaryngology, Chicago, IL, USA

J. Murphy · H. Fung
Division of Oral and Maxillofacial Surgery, Chicago, IL, USA

J. C. Melville et al. (eds.), *Regenerative Strategies for Maxillary and Mandibular Reconstruction*, https://doi.org/10.1007/978-3-319-93668-0_16

Group	N=	Mandible/Maxilla	Benign/Malignant	
Levine et al. [3]	4	Mandible + Maxilla	Benign	USA
Qaisi et al. [2]	3	Mandible	Benign	USA
Rahimov et al. [4]	1	Mandible	Not mentioned	Azerbaijan
Runyan et al. [5]	1	Maxilla	Benign	USA
Yetzer et al. [6]	1	Mandible	Benign	USA
Salman et al. [7]	1	Mandible	Benign	USA

and an immediate prosthesis. These procedures showed good success rates in excess of 95% at 5 years despite immediate loading of the implants [10, 11].

Some of these same concepts were then applied by reconstructive surgeons to the free flap-reconstructed mandible. This neo-mandible was essentially treated as an edentulous mandible and thus allowed for immediate implant placement and provisionalization 3–4 months after the initial surgery once bony healing between the fibular segments has occurred [12, 13]. This improved the time that patients go without a prosthesis. However, this technique is not widely used and can be challenging if the skin paddle is bulky making it difficult to seat the surgical guide in place.

The work that bears the greatest similarity with the current technique is the work by Rohner et al. on prefabricated free flaps [14]. This reconstructive technique is also occlusion driven and has essentially paved the way for the "Fibula Jaw in a Day" procedure. In this technique, implants are placed into the fibula at the leg during phase I surgery. Skin grafting of the fibula is also performed at the same time [15]. The fibula is then harvested 3 months later with the osseointegrated implants, inset at the head, and immediately loaded with a provisional prosthesis.

16.2 Patient Selection

Some of the indications or factors that should be taken into consideration when evaluating patients for the "Fibula Jaw in a Day" procedure include:

- Reconstruction of defects for benign disease: For now, this procedure has been largely limited to the reconstruction of benign disease including the treatment of benign tumors, osteomyelitis, and osteonecrosis [2, 3, 6]. Although the "Fibula Jaw in a Day" technique is likely to be used for the reconstruction of malignant disease in the future in select patients, it is felt that routine use in patients with malignant disease was a relative contraindication [6]. Any delays to sound oncologic management in either the planning or surgical healing phase would be counterproductive to the goal of therapy. In the event of a complication, it could lead to delay of valuable and much-needed radiation therapy, which in turn negatively affects survival.
- Availability of soft tissue for primary closure: Although not a must, this is preferable with "Fibula Jaw in a Day" cases because oral mucosa and specifically attached gingiva are the ideal types of tissue to be used around implants. We try to preserve attached gingiva whenever possible when using this technique. The use of skin paddle around dental implants, although not contraindicated, may be difficult because of the variable thicknesses of fibula skin paddles. Furthermore, due to the thickness, texture, and type of tissue, they are likely to result in pocketing around the implants if used as a long-term solution.
- Single segment reconstruction: Cases that require only a single segment for mandibular reconstruction are ideal cases to start with for the novice surgeon. Conceivably as the number of implanted segments increases, the complexity of the reconstruction also increases.
- Patient finances: Patient finances should generally not be used to dictate treatment plans for patients. Unfortunately, depending on the laws and regulations in one's state, dental

implants and prosthetics may or may not be covered under medical insurance. Thus patients who are not able to afford the cost of the implants and prosthesis may not be able to be treated with this technique.

16.3 Advantages of the "Fibula Jaw in a Day" Procedure

- This technique allows for a single-stage reconstruction of maxillomandibular defects after tumor resection, thus allowing for early restoration of form and function. It essentially reduces the time from resection to dental restoration from a 12-month process to a single day, giving patients a sense of normalcy.
- Another advantage of this procedure is that it decreases the number of procedures and general anesthetics required to fully rehabilitate the patient. Although this has not been investigated, intuitively it is likely to yield some overall cost savings [3].
- One of the main benefits of the "Fibula Jaw in Day" procedure is the psychological impact it has on the patient. As previously discussed, one can imagine the distraught that is caused when one is suddenly faced with reality that they are going to have to lose a significant portion of their jaw and teeth because of a benign tumor. The face, jaws, and teeth are central to our identity and self-image. Going without rehabilitation of this area for a year or so can have a negative psychological impact on patients. The "Fibula Jaw in Day" procedure allows for immediate restoration of form and function, thus giving patients a sense of normalcy.
- Another benefit of this technique is that it provides excellent access to the fibula for implant placement. The excellent access to the fibula at the time of the harvest combined with virtually planned guided implant placement provides the surgeon with an unparalleled opportunity to place and position implants in the most ideal position under direct vision.

16.4 Disadvantages of the "Fibula Jaw in a Day" Procedure

- This is a relatively new procedure, and therefore sufficient long-term data does not exist to compare the safety profile of this procedure relative to conventional microvascular reconstruction. However, the outcomes of initial case series and reports have been good [2, 3, 6, 7].
- As with any new technique, there is usually a learning curve. With an increased number of procedures, one's familiarity and comfort with the procedure increases. One must however be aware that this learning curve exists and take this into consideration. Furthermore, having some knowledge or experience with guided implant placement is very useful for the surgeon attempting to perform this type of surgery. Many of the concepts used in guided implant surgery are applicable and used in "Fibula Jaw in a Day" cases.
- A major downside of this procedure is that in the event of a flap failure, the patient is likely to lose their investment in the implants and prosthesis. The patient must be made aware of this. Patients must also understand that the flap success rate is not 100% and that flap failure may occur in 1–5% of cases with conventional free flaps. Data for safety of this procedure is not available except for small case series and case reports. Safety of this procedure is somewhat extrapolated from previous large series on free flaps with immediate dental implant placement [16–19].

16.5 Step-By-Step Approach

16.5.1 Patient Consultation

Having an honest discussion with the patient about the procedure and having realistic expectations are crucial. All the pros and cons of free flap reconstruction should be discussed in the usual fashion. When discussing the "Fibula Jaw in a Day" portion, it is important to discuss that this is still a relatively new treatment modality with little outcomes data available. Things to explain to the patient should include the understanding that a flap loss could lead to the loss of their investment in the implants and the provisional prosthesis. It also should be explained that should the provisional prosthesis show any signs that could be detrimental to the overall survival of the flap, then the prosthetic portion of the procedure would be aborted. It should also be noted that the implant/prosthesis placement portion of the procedure can add up to 2 h to an already lengthy surgery, especially when being performed early on at a new institution.

At the same time, "Fibula Jaw in a Day" has several advantages as discussed above. It allows for almost immediate restoration of form and function, negating the standard 6–12 month wait required for dental rehabilitation. Additionally, it can result in reduction in the number of subsequent procedures required for dental rehabilitation.

During this visit, the surgeon must be able to assess the compliance of the patient. The surgical team must be assured that the patient will be compliant with prescribed diet and hygiene or substantial or even catastrophic healing complications may arise [6]. It is our protocol to place patients on tube feeds via a thin nasogastric tube for the first 2 weeks or until mucosal healing around the implants has been achieved. A soft non-chew diet is then continued for roughly a total of 12 weeks after the procedure in order to avoid external forces on the fibula and implants during the initial healing phase.

16.5.2 Preoperative Phase

Once it has been decided that the patient is a good candidate for the "Fibula Jaw in a Day" procedure, the preoperative phase of treatment may begin. This phase involves treatment planning for the procedure as well as any other preparatory steps that are needed prior to surgery. For simplicity, this phase has been broken down to the following individual steps.

- Obtain a maxillofacial computed tomography (CT) scan and a CT angiogram (CTA) of the lower extremities with three-vessel runoff with 1 mm cuts to be used during the virtual planning phase.
- Impressions of both upper and lower arches should be obtained. A tooth shade should be obtained at that time to color match the prosthetic teeth to the adjacent remaining dentition. A bite registration of the patient's current bite should also be obtained. Extraoral and intraoral photos should be taken as part of the medical record (Fig. 16.1). These may be referenced back to during surgery, virtual planning phase, or even postoperatively.
- The poured casts are then sent to the virtual planning company so that they can be scanned and superimposed on the 3D images from the CT data. Early on in our experience, we've used fiduciary markers to help increase the accuracy of this process. However, if there are

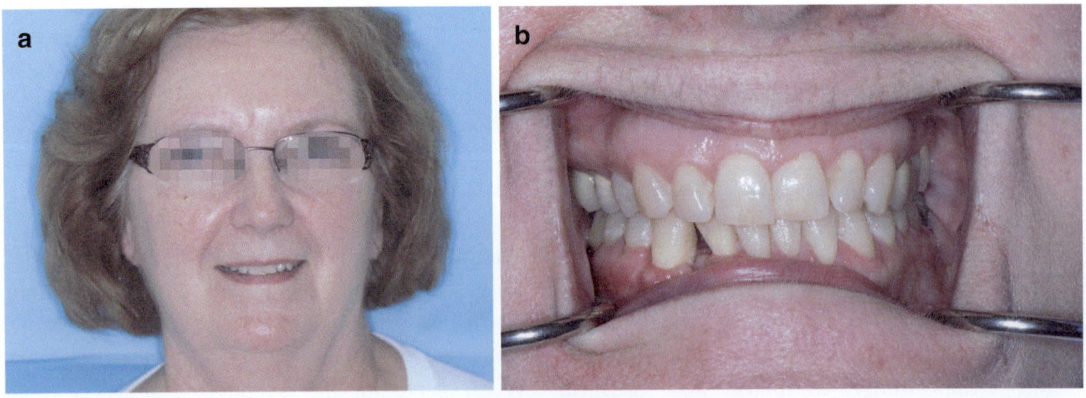

Fig. 16.1 Case example 1. Patient with biopsy-proven ameloblastoma of the anterior mandible. (**a**) Extraoral photo. (**b**) Intraoral photo showing displacement of teeth secondary to the tumor

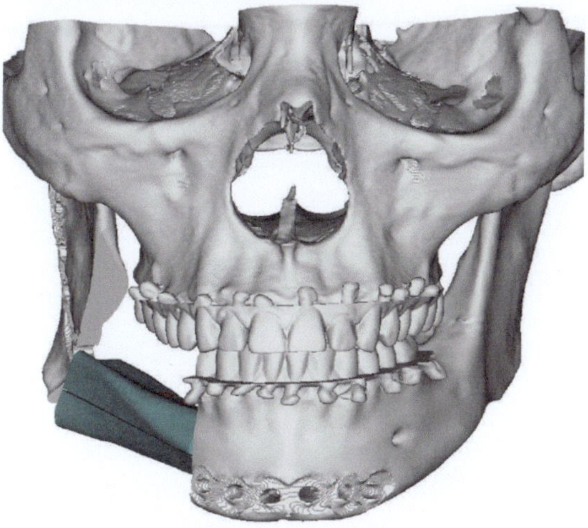

Fig. 16.2 Sowing data from dental casts onto the CT scan allows for greater accuracy at the level of the dentition. A line of demarcation where the two scans have been sown together is noticeable at the level of the dentition

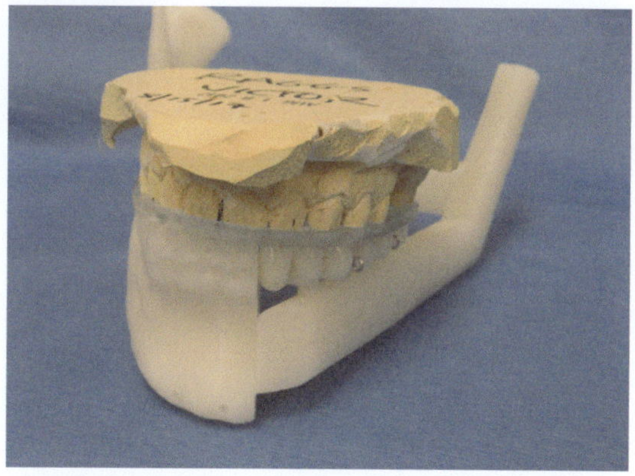

Fig. 16.3 Preoperative models and splint. Sown data from casts allow for the fabrication of accurate occlusal splints

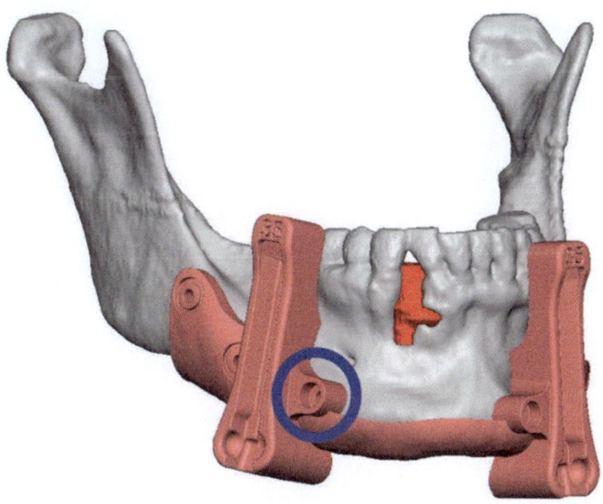

Fig. 16.4 Planned area of mandibular resection based on 1 cm margins from radiographic tumor

enough teeth remaining at the time of the impression, then the data can be superimposed with very high accuracy by overlapping the dentition from both CT scans and the scans of the casts. Sowing the data from the casts onto the CT scans allows much greater resolution at the level of the dentition (Fig. 16.2). This in turn allows for the fabrication of accurate well-fitting occlusal splints that may be used intraoperatively (Fig. 16.3).

- Once scans have been uploaded, a virtual planning session can be set up with the virtual planning company. During the virtual planning phase, the resection is planned in the standard fashion with appropriate margins for the given pathology and extent of tumor.
- -Selection of the lower extremity to be used during the reconstruction depends on patient three-vessel vasculature of the lower extremities and whether the vascular pedicle is planned to come off the anterior or posterior aspect of the reconstruction. Although not necessary, it is the authors' preference to place dental implants in the anterior surface of the fibula in order to avoid interfering with perforators along the posterior aspect of the fibula should a skin paddle be used. All of the abovementioned factors are used in determining which lower extremity is selected for reconstruction.
- Once the tumor is virtually resected and removed, occlusion-driven reconstruction ensues (Fig. 16.4). The desired segment of the fibula to be used for reconstruction is brought into the defect (Fig. 16.5).
- The final positioning of the fibula can be facilitated by bringing the patient's premorbid dentition back into view (Fig. 16.6b). The fibula flap is then positioned in the ideal position for implant rehabilitation of the missing teeth. Alternatively, if the patient's premorbid dentition is not in an ideal position or is missing, one can request

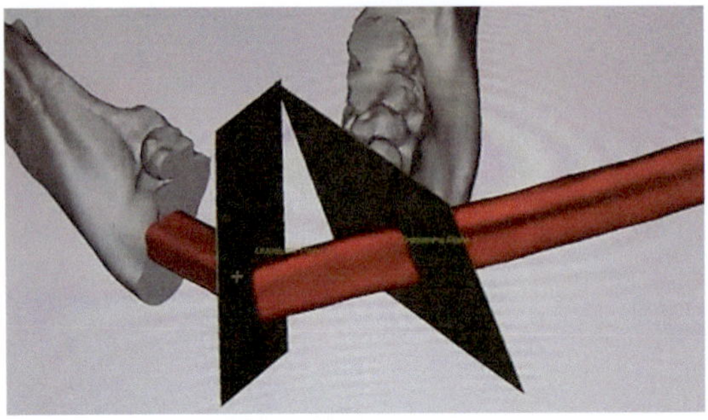

Fig. 16.5 Once mandibular defect is created and the appropriate lower extremity is selected, the fibula is brought up to the defect. Fibular osteotomies are used to help mimic and recreate the premorbid shape

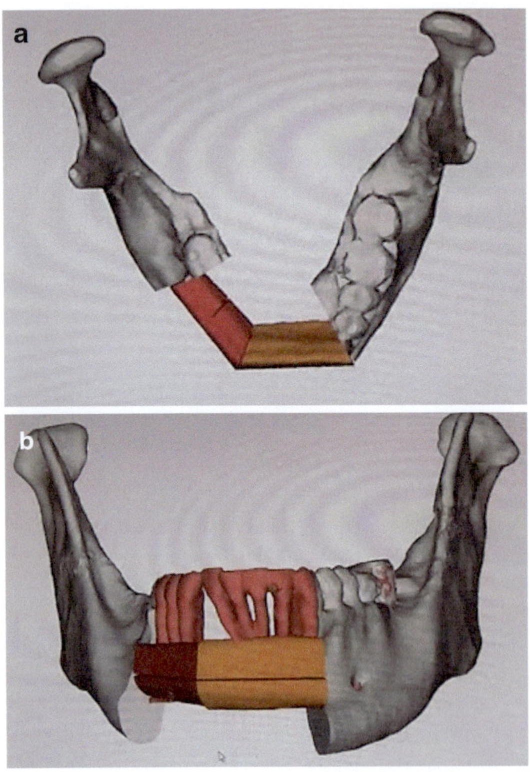

Fig. 16.6 (**a**) After initial fibular osteotomies. (**b**) Premorbid dentition, if not grossly displaced, may be used in helping positioning the fibular reconstruction

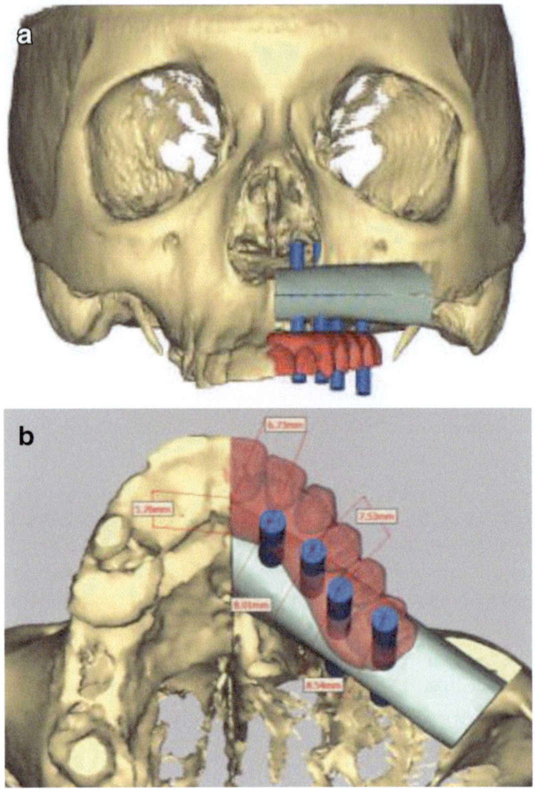

Fig. 16.7 If premorbid dentition is greatly displaced or is not existent, a digital prosthesis may be waxed up in order to help with the position of the fibula and implants. (**a**) Frontal view. (**b**) Occlusal view—prosthesis with fibula and dental implant positioning

a digital wax-up. Based on the models and bite sent in, the dental technologist can perform a digital wax-up of the ideal reconstructed arch. The STL data of this prosthesis can then be imported into the virtual planning software to assist in placement of implants in the proper position (Fig. 16.7).

– During the virtual planning, we use 5 mm diameter cylinders to simulate the dental implants (Fig. 16.7). In traditional implant planning software, such as NobelClinician (Yorba Linda, CA) or Anatomage (San Jose, CA), an implant library exists with different implant shapes, platforms, and lengths to be used in simulating placement of the implants in the mandible or maxilla. However, during virtual planning sessions, we've used 5 mm diameter cylinders instead since the fibula provides adequate bony stock; thus, the use of the exact implant during simulation does not appear to be beneficial. These cylinders are imported into the virtual plan and are

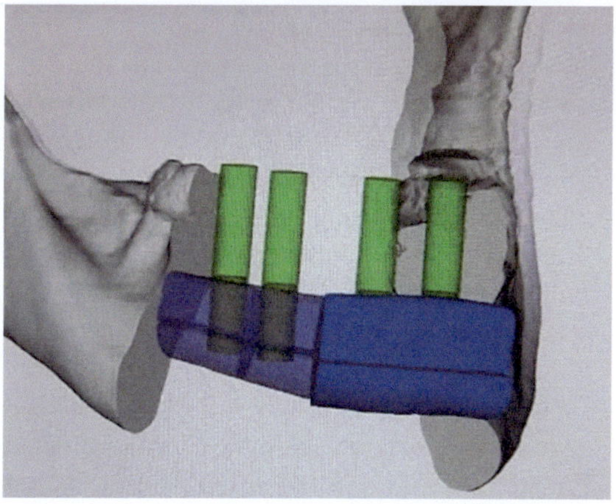

Fig. 16.8 Five millimeter diameter, 25 mm long cylinders are imported into the plan and simulate the dental implant positioning. Ten millimeters are positioned within the bone to simulate 10 mm long implants, and the remaining 15 mm are used to simulate the supra-implant components and needed occlusal clearance

used to simulate implant positioning (Fig. 16.8). The fibular bone and implants are placed in the most ideal position to support the anticipated dentition and prosthesis (Fig. 16.9).

– Similar to “All-on-Four” implant cases, a minimum of 15 mm of occlusal clearance is needed to account for the height of the prosthesis as well as a 5 mm space below the prosthesis to allow for cleanliness. To accomplish this, we customize the implant cylinders to be 25 mm in length. Ten millimeters are placed within the fibula and represent the implant length, and the other 15 mm account for the height of the prosthesis and needed occlusal clearance (Fig. 16.8). The cylinders are moved to ensure they are in bone and parallel to each other. They are usually placed in a way where they exit slightly lingual to the occlusal table whenever possible, to avoid interference between the temporary copings and the facial surface of the prosthesis, which in turn affects access channel placement (Fig. 16.9). The position of the fibula is finalized at this position to make sure that the implants are all located within the bone and that fibula has good bony contact with the adjacent mandible (Fig. 16.10).

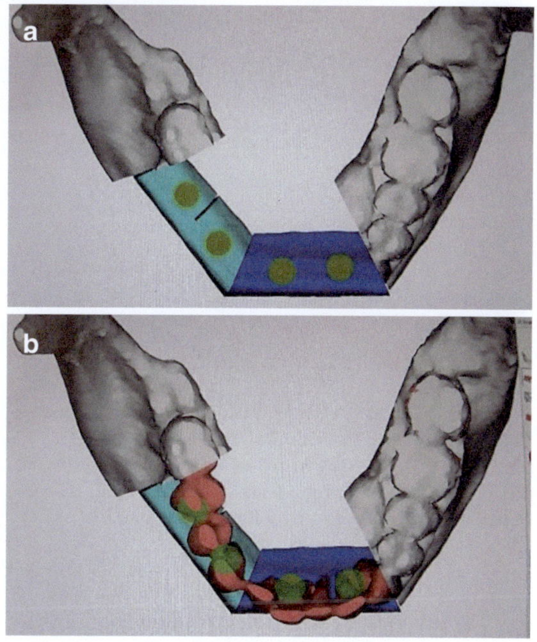

Fig. 16.9 Occlusion-driven reconstruction. (**a**) Shows how implants affect positioning of the fibular segments. (**b**) Positioning of the implants is driven by the anticipated final occlusion

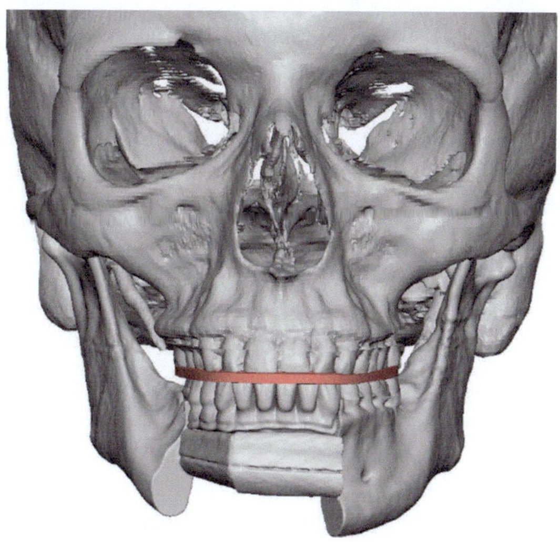

Fig. 16.10 Show final prosthesis and fibula bone positioning, driven by the occlusion

– It has to be understood that using this occlusion-driven technique, it is very likely that the fibula flap may not be at the level of the inferior border of the mandible. In the lateral mandible,

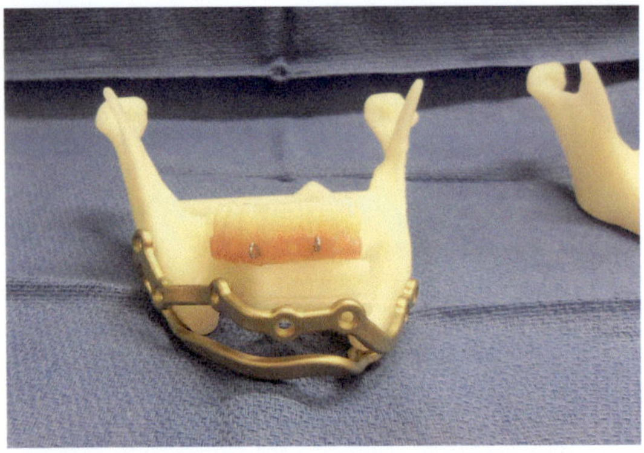

Fig. 16.11 Since the fibula was superiorly positioned, a second bar was used at the inferior border for chin support. This was wrapped with a de-epithelialized skin paddle to prevent thinning of the overlying soft tissues. Alternatively, a double-barrel fibula may be used

this can usually be left alone with no esthetic consequence to facial form. However, in the anterior mandible or when the discrepancy is great, then one must think of ways to augment the inferior border, whether that is using a double-barrel fibula or other techniques (Fig. 16.11).

- Once the fibula and implants have been placed in the desired position, the next step involves designing the custom reconstruction bar. While this is not necessary, it provides numerous advantages that help facilitate the "Fibula Jaw in a Day" procedure. One of the advantages is that the custom-milled plate allows the surgeon to select exactly where each screw hole will go, therefore avoiding interferences between the fixation screws and the dental implants (Fig. 16.12).
- Another advantage of custom plates that is particularly useful in these procedures is the use of predictive holes. These are strategically placed screw holes within the fibula and mandible cutting guides that are used to secure the guides in place during surgery. These same holes are designed to line up with the holes in the custom-milled plate [2] (Fig. 16.13). Therefore, cutting guides can be secured in place during the mandibular resection and fibular osteotomies, and upon removal of the guides, the plate will fit over these holes as virtually planned (Fig. 16.14a). Thus, these holes act as external reference points that help align all the pieces in space to the custom-milled plate and considerably simplify the positioning of all components (Fig. 16.14b).
- The next step during virtual planning session involves designing the cutting guides. For the mandibular cutting guide, we try to use a connected cutting guide whenever possible. This helps increase accuracy of the procedure by maintaining appropriate relationship between predictive holes on either side of the resected mandible (Fig. 16.15). This design is not always feasible especially when trying to minimize external surgical incisions. In those instances, we use separate anatomic cutting guides and try to capture as much unique anatomy within the guide so that it locks accurately in place.
- For the fibula portion, we incorporate the dental implant guide into the fibula guide, therefore preserving the relationship between the fibula osteotomies, the predictive holes, and the dental implant positions within the fibular segments (Fig. 16.16). Since the fibula is a tubular bone, we design the guides so that they anatomic and hug the anatomy of the fibula with a 0.5 mm soft tissue offset (Fig. 16.17a). We additionally design the fibula guide with a proximal 5–10 cm extension in order to allow for engaging as much of the fibular anatomy as possible (Fig. 16.17b). This allows for placement of the guide on the fibula as close as possible to the virtual plan, thus allowing for placement of the implants and the predictive holes in the most accurate position possible. This will increase the accuracy of the reconstruction and decrease the amount of adjustments necessary when insetting the fibula-prosthesis complex at the head.
- Depending on the implant system to be used, the implant guide component can be adjusted accordingly. Guided implant kits have longer drills accounting for the thickness of the tissue and the guide itself. For the Nobel Biocare system (Yorba Linda, CA), the length of the non-cutting portion of the drill is 9 mm

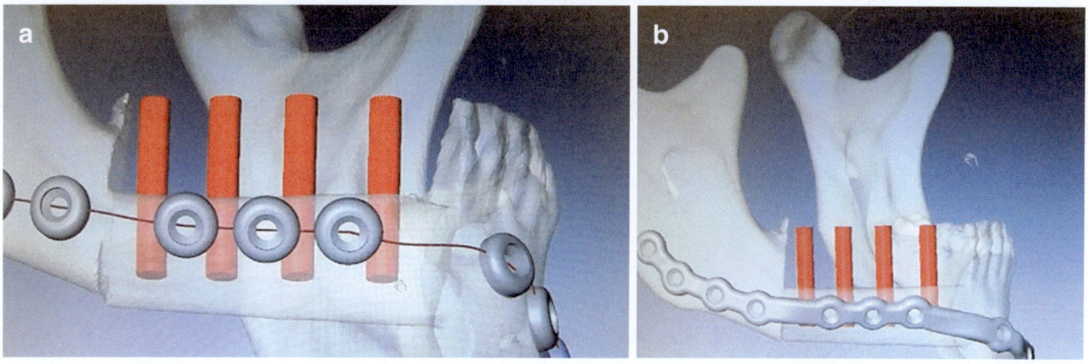

Fig. 16.12 Designing a custom-milled plate. (**a**) The surgeon is able to select where plate holes go, therefore avoiding interference with dental implants. (**b**) Shows milled-bar design after hole positions were selected

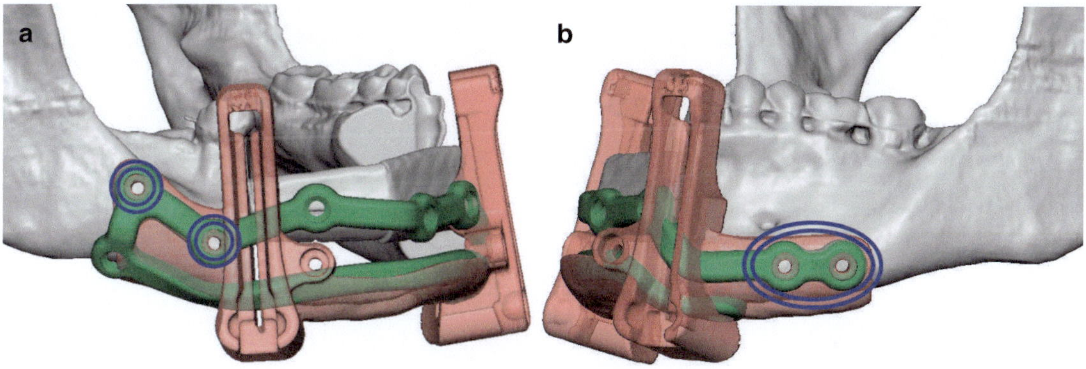

Fig. 16.13 Predictive holes in mandibular resection guide are designed to line up with holes in the milled reconstruction bar. (**a**) Right side. (**b**) Left side. These predictive holes help to align all pieces in space to the custom-milled plate

(Figs. 16.18 and 16.19). Therefore, when designing the fibula guide, the distance from the fibula surface to the top of the implant guide must be roughly 9–10 mm (Fig. 16.20). This allows for accurate preparation of the implant osteotomy sites to the appropriate depth.

- The diameter of the holes in the implant guide portion also depends on the implant system to be used and planned size of implants to be placed. These holes can be fitted with metal sleeves to guide drill placement as is done for standard guided implant surgery. Alternatively, one may use the metal drill guide adaptors that come in the guided kits in lieu of metal sleeves. For the Nobel Biocare system, 5 mm diameter holes allow for the metal drill guide adaptors to slide into the guide with ease (Fig. 16.20). The appropriate adaptor may be used based on the diameter of the implant.
- Although we no longer obtain these at our institution, other models to consider obtaining when first performing this technique include a resected mandibular model and a fibula model. These are briefly discussed in the next section. The fibula model allows for the surgeon to try out the fibula guide on the fibula before surgery and get a feel of the fit of the guide (Fig. 16.21). The resected mandibular model is used intraoperatively and allows for adjustments and simulation of the inset of the fibula-prosthesis complex at the head (Fig. 16.22).
- Design of the prosthesis needs to be approved prior to surgery (Fig. 16.23). The digital laboratory will design a prosthesis based on the virtual plan. Things to consider in the

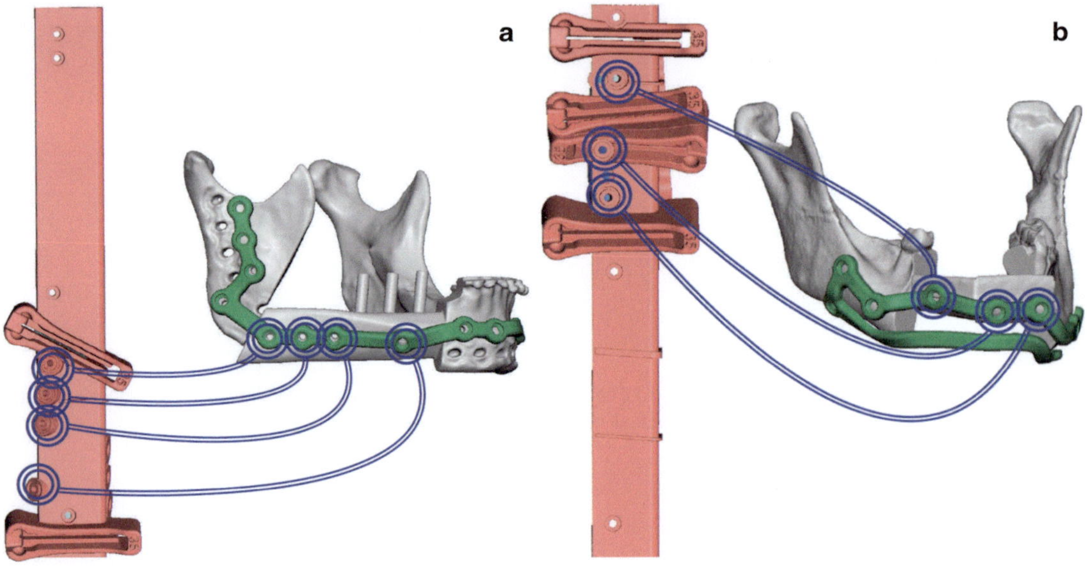

Fig. 16.14 Predictive holes in the fibula cutting guide are designed to line up with holes in the custom-milled bar. Therefore, upon removal of the guide, the predictive holes can be used to line the fibular segments against the custom-milled plate. (**a**) Single segment. (**b**) Positioning of multi-segment fibular reconstruction can be simplified by the use of predictive holes

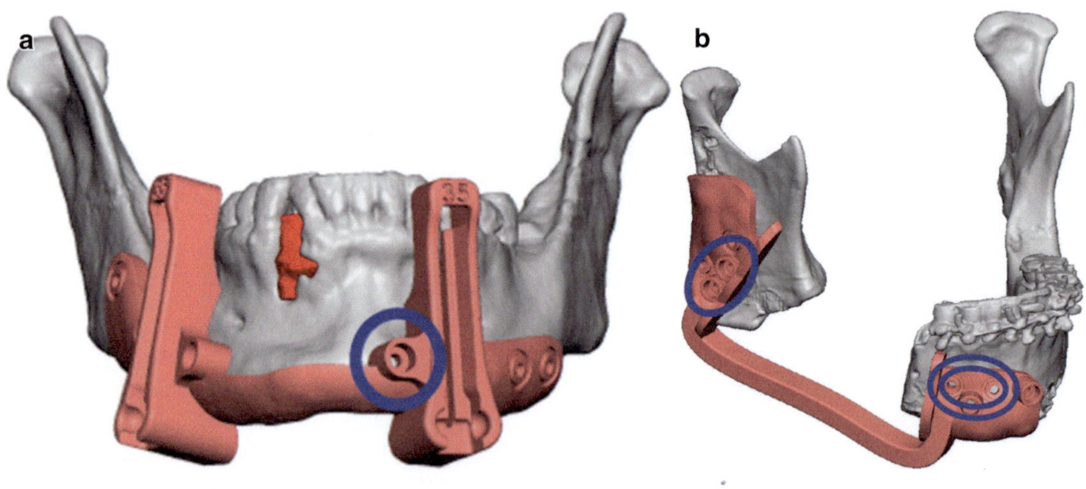

Fig. 16.15 (**a**, **b**) Connected mandibular cutting guides

design are whether lugs or wire loops are needed in the provisional prosthesis to place the patient into maxillomandibular fixation. Additionally, it is usually recommended to have multiple sprue channels incorporated into the prosthesis design to allow for easy access for injecting the flowable luting resin into the prosthesis and securing it to the temporary copings. Finally, we recommend slightly larger access holes for the implant temporary copings as long as it doesn't affect the strength of the prosthesis (Fig. 16.24). We routinely use 5 mm access holes. This allows for some wiggle room and accounts for minimal unavoidable deviations from the virtual plan; it allows for minor movements in the prosthesis so that it may be place in the ideal occlusion.

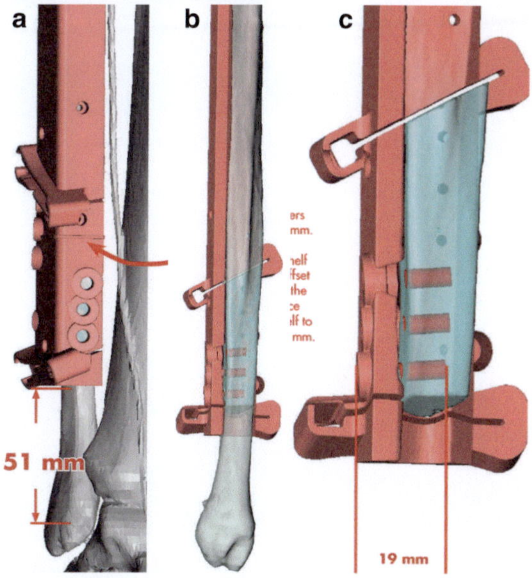

Fig. 16.16 Fibula cutting guide (**a**) with implant guide incorporated into it. (**b**) Notice proximal extension of the guide which helps engage as much of the bony anatomy to increase the accuracy of seating the guide. (**c**) Demonstrates how implant guide component helps position the implants within the bone

- The final step in the preoperative phase involves inspecting all parts, models, splints, and the prosthesis for accuracy prior to surgery. It is also important to have all necessary equipment ready and tested. These include implant drill consoles, hand pieces, and guided implant kits (Fig. 16.18). Implants need to be ordered prior to surgery. Implant parts such as multiunit abutments and temporary copings also need to be ordered. Healing abutments or cover screws are required in the event that the prosthetic part of the case needs to be aborted. Luting resin needs to also be available. In our practice we use Luxatemp (Ridgefield, NJ), which is a bis-acryl composite resin, for securing the prosthesis in place (Fig. 16.25).
- For implants we have come to prefer the NobelReplace narrow platform (3.5 mm) with conical connection. Our preferred length is 10 mm. Due to the density of the fibula, we find that the narrow platform provides excellent engagement of the bone with good torque forces, thus negating the need for using larger diameter implants. Furthermore, the narrower shorter implants theoretically cause less disturbance and interference with the blood supply of the implanted segment. Finally, we find that the machined collar conical connection provides greater strength at the collar portion than the trilobed connection implants. This helps avoid breakage of the collar of the implant under increased torque forces "flowering of the implant." A single regular platform (4.3 mm) implant is made available on reserve in the event it is needed if a narrow platform does not reach adequate torque forces. In our experience, these rarely have to be utilized as the cortical bone provides great implant stability.

Fig. 16.17 (**a**) The fibula guide is designed with an internal surface that engages as much of the fibular anatomy as possible with a 0.5 mm tissue offset. (**b**) The proximal extension also helps to engage additional anatomy and help with reproducibility of seating the guide accurately in place

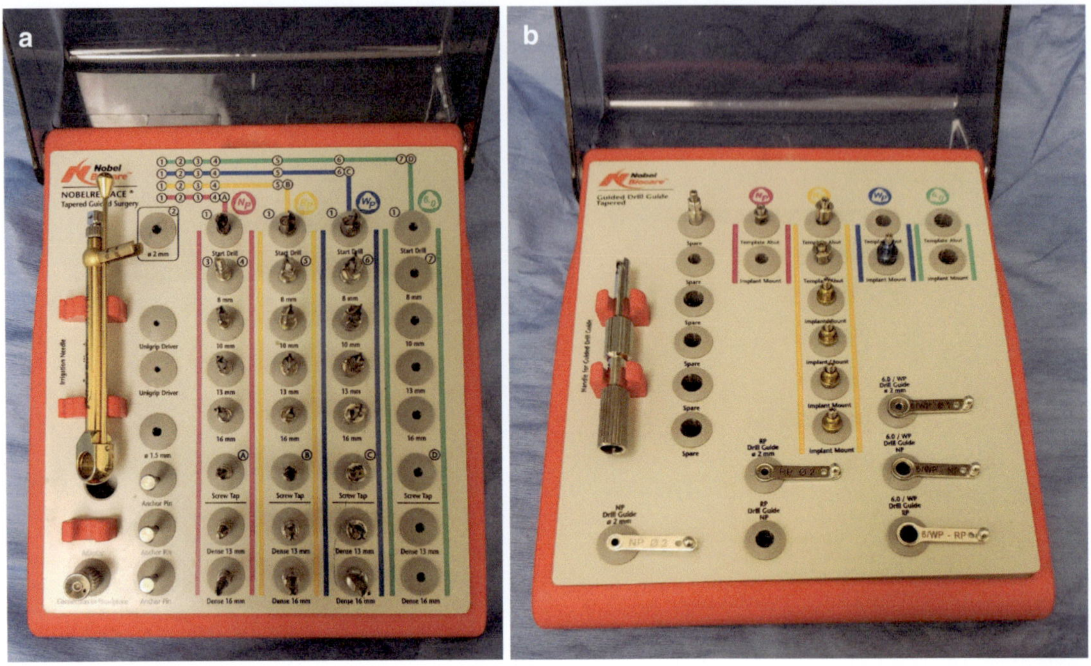

Fig. 16.18 (**a**, **b**) Nobel Biocare guided kit

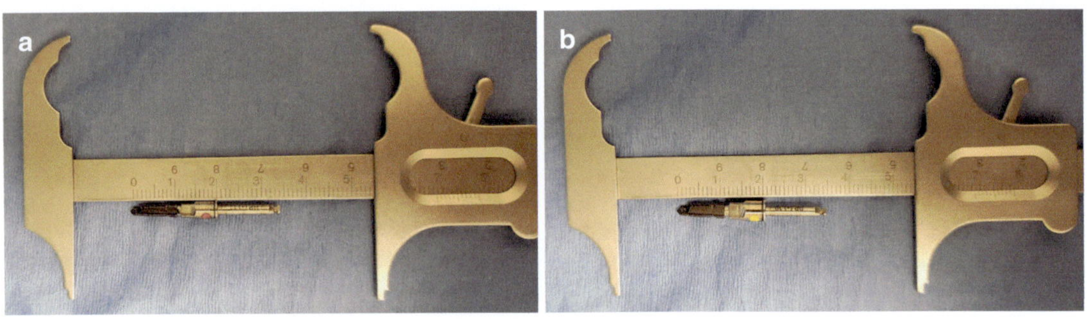

Fig. 16.19 Nobel Biocare guided drills. (**a**) Narrow platform. (**b**) Regular platform. For 10 mm drills, the cutting part of the drill is 10 mm with 9 mm of additional length prior to the stop to accommodate tissue and guide thickness

16.5.3 Intraoperative Phase

- On the day of surgery, one should quickly revise the plan with the patient. All needed parts should have been sterilized and available for surgery (Figs. 16.18 and 16.21).
- Surgery starts in the usual fashion. During tumor resection, we attempt to preserve as much mucosa as possible to allow for primary closure over the fibula flap. Passive mucosal closure over the fibula is crucial in order to avoid a salivary leak postoperatively. Whenever feasible we attempt to preserve keratinized attached mucosa since this is the ideal type of tissue to be placed around dental implants (Figs. 16.15a and 16.26).
- The fibula harvest is carried on in the standard fashion (Fig. 16.27). During the harvest, we measure the distance from the lateral malleolus to the anticipated lower osteotomy, and that area is marked on the fibula. An osteotomy is made below this area to allow for distraction of the fibula bone and application of the fibula cutting guide (Fig. 16.28). This is

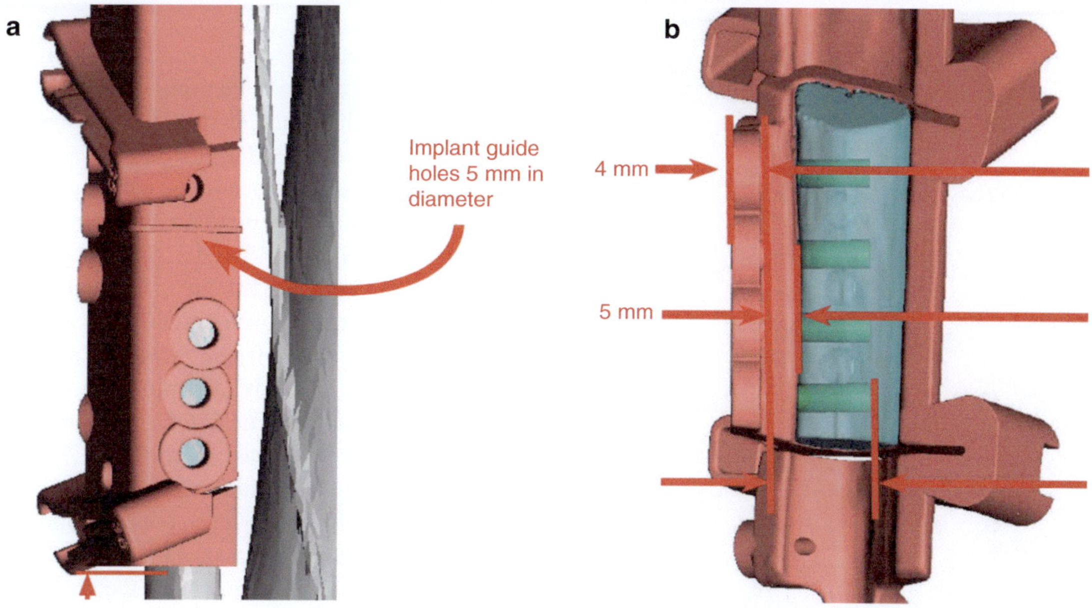

Fig. 16.20 (**a**) Drill component of fibula guide designed with 5 mm diameter holes to allow insertion of Nobel Biocare adaptors (drill guides) into the holes. (**b**) Distance from the surface of the bone to the shelf of the drill guide must be 9–10 mm or equal to the non-cutting portion of the guided drills. This will allow preparation of the implant osteotomy sites to the proper depth

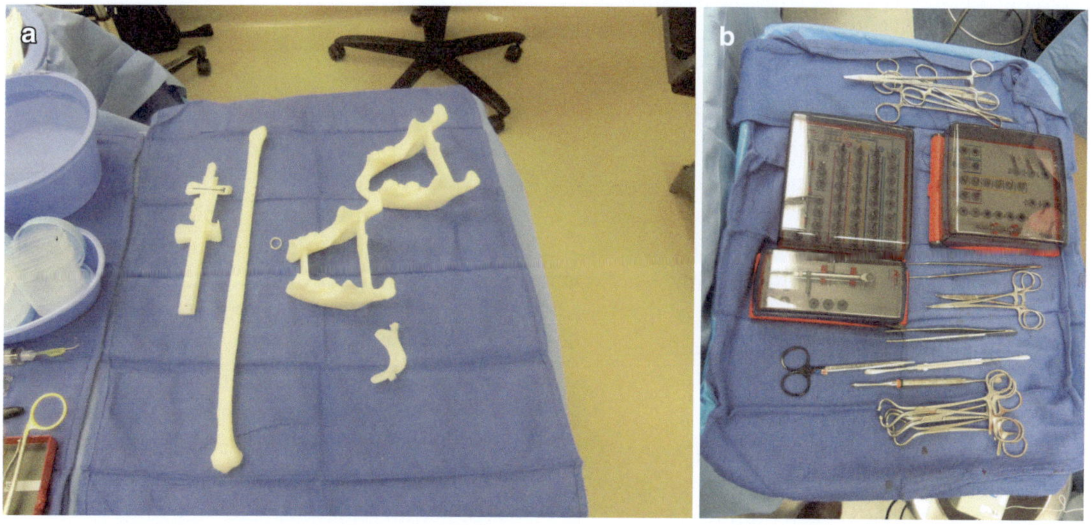

Fig. 16.21 (**a**) Shows OR models and (**b**) OR implant setup. We no longer obtain these models, but they are helpful for surgeons when first trying out this technique. It includes a resected mandibular model on the right and fibula model on the left

secured in place using multiple 2.0 mm screws using both predictive holes and other fixation holes in the guide (Fig. 16.14b). The use of predictive holes will allow these same pre-drilled holes to secure the custom-milled plate in place later on. Accurate seating and positioning of the guide are crucial for replicating the virtual plan and minimizing the amount of adjustments needed to the fibula or prosthesis. To aid with this, in the authors' experience, we

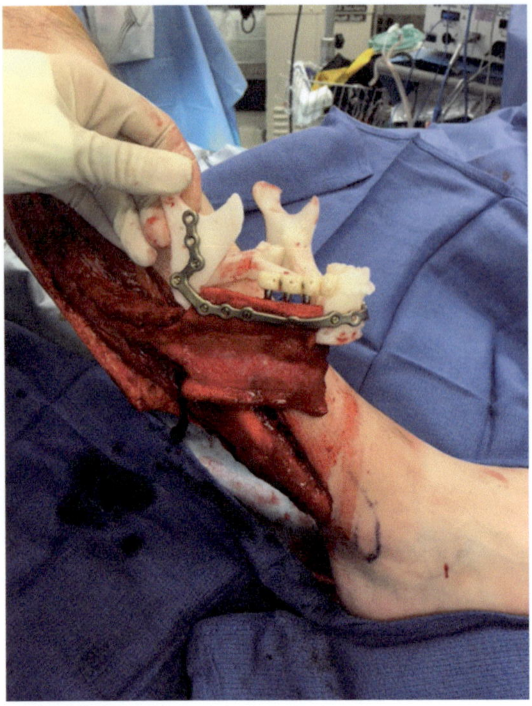

Fig. 16.22 Shows how the resected model is used during surgery to adjust the fibula and prosthesis while still pedicled at the lower extremity

design the fibula guide with a 5–10 cm proximal extension to allow the guide to engage as much of the fibular anatomy as possible (Fig. 16.16).

- Once the guide is secured in place, we then proceed with placement of the dental implants through the implant guide portion of the fibula cutting guide (Fig. 16.28). Implants may be placed using a guided implant kit of the surgeon's choice.
- In the author's experience, it is preferable to place the implants in the fibula first prior to making the fibular osteotomies, especially when implants are being placed in multiple fibular segments. Performing the fibular osteotomies first could theoretically make the fibular segments less stable within the guide as there are fewer screws holding the individual segments to the guide during implant placement.
- The implant osteotomy sites are prepared using sequential drills in the implant kit per the manufacturer recommendations. As previously discussed, the implant guide component may have metal sleeves incorporated into it, or alternatively the metal adaptors for the guided kit may be used instead (Fig. 16.28).
- When using Nobel biocare, we start by placing the 5 mm "wide platform" to pilot drill guide adaptor in place. We start with the 2 mm pilot drill and prepare the site to the appropriate depth. We then move on to the next drill in the sequence, which is the 3.5 mm "narrow platform" drill. This is also prepared to the appropriate depth. The wide platform to narrow platform adaptor (5 mm to 3.5 mm adaptor) is used for this step (Fig. 16.18b). This step is repeated until the appropriate diameter and depth are reached.
- The use of the dense bone drill for the final size drill as well as tapping the implant sites should be strongly considered because of the increased density of the fibular bone compared to the mandible. Not doing so could lead to excessive torque forces on the implants. This increased torque could lead to the breaking of implant at the collar portion "flowering of the implant." implants with conical machined collar connections are less likely to have this problem than implants with trilobed connections. This should be taken into consideration when selecting and ordering implants prior to surgery. Excessive torque forces could also theoretically lead to cracking of the fibula or a fracture along the length of the bone. While we have not encountered this phenomenon during primary implant placement, we have seen this in cases where delayed implant placement was performed several months after the fibula flap reconstruction. This is also more likely to occur with wider diameter implants.
- Once the implant sites have been prepared and tapped, placement of implant usually begins with a handpiece rotating at a torque of 40 Ncm (Fig. 16.28b). The hand wrench may be used to complete the seating of the implant. If excessive torques occur without significant seating of the implant, one should consider backing out the implant and re-preparing the osteotomy site (Fig. 16.29).

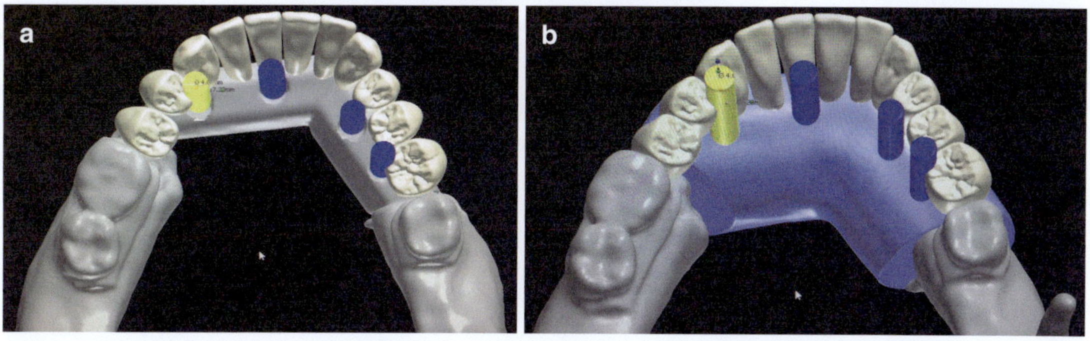

Fig. 16.23 (**a**, **b**) Provisional prosthesis design by digital laboratory based on completed virtual plan

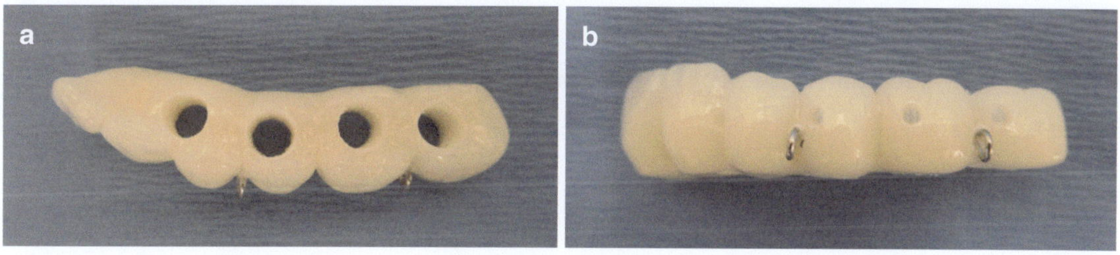

Fig. 16.24 Provisional prosthesis. (**a**) Occlusal view, 5 mm holes all the prosthesis to slide over the temporary copings. (**b**) Buccal view shows wire loops to place the patient in maxillomandibular fixation. Also shows sprue channels which are used for injecting luting material from buccal aspect while prosthesis is up in the occlusion

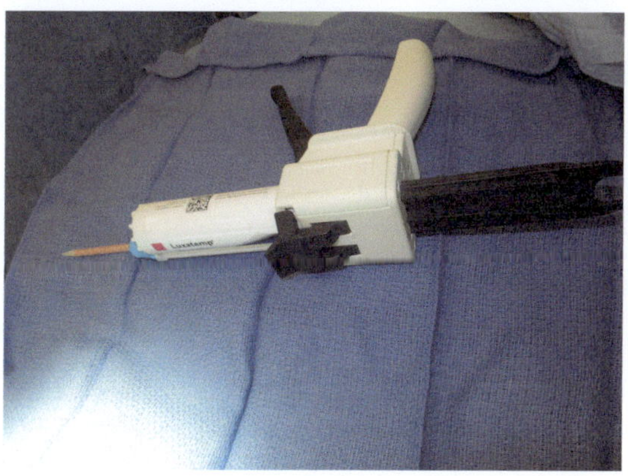

Fig. 16.25 Luxatemp bis-acryl composite resin, used for luting the prosthesis in place

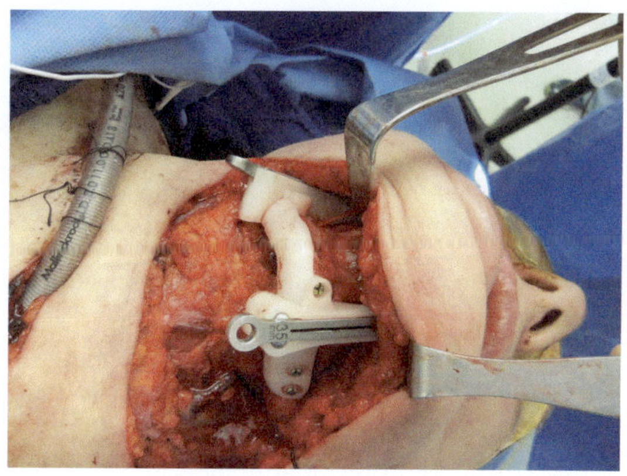

Fig. 16.26 Mandibular cutting guide used in mandibular resection. For benign disease, we attempt to preserve intraoral mucosa as long as it doesn't interfere with margins

- Once all implants have been placed and fully seated, the fibular osteotomies are performed using the osteotomy slots of the guide. A reciprocating saw or a sagittal saw may be used (Fig. 16.30). Care must be taken to avoid injury to the vascular pedicle on the medial aspect of the fibula.
- Once all osteotomies have been completed, we then remove the guide. The proximal segment of the fibula is gently dissected and freed from the vascular pedicle. This bone is removed in order to allow access to the vascular pedicle and also allow better maneuver-

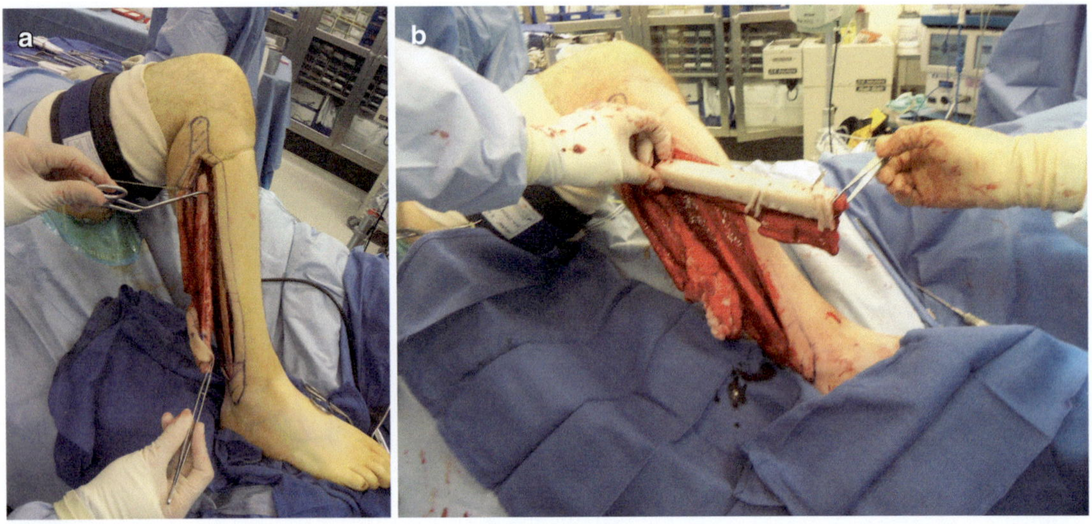

Fig. 16.27 (**a**) Showing fibula free flap harvest. (**b**) Application and securing the fibula guide to the fibula bone

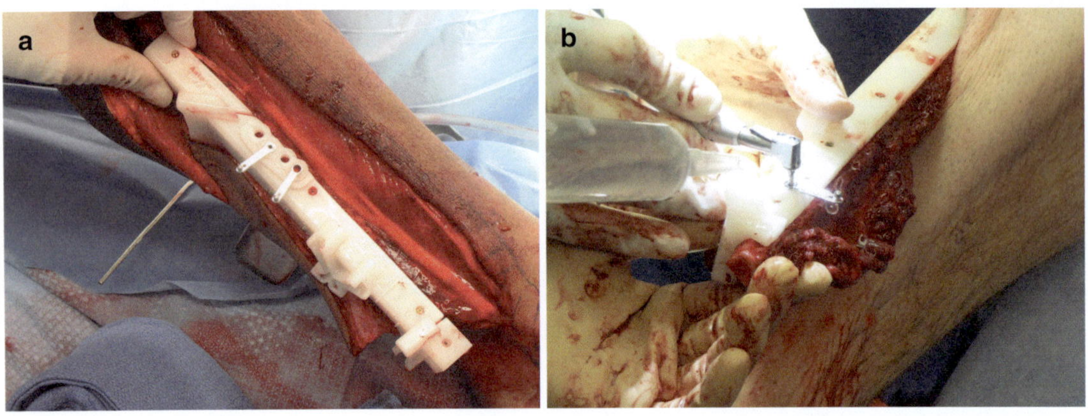

Fig. 16.28 Fibula cutting guide in place (**a**) with drill guide adaptors in place. (**b**) Implant site preparation with sequential drills

ability of the fibular segments while still pedicled at the leg.

- Next, the multiunit abutments are seated on the implants and secured in place (Fig. 16.31). We generally use 3–5 mm healing abutments depending on the diameter of the implants. The multiunit abutments will allow for space underneath the prosthesis for tissue closure and for cleaning under the prosthesis once healing is complete.
- Temporary copings are then placed on top of the multiunit abutments and secured in place (Fig. 16.31). At this point, the prosthesis can be slid over the temporary copings (Fig. 16.32). If the temporary copings extend beyond the occlusal level of the prosthesis, they will need to be trimmed before the prosthesis may be luted in place. The prosthesis may be luted to the temporary copings at this phase or done once flap is transferred to the head. Luting the prosthesis at the head will allow for some movement of the prosthesis before luting due to the built-in 5–6 mm holes in the prosthesis.
- In early stages of performing this procedure, we encourage having a resected mandible model to simulate the inset of the flap-prosthesis construct at the head (Fig. 16.33). This can be done while the flap is still attached

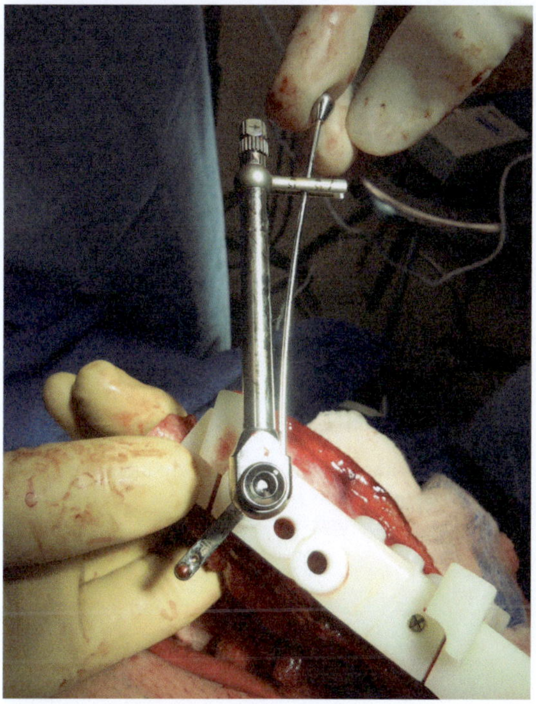

Fig. 16.29 Implant seating started with a drill on implant setting and at 40 Ncm of torque. Once the drill stops, the seating of the implant may be completed with a hand wrench. Excessive torques should be avoided to prevent fractures of the implants or the fibula

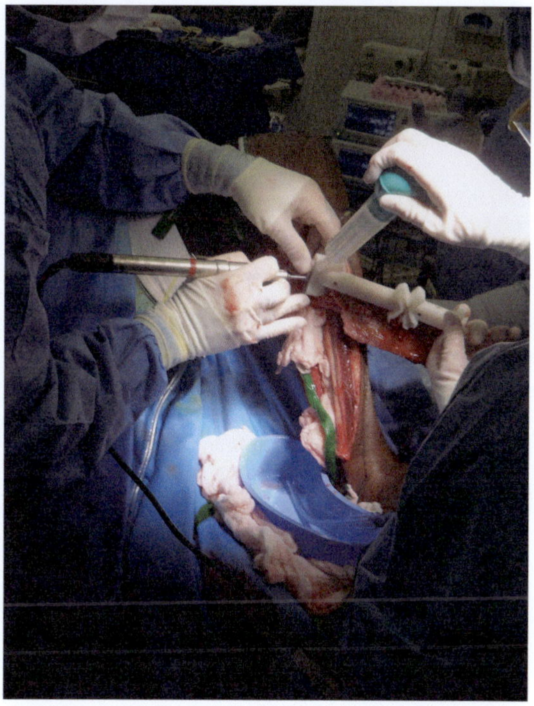

Fig. 16.30 Fibular osteotomies through slots in the fibula guide using a reciprocating saw

to its vascular pedicle and in turn will help to decrease ischemia time and the amount of adjustments needed at the head.

– The resected model is a replica of the anticipated resultant defect from the mandibular resection. The use of this model will allow for positioning of the fibula-implant-prosthesis construct in the resected portion of the model (Fig. 16.33). The occlusal splint in addition to the custom-milled plate will aid in positioning of the fibula construct against the resected model. Because of the impressions and models that were taken preoperatively, the dentition on the resected model has great accuracy allowing the seating of the occlusal splint on the model. The occlusal splint thus helps to align the prosthesis in relation to the remaining dentition (Fig. 16.33). With the fibula seating within the resected model and the prosthesis engaged into the occlusal splint, the prosthesis may be luted in place with crown and bridge composite-based resin. The resin is injected around temporary copings through sprue channels designed in the prosthesis. This allows for injection of the resin, while the occlusal surface of the prosthesis is engaged into the splint in the ideal occlusal position.

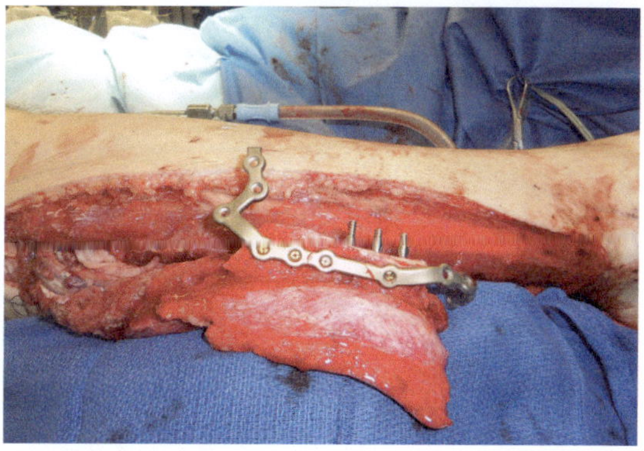

Fig. 16.31 Fibula after guide has been removed, and reconstruction bar has been lined up with and secured using predrilled predictive holes. Notice multiunit abutments and temporary copings that have been secured on top of the implants

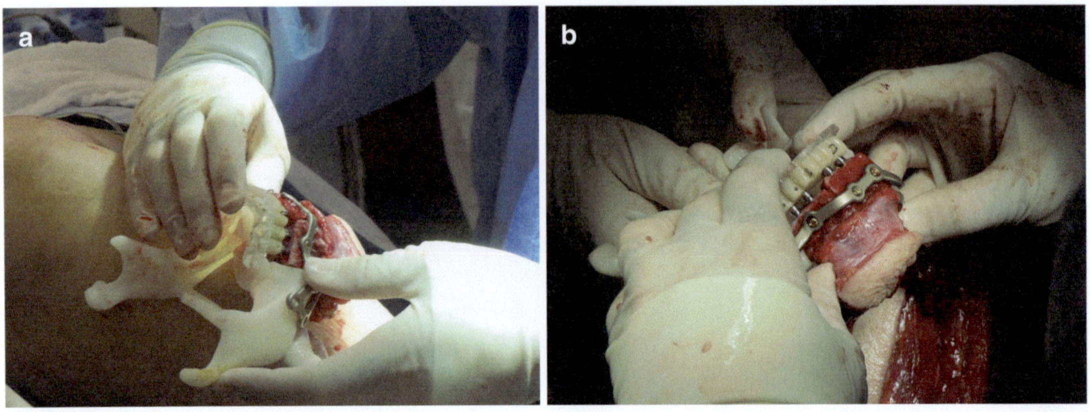

Fig. 16.32 (**a**) Prosthesis has been slid over the temporary copings and plate has been secured. The fibula-prosthesis complex is adjusted and fit into the space within the resected model. The occlusal stent along with predictive holes in the plate helps to align everything in space. (**b**) Same as A. Note how the multiunit abutment inferior to the prosthesis provides for space for tissue closure and eventually oral hygiene

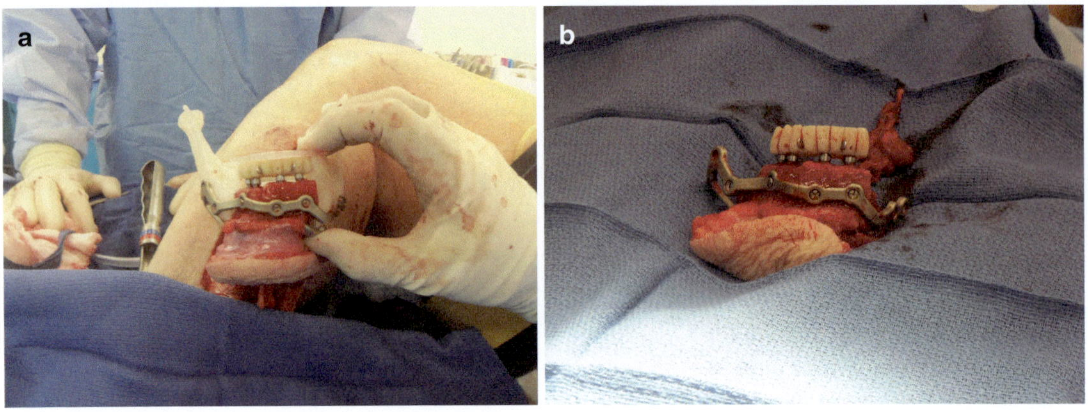

Fig. 16.33 (**a**) Showing the fibula-prosthesis plate complex fitted again the resected model. Note that a skin paddle will be de-epithelialized and used to supplement and pad the chin area. (**b**) Shows the fibula construct upon division of the pedicle and being ready to be transferred to the head

- Prior to luting the prosthesis in place, it is important to place very small pieces of cotton within the temporary copings to avoid resin from flowing on the internal screws holding the copings in place. Doing so may make it impossible to remove the prosthesis later on because of the inability of the screw drivers to engage the screw heads.
- We no longer use resected models during surgery. Once the surgeons feel comfortable with the process, this step is no longer necessary, and adjustments can be expeditiously performed at the head in situ.
- The fibula-prosthesis-plate construct is ready to be transferred to the head at this point after division of the pedicle (Fig. 16.33b).
- At the head the occlusal splint and the reconstruction plate help position the fibula-prosthesis construct in place (Fig. 16.34). The predictive holes act as external reference points that align the fibular segments and the native mandible together (Fig. 16.34a). If the

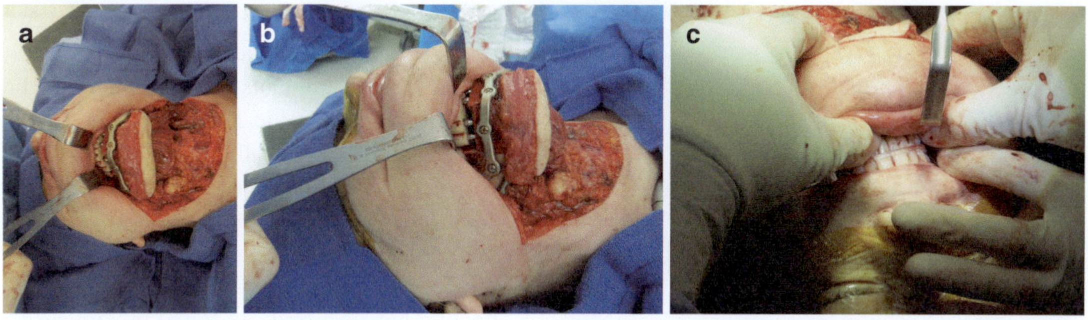

Fig. 16.34 (**a**, **b**) The fibula-prosthesis construct is positioned using predictive holes in the mandible. Upon transfer of the construct, only minimal adjustments should be necessary to get the hole in the plate and mandible to line. (**c**) Patient is wired into the splint. The splint is used simultaneously along with the predictive holes to line up everything in place. Once the prosthesis is up in the splint completely, it may be luted in place

mandible and fibula guides were used correctly, then the plate should fit over all predrilled holes with minimal adjustments (Fig. 16.34b). Positioning is also aided by the use of the occlusal splint which aligns the fibula-prosthesis complex in proper occlusion relative to the adjacent teeth and the opposing dentition (Fig. 16.34c). As indicated previously, it is now our preference to lute the prosthesis in place after the fibula has been transferred to the head. The 5–6 mm holes in the prosthesis allow for slight "wiggle room" around the temporary copings, therefore accommodating minor discrepancies from the virtual plan. With the prosthesis engaged into the splint and the fibula secured in place, the prosthesis is luted in place.

- Another advantage of luting the prosthesis after inset of the fibula is that one is able to close mucosa over the bone and around the implants under direct vision prior to securing the prosthesis in place.
- Mucosal closure must occur in a tension-free manner. Any tension during the wound closure could result in a salivary leak. Mucosa from the buccal and lingual aspects needs to be approximated to each other over the bone and around the implants. Suture may also be suspended around the implants to help take tension off the closure line. If the prosthesis had already been luted in place, then the needle must be passed under the prosthesis during suturing and closure.

Once all wounds have been closed, a feeding tube is placed to be used postoperatively. If the fibular reconstruction includes the condylar unit, then we will usually place the patient in maxillomandibular fixation. Otherwise, the patient is placed with guiding elastics. Occlusion should be checked before leaving the OR. If there are any major occlusal interferences, then those must be corrected. However, these rarely occur if the occlusal splint is used in positioning the prosthesis.

16.5.4 Postoperative Care

- Flap monitoring should follow the same protocol routinely used for buried flaps at the institution. At our institution monitoring is carried out via an implantable Synovis Flow Coupler (Birmingham, AL).
- Feeding postoperatively is done via a Dobhoff tube, which is a special type of nasogastric tube that is flexible and thinner in diameter making it more comfortable for the patient. The routine practice at the authors' institution is to leave the Dobhoff tube in place for 1 week for microvascular cases involving

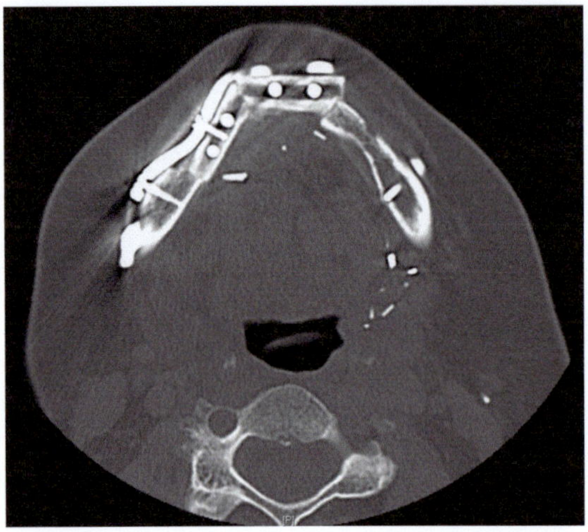

Fig. 16.35 CT scan at 4 months after surgery reveals a good bony union between the two fibular segments and the mandible bilaterally. Notice the relationship between the fixation screws and the implants

oromandibular reconstruction. However, due to the higher risk for salivary leak with the "Fibula Jaw in a Day" procedure, we will normally leave the Dobhoff tube in place for 2 weeks or longer if necessary.

- Once the Dobhoff tube is removed, the patient is started on a soft non-chew diet for 8–12 weeks. This is done so as to avoid chewing forces on the prosthesis, implants, and fibula during the initial healing phase. The patient must be shown how to care for the provisional prosthesis. It is important to clean around the implants and under the prosthesis. A water pick may be used to clean under the prosthesis once the healing is deemed to be adequate.
- Should a salivary leak occur or a wound dehiscence become obvious in the post-op period, the prosthesis must be removed in order to avoid putting the flap at risk. The temporary copings are replaced with either cover screws or short healing abutments, and primary closure over the implants and bone should be obtained. Once tissue healing has occurred, the prosthesis can usually be replaced in about 6 weeks or whenever tissue healing has occurred. This is felt to be only a minor setback, as the prosthesis can be immediately replaced in the same position once soft tissue healing has occurred. As stated previously, in our experience, this occurs when the soft tissues are closed under tension. In those cases, it may be feasible to make the decision intraoperatively to delay the placement of the prosthesis until a later date.
- It is our practice to obtain a postoperative CT scan in 3–4 months after surgery. This is to confirm that bony union has occurred between the native mandible and fibular segments despite early loading (Figs. 16.35, 16.36, and 16.37).
- Once accustomed to the procedure, a maxillofacial prosthodontist is not needed intraoperatively. It may be wise to have a prosthodontist on board during the initial cases in the event that there are significant deviations from the virtual plan. We do, however, have all our patients see a maxillofacial prosthodontist once bony union is confirmed on CT scan, so that they may start the process of fabricating a definitive prosthesis. The final prosthesis utilizes better materials and allows for a better ability to keep the prosthesis clean (Fig. 16.38).

Occasionally, adjunctive surgery may be needed down the road such as vestibuloplasty, tissue recontouring, or palatal grafts, although this has been seldom needed in our experience. Tissue recontouring or grafting may be necessary in the

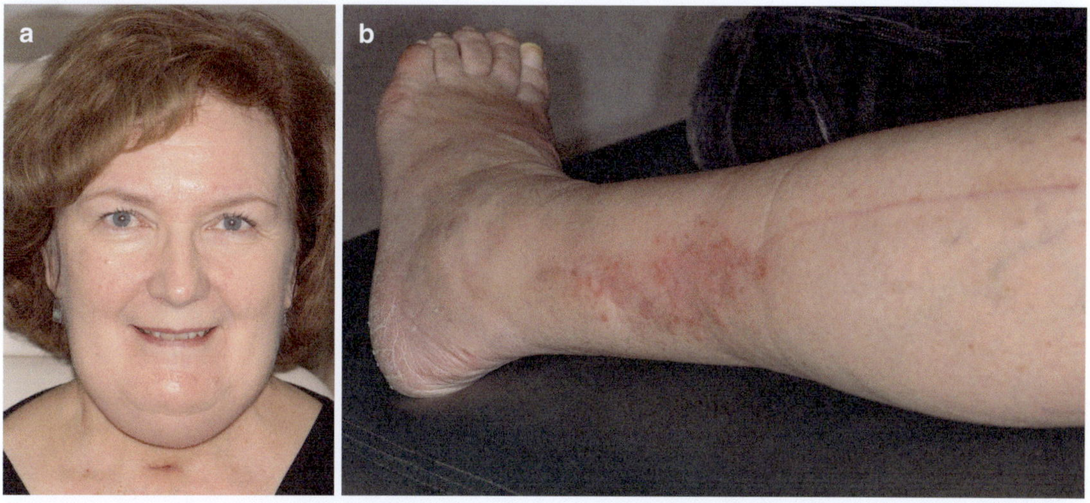

Fig. 16.36 Postoperative photos at 6 months. (**a**) Face frontal. (**b**) Lower extremity wound

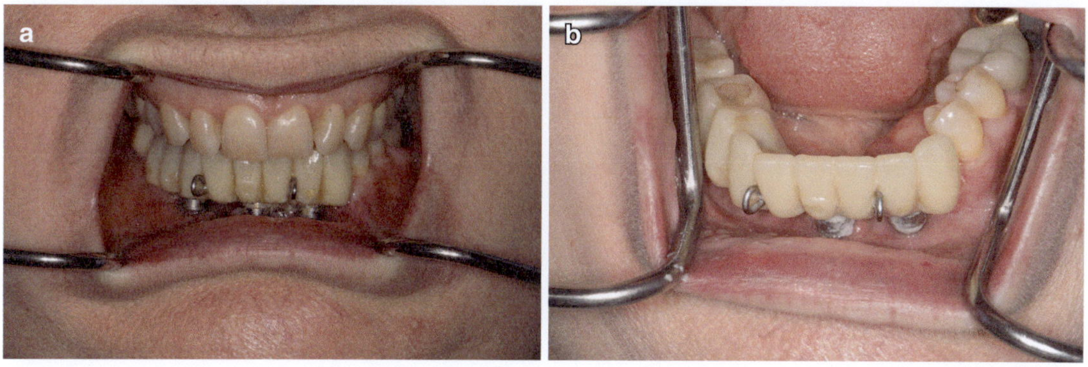

Fig. 16.37 Postoperative intraoral photos at 6 months with provisional prosthesis in place. (**a**) In occlusion and (**b**) open mouth

event the hyperplastic tissue develops around one or some of the implant. In those circumstances we have the prosthodontist adjust the undersurface of the prosthesis so that it may act as a soft tissue stent and hold the soft tissues in place during healing. Our practice has been to use the CO_2 laser for excision of the hyperplastic tissue and performing soft tissue recontouring around the involved implant. The prosthesis is then secured in place. Co-soft or some other reline material may be used in that area of the stent or prosthesis if needed.

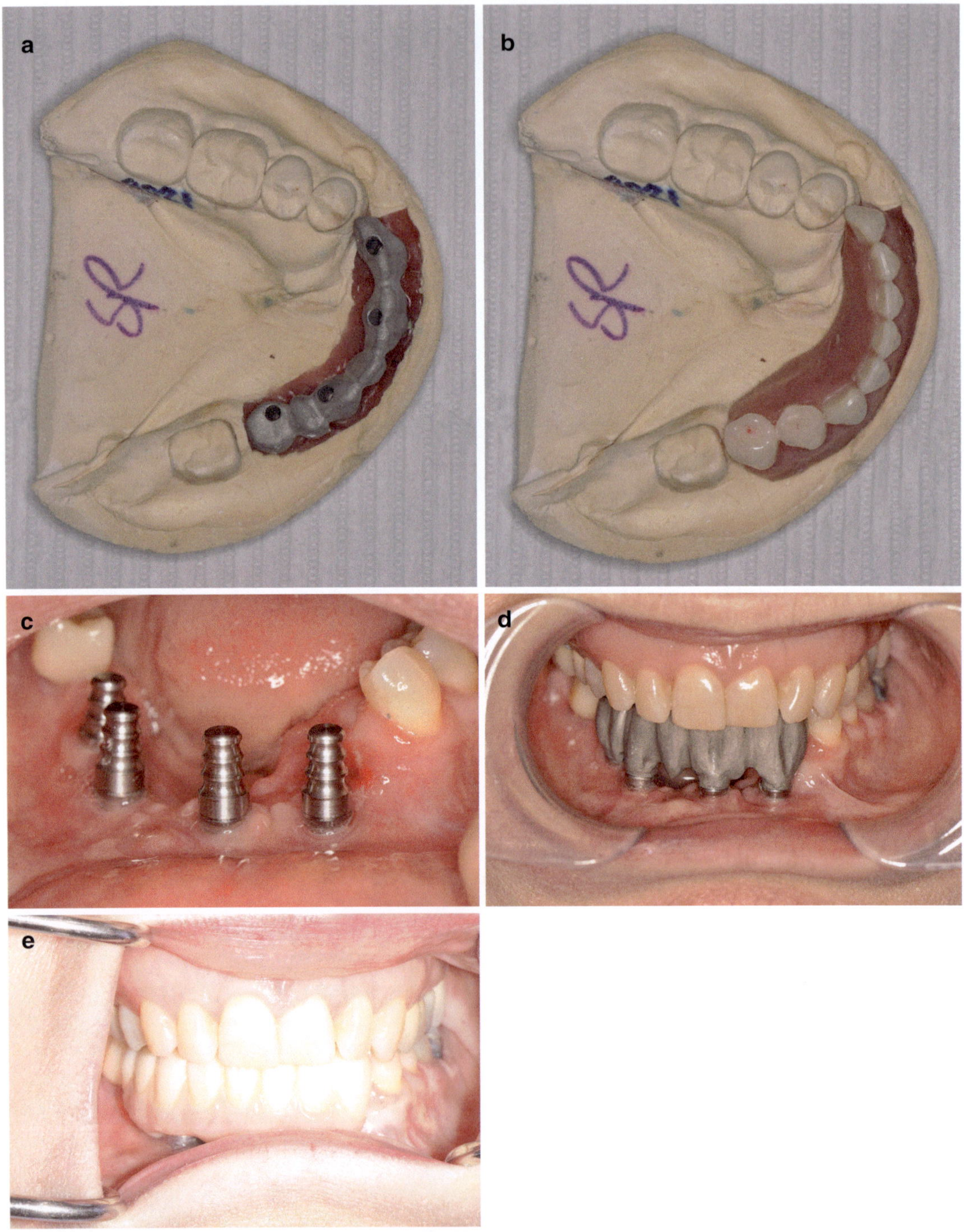

Fig. 16.38 (**a**, **b**) Final prosthesis metal framework and final (**c**) condition of the soft tissues around the implants without the need for any adjunct procedures after surgery. (**d**) Metal framework try in. (**e**) Final prosthesis insertion

16.6 Summary

The Fibula Jaw in a Day procedure has revolutionized our thinking with regard to maxillomandibular reconstruction. Although data from large series do not exist for this procedure, the outcomes from initial series and reports have been good (see above Table). The purpose of this chapter was to provide a step-by-step approach to performing the "Fibula Jaw in a Day" procedure. As one reads the literature, one will notice that there is not one way to executing these types of cases. However, what all cases share is they are based on occlusion-driven reconstruction that begins with the end goal in sight.

References

1. Hanasono MM, Chang DW. Discussion: jaw in a day: total maxillofacial reconstruction using digital technology. Plast Reconstr Surg. 2013;131(6):1392–3.
2. Qaisi M, Kolodney H, Swedenburg G, Chandran R, Caloss R. Fibula jaw in a day: state of the art in maxillofacial reconstruction. J Oral Maxillofac Surg. 2016;74(6):1284.e1–1284.e15.
3. Levine JP, Bae JS, Soares M, et al. Jaw in a day: total maxillofacial reconstruction using digital technology. Plast Reconstr Surg. 2013;131(6):1386–91.
4. Rahimov CR, Farzaliyev IM, Fathi HR, Davudov MM, Aliyev A, Hasanov E. The application of virtual planning and navigation devices for mandible reconstruction and immediate dental implantation. Craniomaxillofac Trauma Reconstr. 2016;9(2):125–33.
5. Runyan CM, Sharma V, Staffenberg DA, et al. Jaw in a day: state of the art in maxillary reconstruction. J Craniofac Surg. 2016;27(8):2101–4.
6. Yetzer JG, Ettinger KS, Arce K, Salinas TJ. Whole-arch single-stage free flap reconstruction and rehabilitation of the mandible: a case report and technical considerations on a new technique. J Oral Maxillofac Surg. 2017;75(2):290–6.
7. Salman SO, Fernandes RP, Rawal SR. Immediate reconstruction and dental rehabilitation of segmental mandibular defects: description of a novel technique. J Oral Maxillofac Surg. 2017;75(10):2270.e1–8.
8. Butura CC, Galindo DF, Jensen OT. Mandibular all-on-four therapy using angled implants: a three-year clinical study of 857 implants in 219 jaws. Oral Maxillofac Surg Clin North Am. 2011;23(2):289–300. vi.
9. Balshi SF, Wolfinger GJ, Balshi TJ. A prospective study of immediate functional loading, following the Teeth in a Day protocol: a case series of 55 consecutive edentulous maxillas. Clin Implant Dent Relat Res. 2005;7(1):24–31.
10. Malo P, Nobre Mde A, Lopes A. Immediate rehabilitation of completely edentulous arches with a four-implant prosthesis concept in difficult conditions: an open cohort study with a mean follow-up of 2 years. Int J Oral Maxillofac Implants. 2012;27(5):1177–90.
11. Maló P, de Araújo Nobre M, Lopes A, Francischone C, Rigolizzo M. "All-on-4" immediate-function concept for completely edentulous maxillae: a clinical report on the medium (3 years) and long-term (5 years) outcomes. Clin Implant Dent Relat Res. 2012;14(Suppl. 1):e139–50.
12. Chiapasco M, Gatti C. Immediate loading of dental implants placed in revascularized fibula free flaps: a clinical report on 2 consecutive patients. Int J Oral Maxillofac Implants. 2004;19(6):906–12.
13. Okay DJ, Buchbinder D, Urken M, Jacobson A, Lazarus C, Persky M. Computer-assisted implant rehabilitation of maxillomandibular defects reconstructed with vascularized bone free flaps. JAMA Otolaryngol Head Neck Surg. 2013;139(4):371–81.
14. Rohner D, Bucher P, Hammer B. Prefabricated fibular flaps for reconstruction of defects of the maxillofacial skeleton: planning, technique, and long-term experience. Int J Oral Maxillofac Implants. 2013;28(5):e221–9.
15. Freudlsperger C, Bodem JP, Engel E, Hoffmann J. Mandibular reconstruction with a prefabricated free vascularized fibula and implant-supported prosthesis based on fully three-dimensional virtual planning. J Craniofac Surg. 2014;25(3):980–2.
16. Urken ML, Buchbinder D, Weinberg H, Vickery C, Sheiner A, Biller HF. Primary placement of osseointegrated implants in microvascular mandibular reconstruction. Otolaryngol Head Neck Surg. 1989;101(1):56–73.
17. Chang YM, Santamaria E, Wei FC, et al. Primary insertion of osseointegrated dental implants into fibula osteoseptocutaneous free flap for mandible reconstruction. Plast Reconstr Surg. 1998;102(3):680–8.
18. Chana JS, Chang YM, Wei FC, et al. Segmental mandibulectomy and immediate free fibula osteoseptocutaneous flap reconstruction with endosteal implants: an ideal treatment method for mandibular ameloblastoma. Plast Reconstr Surg. 2004;113(1):80–7.
19. Shen YF, Rodriguez ED, Wei FC, Tsai CY, Chang YM. Aesthetic and functional mandibular reconstruction with immediate dental implants in a free fibular flap and a low-profile reconstruction plate: five-year follow-up. Ann Plast Surg. 2015;74(4):442–6.

Role of Microvascular Free Flaps Combined with Tissue Engineering

17

Waleed Zaid, Andrew Yampolsky, and Beomjune Kim

17.1 Tissue Engineering

The application of tissue engineering (TE) or regenerative medicine methods for reconstruction of maxillofacial continuity defects remains in its infancy. A number of authors have, however, reported successful use of TE for mandible reconstruction in humans and subhuman primates [1]. TE products typically fall within the concept of the tissue engineering triad including constructs, stem cells, and signaling molecules (Fig. 17.1a) [2]. Although significant TE research is ongoing, only a few TE products are available to enhance mandibular continuity defect reconstruction, and none of them provide all aspects of the triad. In this section, we will share some areas where TE products can be beneficial as an adjunct to traditional reconstruction with vascularized grafts. When choosing adjunctive products, some properties should be considered. Constructs should remodel and be completely replaced by viable bone during the regenerative process. Currently the most useful available constructs are allogeneic mineralized cancellous bone particles between 250 and 1000 microns. These American Association of Tissue Bank process-compliant grafts provide a surface area and geometry that is biomimetic and supports cell attachment [2]. Attachment is essential for mesenchymal stem cells to express their phenotype. Initially osteoclasts demineralize these grafts permitting appositional bone formation on the surface followed by definitive remodeling that in most areas will completely replace the graft with vital bone. Stem cells can be delivered in non-concentrated and concentrated forms [3]. Minimally invasive trephine systems are available to procure autogenous cancellous bone plugs that keep stem cells in their native sub-sinusoidal environment (Fig. 17.1b, c). Concentrated stem cells can be procured in the intraoperative environment using bone marrow aspiration coupled with concentration (BMAC, Harvest Technologies Corp., Lakewood, CO) (Fig. 17.1d). Bone inductive signaling molecules are extensively researched, and although their clinical use is regulated by the standard of care, their marketing is regulated by the Food and Drug Administration (FDA). Infuse (rhBMP-2/ACS 1.5 mg/ml, Medtronic, Minneapolis, MN) is the only inductive bone regenerative cytokine available for clinical use (Fig. 17.1e–g). It has maxillofacial indications related to extraction socket defects [4] and

W. Zaid (✉)
Department of Oral and Maxillofacial Surgery, Louisiana Health Sciences Center in New Orleans, New Orleans, LA, USA

Our Lady of the Lake Regional Medical Center, Baton Rouge, LA, USA
e-mail: Wzaid@lsuhsc.edu

A. Yampolsky · B. Kim
Department of Oral and Maxillofacial Surgery, Louisiana Health Sciences Center in New Orleans, New Orleans, LA, USA

J. C. Melville et al. (eds.), *Regenerative Strategies for Maxillary and Mandibular Reconstruction*,
https://doi.org/10.1007/978-3-319-93668-0_17

posterior maxillary alveolar deficiency requiring a sinus lift bone graft [5]. Thus, the application of this technology to mandibular continuity defects is off-label [6, 7]. The package insert further states that Infuse should not be used at the site of an extant or resected tumor when malignancy is present or when a patient is being treated for malignancy. Based upon the literature sited above and our experience, we currently apply Infuse as an adjunct to non-vascularized cancellous particulate bone grafts used to treat continuity defects present from trauma or infection or after the resection of benign tumors [8]. In addition, we use it to treat post-malignant tumor resection in malignancy-free patients having a delayed reconstruction or those in need of reconstruction as a result of osteoradionecrosis [9, 10]. The advantage of combining Infuse with other reconstructive techniques is the rapid formation of an osteogenic granulation tissue that provides a rich connective tissue and vascular environment to support regeneration (Fig. 17.1h).

17.2 Combining Tissue Engineering Principles with Free Tissue Transfer

Numerous technological advances have opened up a wide range of options for maxillofacial reconstruction. Traditional bone graft stood the test of time and is still widely used with success. Bone morphogenetic protein (BMP) whether used alone or in combination with bone graft has greatly contributed to the field of TE. Distraction osteogenesis has been utilized successfully for augmentation of the alveolar ridge and segmental mandibular defects [2]. Prosthetic total joint replacement made possible functional reconstruction of the missing temporomandibular joint (TMJ). The introduction of free vascularized tissue transfer was a game changer and quickly revolutionized oromandibular reconstruction. This treatment modality gained popularity for composite mandibular defects due to their reliable ability to reconstruct simultaneously both missing soft and hard tissues even in a compromised wound.

In addition, the utilization of virtual surgical planning (VSP) greatly added to surgical precision and efficiency [11]. The virtual environment allowed surgeons to accurately and safely plan surgical osseous resection margins, decrease surgical time by giving the reconstructive team the ability to produce a stereolithographic model and pre-bend reconstruction plates, or be able to create a computer-aided design-computer-aided manufacturing (CAD/CAM) plate which is commonly referred to as a custom-made plate. Finally VSP allows regaining of pre-surgical mandibular contour due to the ability of mirroring contralateral, unaffected mandible or utilizing a generic mandibular form [12, 13].

However, none of the existing technological developments alone can perfectly satisfy the complex anatomical and functional needs of the mandible. Each option has specific indications dictated by the disease, defect, patient, and clinician factors. Combining different reconstructive modalities may bring together the advantages of each option.

17.2.1 Hybridized Mandibular Reconstruction (HMR) with FFF and Tissue-Engineered Graft

In HMR, superiorly positioned FFF is combined with inferiorly positioned tissue-engineered graft, i.e., a mixture of BMP, BMAC, and cortico-cancellous bone graft (Fig. 17.2), in order to precisely recreate the native height of the mandible, to place the dental implants in an ideal restorative relationship to the maxillary arch, to improve the bony union between the fibula-fibula and the fibula-native mandible interface, to recreate the depth of vestibule for manipulation of food and saliva, and to accelerate the bony remodeling of the fibula bone to transform into the shape and quality of the premorbid mandible (Fig. 17.3a–j).

The position of the fibula is dictated by the future prosthetic restoration based on the "reverse engineering" or "prosthetically driven planning" principle [14]. VSP (Medical Modeling Inc., Golden, CO) allows easy translation of this pre-

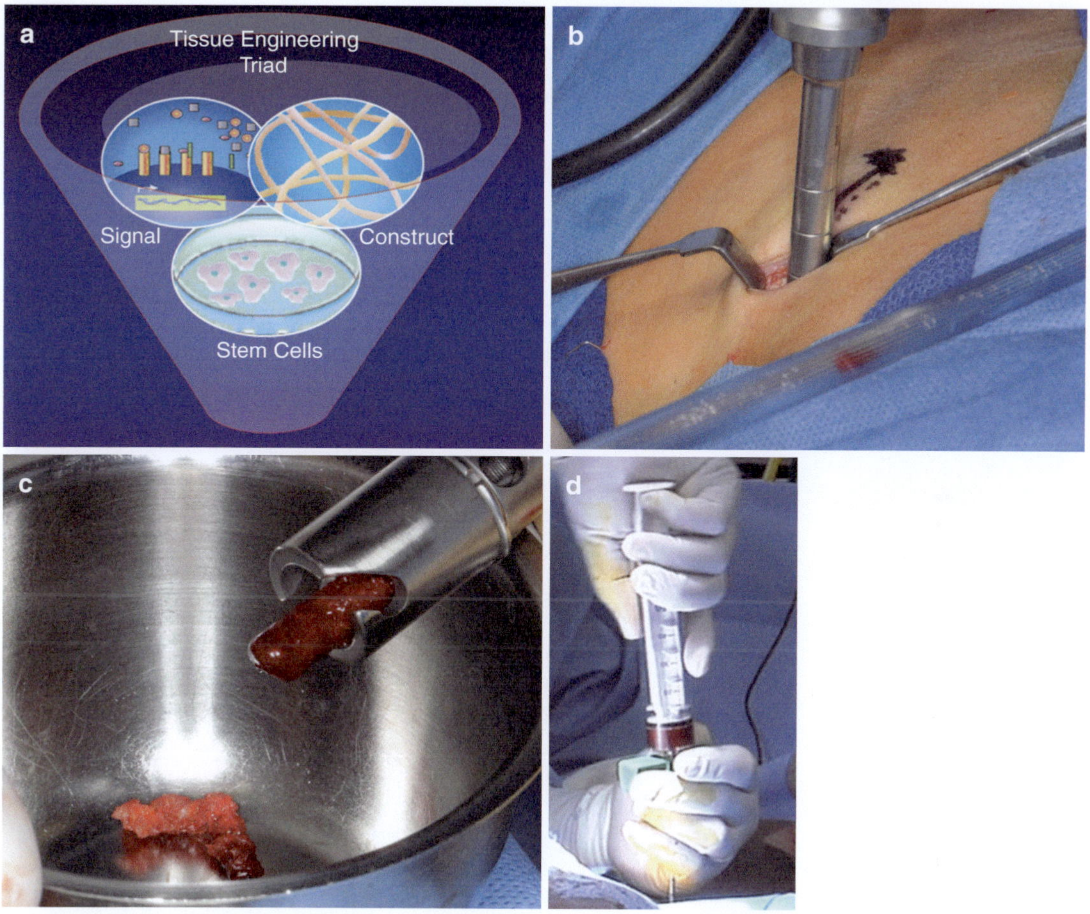

Fig. 17.1 (**a**) The concept of tissue triad with mesenchymal stem cells, scaffolding constructs, and signaling molecules. (**b**, **c**) A trephine system is used to harvest a core of autogenous cancellous bone graft and marrow containing mesenchymal stem cells from the anterior iliac crest. (**d**) Concentrated bone marrow aspirate is collected by the BMAC system from the anterior iliac crest. (**e**) This diagram depicts the molecular structure of rhBMP-2. (**f**) BMP signal transduction leads to osteoblast differentiation of mesenchymal stem cells. (**g**) Microscopic view (40×) of de novo bone formation by BMP 6 months after placement. (**h**) Microscopic view (10×) of the osteogenic granulation created by BMP 10 days after placement in a canine model. (**i**) A 6 cm mandibular continuity defect from osteoradionecrosis resection. Both upper and lower border plates serve as a scaffold for the tissue-engineered graft and define the shape of the mandible. (**j**) Tissue-engineered graft with cancellous bone graft combined with BMP (Infuse, Medtronics). (**k**) The wound bed is first lined with a collagen sponge soaked with BMP. (**l**) Placement of tissue-engineered graft in the defect. (**m**) Postoperative panorex shows adequate alignment of the plates and tissue-engineered graft. (**n**) Postoperative lateral cephalogram shows anatomic recreation of the mandible. (**o**) Dental implants are placed 7 months post-graft placement. Mandibular continuity is reestablished. The superior border plate is removed if necessary for patient comfort. (**p**) An implant bar supra-structure is fabricated. The depth of vestibule on the reconstructed side is similar to the unaffected side. (**q**) The intaglio surface of the denture with locator attachments

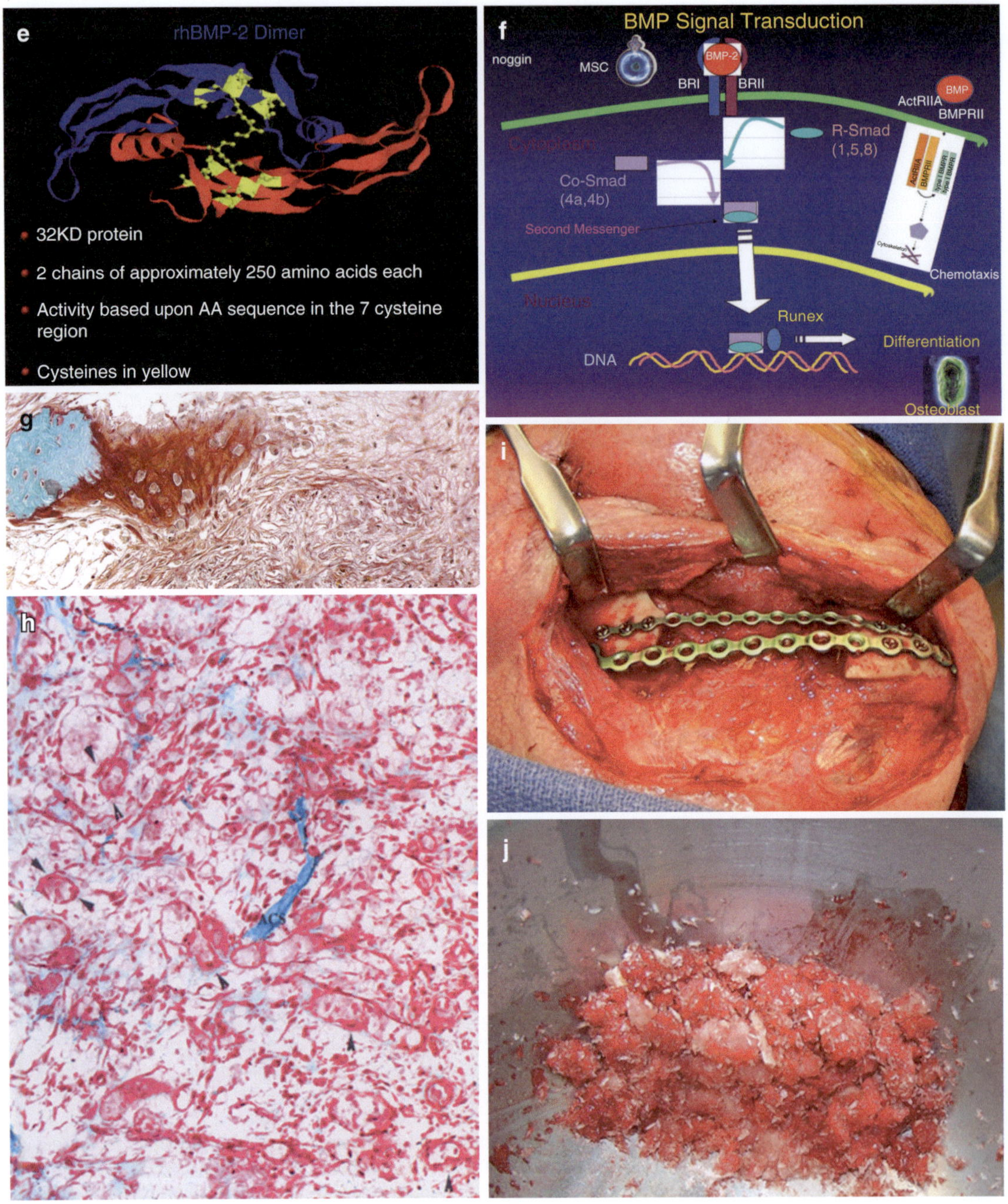

Fig. 17.1 (continued)

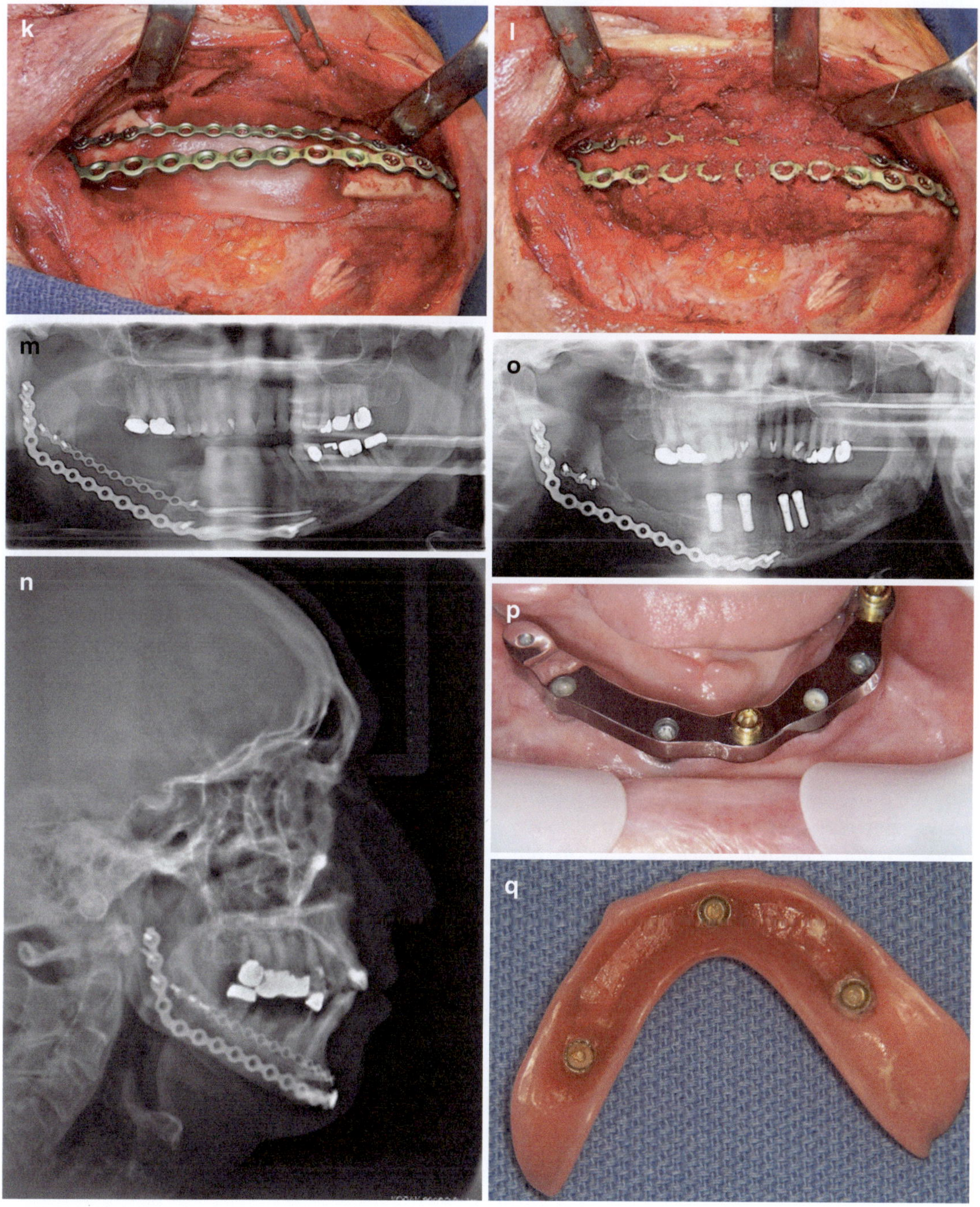

Fig. 17.1 (continued)

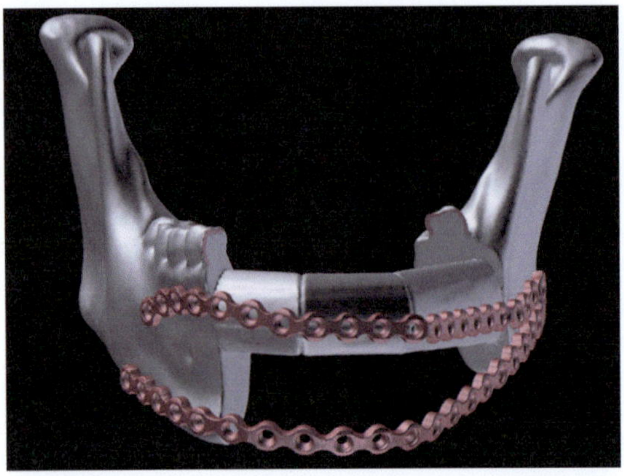

Fig. 17.2 Superiorly positioned fibula provides a sturdy platform for dental implants, while inferiorly placed tissue-engineered graft forms the basal mandibular bone for structural support and lower facial contour. The superior reconstruction plate secures the fibula, while the inferior reconstruction plate provides a scaffold for the tissue-engineered graft

prosthetic plan into the actual surgery (Fig. 17.4). The fibula is placed at an ideal restorative position in relation to the opposing maxillary dentition or a pre-scanned surgical stent in the virtual surgical environment. Even dental implants can be virtually placed if desired. The surgeon is provided with a fibula-cutting guide, a mandibular resection guide, a fibula-positioning guide, and/or a pre-bent plate in order to precisely carry out the plan in the operating room.

One underappreciated downside of FFF is a relatively high incidence of nonunion at the closing osteotomy sites. Due to the straight shape of the fibula bone, the natural contour of the mandible necessitates multiple closing osteotomies of the fibula permitted by its segmental blood supply [2]. Before the VSP technology with cutting guides was available, the osteotomy had to be performed freehandedly and was often less than perfect. Because of the cortical nature of the fibula bone, if bony contact was not adequate, the area often healed with fibrous union. This led to eventual plate fracture and reoperation for the patients. Virtual surgical technology probably has greatly reduced the incidence of nonunion, but many microvascular surgeons agree that achieving perfectly adapted closing osteotomy sites is still challenging even with cutting guides. BMP has shown to improve bony union at the FFF closing osteotomy sites, and HMR with addition of BMAC and cortico-cancellous bone graft may further ameliorate bony healing [9]. The same rule applies to the FFF reconstruction in mandibular reconstruction in the subsigmoid region. After subsigmoid osteotomy, the proximal segment is usually left with a very narrow cut surface (5–10 mm). When fibula is placed against this surface, either weak bony union or nonunion is expected. When osteotomy surfaces are not perfectly adapted, the use of BMP is likely to improve bony union (Fig. 17.5a–g).

HMR is currently contraindicated in cancer population, due to the possibility of tumor proliferation from BMP, albeit controversial [9, 15], and unpredictable success of immediate bone graft in the face of radiation and oral contamination. In these patients, a prosthetic material such as polyetheretherketone (PEEK) (Fig. 17.6a–d) or a thicker titanium reconstruction plate (Fig. 17.7) may be utilized to fill the inferior border defect below the fibula segments. This technique is less ideal than tissue-engineered graft with de novo bone formation and healing but would satisfy both functional and esthetic requirements of mandibular reconstruction.

17.3 Axially Vascularized Tissue-Engineered Constructs

Despite significant advances in reconstructing maxillofacial ablative defects with the advent of microvascular surgery and the popularization of various osteocutaneous free flaps, some limitations exist in current reconstructive techniques. Currently, the microvascular free flaps most commonly used in reconstructing maxillofacial defects include the fibula, the ilium, and the scapula. These flaps all have unique indications as well as limitations.

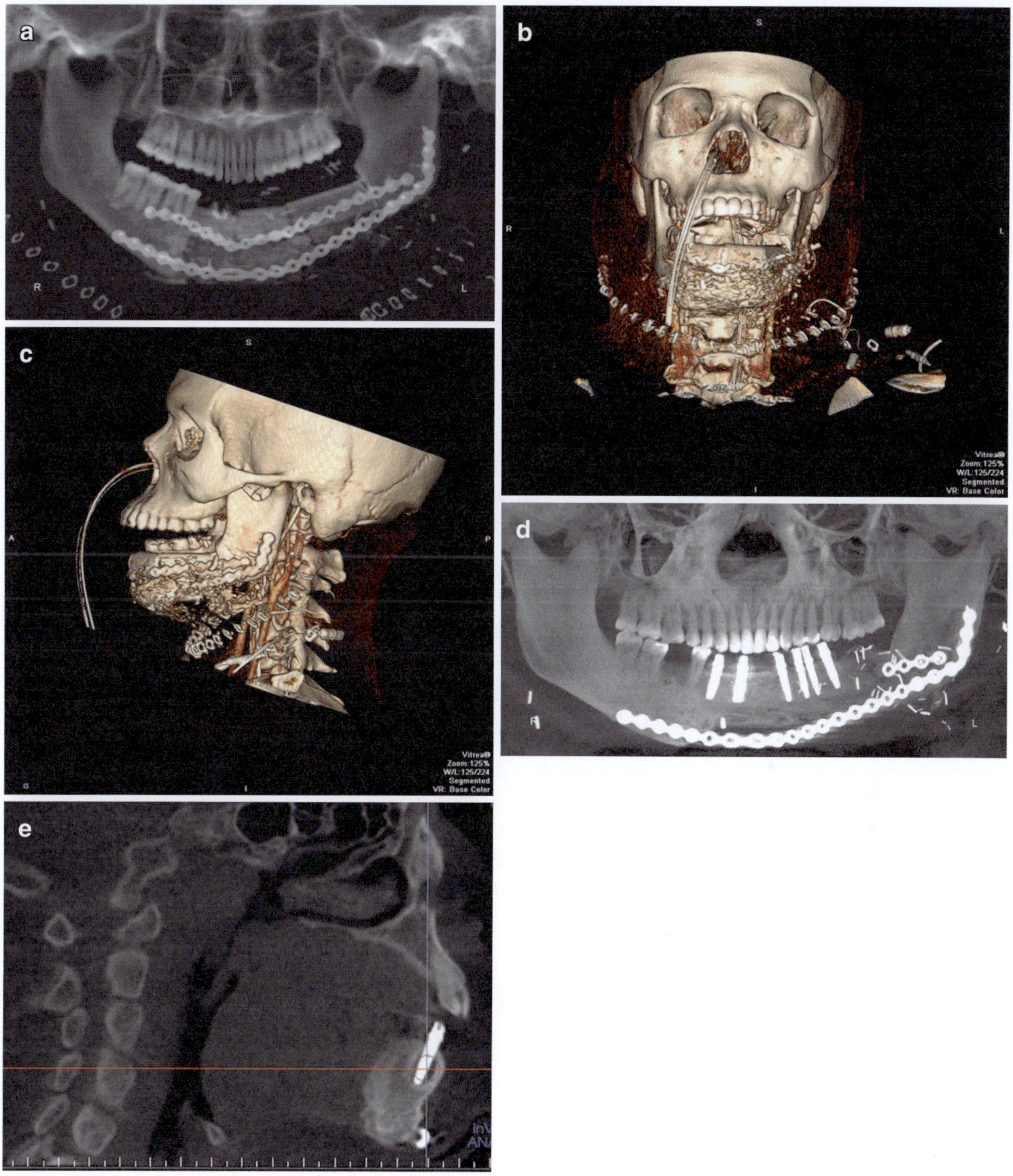

Fig. 17.3 (**a**) This panorex shows HMR after segmental mandibular defect from ameloblastoma. Superiorly positioned fibula segments and inferiorly placed tissue-engineered graft are noted. (**b**, **c**) Three-dimensional images of HMR show excellent anatomic reconstruction of the mandible. (**d**) Complete bony union and remodeling are noted after 6 months. (**e**, **f**) Dental implants are placed at ideal angulations and positions at 9 months, which is possible due to the precise virtual surgical planning. Also, complete remodeling and fusion of the fibula bone with the tissue-engineered graft are noted. The muscular action in the mentolabial region has recreated the B point. (**g**, **h**) Final prosthesis with excellent occlusion is noted in addition to recreation of the premorbid vestibular depth and form. (**i**, **j**) Another HMR case showing recreation of the premorbid vestibular depth and form

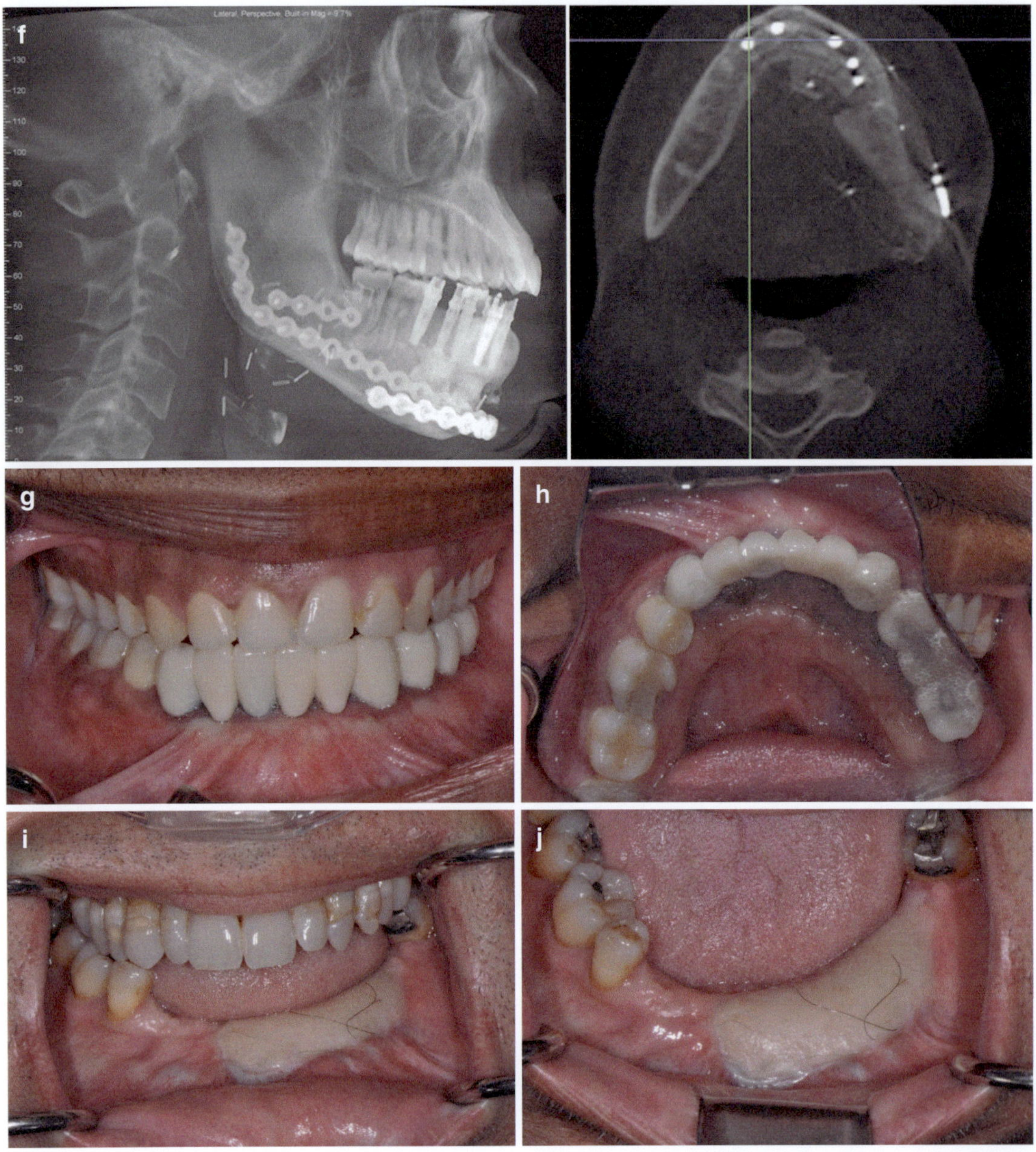

Fig. 17.3 (continued)

A detailed discussion of these flaps is beyond the scope of this text. Briefly, the fibula is able to provide the longest segment of vascularized bone (up to 25 cm) and soft tissue and is thus a popular choice for reconstructing mandibular defects. However, it has significant limitations owing to its limited bone height which makes it difficult to reconstruct mandibular contour while also creating a prosthetically favorable platform for dental reconstruction [2, 11, 16, 17] (Fig. 17.8a–f). Bahr et al. tried to overcome this problem by introducing the double-barrel technique in an attempt to overcome the height mismatch at the interface between the native mandible and fibula free vascularized flap segments. This technique allows improved occlusal relationship with the maxillary dentition but has several limitations as this technique cannot be used for large mandibular

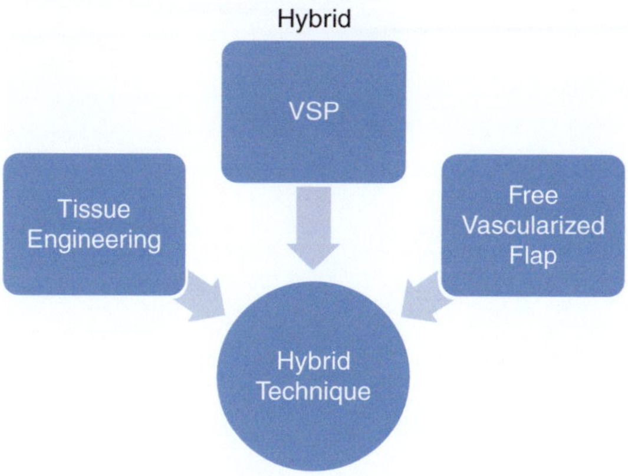

Fig. 17.4 The concept of HMR technique combining FFF, tissue engineering, and virtual surgical planning

segmental defects and leads to significant shortening of the vascular pedicle, which limits your options for microvascular anastomosis [18].

Vascularized iliac crest bone can be used to provide more abundant bony height and width and often has a contour that more accurately recreates the native mandible in comparison to the fibula bone. Nevertheless, it is limited in the length of the bone that can be harvested safely and thus is unsuitable for reconstruction of the long mandibular defects especially with the lack of segmental perforators, which makes closing osteotomies less reliable. Furthermore, it has the potential for significant donor site morbidity such

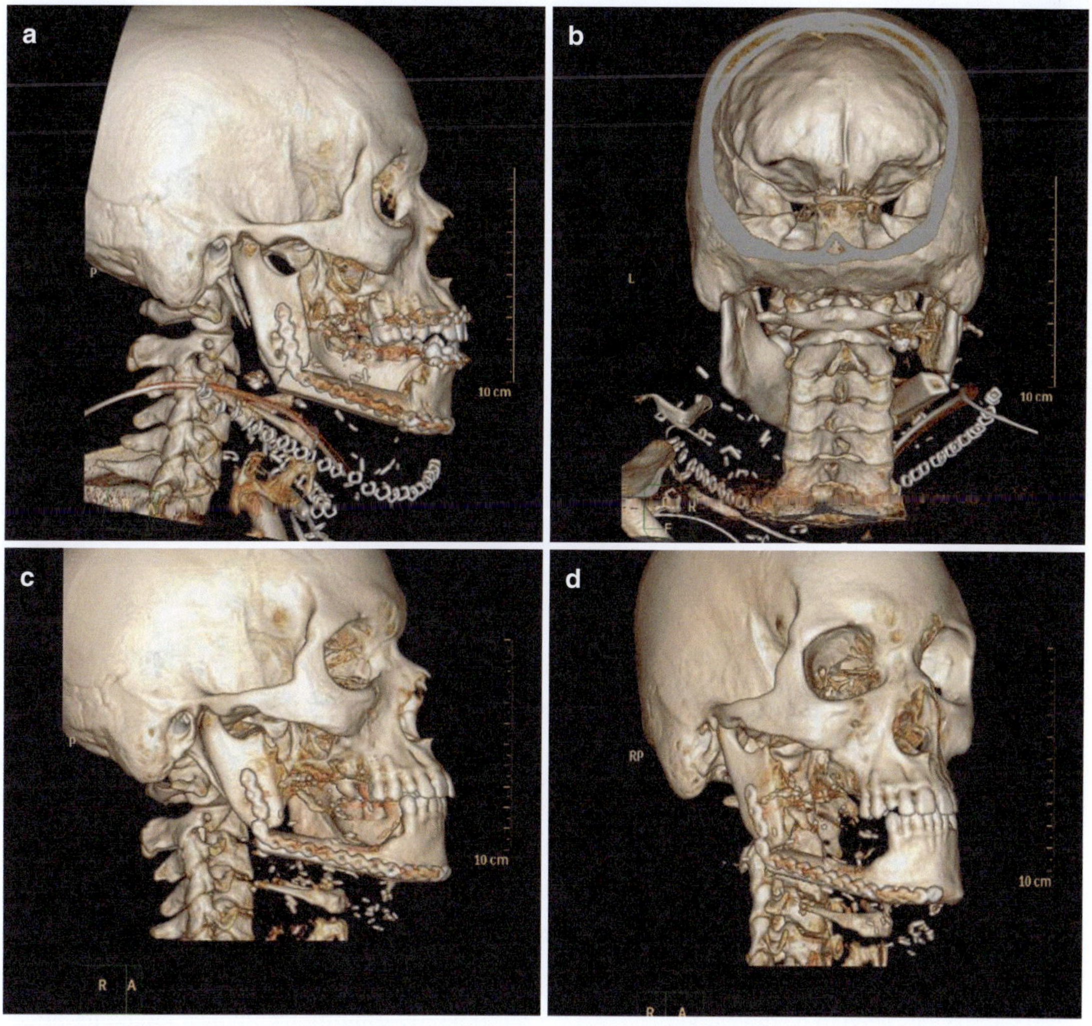

Fig. 17.5 (**a**, **b**) Poor bony contact between the native mandible and the fibula is noted in the subsigmoid region. (**c**, **d**) Nonunion and hardware fracture at 6 months post-op. (**e–g**) Reoperation was undertaken with reconstruction plate replacement in addition to the anterior iliac crest bone graft, BMP, and BMAC

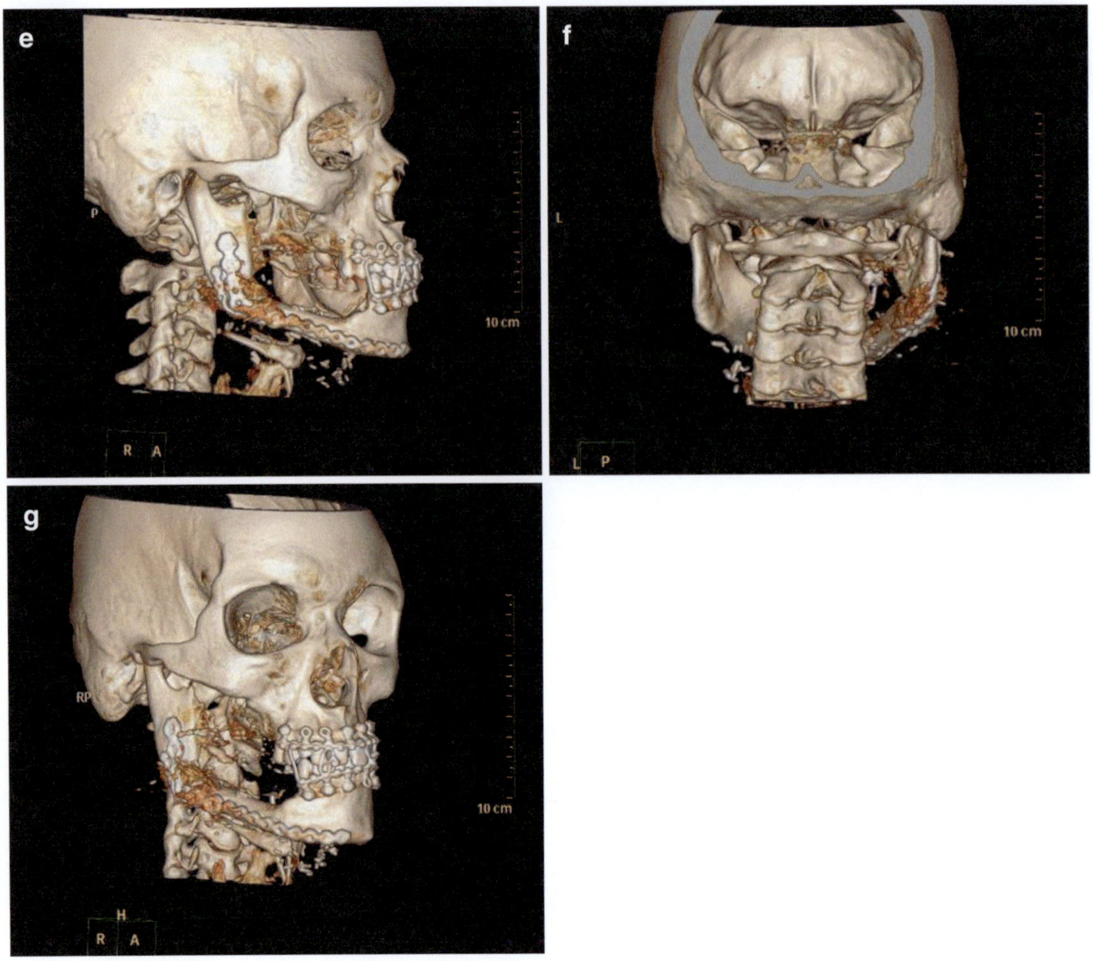

Fig. 17.5 (continued)

as the development of incisional or inguinal hernias and gait disturbance and is limited by a shorter pedicle length [14, 19, 20] (Fig. 17.9a–f).

The osseocutaneous scapula flap has a limited bone stock and is often too thin to accommodate dental implants. The scapular tip is an excellent choice for reconstruction of palatal defects as its shape closely approximates that of the palate (Fig. 17.10) [19, 20]. In summary all the current reconstructive options suffer from limited anatomic compatibility with the defects they intend to reconstruct. Furthermore, the harvest of composite flaps will always result in varying degrees of donor site morbidity. Thus, it is evident that novel approaches need to be developed.

Regenerative medicine and tissue engineering attempt to reconstruct diseased tissue or ablative defects with viable healthy tissue, similar in function and appearance to the native tissue being reconstructed. Consequently, there is a need for a method to create a three-dimensional medium that provides structural support and allows for the migration and differentiation of cells and appropriate release of growth factors [21–23]. In tissue engineering, such a structure is known as a scaffold. Early techniques for creating scaffolds involved solvent casting, particulate leaching, gas foaming, phase separation, melt molding, and emulsion freeze-drying [21–23]. These techniques had limitations as they were unable to produce precise geometry, appropriate pore sizes, and mechanical strength [21–23]. The advent of multiple 3D printing techniques including direct 3D printing, fused deposition modeling, stereolithography, and selective laser sintering has allowed researchers to produce structures with

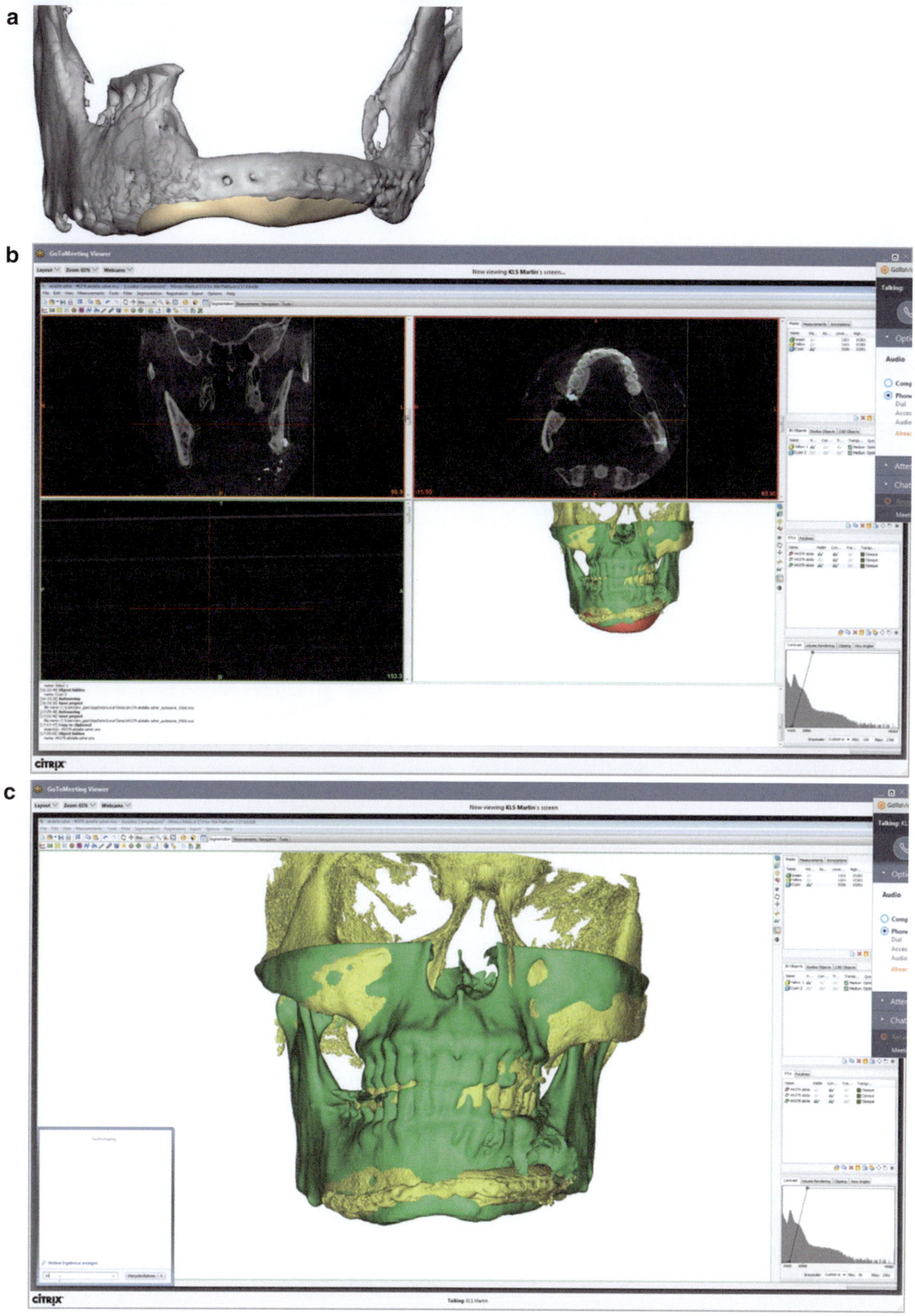

Fig. 17.6 (**a**) A PEEK implant is designed for the inferior border defect due to the patient's refusal of tissue-engineered graft. (**b**, **c**) VSP images. (**d**) Intraoperative picture of the implant, augmenting the inferior border of the fibular reconstruction

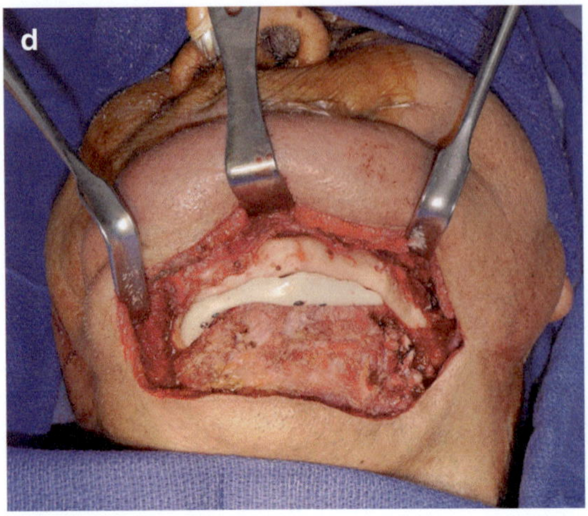

Fig. 17.6 (continued)

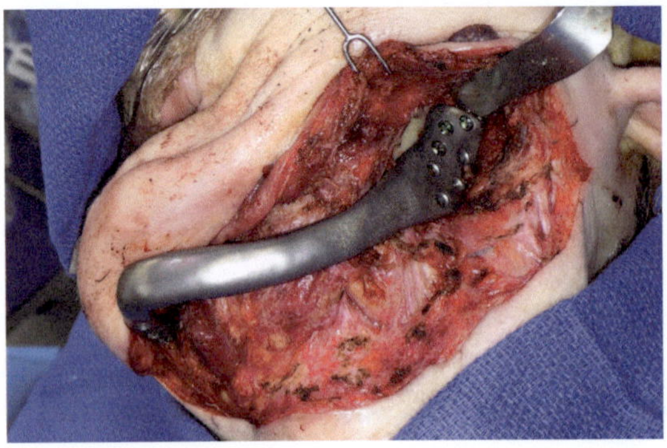

Fig. 17.7 A custom-designed plate with a thicker profile was used in this medically compromised patient in order to prevent hardware fracture in the long term

complex morphologies and precisely controlled pore sizes, on a nanometer scale, that are able to more accurately reproduce the native extracellular matrix, thus creating an ideal environment for tissue regeneration [21–24].

There are a variety of materials used for scaffold creation. Perhaps the most promising in terms of applications for maxillofacial reconstruction is the use of biocompatible ceramics. These are promising because of their osteoconductive potential and mechanical strength, making them an excellent substrate for regeneration of the maxillofacial skeleton. Hydroxyapatite (HA) in particular is the predominant mineral compound in both teeth and the skeleton. Animal studies suggest that scaffolds with pore sizes similar to that of newly formed trabecular bone, in the range of 250–400 μm, are effective at conducting osteoblasts and regenerating bone in critically sized defects [25]. Other researchers have demonstrated that similar scaffolds fabricated out of tricalcium phosphate (TCP) with a pore size in the 300 μm range are able to allow nutrient diffusion to the cells contained within [21, 25].

The advent of 3D printing has allowed researchers to create structures that are precisely designed to promote osseoconduction (Fig. 17.11), release inductive factors, and precisely conform to the complex anatomy of the maxillofacial region. Early research in such grafts relied on the graft obtaining neovascularization through peripheral ingrowth of blood vessels [26]. This understandably limits the theoretical application of such grafts to smaller defects and defects with a healthy soft tissue envelope and precludes use in irradiated tissues [26].

Attempts to overcome limited neovascularization of these structures have involved the addition of growth factors such as vascular endothelial growth factor (VEGF), basic fibroblast growth factor (bFGF), and transforming growth factor-beta (TGF-B) [27–31]. However these techniques have drawbacks, as they rely on recruiting local cells for neo-angiogenesis which may be limited depending on the amount of progenitor cells available in the particular subject [27]. Furthermore, more research is necessary to precisely determine the proper concentrations as well as timing and duration of cytokine release necessary to produce optimum outcomes [27]. The implantation of endothelial progenitor cells (EPCs) into these constructs has shown promise. However, ensuring proper blood supply to tissue-engineered constructs remains a challenge [27].

One novel area of research involves the fabrication of axially vascularized tissue-engineered grafts that can then be harvested and transplanted to the defect site using traditional microsurgical techniques. These techniques aim to implant an osse-

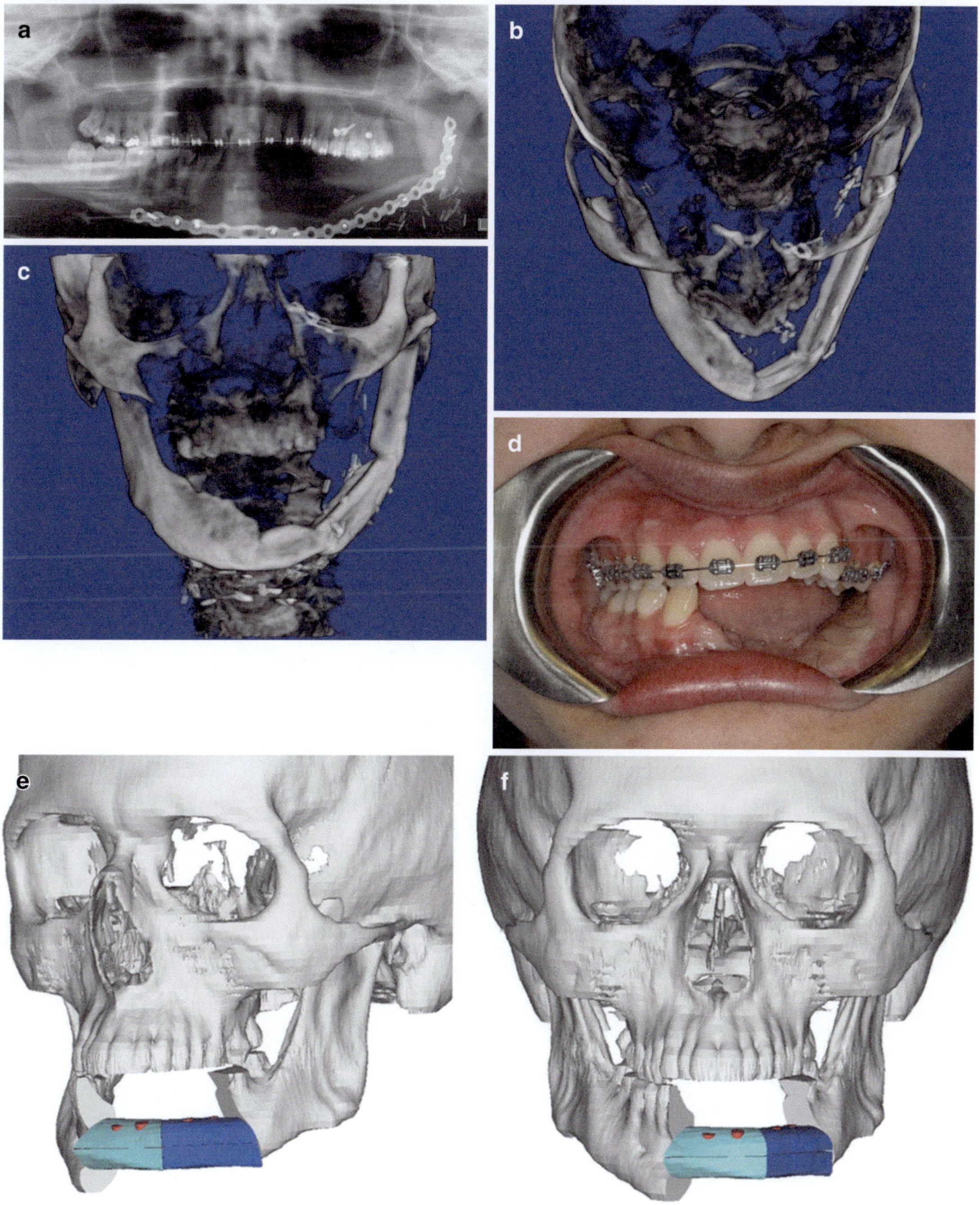

Fig. 17.8 (**a**) The fibula placed at the inferior border. A significant height mismatch with the native mandible leads to a high crown-to-implant fixture ratio if dental implants are placed. (**b**) The fibula is placed too anteriorly in relation to the native maxillary arch. This relation subjects the future prosthesis to a significant cantilever force. (**c**) The fibula is placed too laterally in relation to the native maxillary arch. (**d**) The lack of vestibule results from the inferior positioning of the fibula. (**e**) Superior positioning of the fibula may compromise the lower facial contour

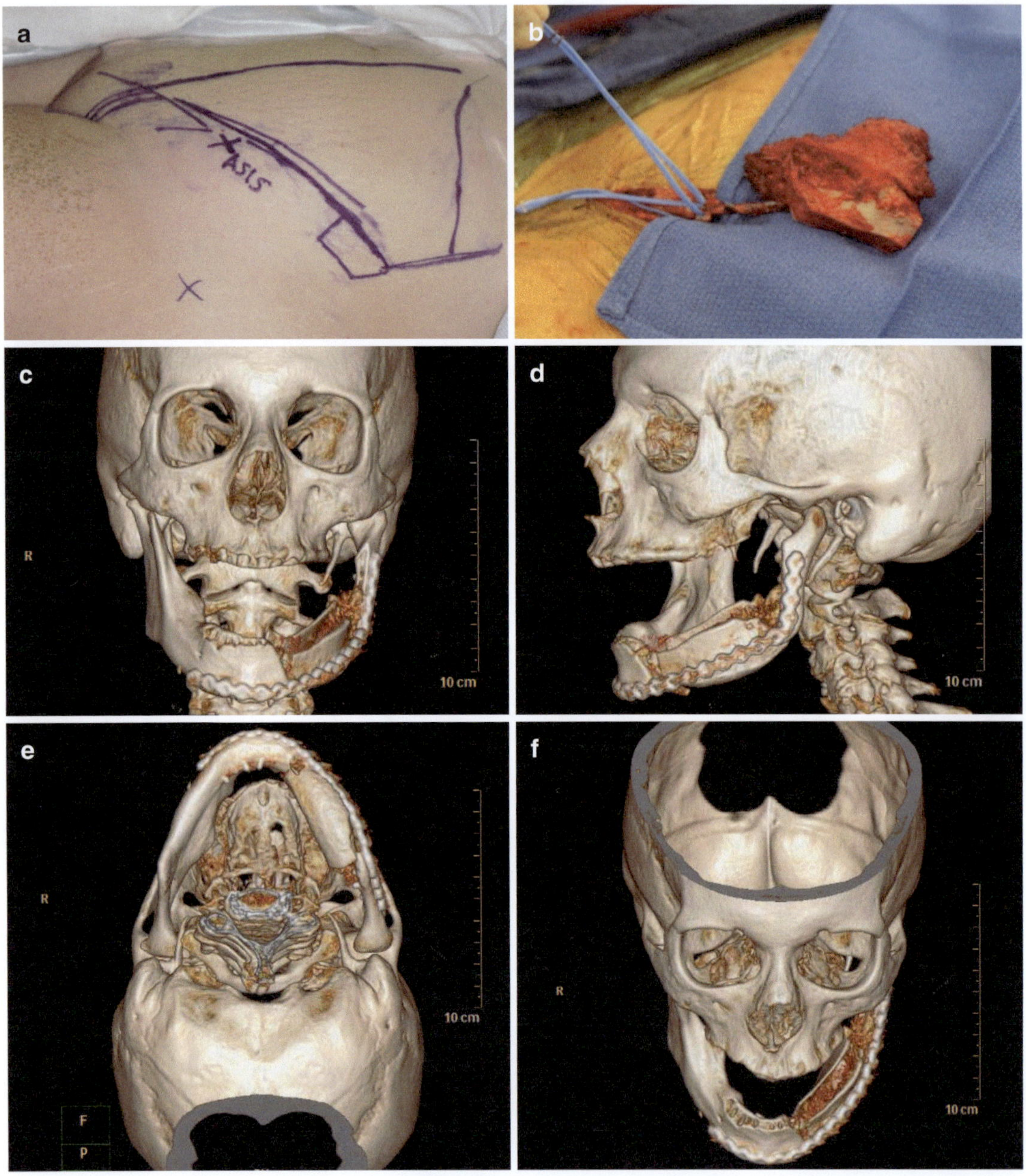

Fig. 17.9 (**a**) Skin markings for the deep circumflex iliac artery flap (DCIAF). (**b**) The iliac crest bone and the internal oblique muscle are harvested in one unit based on the DCIA. Excellent bone stock is noted. The muscle mucosalizes promptly once placed in the oral environment. (**c**, **d**) These three-dimensional images show an excellent height match between the native mandible and DCIAF. (**e**, **f**) The natural curvature of the iliac crest allowed the bone to be placed without closing osteotomies in this case. The width of the bone is more than adequate for dental implantation at the crest level of the neo-mandible

ous scaffold within an axially vascularized territory; thus when transplanted for reconstruction, the construct can be nourished through its axial blood supply in the same manner as a vascularized free flap. Further, these vascularized constructs have the added advantage of perfectly conforming to the defect and avoiding donor site morbidity in a way that a traditional free flap is unable to [26, 32].

There are two main techniques for creating an axially vascularized tissue-engineered construct,

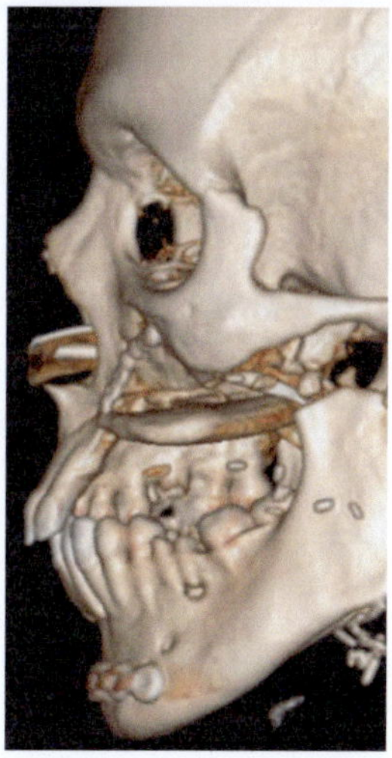

Fig. 17.10 3D CT of a palatal defect reconstructed with a scapular tip flap demonstrating good anatomic conformation but limited bone stock for implant placement

Fig. 17.11 An experimental 3D-printed HA scaffold with controlled porosity

prelamination and prefabrication. Prelamination was first described in 1994 by Pribaz and Fine which most accurately refers to the implantation of a customized structure into the vascular territory of a flap, allowing for extrinsic vascularization of the construct with subsequent transfer to the planned reconstructive site [26, 32, 33]. As an example of this concept, observe Fig. 17.12. In this case a costochondral graft was implanted into the vascular territory of the radial forearm and subsequently used to reconstruct a complete

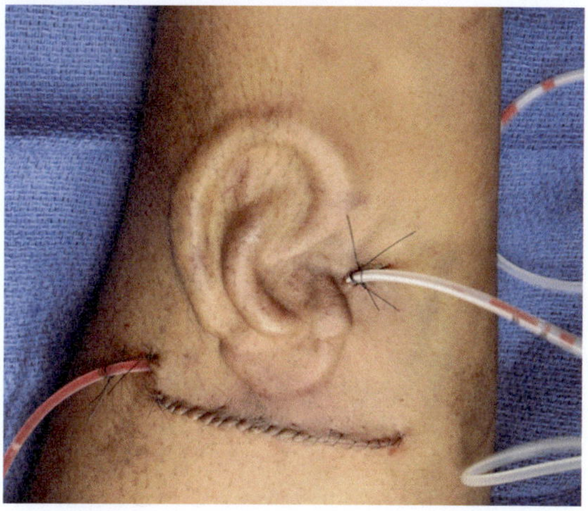

Fig. 17.12 An example of a prelaminated flap designed by implanting a costochondral framework into the vascular territory of a radial forearm free flap

auricular avulsion. There are several reports in the literature where prelamination techniques were used to reconstruct composite mandibular defects. Warnke et al. implanted xenograft blocks with BMP and BMAC into the latissimus dorsi muscle of a patient and then successfully used this construct to reconstruct a composite mandibular defect [26, 32, 34].

Prefabrication of a flap refers to the implantation of a vascular pedicle into a graft construct such as a custom-designed scaffold that aims to reconstruct a composite osseous defect. Recently, several papers have suggested that the creation of arteriovenous loop (AVL) and implantation of the fistula into the scaffold construct provide for improved vascular supply to the graft. This concept has been supported through several promising animal studies [26, 32, 35–38]. Furthermore, recent studies have demonstrated that such AVL-vascularized constructs are able to maintain viability after exposure to ionizing radiation, suggesting that these techniques may 1 day be used to reconstruct oncologic defects in patients who are likely to undergo adjuvant radiation [35].

17.4 Practical Applications

The theoretical workflow for creating an axillary vascularized construct depends on the clinical situation. If one were to create a prelaminated flap, this necessitates a secondary surgical procedure where the construct is implanted into a flap's vascular territory and time must be allotted for integration and neovascularization to take place [37, 38]. A second operation needs to take place to transplant the construct into the surgical defect. Thus, a major advantage of free tissue transfer, the ability to perform immediate reconstruction, is negated. One suggested workflow for prefabricated flaps using an AVL mitigates this issue. In an experiment conducted in a sheep animal model, Eweida was able to demonstrate the viability of creating an AVL utilizing the facial vessels at the time of mandibular ablation, to successfully revascularize an immediately implanted osseous scaffold [36]. Such a technique may show promise in humans. It remains to be seen how such a technique would address the reconstruction of a composite defect involving soft tissue loss as well as bone.

References

1. Spagnoli DB, Marx RE. Dental implants and the use of rhBMP-2. Oral Maxillofac Surg Clin North Am. 2011;23(2):61, vii. https://doi.org/10.1016/j.coms.2011.02.003.
2. Hidalgo DA, Pusic AL. Free-flap mandibular reconstruction: a 10-year follow-up study. Plast Reconstr Surg. 2002;110(2):1.
3. Marx RE, Harrell DB. Translational research: the CD34+ cell is crucial for large-volume bone regeneration from the milieu of bone marrow progenitor cells in craniomandibular reconstruction. Int J Oral Maxillofac Implants. 2014;29(2):201. https://doi.org/10.11607/jomi.te56.
4. Fiorellini JP, Howell TH, Cochran D, et al. Randomized study evaluating recombinant human bone morphogenetic Protein-2 for extraction socket augmentation. J Periodontol. 2005;76(4):605–13. https://doi.org/10.1902/jop.2005.76.4.605.
5. Triplett RG, Nevins M, Marx RE, et al. Pivotal, randomized, parallel evaluation of recombinant human bone morphogenetic protein-2/absorbable collagen sponge and autogenous bone graft for maxillary sinus floor augmentation. J Oral Maxillofac Surg. 2009;67(9):1947–60. https://doi.org/10.1016/j.joms.2009.04.085.
6. Carter TG, Brar PS, Tolas A, Beirne OR. Off-label use of recombinant human bone morphogenetic protein-2 (rhBMP-2) for reconstruction of mandibular bone defects in humans. J Oral Maxillofac Surg. 2008;66(7):1417–25. https://doi.org/10.1016/j.joms.2008.01.058.
7. Herford AS, Boyne PJ. Reconstruction of mandibular continuity defects with bone morphogenetic protein-2 (rhBMP-2). J Oral Maxillofac Surg. 2008;66(4):616–24. https://doi.org/10.1016/j.joms.2007.11.021.
8. Seto I, Asahina I, Oda M, Enomoto S. Reconstruction of the primate mandible with a combination graft of recombinant human bone morphogenetic protein-2 and bone marrow. J Oral Maxillofac Surg. 2001;59(1):3.
9. Sweeny L, Lancaster WP, Dean NR, et al. Use of recombinant bone morphogenetic protein 2 in free flap reconstruction for osteonecrosis of the mandible. J Oral Maxillofac Surg. 2012;70(8):1991–6. https://doi.org/10.1016/j.joms.2011.08.037.
10. Howard BK, Brown KR, Leach JL, Chang CH, Rosenthal DI. Osteoinduction using bone morphogenic protein in irradiated tissue. Arch Otolaryngol Head Neck Surg. 1998;124(9):985–8.
11. Roser SM, Ramachandra S, Blair H, et al. The accuracy of virtual surgical planning in free fibula mandibular reconstruction: comparison of planned and final results. J Oral Maxillofac Surg. 2010;68(11):2824–32. https://doi.org/10.1016/j.joms.2010.06.177.
12. Tarsitano A, Mazzoni S, Cipriani R, Scotti R, Marchetti C, Ciocca L. The CAD-CAM technique for mandibular reconstruction: an 18 patients oncological case-series. J Craniomaxillofac Surg. 2014;42(7):1460–4. https://doi.org/10.1016/j.jcms.2014.04.011.
13. Bao T, He J, Yu C, et al. Utilization of a pre-bent plate-positioning surgical guide system in precise mandibular reconstruction with a free fibula flap. Oral Oncol. 2017;75:133–9.
14. Carini F, Francesconi M, Saggese V, Monai D, Porcaro G. Implant-supported rehabilitation of a patient with mandibular ameloblastoma. Ann Stomatol (Roma). 2012;3(Suppl 2):21–5.
15. Kokorina NA, Zakharkin SO, Krebsbach PH, Nussenbaum B. Treatment effects of rhBMP-2 on invasiveness of oral carcinoma cell lines. Laryngoscope. 2011;121(9):1876–80. https://doi.org/10.1002/lary.21914.
16. Stosic S. Mandibular reconstruction--state of the art and perspectives. Vojnosanit Pregl. 2008;65(5):397–403.
17. Terry Sua RF. Microvascular reconstruction of the mandible: an argumentfor the fibula osteocutaneous free flap. Rev Esp Cir Oral Maxilofac. 2014;2(36):1–8.
18. Bahr W, Stoll P, Wachter R. Use of the "double barrel" free vascularized fibula in mandibular reconstruction. J Oral Maxillofac Surg. 1998;56(1):38–44.
19. Frodel JL, Funk GF, Capper DT, et al. Osseointegrated implants: a comparative study of bone thickness in four vascularized bone flaps. Plast Reconstr Surg. 1993;92(3):8.
20. Urken ML, Buchbinder D, Costantino PD, et al. Oromandibular reconstruction using microvascular composite flaps: report of 210 cases. Arch Otolaryngol Head Neck Surg. 1998;124(1):46–55.
21. Do AV, Khorsand B, Geary SM, Salem AK. 3D printing of scaffolds for tissue regeneration applications.

Adv Healthc Mater. 2015;4(12):1742–62. https://doi.org/10.1002/adhm.201500168.
22. Sachlos E, Czernuszka JT. Making tissue engineering scaffolds work. Review: the application of solid free-form fabrication technology to the production of tissue engineering scaffolds. Eur Cell Mater. 2003;5:40.
23. Wust S, Muller R, Hofmann S. Controlled positioning of cells in biomaterials-approaches towards 3D tissue printing. J Funct Biomater. 2011;2(3):119–54. https://doi.org/10.3390/jfb2030119.
24. Peltola SM, Melchels FP, Grijpma DW, Kellomaki M. A review of rapid prototyping techniques for tissue engineering purposes. Ann Med. 2008;40(4):268–80. https://doi.org/10.1080/07853890701881788.
25. Simon JL, Michna S, Lewis JA, et al. In vivo bone response to 3D periodic hydroxyapatite scaffolds assembled by direct ink writing. J Biomed Mater Res A. 2007;83(3):747–58. https://doi.org/10.1002/jbm.a.31329.
26. Eweida AM, Horch RE, Marei MK, et al. Axially vascularised mandibular constructs: is it time for a clinical trial? J Craniomaxillofac Surg. 2015;43(7):1028–32. https://doi.org/10.1016/j.jcms.2014.10.018.
27. Zan T, Li Q, Dong J, et al. Transplanted endothelial progenitor cells increase neo-vascularisation of rat pre-fabricated flaps. J Plast Reconstr Aesthet Surg. 2010;63(3):474–81. https://doi.org/10.1016/j.bjps.2008.11.076.
28. Li QF, Reis ED, Zhang WX, Silver L, Fallon JT, Weinberg H. Accelerated flap prefabrication with vascular endothelial growth factor. J Reconstr Microsurg. 2000;16(1):45–9.
29. Gurunluoglu R, Meirer R, Shafighi M, Huemer GM, Yilmaz B, Piza-Katzer H. Gene therapy with adenovirus-mediated VEGF enhances skin flap prefabrication. Microsurgery. 2005;25(5):433–41. https://doi.org/10.1002/micr.20142.
30. Haws MJ, Erdman D, Bayati S, Brown RE, Russell RC. Basic fibroblast growth factor induced angiogenesis and prefabricated flap survival. J Reconstr Microsurg. 2001;17(1):4.
31. Iwasawa M. Accelerated maturation in prefabricated flaps by transforming growth factor-beta: an experimental study in the rabbit. Ann Plast Surg. 1993;31(1):72–5.
32. Eweida AM, Nabawi AS, Elhammady HA, et al. Axially vascularized bone substitutes: a systematic review of literature and presentation of a novel model. Arch Orthop Trauma Surg. 2012;132(9):1353–62. https://doi.org/10.1007/s00402-012-1550-3.
33. Pribaz JJ, Fine NA. Prelamination: defining the prefabricated flap--a case report and review. Microsurgery. 1994;15(9):618–23.
34. Warnke PH, Wiltfang J, Springer I, et al. Man as living bioreactor: fate of an exogenously prepared customized tissue-engineered mandible. Biomaterials. 2006;27(17):3163–7.
35. Eweida A, Frisch O, Giordano FA, et al. Axially vascularized tissue-engineered bone constructs retain their in vivo angiogenic and osteogenic capacity after high-dose irradiation. J Tissue Eng Regen Med. 2016;12(2):e657–68. https://doi.org/10.1002/term.2336.
36. Kokemuller H, Jehn P, Spalthoff S, et al. En bloc prefabrication of vascularized bioartificial bone grafts in sheep and complete workflow for custom-made transplants. Int J Oral Maxillofac Surg. 2014;43(2):163–72. https://doi.org/10.1016/j.ijom.2013.10.013.
37. Eweida AM, Nabawi AS, Abouarab M, et al. Enhancing mandibular bone regeneration and perfusion via axial vascularization of scaffolds. Clin Oral Investig. 2014;18(6):1671–8. https://doi.org/10.1007/s00784-013-1143-8.
38. Leibig N, Wietbrock JO, Bigdeli AK, et al. Flow-induced axial vascularization: the arteriovenous loop in angiogenesis and tissue engineering. Plast Reconstr Surg. 2016;138(4):825–35. https://doi.org/10.1097/PRS.0000000000002554.

Part IV

Future Trends for Maxillomandibular Regeneration and Reconstruction

3D Printing and Adenosine Receptor Activation for Craniomaxillofacial Regeneration

18

Christopher D. Lopez, Lukasz Witek, Roberto L. Flores, Andrea Torroni, Eduardo D. Rodriguez, Bruce N. Cronstein, and Paulo G. Coelho

18.1 Background

Maxillofacial reconstruction has evolved over the past 15 years toward customized treatment plans which fulfill both functional and aesthetic outcomes of facial restoration. The merger of craniomaxillofacial and microvascular principles has given rise to tailored-specific free tissue flaps which reconstruct bony maxillofacial defects while rebuilding lining, soft tissue mass, and facial subunits—all of which are key to achieving outcomes that approach normalcy. In addition, surgical planning technology has been able to expedite functional recovery: flap-based reconstruction of the mandible and maxilla with preplanned dental prosthesis can be completed in one operation instead of requiring rehabilitation periods that can take up to a year [1, 2].

While only a small subset of facial trauma requires free tissue transfer, the severity of these cases has been the impetus for the recent advances in craniomaxillofacial and microvascular principles. Unfortunately, the contemporary techniques for complex maxillofacial reconstruction remain limited: autologous bone transfer is complicated by limited bone stock and shape, donor site morbidity, surgical site infection, delayed healing, long operative times, and cost [3, 4]. In appreciation of these limitations, advances in bone tissue engineering have sought to restore maxillofacial deficiencies using customizable constructs commonly augmented with osteogenic agents and/or cellular therapy. The numerous bone constructs described, coupled with the paucity of clinically available tissue engineering-based therapies, highlight the importance of highly translational approaches that consider the challenges faced when attempting bench-to-bedside deployment.

C. D. Lopez · P. G. Coelho (✉)
Hansjörg Wyss Department of Plastic Surgery, NYU Langone Health, New York, NY, USA

Department of Biomaterials and Biomimetics, NYU College of Dentistry, New York, NY, USA
e-mail: christopher.lopez2@nyumc.org; Pc92@nyu.edu

L. Witek
Department of Biomaterials and Biomimetics, NYU College of Dentistry, New York, NY, USA
e-mail: Lw901@nyu.edu

R. L. Flores · A. Torroni · E. D. Rodriguez
Hansjörg Wyss Department of Plastic Surgery, NYU Langone Health, New York, NY, USA
e-mail: Roberto.Flores2@nyumc.org; Andrea.Torroni@nyumc.org; Eduardo.rodriguez@nyumc.org

B. N. Cronstein
Department of Medicine, NYU Langone Health, New York, NY, USA
e-mail: Bruce.cronstein@nyumc.org

J. C. Melville et al. (eds.), *Regenerative Strategies for Maxillary and Mandibular Reconstruction*,
https://doi.org/10.1007/978-3-319-93668-0_18

In this chapter, we will review recent developments our laboratory has made in materials science, osteogenic stimulation, and the integration of these components into an informed tissue engineering construct for maxillofacial bony restoration.

18.2 Advances in Material Science

Calcium phosphate-based biomaterials have been used clinically for decades and have well-established biocompatibility and safety properties [5, 6]. As a result, these biomaterials are used as bone graft substitutes in over half of all grafting procedures in the United States annually [7]. While outcomes are generally favorable for small-scale bony defects, large defects require an interface between the implant and bone margin that facilitates osteoconduction and implant absorption. This is not possible with granular or en bloc materials which are easily manufactured in high quantities and, consequently, cannot be customized to address unique anatomic defects in a personalized fashion. Additionally, they lack both the geometric stability, spatial coordination, and degradation kinetics necessary to maximize bone formation through osteogenesis and osseoconduction.

18.2.1 Geometric Considerations and the Role of 3D Printing in Scaffold Design

The concept that geometric design of biomaterials can alter bone healing is relatively new. Over the last decade, reports investigating bone healing in endosteal metallic implants have determined that osseoconduction is dependent on many physicochemical factors that can be manipulated. This includes macrogeometric-, micrometer-, and nanometer-level changes in implantable device physical dimensions as well as its relationship to the osteotomy walls created by surgical instrumentation [8–11]. These analyses of implant geometry have elucidated that early and late stages of bone healing are directly related to implantable device design and can be leveraged to maximize bone formation. More recently, these principles have seen successful translation to other biomaterials that are even more suited to facilitate bone formation: calcium phosphate-based bioactive ceramics [5, 12–14].

Although calcium phosphate-based bio-ceramics typically exist in powder form, the advent of biomedical applications for 3D printing has yielded modalities capable of printing these bio-ceramics as an ink/slurry capable of holding form. As a result, 3D printing has become a powerful tool capable of printing calcium phosphate-based bio-ceramics in any form, making it a tool of significant interest to tissue engineering and surgery communities.

Three-dimensional printing can create a personalized scaffold macrogeometry based on clinical imaging and thus offers a valuable approach to patient-specific bony defects. By using computer-aided design and manufacturing (CAD/CAM), scaffold constructs can be designed to any bulk shape and size as well as a porous design (Fig. 18.1a–c). Despite the existence of 3D printing for biomedical applications for over a decade, bone

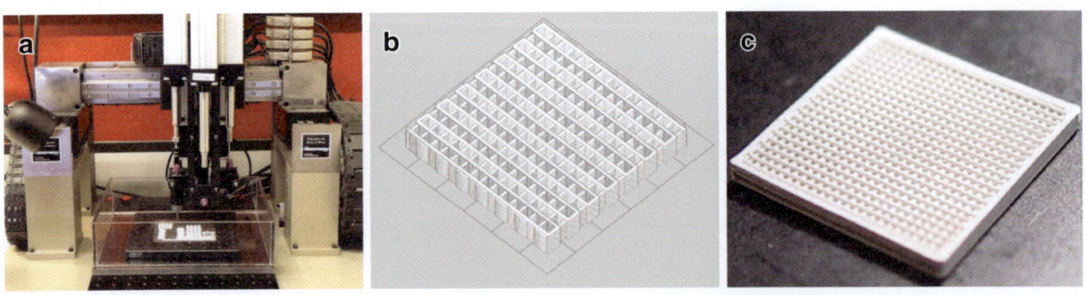

Fig. 18.1 3D printing. (**a**) Printer, (**b**) CAD/CAM software design of scaffold, and (**c**) scaffold after printing

regeneration using 3D printing has remained mixed in outcomes until recently [15–29]. It is suggested that these results are consequent to approaches to 3D printing that do not take advantage of the lessons learned from endosteal implant bone healing. A review of factors that affect bone healing in metallic biomaterials clearly cites that dimensions ranging from macrogeometric to nanogeometric all merit consideration. Only recently have these dimensions have been investigated through 3D printing [5, 9] in a stepwise fashion.

18.2.2 β-Tricalcium Phosphate

Historically, the investigation of bio-ceramics for bone healing has been predominantly focused on hydroxyapatite (HA). HA is the main inorganic component of the bone, making its study a logical choice. Furthermore, HA has favorable bone regenerative properties such as biocompatibility and osseoconductivity. However, it also has less desirable properties, such as unfavorable resorption kinetics. When designing scaffolds for bone regeneration, an ideal construct should ideally facilitate bone formation at a defect and then resorb/degrade without operative intervention. Unfortunately, HA has a low resorption rate in vivo of approximately 1–2% per year at 5 years post-implantation [30], thereby limiting complete bony regeneration.

β-Tricalcium phosphate (β-TCP), on the other hand, has far more rapid resorption/degradation kinetics both in vitro and in vivo when compared to HA. A biomaterial that is also highly osseoconductive and biocompatible, β-TCP can facilitate bone regeneration while also demonstrating favorable resorption that happens in tandem with bone formation. This unique material property can be manipulated to a significant degree with modalities such as 3D printing: when designing scaffold interstices, the rate of degradation can be altered by mesogeometric changes in lattice porosity size, strut circumference, or any other change that alters the surface area. At the nanogeometric level, alterations in ink formulation or sintering that alter porosity can also be engineered to tailor the scaffold degradation rate.

18.3 Bioactive Molecules

18.3.1 Adenosine Receptors

A wide array of biologic agents have been investigated in order to alter bone metabolism in favor of bone formation/healing. Of note, the recently described purinergic receptors have demonstrated promising bone regenerative potential [13, 31–35].

Adenosine has been termed the "retaliatory metabolite" for its unique physiologic role: it is secreted in response to ATP depletion, serves as a marker of metabolic status at the cellular level, and is known to attenuate activity for a wide array of cell types as a protective mechanism. For example, unstressed tissues have an extracellular adenosine concentration of ~1 μM/L, but in septic patients, this number can increase tenfold [36]. When released, adenosine binds to receptors on the surface of immune cells and regulates exuberant immune responses. A practical example is adenosine binding to receptors on neutrophils and inducing the suppression of superoxide anion [37]. These regulatory effects have been reported for almost a century: Drury and Szent-Gyorgyi reported in 1929 that adenosine induced a protective vasodilatory and negative inotropic effect on stressed cardiac vessels and tissue. Today, this mechanism is taken advantage of for cardiac stress testing.

Adenosine bioavailability is determined by regulated, interdependent cellular processes [38]. For example, in the setting of tissue hypoxia/ischemia, purinergic metabolic pathways increase dephosphorylation of ATP to adenosine inside the cell via 5′ nucleotidase and suppress adenosine kinase activity, the enzyme needed to rephosphorylate adenosine [38]. Extracellular adenosine accumulates by two suggested mechanisms: the first involves precursor nucleotides (e.g., ATP, ADP, AMP) released from cells and catabolized into adenosine by several ectonucleosidases. The latter involves intracellular adenosine being shunted extracellularly by specialized nucleoside transporters [38].

There are four adenosine receptors: A1, A2A, A2B, and A3. A1 and A3 receptors are coupled to either Gi-coupled signal transduction proteins or

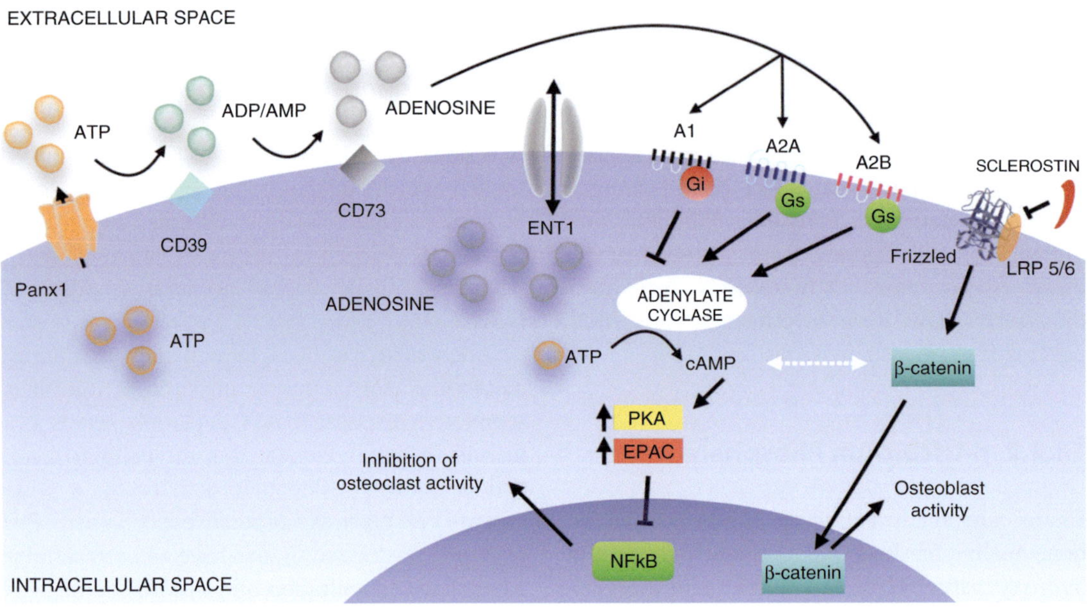

Fig. 18.2 Illustration of adenosine receptor intracellular pathway

ion channels, and A2 receptors are G-alpha-s-linked receptors that stimulate adenyl cyclase/cAMP [39]. Adenosine receptors have effects on both osteoblast and osteoclast syntheses (Fig. 18.2) [40]. In brief, the following have been reported:

1. A1R ligation plays a role in osteoclast formation: In its absence or antagonism, defects have been reported to arise in CSF-1 (colony-stimulating factor 1) and RANK-L (receptor activator of nuclear factor κ-B ligand), thus A1R activation induces osteoclastogenesis both in vitro and in vivo [41].
2. A2AR activation inhibits osteoclastogenesis and A2AR antagonism activates it [42]. Also, osteoblast function is positively modulated by A2AR activation [43]. A2ARs seem to play a role in establishing osteoblast maturation and the maintenance of osteoblast phenotype [44]. A2AR gene and protein expression is upregulated at later stages of osteoblast differentiation [44] but is less present at earlier stages (e.g., undifferentiated stages of mesenchymal stem cells)—Furthermore, A2AR ligation does not alter the gene expression of markers of bone formation, such as *RUNX-2* or *ALP* at earlier stages, but it does upregulate osteogenesis at later stages of differentiation [44].
3. A2BR activation likely plays the most prevalent role in human primary osteoblast-like cells. Osteogenic differentiation seems to be most dramatic under the influence of A2B agonism [45], and it has been suggested that the significance of A2BRs lies in establishing commitment and differentiation of stem cells toward osteoblast lineage. A2BRs are the predominant adenosine receptor subtype expressed in undifferentiated mesenchymal stem cells (MSCs) and upon activation promote osteoblast lineage marker upregulation, osteoblastogenesis, and increases in mineralization.
4. A3AR has not been shown to directly affect osteoclastogenesis or osteoblast bone deposition when stimulated or blocked. However, A3 agonism has been shown to downregulate bone resorption and the number of osteoclasts in murine arthritis [46].

It is critical to note that the adenosine receptor-derived changes in bone homeostasis only occur at supraphysiologic concentrations [45]. Non-stressed cellular environments do not accumulate sufficient adenosine extracellularly to activate any of the aforementioned receptors—even with continuous blockage of the adenosine deaminase enzyme.

These supraphysiologic concentrations can be achieved with pharmacologic agents. An example that has garnered significant interest is dipyridamole, because of its well-established, decades-long safe clinical use in cardiac stress testing and antiplatelet therapy, even in pediatric patient populations [47–49]. Dipyridamole blocks adenosine uptake into the cell via type 1 equilibrative nucleoside transporter (ENT1) and thereby ensures extracellular accumulation.

18.4 The Integration of Tissue Engineering Principles

The bone regenerative capacity of adenosine receptor ligation is ideally leveraged via delivery to defect sites while avoiding systemic effects. This principle of localized drug delivery directly to injury sites is critical in order to avoid unintended effects resulting from activation of other systems—a likely event given the prevalent expression of the adenosine receptor. Furthermore, local delivery simplifies release kinetics and circumvents issues such as patient compliance with medication. Finally, the concentration of dipyridamole needed to affect alterations of bone metabolism is much lower when delivered locally as opposed to systemically, and this local delivery dosage is much lower than systemic concentrations already approved for adult and pediatric patients. Therefore, the local delivery model of dipyridamole may be well suited for future clinical deployment in conjunction with a bone tissue engineering scaffold.

A small but growing body of work demonstrates the effects of localized adenosine receptor activation via 3D-printed, geometrically tailored osseoconductive scaffold carriers.

18.4.1 Calvarial Defect Regeneration: Translation Between Species

The earliest *in vivo* report of localized adenosine receptor delivery for bone healing was by Mediero and colleagues in 2015. A collagen sponge replaced a murine calvarial defect, and it was found that daily administration of dipyridamole regenerated the bone as well as BMP-2 [34]. Although these sponges supported bioactive molecule effects, they lacked the structural support necessary to fit and fill large bony defects and facilitate cellular communication and did not contribute regenerative properties alone. As a result, dipyridamole was more recently combined with 3D-printed bio-ceramic scaffolds which independently demonstrated osseoconductive properties and could be customized to geometrically restore any bony defect. A2A receptor activation was determined to be a critical mechanism of action for bone: in these same murine models, A2A receptor knockouts failed to regenerate cranial bone defects, while controls healed well, highlighting the critical role the A2A receptor plays in bone regeneration [50].

These same principles were also successfully demonstrated in even larger, more translational models: the sheep calvaria. The sheep had two ipsilateral trephine-induced (11-mm-diameter) calvarial defects at $t = 0$, and a second pair of trephines induced contralateral at $t = 3$ weeks. Scaffolds were coated in either 100 μM dipyridamole and a collagen carrier bound to the scaffold via cross-linking or the collagen carrier alone. Each side received one of each scaffold. Following euthanization for two end points ($t = 3$ weeks and $t = 6$ weeks), scaffolds were evaluated through micro CT and non-decalcified histology. Both uncoated and dipyridamole-coated scaffolds exhibited regenerative properties, but dipyridamole significantly augmented this healing at both time points. Interestingly, A2AR activation seemed to have recruited the intrinsic osteogenic capacity of the local dura mater [13]. This capacity was previously observed by others [51].

18.4.2 Load-Bearing Translational Preclinical Models

The regenerative capacity of uncoated bioceramic scaffolds with deliberate geometric design has been demonstrated. Full-thickness, unilateral mandibulectomies ~12 mm in length (compared to 10 mm partial-thickness defects qualifying as critical-sized) were performed on adult New Zealand white rabbits at the segmental mandibular body and replaced with 3D-printed scaffolds that replaced the defect in size and shape (Fig. 18.3a). At 8 weeks in vivo, β-tricalcium phosphate scaffolds demonstrated bone regeneration on gross examination (Fig. 18.3b). This significant degree of osseoconduction bridged the defect span through initial extensive woven bone formation despite immediate return to masticatory function. Micro-CT imaging confirmed significant amounts of bone growth, and histology confirmed that at 8 weeks, intramembranous-like healing pathways are involved, depicting highly cellular, vascularized woven bone structure directly in contact with scaffold struts and lumen (Fig. 18.4a–c). Higher magnification imaging depicts resorbing scaffold along with areas of lamellar bone formation around primary osteonic structures. Electron microscopy illustrates intramembranous-like healing with entire woven to lamellar reorganization spectrum, scaffold resorption (Fig. 18.5), and formation of concentric rings. Microcracks that correspond to bone remodeling are also seen.

These 3D-printed scaffolds have also demonstrated regenerative capacity, both with and without dipyridamole coating, at another part of the rabbit mandible: the load-bearing, relatively avascular ramus. Critical-sized, full-thickness partial mandibulectomies of NZWR rami were performed and replaced with one of the three scaffold types: uncoated, 100 μM dipyridamole carried via collagen cross-linking to scaffold, and collagen carrier alone. Like the segmental body scaffold, these scaffolds were designed to precisely fit and fill the defect made (Fig. 18.6). At a point of $t = 8$ weeks, bone regeneration was markedly enhanced with the addition of dipyridamole at the ramus, though all uncoated scaffolds and collagen-coated scaffolds did have bone growth with remodeling sites, vascular supply, and both immature woven bone and reorganized lamellar bone (Fig. 18.7). Scaffold degradation was also most extensive with dipyridamole coating, likely because of the rebounding effects after the transient changes dipyridamole has on the bone. Since normal bone homeostasis must eventually be reestablished, the effects of A2AR activation on osteoclast activity attenuation is likely followed by an increase in activity once dipyridamole concentrations become too low to affect bone metabolism [34]. Of note, while collagen coating functions to carry dipyridamole, collagen alone seems to limit the early stages of bone remodeling. It is likely that collagen impedes cellular communication between osteogenic cells that induce regeneration and the highly osseoconductive scaffold material. It has also been observed

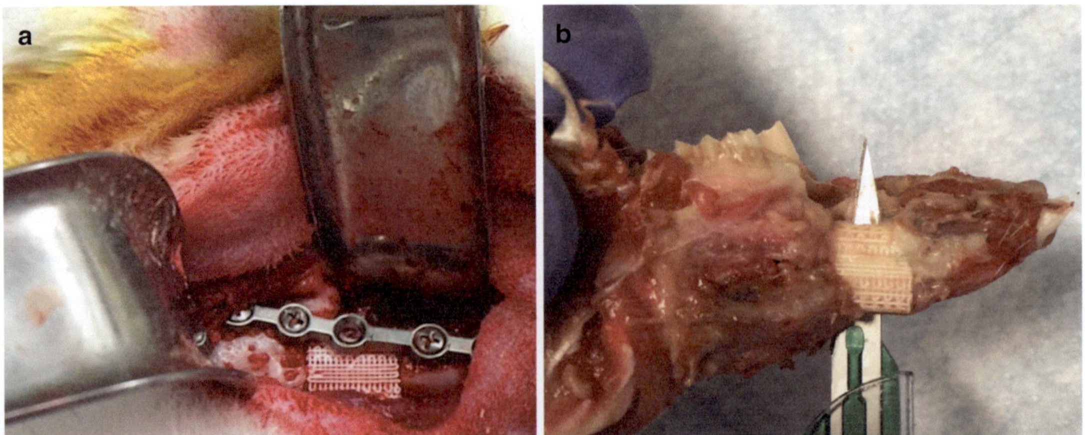

Fig. 18.3 (**a**) Intraoperative scaffold placement with plating for support and (**b**) scaffold integration with rabbit mandible at $t = 8$ weeks after necropsy

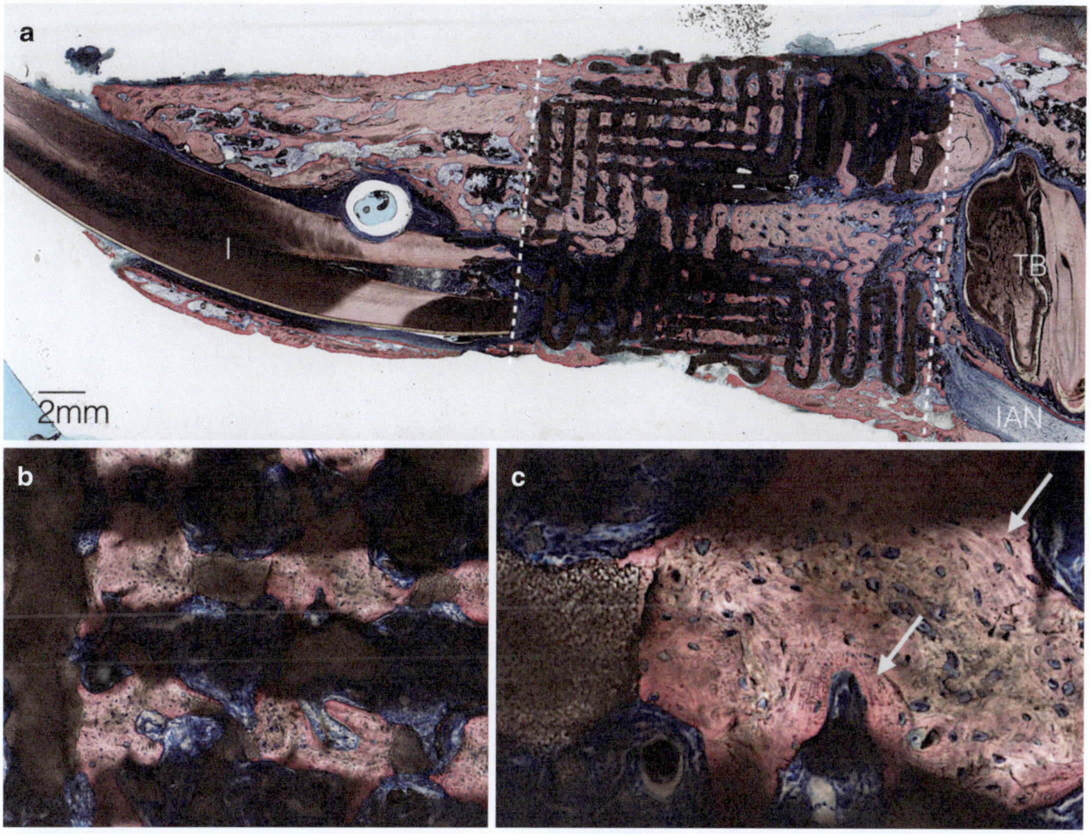

Fig. 18.4 (**a**) Sagittal histologic slice of scaffold in continuity with rabbit mandible. (**b**) High magnification from (**a**) demonstrating porous ingrowth. (**c**) Highly cellular and vascularized woven bone structure, as well as newer, organized lamellar bone formation depicted by arrows (*I* Incisor, *T* tooth, *IAN* inferior alveolar nerve)

that dipyridamole actually facilitates a different pattern of healing than uncoated or collagen-coated scaffolds at the ramus: while it is expected that partial ramus mandibulectomies have the greatest regenerative potential at sites closest in proximity to three-wall defects, only dipyridamole-coated scaffolds demonstrated this healing pattern. Without dipyridamole, uncoated scaffolds and collagen-coated scaffold induced bone formation in uncoordinated pockets.

18.4.3 Skeletally Immature Translational Models

The challenges for regenerating the skeletally immature bone are different than that of a skeletally mature bone. Alveolar cleft defects of the primary palate create structural instability of the maxillary arch, an inability to support tooth eruption, and facial asymmetry. Furthermore, current tissue engineering-based therapies must consider the growing maxillofacial skeleton and therefore cannot introduce therapies that will potentially induce premature suture fusion.

Skeletally immature NZWRs underwent unilateral, ~3.5 mm × ~3.5 mm alveolar cleft defect injury visualized via a direct transfacial approach (Fig. 18.8a, b). Defects created without scaffold intervention served as negative controls, and scaffolds were coated with different concentrations of dipyridamole, 100, 1000, or 10,000 μm, all carried by collagen cross-linked to scaffolds. After $t = 8$ weeks, animals were euthanized, and it was observed that the regenerative properties of 3D-printed scaffolds with dipyridamole induce

Fig. 18.5 Electron microscopy depicting scaffold struts (green arrows) and new, intramembranous like bone formation throughout scaffold interstices (red arrows). New, woven bone is seen filling sites of scaffold degradation (yellow arrows), and regions of lamellar reorganization juxtaposed with immature woven bone are evident (blue arrows, lamellar as brighter bone, woven is darker)

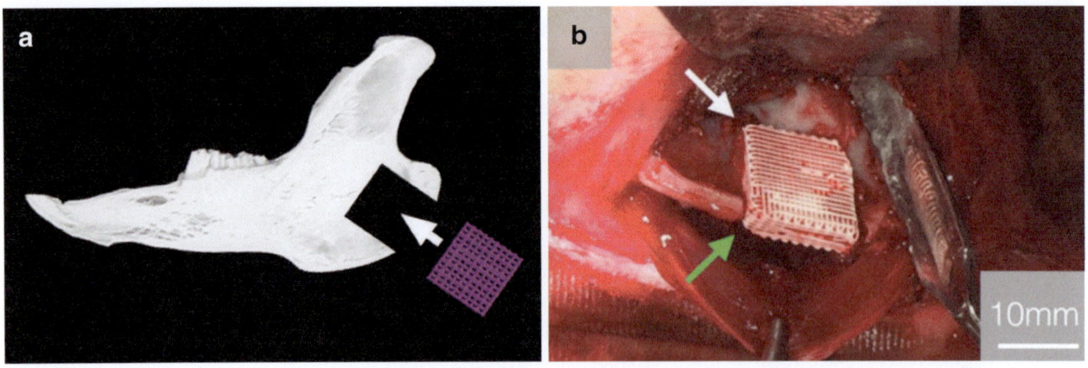

Fig. 18.6 (**a**) Surgical injury model depicting scaffold (purple) insertion site with arrow and (**b**) intra operative photo of scaffold placement at ramus

dose-dependent bone formation at the alveolar cleft in rabbits (Fig. 18.8c). Perhaps most importantly, even at doses of 1–2 logarithmic increases more than needed to increase bone formation, maxillary and sutures remained patent (Fig. 18.8c, d). Healing was both intramembranous-like and endochondral-like, with highly cellular and vascularized structure throughout scaffold porosity (Fig. 18.9).

18.5 Conclusions and Future Directions

The field of surgery arose from the need to visualize the treatment of disease, and its initial challenges were predominantly based on ensuring technical ability. Today's clinicians face a new challenge: the explosion of medical knowledge and innovation that continues to grow every day.

Fig. 18.7 Optical micrographs of scaffold inserted into ramus after 8 weeks. Bone (pink) on either side of scaffold depicts original ramus bone thickness and scaffold (black) demonstrates directional bone growth throughout porosity. (**a**) Control group; (**b**) COLL group; (**c**) DIPY group

In a world where a plethora of biomaterials are readily available for reconstructive purposes, it is no longer sufficient to be a technically skilled surgeon. A strong understanding of regenerative materials available clinically, as well as those with significant translational potential, is essential for sound practice, surgical judgment, and innovative patient care. From linen sutures used in Ancient Egypt to custom, patient-specific 3D-printed bone-cutting guides used in facial transplantation today, biomaterials have a long, rich history in aiding human healing, and boundaries continue to be pushed.

Tissue engineering craniomaxillofacial bone has been the dream of reconstructive surgeons for over a quarter of a century. Unfortunately, early efforts yielded disappointing results of either limited volume or poor bone quality. The clinical potential of tissue-engineered craniomaxillofacial regeneration has yet to be realized, but highly translational constructs are under active investigation. Three-dimensional printing bioactive ceramic scaffolds with deliberate geometries for osseoconduction has provided regenerative cells with physical pathways to communicate with and fill defects by healing, and the porosity is also permissive to other variables that contribute to bone healing, such as blood clots rich in osteogenic factors [52]. For the first time, significant bone healing responses to load-bearing bones such as the body and ramus of the mandible are being reported. A2AR activation appears to stimulate local dura tissue to contribute to regeneration, and alveolar clefts respond to dipyridamole delivery in a dose-dependent manner without any changes to maxillary suture biology at 8 weeks.

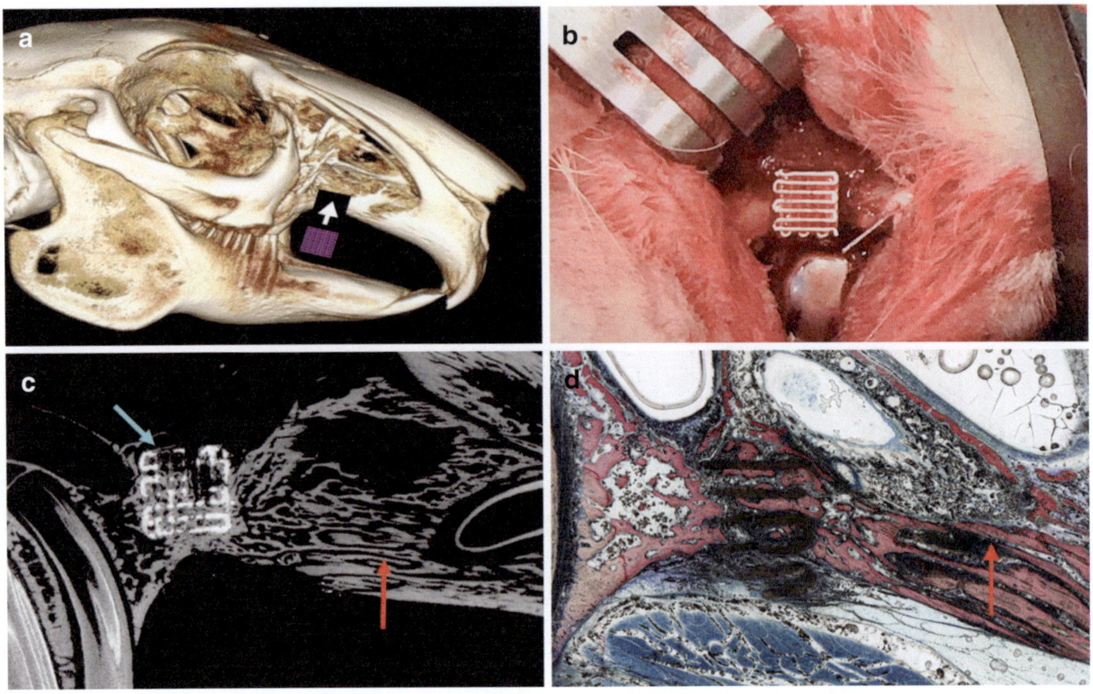

Fig. 18.8 Alveolar cleft model in immature rabbit maxilla at $t = 8$ weeks. (**a**) Schematic depicting surgical injury model with scaffold (purple) insertion site. (**b**) Intraoperative photo of scaffold insertion. (**c**) MicroCT slice at $t = 8$ weeks with scaffold (blue arrow) and patent sutures (red arrow). (**d**) Non-decalcified histologic section of scaffold (black), new bone growth within scaffold (pink) and patent sutures (red arrow)

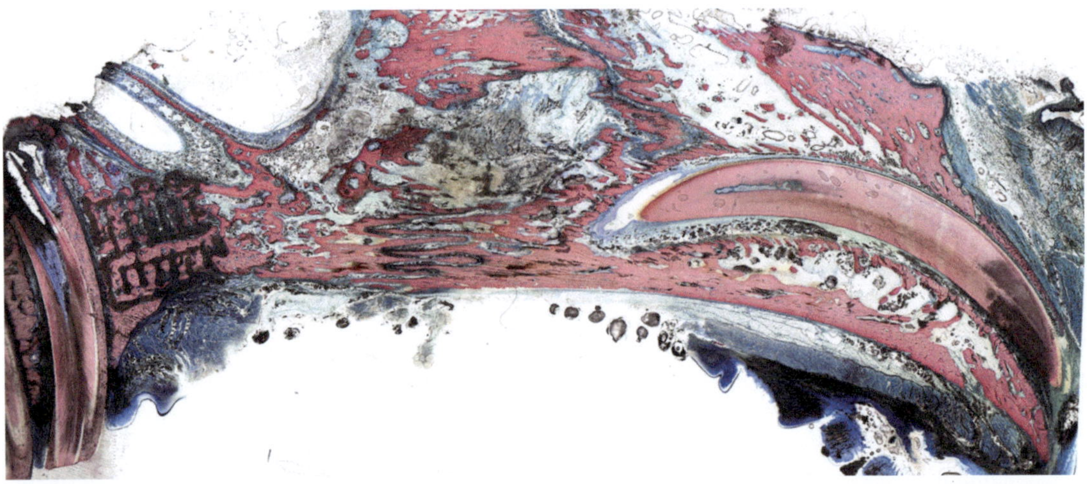

Fig. 18.9 Alveolar cleft with scaffold inserted and bone regeneration at $t = 8$ weeks. Scaffold (black) lattice with bone (pink) growth throughout porosity demonstrating osseoconduction via intramembranous and endochondral-like healing. Suture potency is noted as well in higher magnification (yellow arrow)

Favorable outcomes have been demonstrated by utilizing principles of biomaterial design alone, as well as with the addition of A2AR activation. While groups have also attempted to address the challenge of large bony defect regeneration through stem cell investigation and molecules such as bone morphogenetic proteins [53, 54], outcomes have been inconsistent. Unfortunately, clinical deployment of stem cell therapies has not yet demonstrated desired outcomes, and from a regulatory standpoint, they remain challenging to use and ethically debated. An approach to regeneration such as that presented in this chapter, with both biomaterials and bioactive molecules proven to be safe, which capitalize on endogenous healing mechanisms to work, has significant promise for clinical use and warrants further investigation.

Acknowledgments We would like to wholeheartedly thank Dr. Joseph G. McCarthy for his endorsement of our work and contributions to this chapter. His insight on advances and setbacks in craniomaxillofacial surgery over the last quarter of a century provides us with the unique opportunity to focus on clinically relevant challenges that have challenged reconstructive surgeons for decades.

The work presented in this chapter was supported by NIH/NIAMS 5R01AR068593-02, 3R01AR068593-02S1, 5R01AR068593-03, & 3R01AR068593-03S1, NIH/NICHD R21HD090664-01, and DoD W81XWH-16-1-0772. Drs. Coelho and Cronstein are co-inventors of the 3D printing technology presented in this chapter.

References

1. Levine JP, Bae JS, Soares M, Brecht LE, Saadeh PB, Ceradini DJ, et al. Jaw in a day: total maxillofacial reconstruction using digital technology. Plast Reconstr Surg. 2013;131(6):1386–91.
2. Runyan CM, Sharma V, Staffenberg DA, Levine JP, Brecht LE, Wexler LH, et al. Jaw in a day: state of the art in maxillary reconstruction. J Craniofac Surg. 2016;27(8):2101–4.
3. Hidalgo DA. Condyle transplantation in free flap mandible reconstruction. Plast Reconstr Surg. 1994;93(4):770–81.
4. Hidalgo DA, Rekow A. A review of 60 consecutive fibula free flap mandible reconstructions. Plast Reconstr Surg. 1995;96(3):585–96.
5. Simon JL, Michna S, Lewis JA, Rekow DE, Thompson VP, Smay JE, et al. In vivo bone response to 3D periodic hydroxyapatite scaffolds assembled by direct ink writing. J Biomed Mater Res A. 2007;83A(3):747–58.
6. Inzana JA, Olvera D, Fuller SM, Kelly JP, Graeve OA, Schwarz EM, et al. 3D printing of composite calcium phosphate and collagen scaffolds for bone regeneration. Biomaterials. 2014;35(13):4026–34.
7. Roberts TT, Rosenbaum AJ. Bone grafts, bone substitutes and orthobiologics: the bridge between basic science and clinical advancements in fracture healing. Organogenesis. 2012;8(4):114–24.
8. Steigenga JT, Al-Shammari KF, Nociti FH, Misch CE, Wang H-L. Dental implant design and its relationship to long-term implant success. Implant Dent. 2003;12(4):306–17.
9. Wilson CE, de Bruijn JD, van Blitterswijk CA, Verbout AJ, Dhert WJA. Design and fabrication of standardized hydroxyapatite scaffolds with a defined macro-architecture by rapid prototyping for bone-tissue-engineering research. J Biomed Mater Res A. 2004;68A(1):123–32.
10. Jimbo R, Anchieta R, Baldassarri M, Granato R, Marin C, Teixeira HS, Tovar N, Vandeweghe S, Janal MN, Coelho PG. Histomorphometry and bone mechanical property evolution around different implant systems at early healing stages: an experimental study in dogs. Implant Dent. 2013;22(6):596–603.
11. Coelho PG, Jimbo R. Osseointegration of metallic devices: current trends based on implant hardware design. Arch Biochem Biophys. 2014;561:99–108.
12. Ishack S, Mediero A, Wilder T, Ricci JL, Cronstein BN. Bone regeneration in critical bone defects using three-dimensionally printed β-tricalcium phosphate/hydroxyapatite scaffolds is enhanced by coating scaffolds with either dipyridamole or BMP-2. J Biomed Mater Res B Appl Biomater. 2017;105(2):366–75.
13. Bekisz JM, Flores RL, Witek L, Lopez CD, Runyan CM, Torroni A, et al. Dipyridamole enhances osteogenesis of three-dimensionally printed bioactive ceramic scaffolds in calvarial defects. J Craniomaxillofac Surg. 2018;46(2):237–44.
14. Lopez CD, Diaz-Siso JR, Witek L, Bekisz JM, Cronstein BN, Torroni A, et al. Three dimensionally printed bioactive ceramic scaffold osseoconduction across critical-sized mandibular defects. J Surg Res. 2018;223:115–22.
15. Seitz H, Rieder W, Irsen S, Leukers B, Tille C. Three-dimensional printing of porous ceramic scaffolds for bone tissue engineering. J Biomed Mater Res B Appl Biomater. 2005;74((2):782–8.
16. Cai S, Xi J, Chua CK. A novel bone scaffold design approach based on shape function and all-hexahedral mesh refinement. Methods Mol Biol. 2012;868:45–55.
17. Castilho M, Dias M, Gbureck U, Groll J, Fernandes P, Pires I, et al. Fabrication of computationally designed scaffolds by low temperature 3D printing. Biofabrication. 2013;5(3):035012.
18. Wang X, Schröder HC, Müller WE. Enzymatically synthesized inorganic polymers as morphogenetically active bone scaffolds: application in regenerative medicine. Int Rev Cell Mol Biol. 2014;313:27–77.
19. Zhou Z, Buchanan F, Mitchell C, Dunne N. Printability of calcium phosphate: calcium sulfate powders for the application of tissue engineered bone scaffolds using the 3D printing technique. Mater Sci Eng C. 2014;38:1–10.
20. Farzadi A, Solati-Hashjin M, Asadi-Eydivand M, Osman NAA. Effect of layer thickness and printing

orientation on mechanical properties and dimensional accuracy of 3D printed porous samples for bone tissue engineering. PLoS One. 2014;9(9):e108252.
21. Yao Q, Wei B, Guo Y, Jin C, Du X, Yan C, et al. Design, construction and mechanical testing of digital 3D anatomical data-based PCL–HA bone tissue engineering scaffold. J Mater Sci Mater Med. 2015;26(1):1–9.
22. Cox SC, Thornby JA, Gibbons GJ, Williams MA, Mallick KK. 3D printing of porous hydroxyapatite scaffolds intended for use in bone tissue engineering applications. Mater Sci Eng C. 2015;47:237–47.
23. Hollister SJ, Flanagan CL, Zopf DA, Morrison RJ, Nasser H, Patel JJ, et al. Design control for clinical translation of 3D printed modular scaffolds. Ann Biomed Eng. 2015;43(3):774–86.
24. Luo Y, Zhai D, Huan Z, Zhu H, Xia L, Chang J, et al. Three-dimensional printing of hollow-struts-packed bioceramic scaffolds for bone regeneration. ACS Appl Mater Interfaces. 2015;7(43):24377–83.
25. Brunello G, Sivolella S, Meneghello R, Ferroni L, Gardin C, Piattelli A, et al. Powder-based 3D printing for bone tissue engineering. Biotechnol Adv. 2016;34(5):740–53.
26. Xinning Y, Jinghua F, Jianyang L, Xianyan Y, Dongshuang H, Zhongru G, et al. [Fabrication of bioactive tissue engineering scaffold for reconstructing calcified cartilage layer based on three-dimension printing technique]. Zhejiang da xue xue bao Yi xue ban= Journal of Zhejiang University Medical. Sciences. 2016;45(2):126–31.
27. Cooke MN, Fisher JP, Dean D, Rimnac C, Mikos AG. Use of stereolithography to manufacture critical-sized 3D biodegradable scaffolds for bone ingrowth. J Biomed Mater Res B Appl Biomater. 2003;64(2):65–9.
28. Gonçalves EM, Oliveira FJ, Silva RF, Neto MA, Fernandes MH, Amaral M, et al. Three-dimensional printed PCL-hydroxyapatite scaffolds filled with CNTs for bone cell growth stimulation. J Biomed Mater Res B Appl Biomater. 2016;104(6):1210–9.
29. Provaggi E, Leong JJ, Kalaskar DM. Applications of 3D printing in the management of severe spinal conditions. Proc Inst Mech Eng H J Eng Med. 2017;231(6):471–86.
30. Moore WR, Graves SE, Bain GI. Synthetic bone graft substitutes. ANZ J Surg. 2001;71(6):354–61.
31. Costa MA, Barbosa A, Neto E, Sa-e-Sousa A, Freitas R, Neves JM, et al. On the role of subtype selective adenosine receptor agonists during proliferation and osteogenic differentiation of human primary bone marrow stromal cells. J Cell Physiol. 2011;226(5):1353–66.
32. Mediero A, Frenkel SR, Wilder T, He W, Mazumder A, Cronstein BN. Adenosine A2A receptor activation prevents wear particle-induced osteolysis. Sci Transl Med. 2012;4(135):135ra65.
33. Mediero A, Cronstein BN. Adenosine and bone metabolism. Trends Endocrinol Metab. 2013;24(6):290–300.
34. Mediero A, Wilder T, Perez-Aso M, Cronstein BN. Direct or indirect stimulation of adenosine A2A receptors enhances bone regeneration as well as bone morphogenetic protein-2. FASEB J. 2015;29(4):1577–90.
35. Mediero A, Wilder T, Reddy VS, Cheng Q, Tovar N, Coelho PG, et al. Ticagrelor regulates osteoblast and osteoclast function and promotes bone formation in vivo via an adenosine-dependent mechanism. FASEB J. 2016;30(11):3887–900.
36. Martin C, Leone M, Viviand X, Ayem ML. High adenosine plasma concentration as a prognostic index for outcome in patients with septic shock. Crit Care. 2000;28(9):3198–202.
37. Cronstein BN, Kramer SB, Weissmann G, Hirschhorn R. A new physiological function for adenosine: regulation of superoxide anion production. Trans Assoc Am Phys. 1983;96:384–91.
38. Haskó G, Cronstein BN. Adenosine: an endogenous regulator of innate immunity. Trends Immunol. 2004;25(1):33–9.
39. Haskó G, Pacher P, Deitch EA, Vizi SE. Shaping of monocyte and macrophage function by adenosine receptors. Pharmacol Ther. 2007;113(2):264–75.
40. Evans BAJ, Elford C, Pexa A, Francis K, Hughes AC, Deussen A, et al. Human osteoblast precursors produce extracellular adenosine, which modulates their secretion of IL-6 and osteoprotegerin. J Bone Miner Res. 2006;21(2):228–36.
41. Kara FM, Doty SB, Boskey A, Goldring S, Zaidi M, Fredholm BB, et al. Adenosine A1 receptors regulate bone resorption in mice: adenosine A1 receptor blockade or deletion increases bone density and prevents ovariectomy-induced bone loss in adenosine A1 receptor–knockout mice. Arthritis Rheum. 2010;62(2):534–41.
42. Mediero A, Kara FM, Wilder T, Cronstein BN. Adenosine A(2A) receptor ligation inhibits osteoclast formation. Am J Pathol. 2012;180(2):775–86.
43. Mediero A, Wilder T, Perez-Aso M, Cronstein BN. Direct or indirect stimulation of adenosine A2A receptors enhances bone regeneration as well as bone morphogenetic protein-2. FASEB Journal. 2015;29(4):1577–90.
44. Gharibi B, Abraham AA, Ham J. Adenosine receptor subtype expression and activation influence the differentiation of mesenchymal stem cells to osteoblasts and adipocytes. J Bone Miner Res. 2011;26(9):2112–24.
45. Costa AM, Barbosa A, Neto E, Sousa SA, Freitas R, Neves JM, et al. On the role of subtype selective adenosine receptor agonists during proliferation and osteogenic differentiation of human primary bone marrow stromal cells. J Cell Physiol. 2011;226(5):1353–66.
46. Rath-Wolfson L, Bar-Yehuda S, Madi L, Ochaion A, Cohen S, Zabutti A, et al. IB-MECA, an A3 adenosine receptor agonist prevents bone resorption in rats with adjuvant induced arthritis. Clin Exp Rheumatol. 2006;24(4):400–6.
47. FitzGerald GA. Dipyridamole. N Engl J Med. 1987;316(20):1247–57.

48. Patrono C, Coller B, Dalen JE, Fuster V, Gent M, Harker LA, et al. Platelet-active drugs: the relationships among dose, effectiveness, and side effects. Chest. 1998;114(5 Suppl):234S–64S.
49. Monagle P, Chan AKCKC, Goldenberg NA, Ichord RN, Journeycake JM, Nowak-Göttl U, et al. Antithrombotic therapy in neonates and children: Antithrombotic Therapy and Prevention of Thrombosis, 9th ed: American College of Chest Physicians Evidence-Based Clinical Practice Guidelines. Chest. 2012;141(2 Suppl):e737S–801S.
50. Ishack S, Mediero A, Wilder T, Ricci JL, Cronstein BN. Bone regeneration in critical bone defects using three-dimensionally printed β-tricalcium phosphate/hydroxyapatite scaffolds is enhanced by coating scaffolds with either dipyridamole or BMP-2. J Biomed Mater Res B Appl Biomater. 2017;105(2):366–75.
51. Slater BJ, Kwan MD, Gupta DM, Lee JK, Longaker MT. The role of regional posterior frontal dura mater in the overlying suture morphology. Plast Reconstr Surg. 2009;123(2):463–9.
52. Coelho PG, Jimbo R, Tovar N, Bonfante EA. Osseointegration: hierarchical designing encompassing the macrometer, micrometer, and nanometer length scales. Dent Mater. 2015;31(1):37–52.
53. Weinand C, Neville CM, Weinberg E, Tabata Y, Vacanti JP. Optimizing biomaterials for tissue engineering human bone using mesenchymal stem cells. Plast Reconstr Surg. 2016;137(3):854–63.
54. Cooper GM, Miller ED, DeCesare GE, Usas A, Lensie EL, Bykowski MR, et al. Inkjet-based biopatterning of bone morphogenetic protein-2 to spatially control calvarial bone formation. Tissue Eng Part A. 2010;16(5):1749–59.

19 The Future of Bioengineering for Head and Neck Reconstruction: The Customized Free Flap

Huy Tran, James C. Melville, Jonathan W. Shum, F. Kurtis Kasper, Mark E. Wong, and Simon Young

19.1 Current Methods of Maxillofacial Reconstruction

The aim of maxillofacial reconstruction is to restore the esthetic form of the affected region while concurrently imparting satisfactory functional ability. This objective has been the main driving force behind the advancements of surgical techniques and biomaterials, pushing through many boundaries over the years. While many options exist for the reconstructive surgeon, they can generally be classified into one of two categories: graft and flap reconstructions. Graft tissue relies on the existing vascular network from the recipient site for neovascularization, healing, and integration. It can be taken from donors of different species (xenograft), from donors of the same species (allograft), or from a different site of the same host (autograft). On the other hand, flap tissue retains its existing vascular network and may require anastomosis to the recipient vessels (free flap). In general, flap surgery can be further divided into three categories: local, regional, and free flaps. While a vast majority of reconstructive options exist, the final selection must consider the characteristics of an existing or prospective maxillofacial defect, the patient's demographics and overall health, and the surgeon's experience and resources.

Traditionally, maxillofacial defects have been commonly reconstructed with autogenous particulate bone contained within an allogeneic crib. Under the ideal conditions, this technique is effective and provides excellent functional and esthetic results. Autogenous bone, in the form of a non-vascularized graft or vascularized flap, is considered the gold standard among biomaterials for reconstructive purposes. It is the only bone graft endowed with osteogenic, osteoinductive, and osteoconductive properties. Allogeneic bone and bone substitutes have been extensively studied over the years, although no single material is yet capable of matching the regenerative potential of autogenous grafts to date. A composite system of biomaterials can often be utilized to augment bone regeneration. In Chap. 13, we discussed the use of particulated allogeneic bone supplemented with bone morphogenetic protein (rhBMP-2) and bone marrow aspirate concentrate (BMAC) to reconstruct maxillofacial

H. Tran · J. C. Melville · J. W. Shum · M. E. Wong
S. Young (✉)
Department of Oral and Maxillofacial Surgery, University of Texas Health Science Center
Department of Oral and Maxillofacial Surgery, Oral, Head & Neck Oncology and Micro-vascular Reconstructive Surgery, Houston, TX, USA
e-mail: Huy.Q.Tran@uth.tmc.edu; James.C.Melville@uth.tmc.edu; Jonathan.Shum@uth.tmc.edu; Mark.E.Wong@uth.tmc.edu; Simon.Young@uth.tmc.edu

F. Kurtis Kasper
Department of Orthodontics, The University of Texas School of Dentistry at Houston, Houston, TX, USA
e-mail: Fred.K.Kasper@uth.tmc.edu

J. C. Melville et al. (eds.), *Regenerative Strategies for Maxillary and Mandibular Reconstruction*,
https://doi.org/10.1007/978-3-319-93668-0_19

defects. Non-vascularized autogenous or allogeneic bone grafts require a healthy soft tissue envelope to prevent oral contamination and subsequent graft failure. When such requirements cannot be fulfilled, the surgeon can elect for staged reconstruction where a vascularized soft tissue flap can first be transferred into the defect prior to bone grafting at a later time.

In 1989, the fibula free flap was first reported in the literature as a novel method for mandible reconstruction [1]. It has since remained a workhorse for clinicians by providing consistent, excellent surgical outcomes. In addition to the fibula, osteocutaneous composite flaps such as the deep circumflex iliac artery bone or scapula flaps provide the advantage of simultaneously reconstructing both hard and soft tissue defects in a single procedure. Wounds requiring significant soft tissue augmentation can be reconstructed with the radial forearm or anterolateral thigh free flaps. The option for free tissue transfer provides an excellent solution to maxillofacial defects with compromised soft tissue health as a result of injury, radiation, or postsurgical scarring.

19.2 Limitations Associated with Current Reconstructive Options

While various reconstructive options exist, it is important to understand the goals and limitations of each technique. Mandibular continuity defects require stabilization of the non-resected segments in order to establish a functional occlusion and provide a good esthetic outcome. This can be achieved by placement of a reconstruction plate between the segments, with or without bone grafting. The placement of a reconstruction plate in the absence of a bone graft may be necessary in recipient sites that have active infection or inadequate soft tissue coverage. However, this method has demonstrated a high failure rate in the postoperative period. Kim et al. reported a failure rate of 52% when anterior mandible defects were reconstructed with plates alone. In this study, lateral mandibular reconstruction yielded a 7.7–12.5% failure rate. Additionally, 17% of cases resulted in dehiscence of soft tissue overlying the plates [2]. Evidently, it is important to augment mandibular reconstruction with bone grafting when possible.

When primary closure can be achieved, the maxillomandibular defect can be reconstructed with autogenous bone alone, autogenous and allogeneic composite systems, or allogeneic bone augmented with mesenchymal cells and growth factors. The non-vascularized graft is extremely vulnerable to intraoral bacterial contamination and mobility during the healing period. The ideal conditions of autogenous bone grafting involve stable graft fixation and adequate vascularized soft tissue coverage, which is often insufficient as a result of traumatic injury or post-oncologic ablation. When exposed to the oral cavity and salivary contamination, the graft can experience failure rates of up to 50% [3]. Additionally, some oncologic patients require adjuvant radiation therapy that results in hypovascularized and hypocellular local tissue. These challenges increase the risk of wound dehiscence and graft failure. To mitigate these issues, the surgeon can elect for staged reconstruction or application of free tissue transfer.

While vascularized free tissue transfer has become the preferred method of reconstruction in complex maxillofacial defects, it too presents with limitations that need to be addressed. Satisfactory surgical results when utilizing microvascular free flaps in part depend on the elected donor site's anatomical shape and size. The surgeon takes what is available and tries to mold it to what is needed. With regard to the esthetic and functional outcomes, the raised flap is often too small or too bulky and may not align to the defect fittingly. Most cases require a revision surgery to debulk, contour, or augment the reconstructed site. The risks of donor site morbidity, prolonged surgical time, and increased hospital stay may render vascularized free tissue transfer unsuitable for select patients. While the anastomosed free flap is more resilient to the damaging effects of radiation or scarring from prior surgery compared to non-vascularized graft, these factors may still play a considerable role in the predictability of a successful reconstruction.

To conform with the reconstructive objective discussed previously, we continue to explore novel methods to restore maxillofacial defects to their inherent form and function. It is logical to believe that in order to maximize such effort, like tissue needs to be replaced with like tissue. Additionally, we aim to increase surgical efficacy, to decrease morbidity associated with the operation, and to optimize healing and flap integration in the postoperative period. These notions led to a new concept of tissue engineering a prevascularized free flap that is customized to the recipient site in shape and size prior to harvesting. By eliminating the anatomical and geometric problems associated with conventional vascularized free flaps, we can provide an optimal reconstructive option and result for the patient.

19.3 Staged Mandibular Reconstruction Using the Space Maintenance, Wound Optimization, and Osseous Reconstruction (SWOR) Technique

19.3.1 Space Maintenance and Wound Optimization

As surgical modalities continue to improve, immediate reconstruction is favored by both patients and the clinicians due to the availability of free tissue transfer. Compared to autogenous or allogeneic grafts, vascularized free tissue transfer is less reliant on the recipient bed's vascular network and soft tissue volume, which may be compromised by radiation, trauma, or scarring from a prior surgery. Despite its versatility and resiliency, immediate reconstruction with a free flap often sacrifices esthetics and function because of the anatomical and geometric differences. In complex maxillofacial defects where the margins are not established (medication- or radiation-induced osteonecrosis), infected wounds (osteomyelitis), or with medically ill patients unable to withstand a prolonged operation, it may be prudent to optimize the defect (by ablation or debridement) and the patient medically prior to the final reconstruction. In these cases, staged reconstruction increases surgical efficiency and can provide the patient with a better and more predictable outcome.

If staged surgery is elected, care must be taken to physiologically optimize the patient and the targeted site prior to the final reconstruction. In the postoperative period, wound contracture and soft tissue prolapse are mechanisms that effectively reduce the volume in the space between the non-resected segments. The associated anatomical structures including nerves and blood vessels are intertwined with granulation tissue during the healing period and eventually become closely approximated with the formed scars. Consequently, accessing the site to be reconstructed will be difficult due to the increased risks of nerve damage and bleeding. Any tissue being excised or dissected during the second surgery must be carefully examined to avoid accidental injuries to vital structures. One solution to this problem is to place an alloplastic material into the defect at the conclusion of the initial surgery. The material will serve to maintain the defect volume and prevent prolapse of the soft tissue during the healing period. This approach essentially gives the wound site and the oral cavity time to heal while being primed for definitive reconstruction.

By taking advantage of computer-aided modeling and current tissue engineering technology, we have introduced the concept of "Space Maintenance, Wound Optimization, and Osseous Reconstruction." (SWOR) (Fig. 19.1) that can further advance maxillofacial reconstruction. In the first stage of this approach, a porous polymethyl methacrylate (PMMA) space maintainer can be fabricated ex vivo to conform to the configuration of the defect. The porous property allows inward tissue migration and interdigitation with fibrous tissue, effectively improving the mechanical attachment between the space maintainer and overlying soft tissue. This relationship enhances the soft tissue's resiliency to the shearing forces of mastication and lowers the rate of wound dehiscence. In a rabbit mandibular defect model, we have demonstrated a lower rate of wound dehiscence when a porous space

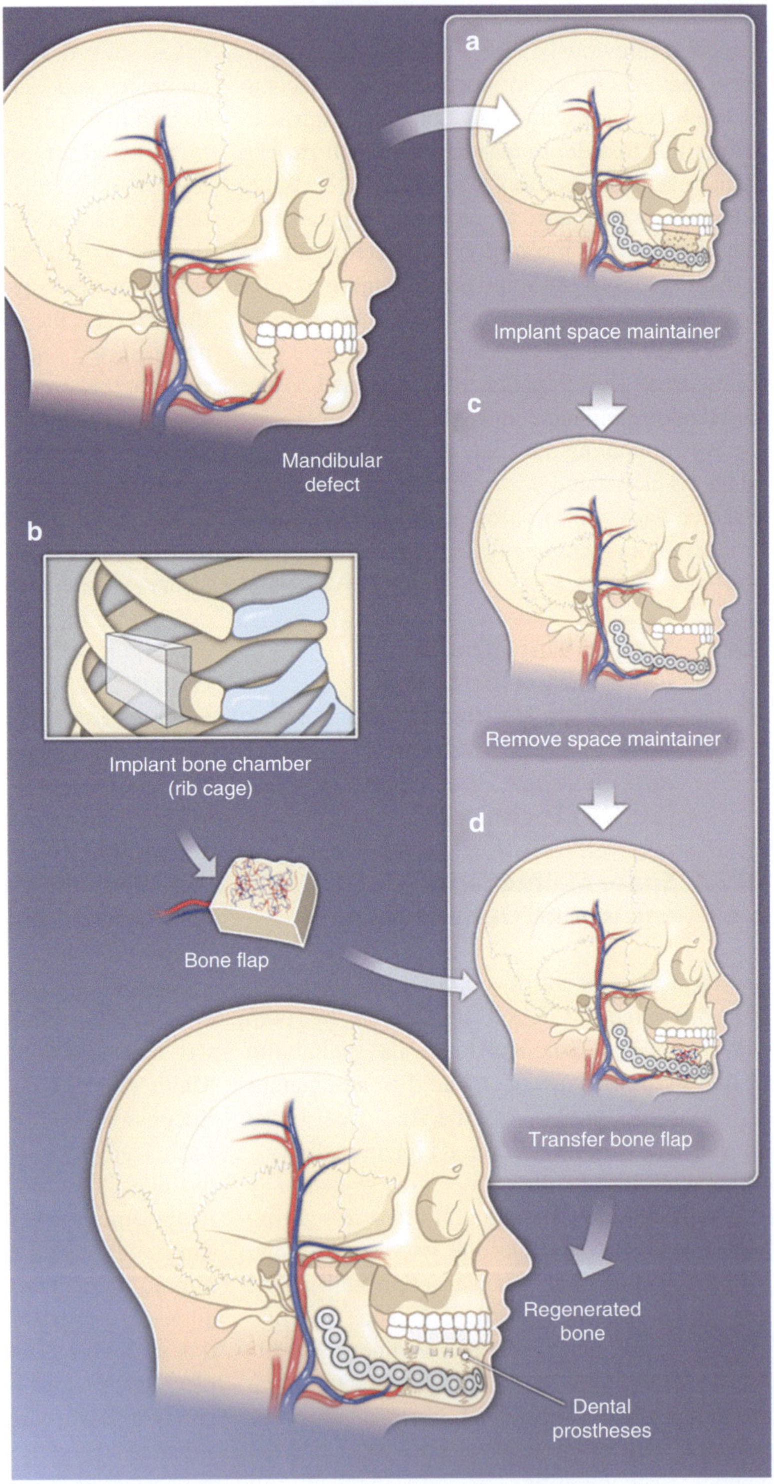

Fig. 19.1 Stage I (A and B) involves placement of an alloplastic implant to prevent volumetric shrinkage of the defect space and optimize the wound site for repair (A). Simultaneously, a chamber filled with bone graft is implanted in a distal site (such as a rib) in approximation to periosteum, in order to form an autologous bone flap (B). Stage II (C and D) involves removal of the alloplastic space maintainer (C) and harvest of the autologous bone flap for vascularized free tissue transfer (D). Microsurgical anastomosis is performed between the pedicle and the recipient vessels to enable continued viability of the transferred bone

maintainer was used compared to a smooth-surfaced, nonporous space maintainer [4, 5].

Taking this concept one step further, we can utilize patient-specific 3D models to prefabricate the porous PMMA space maintainer in the surgical planning stage. Having a space maintainer already approximated to the defect's size and shape, less time is spent intraoperatively for mixing, molding, and trimming. This strategy of prefabricating patient-specific space maintainers was demonstrated in a case series of patients with benign mandibular pathologic lesions [6]. A group of five patients with the diagnosis of either ameloblastoma, keratocystic odontogenic tumor, or glandular odontogenic cyst underwent resection of their associated lesions. At the conclusion of the resection, a preformed space maintainer was placed into the defect, secured by a bone plate and positional screws, and finally closed primarily (Figs. 19.2 and 19.3). Three patients successfully retained their space maintainers and subsequently proceeded to bony reconstruction. The other two patients had intraoral exposure of their space maintainers during the healing period requiring surgical removal. Two factors may have contributed to the wound dehiscence and space maintainer failure in these patients: (1) the space maintainer was placed in the anterior midline location, which suffers from greater mechanical stress, and (2) the overlying soft tissue was thin as a result of supraperiosteal dissection in the initial surgery. In these situations, soft tissue augmentation with allogeneic dermal graft (allograft) or xenogeneic collagen graft (Mucograft) should be considered.

Prefabricated space maintainers can be an effective and cost-efficient adjunct in staged maxillofacial reconstruction. It is important to note that for this approach, preoperative assessment is critical in the predictability of a successful reconstruction. The patient must be reliable and understand the need for multiple operations. The overlying soft tissue should be carefully examined with regard to volume (for primary coverage of space maintainer) and quality (adequate thickness to prevent wound dehiscence). Intraoperatively, the surgeon should strive for conservative preservation of defect dimensions, especially in the vertical (height) plane. The inserted space maintainer should be in contact with the non-resected bone, fixated with a rigid bone plate and adequate number of bone screws, and must not have rough or sharp edges. In select cases at high risk of infection, antibiotics potentially could be added to the space maintainers for controlled local release to improve outcome [7].

19.3.2 In Vivo Bioreactor Strategy for Customized Autologous Flap Generation

While the advancement of space maintainer technology aims to optimize the defect site in the initial surgery, efforts have also been made to explore and expand a variety of reconstructive

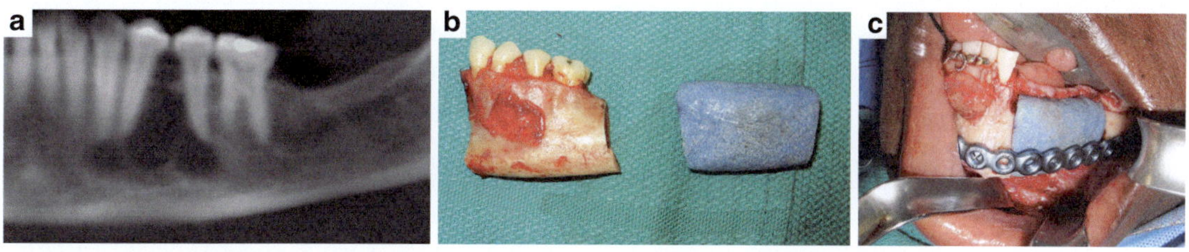

Fig. 19.2 (**a**) Panoramic radiograph of a patient with a left mandibular ameloblastoma. (**b**) Customized porous PMMA spacer (right) approximating the geometry of the resection specimen (left). (**c**) Implantation of porous PMMA spacer into mandibular resection defect

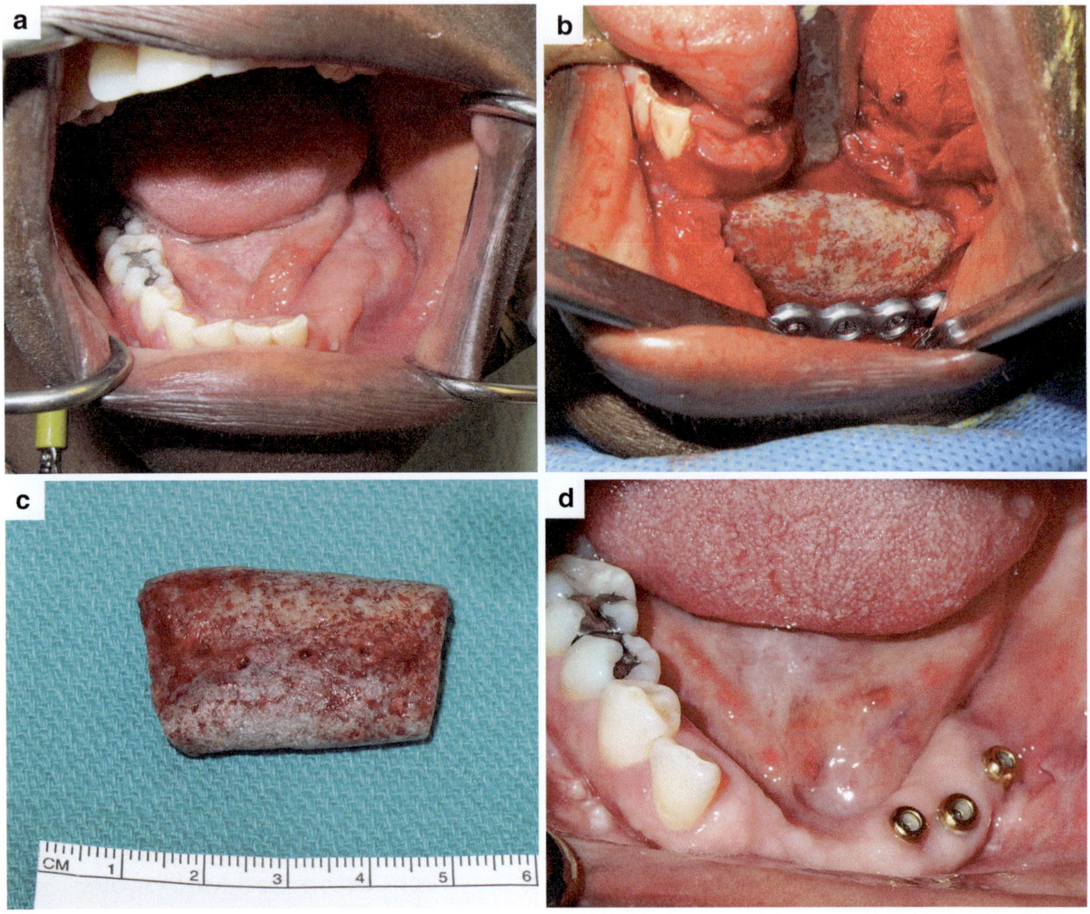

Fig. 19.3 (**a**) Postoperative healing of mandibular resection site with spacer in place at 4 months. (**b**) Exposure of porous PMMA spacer at the time of reconstruction. (**c**) Removal of porous PMMA spacer from mandibular defect showing gross evidence of soft tissue ingrowth. (**d**) Mandibular defect reconstruction 3 years status post bone grafting and dental implant placement

options. Non-vascularized autogenous graft or allogeneic graft that is supplemented with growth factors can produce excellent results when a sufficient healthy soft tissue volume exists. On the other hand, versatile and resilient free flaps give the surgeon a reliable tool for reconstruction of complex craniofacial defects. With regard to free tissue transfer, we previously discussed in this chapter the frequent need for revision surgery in order to improve flap esthetics. By utilizing the concept of tissue engineering, our group (and others) is attempting to explore a novel method to eliminate this geometric limitation between donor and recipient sites. Specifically, we believe that the human body can be harnessed as a bioreactor capable of generating autologous tissue flaps or grafts that are customizable to the target defect site [8].

Traditionally, the bioreactor model is reserved for the generation of cells and tissues in the laboratory or in vitro. The generated tissues are grown and maintained in optimized laboratory conditions, but they typically are found to lack certain key components such as the associated vasculature. This is hypothesized to be due to the inability to replicate the complexities of the in vivo environment [8]. As an alternative, we sought to utilize the human body as a natural bioreactor, encouraging it to generate its own autogenous tissues by providing it with essential growth factors and resources. A chamber molded to a specific shape and size can be filled with osteoconductive or osteoinductive biomaterial and implanted at a body site that is distant from the defect. Cellular migration and ingrowth of local vasculature subsequently occur that initiate the formation of vascularized autogenous bone with the desired size and shape. The bone formed in this chamber can then be harvested as a graft (without a vascular supply) or flap (with a vascular supply) and transferred to the target defect. If needed, local soft tissue such as the muscle or connective tissue can also be harvested for definitive reconstruction.

Early research of this concept was done in sheep models using PMMA chambers implanted in apposition to the cambium layer of the periosteum of the ribs [9, 10]. In this study, the rectangular-shaped chambers were filled with morcellized bone graft and/or porous poly(lactic-*co*-glycolic acid) (PLGA) polymer wafers prior to implantation. After 6–13 weeks, the chambers were surgically removed and analyzed. Histologically, the bone extracted from the chamber was well-vascularized and approximated the geometry of the PMMA chamber, demonstrating that the generated tissue can be harvested as a non-vascularized graft or flap with the associated intercostal vessels [9]. This large animal study was the first of many that aimed to prove the practicality of the in vivo bioreactor strategy.

Subsequent studies aimed to characterize factors associated with in vivo bioreactor efficacy, including types of scaffold materials, optimal implantation duration, and stability of tissues within the chambers over time. In one animal study, PLGA wafers were implanted with or without morcellized bone graft. The result was that bone formation only occurred in areas containing the bone graft, emphasizing the importance of providing an osteoinductive signal to facilitate new bone generation [10]. In a follow-up study, the PMMA chambers filled with morcellized bone graft were either autoclaved to denature the growth factors or left as is prior to implantation in the sheep model [11]. Bone level was significant only in chambers with the native morcellized bone graft, again demonstrating the effects of osteogenic cues and osteoconductive scaffold to support bone formation. Optimal duration of implantation was found to be at 9 weeks in order to generate quality bone. At 12 and 24 weeks, significant decrease in bone volume was observed [11, 12].

While initial animal studies yielded excellent results when the chambers were implanted in orthotopic site, other sites in the human body are continually being explored as bioreactor candidates. A study led by Brey et al. in 2007 compared the level of bone formation when the chambers are implanted in an orthotopic site with appositional contact to the periosteum versus those implanted in contact with the fascia of the latissimus dorsi [13]. While both groups demonstrated vascularized tissue formation, significant bone formation was observed in the chambers in contact with the periosteum. Brey and his group concluded that the orthotopic site is more suitable for engineering bone tissue than ectopic sites. Other studies in small [14, 15] and large [16] animal models have demonstrated that ectopic sites can be utilized effectively in engineering vascularized bony tissue by incorporating exogenous osteogenic growth factor in the chamber.

Based on the current evidence, the in vivo bioreactor approach has demonstrated to be capable of generating vascularized autologous tissue configured to the desired geometrical shape and size. Nevertheless, the success of this technique lies in the viability of the engineered tissue after the transfer of the flap to the recipient site. This was only recently explored and evaluated in a sheep model. The chambers were initially filled with morcellized bone graft, a synthetic ceramic-based bone graft material, or a combination of both materials. They were then implanted in the ribs of the sheep and placed in contact with the periosteum. After 9 weeks, the vascularized bone flaps were harvested and transferred to a mandibular defect [17]. The tissue remained viable for 12 weeks in all three sheep that received the transfers. In this study, the successful use of ceramic-based bone graft also suggested the potential role of synthetic material in the in vivo bioreactor strategy, eliminating the requirement of autologous bone and thus decreasing associated donor site morbidity.

Current (unpublished) preclinical work by our group has advanced the SWOR treatment concept further by utilizing 3D printing technology to generate high-fidelity patient-specific vascularized flaps. Sheep with a challenging superior marginal mandibular defect received a porous space maintainer at the defect site immediately following the marginal mandibulectomy. A separate surgical team simultaneously implanted 3D printed bioreactors filled with either synthetic (Mastergraft® Granules) or autologous (morcellized rib) scaffold material into orthotopic rib resection sites. After a 9-week healing period, the mandibular space maintainers were removed, and the tissues generated from the 3D printed in vivo bioreactors were transferred as both vascularized bone flaps and free bone grafts to the sheep mandibular defect (Fig. 19.4). Twelve weeks after reconstruction, the mandibles were harvested for evaluation. Utilizing both surgical and radiological outcomes, there were no statistically significant differences in radiological markers of bone architecture between tissues generated in bioreactors filled with either autologous or synthetic materials. Furthermore, this customized in vivo bioreactor strategy was successful in facilitating reconstruction of the challenging large tissue defect in 83% of animals, illustrating the potential of this "personalized medicine" type of approach to the reconstruction of patients with complex maxillofacial defects.

The in vivo bioreactor strategy was first applied to humans in 2006 through a case of mandibular augmentation [18]. In this case, the iliac crest was elected to be the bioreactor site. A PMMA chamber that was filled with autograft bone was implanted in contact with the periosteum of the iliac crest. After 8 weeks, the generated tissue was extracted and transferred along with the periosteum to the donor mandibular site. Follow-up at 16 months showed adequate mandibular height for the insertion and maintenance of osseointegrated dental implants. While the patient eventually died from hepatocellular carcinoma, this case demonstrated successful application of the in vivo bioreactor strategy in mandibular augmentation. Other case reports have also been published that demonstrated craniofacial reconstruction using the in vivo bioreactor strategy with a variety of scaffold materials implanted in both orthotopic and ectopic sites. In these studies, the duration of implantation ranged from 7 weeks to over 6 months. Additionally, growth factors or bone marrow aspirate was added to augment bone formation in some cases [19–23]. Despite short-term successes in most of these cases, the variability in regard to biomaterial used, duration of implantation, and role of growth factors stress the importance of continued investigation in order to develop a standardized surgical approach [8]. The in vivo bioreactor strategy aims to generate autologous vascularized bone flaps, with or without associated soft tissue, that is shaped to match the geometry of a craniofacial bony defect. At the same time, the space maintainer serves to prime and preserve the geometry and physiology of the defect for future reconstruction [24].

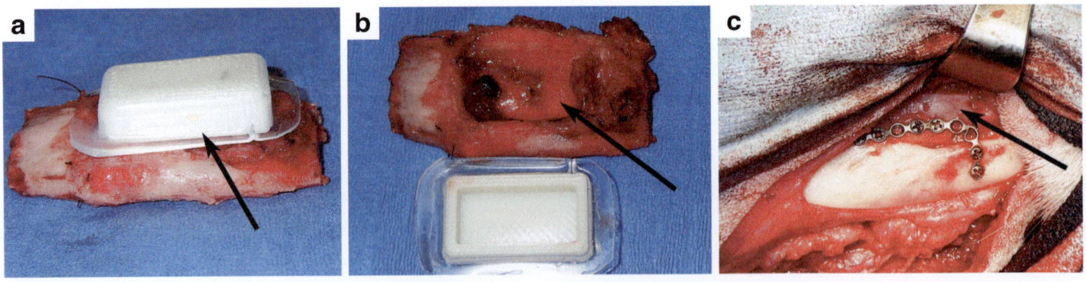

Fig. 19.4 (**a**) 3D printed bioreactor (black arrow) attached to the underlying rib removed at 9 weeks postimplantation. (**b**) After 9 weeks of implantation, the tissue had grown within and conformed to the dimensions of the 3D printed bioreactor (black arrow). (**c**) The tissue (black arrow) was transferred to the mandibular defect site and found to have appropriate volume and geometry for reconstruction

19.4 Conclusion

The reconstruction of composite defects of the head and neck region poses a number of challenges. The ability to replace multiple tissue types in an inhospitable environment is one of them. Additional challenges include the need to develop reconstructive systems that are both functional and esthetically acceptable, considering the prominence of the face and the dynamic roles played by its components. To this end, a variety of tissue engineering and virtual surgical planning tools are being developed to facilitate functional and esthetic reconstruction targeted to the patient-specific geometry of the defect, with a particular focus on the skeletal component (see Chap. 18) [25–29]. Indeed, 3D printing techniques are being applied in the fabrication of custom-shaped scaffolds to support mandibular regeneration in preclinical models [26]. While the proof of concept of translation of 3D printing in the in vivo bioreactor strategy has been established, several aspects of the approach remain to be explored, including the ability to generate and harvest bone of complex geometries. Nevertheless, emerging bioengineering strategies present exceptional potential to advance the options available to the informed surgeon for reconstruction of complex maxillofacial defects.

References

1. Hidalgo DA. Fibula free flap: a new method of mandible reconstruction. Plast Reconstr Surg. 1989;84(1):71–9.
2. Kim M-R, Donoff RB. Critical analysis of mandibular reconstruction using AO reconstruction plates. J Oral Maxillofac Surg. 1992;50(11):1152–7.
3. Lawson W, Loscalzo LJ, Baek SM, Biller HF, Krespi YP. Experience with immediate and delayed mandibular reconstruction. Laryngoscope. 1982;92(1):5–10.
4. Kretlow JD, Shi M, Young S, Spicer PP, Demian N, Jansen JA, Wong ME, Kasper FK, Mikos AG. Evaluation of soft tissue coverage over porous polymethylmethacrylate space maintainers within nonhealing alveolar bone defects. Tissue Eng Part C Methods. 2010;16(6):1427–38.
5. Nguyen C, Young S, Kretlow JD, Mikos AG, Wong M. Surface characteristics of biomaterials used for space maintenance in a mandibular defect: a pilot animal study. J Oral Maxillofac Surg. 2011;69(1):11–8.
6. Henslee AM, Spicer PP, Shah SR, Tatara AM, Kasper FK, Mikos AG, Wong ME. Use of porous space maintainers in staged mandibular reconstruction. Oral Maxillofac Surg Clin North Am. 2014;26(2):143–9.
7. Spicer PP, Shah SR, Henslee AM, Watson BM, Kinard LA, Kretlow JD, Bevil K, Kattchee L, Bennett GN, Demian N, et al. Evaluation of antibiotic releasing porous polymethylmethacrylate space maintainers in an infected composite tissue defect model. Acta Biomater. 2013;9(11):8832–9.
8. Tatara AM, Wong ME, Mikos AG. In vivo bioreactors for mandibular reconstruction. J Dent Res. 2014;93(12):1196–202.
9. Miller MJ, Goldberg DP, Yasko AW, Lemon JC, Satterfield WC, Wake MC, Mikos AG. Guided bone growth in sheep: a model for tissue-engineered bone flaps. Tissue Eng. 1996;2(1):51–9.
10. Thomson RC, Mikos AG, Beahm E, Lemon JC, Satterfield WC, Aufdemorte TB, Miller MJ. Guided tissue fabrication from periosteum using preformed biodegradable polymer scaffolds. Biomaterials. 1999;20(21):2007–18.
11. Cheng MH, Brey EM, Allori A, Satterfield WC, Chang DW, Patrick CW Jr, Miller MJ. Ovine model for engineering bone segments. Tissue Eng. 2005;11(1–2):214–25.
12. Cheng MH, Brey EM, Allori AC, Gassman A, Chang DW, Patrick CW Jr, Miller MJ. Periosteum-guided prefabrication of vascularized bone of clinical shape and volume. Plast Reconstr Surg. 2009;124(3):787–95.
13. Brey EM, Cheng MH, Allori A, Satterfield W, Chang DW, Patrick CW Jr, Miller MJ. Comparison of guided bone formation from periosteum and muscle fascia. Plast Reconstr Surg. 2007;119(4):1216–22.
14. Kusumoto K, Bessho K, Fujimura K, Akioka J, Ogawa Y, Iizuka T. Prefabricated muscle flap including bone induced by recombinant human bone morphogenetic protein-2: an experimental study of ectopic osteoinduction in a rat latissimus dorsi muscle flap. Br J Plast Surg. 1998;51(4):275–80.
15. Roldán JC, Jepsen S, Miller J, Freitag S, Rueger DC, Açil Y, Terheyden H. Bone formation in the presence of platelet-rich plasma vs. bone morphogenetic protein-7. Bone. 2004;34(1):80–90.
16. Geuze RE, Theyse LF, Kempen DH, Hazewinkel HA, Kraak HY, Oner FC, Dhert WJ, Alblas J. A differential effect of bone morphogenetic protein-2 and vascular endothelial growth factor release timing on osteogenesis at ectopic and orthotopic sites in a large-animal model. Tissue Eng Part A. 2012;18(19–20):2052–62.
17. Tatara AM, Kretlow JD, Spicer PP, Lu S, Lam J, Liu W, Cao Y, Liu G, Jackson JD, Yoo JJ, et al. Autologously generated tissue-engineered bone flaps for reconstruction of large mandibular defects in an ovine model. Tissue Eng Part A. 2015;21(9–10):1520–8.
18. Cheng MH, Brey EM, Ulusal BG, Wei FC. Mandible augmentation for osseointegrated implants using

tissue engineering strategies. Plast Reconstr Surg. 2006;118(1):1e–4e.
19. Heliotis M, Lavery KM, Ripamonti U, Tsiridis E, di Silvio L. Transformation of a prefabricated hydroxyapatite/osteogenic protein-1 implant into a vascularised pedicled bone flap in the human chest. Int J Oral Maxillofac Surg. 2006;35(3):265–9.
20. Kokemueller H, Spalthoff S, Nolff M, Tavassol F, Essig H, Stuehmer C, Bormann KH, Rucker M, Gellrich NC. Prefabrication of vascularized bioartificial bone grafts in vivo for segmental mandibular reconstruction: experimental pilot study in sheep and first clinical application. Int J Oral Maxillofac Surg. 2010;39(4):379–87.
21. Orringer JS, Shaw WW, Borud LJ, Freymiller EG, Wang SA, Markowitz BL. Total mandibular and lower lip reconstruction with a prefabricated osteocutaneous free flap. Plast Reconstr Surg. 1999;104(3):793–7.
22. Warnke PH, Springer ING, Wiltfang J, Acil Y, Eufinger H, Wehmöller M, Russo PAJ, Bolte H, Sherry E, Behrens E, et al. Growth and transplantation of a custom vascularised bone graft in a man. Lancet. 2004;364(9436):766–70.
23. Warnke PH, Wiltfang J, Springer I, Acil Y, Bolte H, Kosmahl M, Russo PA, Sherry E, Lutzen U, Wolfart S, et al. Man as living bioreactor: fate of an exogenously prepared customized tissue-engineered mandible. Biomaterials. 2006;27(17):3163–7.
24. Atala A, Kasper FK, Mikos AG. Engineering complex tissues. Sci Transl Med. 2012;4(160):160rv12.
25. Levine JP, Bae JS, Soares M, Brecht LE, Saadeh PB, Ceradini DJ, Hirsch DL. Jaw in a day: total maxillofacial reconstruction using digital technology. Plast Reconstr Surg. 2013;131(6):1386–91.
26. Lopez CD, Diaz-Siso JR, Witek L, Bekisz JM, Cronstein BN, Torroni A, Flores RL, Rodriguez ED, Coelho PG. Three dimensionally printed bioactive ceramic scaffold osseoconduction across critical-sized mandibular defects. J Surg Res. 2018;223:115–22.
27. Monaco C, Stranix JT, Avraham T, Brecht L, Saadeh PB, Hirsch D, Levine JP. Evolution of surgical techniques for mandibular reconstruction using free fibula flaps: the next generation. Head Neck. 2016;38(Suppl 1):E2066–73.
28. Qaisi M, Kolodney H, Swedenburg G, Chandran R, Caloss R. Fibula jaw in a day: state of the art in maxillofacial reconstruction. J Oral Maxillofac Surg. 2016;74(6):1284.e1–1284.e15.
29. Runyan CM, Sharma V, Staffenberg DA, Levine JP, Brecht LE, Wexler LH, Hirsch DL. Jaw in a day: state of the art in maxillary reconstruction. J Craniofac Surg. 2016;27(8):2101–4.

MIX
Papier aus verantwortungsvollen Quellen
Paper from responsible sources
FSC® C105338

If you have any concerns about our products,
you can contact us on
ProductSafety@springernature.com

In case Publisher is established outside the EU,
the EU authorized representative is:
Springer Nature Customer Service Center GmbH
Europaplatz 3, 69115 Heidelberg, Germany

Printed by Libri Plureos GmbH
in Hamburg, Germany